MARE Publication Series

Volume 21

The MARE Publication Series is an initiative of the Centre for Maritime Research (MARE). MARE is an interdisciplinary social-science network devoted to studying the use and management of marine resources. It is based jointly at the University of Amsterdam and Wageningen University (www.marecentre.nl).

The MARE Publication Series addresses topics of contemporary relevance in the wide field of 'people and the sea'. It has a global scope and includes contributions from a wide range of social science disciplines as well as from applied sciences. Topics range from fisheries, to integrated management, coastal tourism, and environmental conservation. The series was previously hosted by Amsterdam University Press and joined Springer in 2011.

The MARE Publication Series is complemented by the Journal of Maritime Studies (MAST) and the biennial People and the Sea Conferences in Amsterdam.

More information about this series at http://www.springer.com/series/10413

Ratana Chuenpagdee • Svein Jentoft
Editors

Transdisciplinarity for Small-Scale Fisheries Governance

Analysis and Practice

Springer

Editors
Ratana Chuenpagdee
Department of Geography
Memorial University of Newfoundland
St. John's, NL, Canada

Svein Jentoft
Norwegian College of Fishery Science
UiT The Arctic University of Norway
Tromsø, Norway

Too Big To Ignore (TBTI; toobigtoignore.net) is a global research network and knowledge mobilization partnership, funded by the Social Sciences and Humanities Research Council of Canada, and supported by 15 partner organizations and over 500 members from around the world. The network aims at elevating the profile of small-scale fisheries, arguing against their marginalization in national and international policies, and developing research and governance capacity to address global fisheries challenges.

ISSN 2212-6260 ISSN 2212-6279 (electronic)
MARE Publication Series
ISBN 978-3-030-06941-4 ISBN 978-3-319-94938-3 (eBook)
https://doi.org/10.1007/978-3-319-94938-3

Cover illustration: Lobster traps of Newfoundland, Photo by Jaehong Jin, Portugal Cove-St. Philip's, NL, Canada; July 2016

This Springer imprint is published by the registered company Springer Nature Switzerland AG
The registered company address is: Gewerbestrasse 11, 6330 Cham, Switzerland

Foreword

Ignorance about small-scale fisheries has prevailed way too long in this water world of ours. Ironically, small-scale fisheries are responsible for feeding vast numbers of people, and, though often theoretically recognized as important, in practice, they are typically marginalized. Instead, large-scale fisheries draw attention and help. While it may be true that there have been a plethora of studies of fisheries and fisheries governance, rarely have researchers looked beyond the questions raised by their own singular disciplines. The amazing, multiyear project entitled, "Too Big To Ignore (TBTI): Global Partnership for Small-Scale Fisheries Research," funded by the Social Sciences and Humanities Research Council of Canada (SSHRC), has taken steps to remedy our collective ignorance. This book is a synthesis of 6 years of workshops, congresses, and research focused on the potential for taking an innovative, transdisciplinary approach to the governance of small-scale fisheries. Case studies reflecting local, national, and global contexts illuminate challenges in governance, stewardship, food security, alleviation of poverty, livelihoods, human rights, and social justice associated with small-scale fisheries. Importantly, a path forward is laid out to help build capacity for the practical application of the ideals of transdisciplinary governance, research, and development.

As a graduate student, I learned about ethnographic fieldwork by spending a summer on the fleet of trawlers that fished out of Provincetown, Massachusetts (USA). Later, I spent a year in Saint-Louis, Senegal (West Africa), learning about the essential role of small-scale earnings derived from the marketing of fish by women. Despite the differences in the scale of fishing operations (including quantities of landings, vessel sizes, gear used, processing, and markets), similarities in challenges, as well as values, norms, and the significance of local knowledge, were striking. These studies led to a 30+-year career primarily focused on the fishing industry, writ large, of the northeast United States.

While the region's industry is dominated by what would be considered large-scale fisheries in other parts of the world, the challenges of governance share many of the same issues faced by small-scale fisheries around the globe. As the researchers in this volume found in their work with small-scale fisheries, problem-solving in the US fisheries has been most effective in the context of collaboration among a whole array of stakeholders including researchers from a variety of disciplines, fishers and their representatives (often wives), civil society organizations, and government managers. Though these efforts may not be characterized as fully transdisciplinary, they could be considered as moving in that direction. The work described herein is, I would argue, valuable to all interested in fisheries, regardless of scale. Nevertheless, given the well-documented tendency to turn a blind eye to the small-scale, this book is imperative for understanding and aiding the multitude of fisheries around the world whose value is not easily measured by quantities landed and money earned.

The editors of this volume, Ratana Chuenpagdee and Svein Jentoft, have long been clear-eyed advocates for small-scale fisheries, recognizing their significance and working on theoretical frameworks that could lead to the engagement of all in the effort to solve the "wicked" problems associated with these fisheries and their communities. Their research, writing, teaching, and outreach created the foundation for the project and this book.

My first involvement with TBTI was participation in a workshop on "Fishing Futures" that explored case studies from a variety of locales around the world. Later, I was able to attend the Second World Small-Scale Fisheries Congress held in Merida, Mexico in 2014. The latter provided a wealth of information, rich descriptions, and thoughtful analyses of diverse fisheries. An innovative aspect was the provision of translators for the fishers who attended thus permitting first-person commentary and a modicum of empowerment. Small group presentations and discussions encouraged attendees to delve into particular topics at greater length and with more attention to a transdisciplinary perspective than is usually the case in academic conferences. This book reflects some of the thinking that served as the basis for these in-person meetings, carrying it forward to identify what has and hasn't worked in implementation.

I am very pleased to have the opportunity to recommend this book as an excellent compilation of essays offering a window into the world of small-scale fisheries and governance. The book makes a strong case for transdisciplinary approaches to both research and action in order to assure that these fisheries, the humans that rely on them, and their communities, survive.

Anthropologist (retired)
Massachusetts Institute of Technology's
Sea Grant Program
Cambridge, MA, USA

Madeleine Hall-Arber

Series Editors' Preface

The majority of volumes that have been published in the MARE Publication Series have been devoted to capture fisheries, with many highlighting the fate of small-scale fishing populations in the world. There were good reasons for authors to pay attention to the latter topic: small-scale fisheries provide an important source of employment and food security and are argued to be more environmentally appropriate than the industrial fisheries that have come about in the past century. Nevertheless, small-scale fishers are becoming increasingly marginalized. It is for these reasons that FAO has recently taken up their cause and spearheaded the *Voluntary Guidelines for Securing Sustainable Small-Scale Fisheries* (2014). The Too Big To Ignore (TBTI): Global Partnership for Small-Scale Fisheries Research, of which the current volume is a product, has made a very important contribution to the international effort in support of small-scale fishers (men and women), mainly by means of research activities and advocacy. We are proud that the five edited volumes produced by TBTI in the course of 5 years, including this one, have all been published in the MARE Publication Series.

The present volume is organized according to the key questions that were formulated to guide TBTI research. The interactive governance perspective that has imbued the work of TBTI throughout also assists in structuring this volume, the main topic of which is the promotion of transdisciplinary knowledge and action. Transdisciplinary science integrates the scientific knowledge of academics with the experiential knowledge of practitioners, while transdisciplinary governance recognizes that policy is the result not only of rational deliberation but of agreement on values that are pertinent in specific societal contexts. This volume is intended to give the process of realizing sustainable small-scale fisheries another useful impulse. We wish to congratulate Ratana Chuenpagdee and Svein Jentoft, who have provided strong leadership to the TBTI project, with the completion of yet another high quality manuscript!

The MARE Publication Series commenced in 2004 with Amsterdam University Press but moved to Springer Academic Publishers in 2012. It has hitherto contained 20 edited and single-authored volumes on a variety of regions and topics in the topical field of people, coasts, and seas. Fritz Schmuhl and other staff of Springer have facilitated the production process, for which we again say thanks!

On behalf of the series editors,

University of Amsterdam Maarten Bavinck
Amsterdam, The Netherlands

Acknowledgments

This book results from the research that has been carried out by members of the Too Big To Ignore (TBTI): Global Partnership for Small-Scale Fisheries Research and colleagues around the world to help reveal the importance of small-scale fisheries to food security, well-being, livelihoods, community viability, environmental stewardship, and fisheries sustainability. We are truly grateful for the strong commitment and contributions of all the authors, many of whom have been the long-time support and leaders of TBTI. We appreciate the time and constructive comments from the many people who have reviewed individual chapters. We are indebted to Dr. Madeleine Hall-Arber, Massachusetts Institute of Technology's Sea Grant Program, for the ardent support to TBTI and for kindly writing the book foreword. Our thanks go also to Nicole Franz of FAO and Sebastian Mathew of International Collective in Support of Fishworkers (ICSF) for their endorsement.

A very special thanks to Vesna Kereži, TBTI Project Manager, for her dedication to TBTI in general and for her fervent effort to bring this book, as well as the previous ones, to completion. Once again, we are happy to include this book in the Springer MARE Book Series, which was enthusiastically received by the series editor, Maarten Bavinck, and well supported by Springer staff, Fritz Schmuhl and Joseph Daniel. Svein Jentoft wishes to thank Daniela Kalikoski for her support during his sabbatical stay with FAO while working on this book.

TBTI is funded by the Social Sciences and Humanities Research Council of Canada (grant number 895-2011-1011), through the partnership program. We thank Memorial University of Newfoundland, St. John's, Canada, for hosting TBTI and for the generous support that it provides, in terms of cash and in-kind contribution, which has made it possible for us to conduct our research and to build partnerships for sustainable small-scale fisheries worldwide.

Contents

Contributors

Melinda Agapito Too Big To Ignore Global Partnership for Small-Scale Fisheries Research, Memorial University of Newfoundland, St. John's, NL, Canada

Alfonso Aguilar-Perera Departamento de Biología Marina, Universidad Autónoma de Yucatán, Mérida, Mexico

Minerva Arce-Ibarra Departmento de Sistemática y Ecología Acuática de El Colegio de la Frontera Sur, Chetumal, Mexico

Milena Arias-Schreiber Swedish Institute for the Marine Environment, Gothenburg University, Gothenburg, Sweden

Lovasoa Cédrique Augustave Norwegian College of Fishery Science, UiT-The Arctic University of Norway, Tromsø, Norway

Maarten Bavinck Amsterdam Institute for Social Science Research, University of Amsterdam, Amsterdam, The Netherlands

Fikret Berkes University of Manitoba, Winnipeg, MB, Canada

Bonnie Bishop Education Coordinator, Vancouver, BC, Canada

Shannon D. Bower Fish Ecology and Conservation Physiology Laboratory, Department of Biology, Carleton University, Ottawa, ON, Canada

Catalina García Chavez Erigaie Foundation, Amsterdam Institute for Social Science Research, University of Amsterdam, Amsterdam, The Netherlands

Ratana Chuenpagdee Department of Geography, Memorial University of Newfoundland, St. John's, NL, Canada

Andrés M. Cisneros-Montemayor Nereus, Institute for the Oceans and Fisheries, The University of British Columbia, Vancouver, BC, Canada

Steven J. Cooke Fish Ecology and Conservation Physiology Laboratory, Department of Biology, Carleton University, Ottawa, ON, Canada

Alyne E. Delaney Innovative Fisheries Management, Aalborg University, Aalborg, Denmark

Rodolphe Devillers Department of Geography, Memorial University of Newfoundland, St. John's, NL, Canada

David V. Fazzino II Department of Anthropology, Bloomsburg University, Bloomsburg, PA, USA

Katia Frangoudes Univ Brest, Ifremer, CNRS, UMR 6308, AMURE, IUEM, Plouzané, France

Glenna Gannon School of Environment and Sustainability, University of Saskatchewan, Saskatoon, SK, Canada

João Garcia Rodrigues Campus Do*Mar – International Campus of Excellence Galicia-North Portugal, Faculty of Political and Social Sciences, University of Santiago de Compostela, Santiago de Compostela, Spain

Jennifer Gee Food and Agriculture Organization (FAO) of the United Nations, Rome, Italy

Siri Gerrard The Arctic University of Norway, KVINNFORSK, UiT – The Arctic University of Norway, Tromsø, Norway

Miguel González Department of Social Science, York University, Toronto, ON, Canada

Tek Bahadur Gurung Nepal Agricultural Research Council, Kathmandu, Nepal

Gordon M. Hickey Department of Natural Resource Sciences, McGill University, Sainte-Anne-de-Bellevue, QC, Canada

Moenieba Isaacs Institute for Poverty, Land and Agrarian Studies, University of the Western Cape, Cape Town, South Africa

Svein Jentoft Norwegian College of Fishery Science, UiT-The Arctic University of Norway, Tromsø, Norway

Andrew F. Johnson Marine Biology Research Division, Scripps Institution of Oceanography, San Diego, CA, USA

Derek S. Johnson Department of Anthropology, University of Manitoba, Winnipeg, MB, Canada

Helga Josupeit INFOPESCA, Montevideo, Uruguay

Kungwan Juntarashote Department of Fisheries Management, Kasetsart University, Bangkok, Thailand

Jeppe Kolding Department of Biology, University of Bergen, Bergen, Norway

Annie Lalancette St. Mary's University, Halifax, NS, Canada

Mimi E. Lam Centre for the Study of the Sciences and the Humanities, University of Bergen, Bergen, Norway

Marta Leite Natural Resources Institute, University of Manitoba, Winnipeg, MB, Canada

Marc Léopold Institut de Recherche pour le développement (IRD) c/o Institut Halieutique et des Sciences Marines (IH.SM), Université de Toliara, Toliara, Madagascar

Charles Z. Levkoe Canada Research Chair in Sustainable Food Systems, Department of Health Sciences, Lakehead University, Thunder Bay, Canada

Sebastian Linke School of Global Studies, Gothenburg University, Gothenburg, Sweden

Philip A. Loring Department of Geography, Environment, and Geomatics and the Arrell Food Institute, University of Guelph, Guelph, Canada

Kristen Lowitt Department of Geography, Brandon University, Brandon, MB, Canada

Ana Isabel Márquez Pérez Universidad Nacional de Colombia, San Andrés, Colombia

Sérgio M. Gomes de Mattos Ministry of Planning, Recife, Brazil

Patrick McConney University of the West Indies, Cave Hill, Barbados

Beatriz Mesquita P. Ferreira Fundação Joaquim Nabuco – FUNDAJ, Recife, Brazil

Mbachi Ruth Msomphora Research and Publishing Support - Department of Library services, Science and Health Library, UiT – The Arctic University of Norway, Tromsø, Norway

Prateep K. Nayak School of Environment, Enterprise and Development, University of Waterloo, Waterloo, ON, Canada

Connie Nelson Food Security Research Network, Lakehead University, Thunder Bay, ON, Canada

Paul Onyango Department of Aquatic Sciences and Fisheries, University of Dar es Salaam, Dar es Salaam, Tanzania

Sölmundur K. Pálsson Department of Anthropology, University of Manitoba, Winnipeg, MB, Canada

José J. Pascual-Fernández Instituto Universitario de Investigación Social y Turismo, Universidad de La Laguna, Santa Cruz de Tenerife, Spain

Maria Pena University of the West Indies, Cave Hill, Barbados

Rodrigo Pereira Medeiros Universidade Federal do Paraná, Apucarana, Brazil

Graham J. Pierce Departamento de Ecología y Recursos Marinos, Instituto de Investigaciones Marinas (CSIC), Vigo, Spain

Evelyn Pinkerton School of Resource & Environmental Management, Simon Fraser University, Burnaby, BC, Canada

Cristina Pita Department of Environment and Planning & Centre for Environmental and Marine Studies (CESAM), University of Aveiro, Aveiro, Portugal

Alicia Said Too Big To Ignore Global Partnership for Small-Scale Fisheries Research, Memorial University of Newfoundland, St. John's, NL, Canada

Suvaluck Satumanatpan Faculty of Environment and Resources Studies, Mahidol University, Nakorn Pathom, Thailand

Joeri Scholtens Amsterdam Institute for Social Science Research, University of Amsterdam, Amsterdam, The Netherlands

Anna Schuhbauer Fisheries Economic Research Unit, Institute for the Oceans and Fisheries, The University of British Columbia, Vancouver, BC, Canada

Andrew M. Song Australian Research Council Centre of Excellence for Coral Reef Studies, James Cook University, Townsville, Australia

WorldFish, Honiara, Solomon Islands

Merle Sowman Department of Environmental and Geographical Science, University of Cape Town, Cape Town, South Africa

Natasha Stacey Research Institute for the Environment and Livelihoods, Charles Darwin University, Darwin, Australia

Johny Stephen Tata Institute of Social Science, Hyderabad, India

Wichin Suebpala Marine Biodiversity Research Group, Faculty of Sciences, Ramkhamhaeng University, Bangkok, Thailand

U. Rashid Sumaila Fisheries Economic Research Unit, Institute for the Oceans and Fisheries, The University of British Columbia, Vancouver, BC, Canada

The Liu Institute for Global Issues, The University of British Columbia, Vancouver, BC, Canada

Jackie Sunde ABALOBI, Cape Town, South Africa

Brice Trouillet Université de Nantes, CNRS, UMR LETG, Nantes, France

Daniele A. Vila-Nova Independent researcher, Ecology and Conservation, Recife, Brazil

Matias John Wojciechowski World Fisheries Trust – WFT, Victoria, BC, Canada

Thamasak Yeemin Marine Science Association of Thailand, Faculty of Sciences, Ramkhamhaeng University, Bangkok, Thailand

Part I
Introduction

Chapter 1
The Quest for Transdisciplinarity in Small-Scale Fisheries Governance

Svein Jentoft and Ratana Chuenpagdee

Abstract This chapter introduces the contributions and topics raised in this book by authors from around the world. It stresses the relevance of small-scale fisheries for sustainable livelihoods and communities, food security, and poverty eradication, as also emphasized in the 'Voluntary Guidelines for Securing Sustainable Small-Scale Fisheries' endorsed by FAO member states in 2014. The chapter advocates the need for an analytical lens and a theoretical approach to small-scale fisheries governance. 'Interactive governance' is held up as a useful tool for recognizing the 'wicked problems' that are commonly associated with the governance and governability of small-scale fisheries. The chapter lays out an argument for why it is essential to cross the disciplinary boundaries of sciences, and to ground it in the local knowledge and practice of small-scale fisheries stakeholders. Thus, the governance of small-scale fisheries must be founded on transdisciplinary perspectives in order to effectively address the multiple concerns associated with the wellbeing, viability and sustainability of small-scale fisheries and fisheries communities, which are the basic conditions for their ability to fulfill their potentials and contributions to society. The chapter also includes an overview and summary of the book contents.

Keywords Interactive governance · TBTI · SSF Guidelines · Wicked problems · Transdisciplinarity · Implementation

---

S. Jentoft (✉)
Norwegian College of Fishery Science, UiT- The Arctic University of Norway, Tromsø, Norway
e-mail: svein.jentoft@uit.no

R. Chuenpagdee
Department of Geography, Memorial University of Newfoundland, St. John's, NL, Canada
e-mail: ratanac@mun.ca

R. Chuenpagdee, S. Jentoft (eds.), *Transdisciplinarity for Small-Scale Fisheries Governance*, MARE Publication Series 21,
https://doi.org/10.1007/978-3-319-94938-3_1

1.1 Introduction

With the millions of people employed in small-scale fisheries globally, and considering their contribution to food security and poverty eradication, their importance for sustaining ecosystems, local communities, and livelihoods, and the cultural inheritance they represent, small-scale fisheries are both 'too big to ignore' and 'too important to fail'. FAO member states endorsed this view in 2014 with the adoption of the *Voluntary Guidelines for Securing Sustainable Small-Scale Fisheries in the Context of Food Security and Poverty Eradication* (SSF Guidelines; FAO 2015). The relevance of small-scale fisheries is now commonly accepted, and the main challenge is to put this shared vision into concrete policies and strategies (Jentoft 2014).

This is ultimately a governance issue of considerable complexity and urgency. With the SSF Guidelines, it has also become an issue of realizing the human rights and dignity of the people who are engaged in small-scale fisheries. Thus, supporting small-scale fisheries, and making it possible for them to thrive, is not just about the services they deliver to society at large, but also about social values and ethics. In other words, it is not only smart politics from ecological, economic, and social perspectives, but also 'the right thing to do'. By approving the SSF Guidelines, FAO member states agree on what constitutes progress and sustainability in small-scale fisheries, and, therefore, what policy-makers should aim at. They also define a role for the academic community to contribute the knowledge needed to make these ideals become a reality, and to build the much needed capacity within the fisheries communities and among policy-makers for the implementation of the SSF Guidelines.

As a 'Global Partnership for Small-Scale Fisheries Research', Too Big To Ignore (TBTI) has taken on this responsibility. In many publications, seminars, and conferences, TBTI partners and members have actively promoted this agenda by arguing theoretically and showing empirically why and how small-scale fisheries are important in their particular contexts, and what social and political conditions are conducive to their wellbeing and functioning. The current volume is another such effort, which attempts to answer the following question: if small-scale fisheries are indeed too big to ignore, how exactly do we recognize them?

Recognition is, notably, a concept with a double meaning. First, recognition is about acknowledging the existence of small-scale fisheries and the value of their contribution, which FAO member states did with the SSF Guidelines. Now, the same acknowledgement must take place at regional, national, and local levels. There is no guarantee that governing actors in countries and relevant organizations will do that, at least not on their own without pressure from anyone. Given their progressive agenda, it is highly likely that the SSF Guidelines will meet resistance at all levels, and that the ground is not always fertile to them, as demonstrated in our previous publication in the Mare Series (Jentoft et al. 2017). Policy-makers and governors at these levels may chose to ignore them. TBTI is an effort to reduce this risk through research into the realities and prospects of small-scale fisheries. To recognize small-scale fisheries, we need to know about them. Not knowing is no excuse for policy inaction TBTI thus has a normative foundation, based on the idea that small-scale fisheries are worth supporting, that their problems can be resolved, and that transdisciplinary research is an essential contribution to make this happen.

The other meaning of recognition is about seeing. By what do we recognize small-scale fisheries? Whereas acknowledgement is about social and ethical values, and hence a political issue, recognition is analytical and empirical. Recognizing small-scale fisheries as an empirical phenomenon requires a conceptual tool: to be able to see them, you need to have something to see them *with*. Without a proper analytical framework, you may not recognize and, therefore, ignore them without wanting to. For TBTI, it has been essential to be able to provide such a lens to a wide range of audiences. Academically, we have offered 'interactive governance' as a theory and concept to see small-scale fisheries, as outlined in a number of publications (such as Kooiman et al. 2005; Bavinck et al. 2013; Jentoft and Chuenpagdee 2015). We believe, however, that interactive governance theory has more to offer, as we apply it to examine why small-scale fisheries continue to be ignored despite the recognition of their importance.

Thus, interactive governance recommends that the problems associated with the governance of small-scale fisheries should be sought within the system-to-be-governed, the governing system, and in the interactions that occur between the two. Where the governability problem is to be found cannot be determined *ex ante*, but must be subject to empirical examination from both 'etic' and 'emic' perspectives (Harris 1976). Interactive governance provides a lens, a set of concepts, by which governance problems can be examined. But in addition to this, the conceptual frameworks that small-scale fisheries people apply themselves, and the categories they apply to make sense of their world, must also be recognized. Otherwise, governance failure, in technical, social, and ethical senses, is a likely outcome.

In line with Rittel and Webber's (1973) description of the nature of wicked problems, many of the issues and challenges facing small-scale fisheries needing governance intervention are indeed 'wicked'. This means that the problems are complex, hard to define, and delineate from other problems, and are not solved once and for all. It is often not clear if, and when, the problems have been solved or whether they will reappear at a later stage. Such problems cannot be unilaterally defined and addressed by some higher authority, but require a collective and interactive approach to identify what they are in the first place. Government has a role to play, but interactive governance is also inclusive of other actors. It involves building partnerships with civil society organizations, academia, and small-scale fisheries stakeholders in a way that helps to level the playing field and broaden the knowledge base of decision making. Involving different actors and groups in an interactive and iterative process of problem identification, and eventually problem solving, is the premise for a transdisciplinary approach to fisheries governance.

Good small-scale fisheries governance calls for transdisciplinary knowledge, which involves more than the universal knowledge of natural science and the context specific knowledge that social science offers. Transdisciplinary science integrates knowledge of multiple academic disciplines and the contextual and ethically founded 'phronetic knowledge' of stakeholders (Flyvbjerg 2001; Jentoft 2006). Transdisciplinary governance, on the other hand, is about understanding how the wicked problems raise issues that are about social values, on which scientists have no supreme authority, and are, therefore, not the only relevant voice. The SSF Guidelines emphasize the importance of securing the interests, rights, and participation of small-scale fishers, especially those who are poor and marginalized, and who from their own experience know what this means. Interactive small-scale fisheries governance, therefore, involves facilitating participation of definitive stakeholders, empowerment of marginalized groups, and a process that is transparent and socially just, respecting human rights. In effect, addressing major concerns in small-scale fisheries requires both transdisciplinary science and transdisciplinary governance.

With this book, we are asserting the usefulness of the transdisciplinary approach. The book illustrates that the problems and priorities in small-scale fisheries require a broadening of perspectives that cuts across academic disciplines, that bridges division between scientific and local knowledge, and that bring about innovation in teaching and learning, and ultimately about the governance of small-scale fisheries. It aims at practitioners, researchers, government and non-governmental organizations, community groups, and civil society organizations involved directly or indirectly in the governance of small-scale fisheries at various levels. The book centers around main concerns and key policy priorities and interventions in small-scale fisheries, which TBTI identifies, related to economic viability, sustainable livelihoods, conservation and stewardship, allocation of rights and access, and governance performance.

Through a transdisciplinary lens, the book contributes to the creation of enabling environments for the implementation of the SSF Guidelines, including capacity building and empowerment through research, knowledge acquisition and integration, and communication. Readers of the book will learn to recognize the wicked problems of small-scale fisheries governance and the need to examine small-scale fisheries in their diversity, complexity, and dynamics, and the scale issues associated with their governance. These features make it difficult, if not impossible to arrive at some unified, agreed upon definition of what small-scale fisheries are. For this reason also, the SSF Guidelines refrain from defining them. Instead, definitions are left for member states to determine, based on the particular context within which small-scale fisheries operate, and the multiple characteristics of small-scale fisheries. Globally, these characteristics vary enormously, which governance actors must take into account. However, recognizing the diversity of small-scale fisheries, and what is unique about them at the place where they exist, does not detract from the need for small-scale fisheries governance to be principled, for example based on the human rights and other principles pertaining to good governance, as pointed out in the SSF Guidelines.

The SSF Guidelines provide a set of governance principles, all deliberated, negotiated, and endorsed. At the pinnacle are respect for human rights and dignity of participants within the small-scale fisheries value chain. These are universal principles that are applicable regardless of local context, but they must still, as part of the governance of implementation, be compared to those governance principles that are practiced in the fishing industry. Closing the gap between these high ideals and governance routine is integral to the implementation process. But as the gap varies from country to country and needed fisheries reforms may range from marginal to fundamental, as the case may be, so must the governance approach. From a governance perspective, the bigger the gap, the bigger the governability challenge. The moral and ethical foundation of these principles, and the persistent challenge that their implementation involves, are evidences of the wicked nature. The governance system and process must allow the problem definition and identification to emerge through an interactive process of collective reflection based on knowledge that transcends science (Rittel and Webber 1973). Small-scale fisheries governance is a process, which science can inform but not rule.

1.2 About this Book

TBTI set out to answer five key questions essential for the sustainability of small-scale fisheries globally, for which a range of more focused research clusters were established. The questions stemmed from literature review, discussion at the first small-scale fisheries congress in Bangkok in 2010, a synthesis of submitted questions to an online survey by small-scale fisheries researchers and stakeholders, a workshop in Cape Town in 2011, and various exchanges among network members. The questions were given the titles (*1*) *Strengthening the base,* (*2*) *Broadening the scope,* (*3*) *Enhancing stewardship,* (*4*) *Defending the beach,* and (*5*) *Governing the governance.* This volume is organized around these topics, with chapter contributions from cluster leaders and participants. These are also the headings for each section of this book. Crosscutting is the basic notion that all five questions refer to challenges and problems that are inherently 'wicked'. These questions point to complex problems that require in-depth examination, thick description, and broad perspectives to analyze and address. As such, they need a knowledge that integrates academic disciplines and transcends science by drawing on experience-based knowledge, ethical principles, and values of small-scale fisheries stakeholders along the value chain and gender divides. Such transdisciplinary knowledge must be generated and implemented through a governance approach that is inclusive and interactive, in a way that is empowering, socially just, and dynamic, and in accordance with human rights principles, as also stressed in the SSF Guidelines. Arguments for this approach and discussions for what it involves, are given in the first and last sections of this book with illustrations in between, in individual chapters. The goal of TBTI has been to elevate the profile of small-scale fisheries, by addressing their marginalization, and promoting their viability and sustainability. Chapter 2, titled **Too Big To Ignore – a transdisciplinary journey** by Ratana Chuenpagdee, provides an overview of the TBTI project as a whole, and how it has evolved with various publications until this book, which can also be read as a synthesis of current knowledge about small-scale fisheries.

Part 2: ***Broadening the scope*** has chapters devoted to the essential aspects of small-scale fisheries that need to be accounted for and emphasized in order to increase the awareness of their social contributions and overall importance to community viability, food security, and wellbeing. Included here are the understandings of public and private values through inquiry and reflection, going beyond present economic benefits to consider other aspects associated with small-scale fisheries, such as cultural, historical, and intergenerational values. Chapter 3, titled **The value of values for understanding transdisciplinary approaches to small-scale fisheries,** by Derek S. Johnson, Annie Lalancette, Mimi E. Lam, Marta Leite, and Sölmundur K. Pálsson, compares six approaches to study values in small-scale fisheries: economic valuation, ecosystem services, political economy, social wellbeing, interactive governance, and post-normal science. With this, they demonstrate what the benefits and challenges are in a value-sensitive transdisciplinary approach to small-scale fisheries. They argue that transdisciplinary approaches are not the same as synthesis, but a process of translation through communication between varying perceptions, paradigms, and priorities of academic disciplines and stakeholder knowledge. Chapter 4, by Philip A. Loring, David V. Fazzino II, Melinda Agapito, Ratana Chuenpagdee, Glenna Gannon, and Moenieba Isaacs, titled **Fish and food security in small-scale fisheries,** is about the role this sector makes in feeding the world. Fish remains among the most eaten foods and traded commodities in the world, and here small-scale fisheries play an essential role. Yet, the authors argue that small-scale fisheries are often left out of discussions about food security and the study of food systems, and, therefore, risk being undermined. Their chapter provides an overview of the various ways that small-scale fisheries currently provide for food security in regions around the world, and addresses aspects that are critical for food security, such as access, availability, and affordability of fish-as-food, and the degree to which people in fishing communities have the flexibility to respond to variability and change in any specific fishery. Chapter 5, co-authored by Kristen Lowitt, Charles Z. Levkoe, Andrew M. Song, Gordon M. Hickey, and Connie Nelson, is a case study of the Great Lakes region in North America, named **Broadening the knowledge base of small-scale fisheries through a food systems framework: a case study of the Lake Superior region.** They find that a dominating 'resourcist' management practice tends to leave out the diversity of local social and cultural values. They illustrate their point in three situations: the traditional fisheries of Batchewana First Nation; Eat the Fish, a small business marketing local fish through alternative food networks in Northwestern Ontario; and, Bodin's Fisheries in Wisconsin, a regional fish processor and retail outlet. The chapter concludes by offering a set of three key transdisciplinary questions on fishery-food system linkages that suggest the need for 'broadening the scope' of fisheries governance.

Part 3: ***Strengthening the base*** focuses on the existing and potential options for improving economic viability of small-scale fisheries and reducing their vulnerability by increasing their resilience to large scale processes of change that compromise small-scale fisheries resources and markets, through, for example, climate change and other eco-system stressors. In Chap. 6, Anna Schuhbauer, Andrés M Cisneros-Montemayor, and U. Rashid Sumaila argue that economic viability is often confused with financial viability, thus making profits the only relevant goal. In their chapter, **Economic viability of small-scale fisheries: a transdisciplinary evaluation approach,** they make a case for a more inclusive concept of viability as "the achievement of nonnegative net benefits to society over time", and explain how an assessment according to this definition can be carried out, by focusing on the total revenue and costs of fishing and fisheries subsidies, and by integrating attributes drawn from economic, social, ecological, and institutional aspects of fishing. Mexican fisheries are used as a case-study. To improve economic viability, policy-makers need to ensure balanced management in regard to social, economic, ecological, and institutional concerns. One key dimension seeking balance is addressed by Katia Frangoudes and Siri Gerrard in Chap. 7, **Gender perspective in fisheries: examples from the South and the North.** Women play an important role in small-scale fisheries all over the world. However, they often have few, if any, formal rights and gain little political attention. Therefore, there is insufficient knowledge about women's roles, gender relations and women's way of life, and how women are affected by ecological, economic, and social change. In their chapter, the authors review literature on women's roles, relations, and identities in small-scale fisheries, drawing from experiences in multiple countries in the global North and South. They are also interested in examining the extent to which the nationality, discipline, and gender of the researchers influence how women are portrayed. Chapter 8, titled **Markets, distribution, and value chains in small-scale fisheries: a special focus on Europe** by José J. Pascual-Fernández, Cristina Pita, Helga Hosupeit, Alicia Said, and João Garcia Rodrigues, focuses on the post-harvesting side of small-scale fisheries, in their capacity to sell their fish, receive fair prices, and add value to their catches. The authors observe that local fishing catches are not sufficiently differentiated in the market from large-scale fisheries and aquaculture. In their chapter, the authors describe what small-scale fishers do to add value and be more effective and innovative in the market place, and what transdisciplinary knowledge they need in order to pursue such strategies, using, as an example, the experience in Tenerife (Spain), where a particular branding of fish from the small-scale sector is promoted through collaboration between scientists, local government, and the industry. Chapter 9, **Governing for viability: the case of Velondriake Locally Managed Marine Area in Madagascar,** by Lovasoa Cédrique Augustave, takes the reader to Africa. The fisheries in the region of Madagascar went through a change when an export company focusing on the fresh octopus trade arrived. Overexploitation and degradation of the coral reef ecosystem and concern about the viability of the fishing communities resulted in a temporary no-take zone, as part of the locally managed marine area (LMMA). Behind this initiative were governments, research institutions, environmental organizations, and villagers entering into a partnership arrangement. The chapter performs a governability assessment to examine the natural and social characteristics of the fishing communities and the institutional and policy requirements for the successful implementation of the LMMA, which in the current situation is facing a number of challenges. The author argues that transdisciplinary perspectives are instrumental for the long-term sustainability and viability of the fishing communities.

Part 4: *Enhancing the stewardship* examines ways of minimizing environmental impacts and fostering stewardship within small-scale fisheries. Several questions are related to this big question, including identifying the carrying capacity of the ecosystem and describing gear and fleet interactions among small-scale fisheries, and between them and large-scale fisheries. The importance of promoting conservation and stewardship initiatives that make it possible for small-scale fisheries people to maintain their livelihoods without eroding their ecological base is well recognized. To achieve this, it is essential to know more about small-scale fisheries, where they are and what their situation is. Today, such data is largely missing at global and national levels. In Chap. 10, **Stewardship and sustainable practices in small-scale fisheries,** Patrick McConney, Rodrigo Pereira Medeiros, José J. Pascual-Fernández, and Maria Pena provide an overview of the many factors associated with stewardship and the sustainability of small-scale fisheries from a transdisciplinary applied science perspective. From the combined perspective of ecology, socioeconomics, and governance, they examine some promising multi-scale initiatives, which include diverse stakeholders. Topics addressed relate to fishing gear and methods; marine protected areas; livelihoods; illegal, unreported and unregulated fishing; postharvest and trade; and organizations ranging from producer cooperatives to the Marine Stewardship Council. Prateep K. Nayak and Fikret Berkes in Chap. 11, titled **Interplay between local and global: change processes and small-scale fisheries,** discuss how local and global drivers work together to affect the vulnerabilities of small-scale fisheries. Situated empirically in the Bay of Bengal, East Coast of India, their chapter addresses the conflict with aquaculture and how small-scale fisheries are using a conservation-development strategy to overcome the challenges. For these authors, vulnerability is multidimensional, complex, highly dynamic, and a relative concept, which they hold are topics in need of a transdisciplinary perspective. They believe that, because of the complexity associated with global change, small-scale fisheries will continue to be vulnerable to multiple challenges; but their inherent flexibility is a real asset that could ensure their viability. Chapter 12 remains in the same region with a contribution by Suvaluck Satumanatpan, Ratana Chuenpagdee, Wichin Suebpala, Thamasak Yeemin, and Kungwan Juntarashote, titled **Enhancing the stewardship in Trat Bay, Thailand: a transdisciplinary exercise.** The chapter applies a transdisciplinary perspective to examine issues and opportunities in an ecologically sensitive area, where small-scale fisheries compete with large-scale fisheries and other activities including urban development, tourism industry, and transportation. This makes a particularly complex governance challenge, exacerbated by a general lack of data to determine the levels of impacts. Small-scale fisheries are not well organized and have little political influence on decisions that affect the viability of their livelihoods. The authors argue that innovative thinking is required to level the playing field for all stakeholders, and to encourage collaboration and synergies among all sectors, with due recognition of the social and cultural importance that small-scale fisheries have in the context of conservation and stewardship.

Part 5: *Defending the beach* addresses one of the most challenging questions concerning the existence of small-scale fisheries: what mechanisms are required to secure physical space and rights for small-scale fishing people? As in the case study from Thailand, small-scale fishing is usually not the only activity taking place in coastal areas. Population expansion, a growing tourism sector, industrial fishing, aquaculture, tourism, and no-take marine protected areas, among others, all compete for space with small-scale fisheries. Thus, small-scale fisheries risk being pushed aside and denied access. These developments raise the issue of territorial rights and ownership and, hence, power relations. Evelyn Pinkerton, in Chap. 13, named **Strategies and policies supporting small-scale fishers' access and conservation rights in a neoliberal world,** takes a global look on how market-based quota systems are affecting small-scale fisheries. She argues that small-scale fisheries are being systematically excluded from access to, and benefits from, fisheries. Still, they persist, in many ways, through clever innovations at various scales, such as forming organizations that buy up, or are allocated, preexisting privatized fishing quotas and then lease them out to members at affordable prices. In some instances, government is regulating the quota market in ways that support small-scale fisheries. She argues that these strategies suggest the important role of transdisciplinary thinking that goes beyond economics to consider issues of distribution and community survival. Concerns of access and rights are even more prominent in the indigenous context. Svein Jentoft, Natasha Stacey, Jackie Sunde, and Miguel González illustrate this by exploring the political legal frameworks for securing the tenure rights of indigenous small-scale fisheries in Norway, Australia, South Africa, and Nicaragua. In their Chap. 14, titled **The small-scale fisheries of indigenous peoples: A struggle for secure tenure rights,** indigenous peoples are protected by international and domestic law; but such frameworks have yet to bring them out of the marginalized and impoverished situation they often find themselves in, not only within their countries, but also within their industry. The authors reflect on which institutional reforms would facilitate the self-determination and sustainable economic development of indigenous small-scale fisheries. Chapter 15, **Defending the beach: transdisciplinary approaches in small-scale fisheries in Pernambuco, Brazil,** by Matias John Wojciechowski, Beatriz Mesquita P. Ferreira, Daniele A. Vila-Nova, and Sérgio M. Gomes de Mattos, brings the reader to an area where small-scale fishing activities are targeting shellfishes, either as a full or part time livelihood activity and food source. However, conflicts related to access to beaches, estuaries and seas due to privatization of land, a tourism industry in demand of space, aquaculture, and increasing urbanization, involve fishers into an emblematic struggle to defend their territorial interests and rights.

Part 6: *Governing the governance* provides insights into the question of what institutions and principles are suitable in the governance of small-scale fisheries. The underlying assumption is that, for the most part, current governance systems do not sufficiently address the interests of small-scale fishing people. Neither do they enable them to become directly involved in the act of governance. The diversity, complexity, and dynamics of small-scale fisheries worldwide pose major challenges to governance. Appropriate institutions need to be designed and established in order to provide action spaces for small-scale fishers to manoeuvre in the changing economic, social, and political landscape within which they operate. Principles that align well with small-scale fisheries' nature, and the images governing them, are likely different from those for large-scale fisheries. Innovative governance thinking is therefore needed. Chapter 16, **Governing change in small-scale fisheries: theories and assumptions,** by Svein Jentoft, explores, at a meta-level, the ideas of how governance can generate the change needed to bring small-scale fisheries out of poverty, while adapting to climate change related impacts. He argues that the two problems, which are both wicked, are closely related. Climate change adaptation would need to help bring people out of poverty, as poverty is a hindrance to their adaptive capacity. He argues that change is not the same as progress, and adaptation to climate change is not synonymous with development. Small-scale fisheries would need to undergo change that people in the industry can associate with progress, and for that governors must recognize the ethical implications and demands of fisheries development initiatives and strategies. In Chap. 17, **Transdisciplinary engagement to address transboundary challenges for small-scale fishers,** by Joeri Scholtens, Andrew M. Song, Johny Stephen, Catalina García Chavez, Maarten Bavinck, and Merle Sowman, the focus is on how transboundary maritime issues complicate the operation and welfare of small-scale fishers, and may further their marginalization. Seeing this as a transdisciplinary issue in the pursuit of empowering small-scale fishers vis-à-vis external dynamics that challenge their survival, the interventions needed would be partly about building organizational capacity and government interventions aimed at the system interactions as a whole. They argue that, regardless of having different perspectives, incentives, and accountabilities, transdisciplinary engagement can be empowering for small-scale fishers in transboundary governance. Chapter 18 is **Using transdisciplinary research solutions to support governance in inland fisheries,** written by Shannon D. Bower, Andrew M. Song, Paul Onyango, Steven J. Cooke, and Jeppe Kolding. Inland and marine small-scale fisheries face many of the same challenges, which require long-term solutions to governance issues as well as engagement and cooperation from multiple disciplines and actors. Pollution, habitat alteration, invasive species, and hydropower development are often external threats imposed on inland small-scale fisheries and are among the pressures impacting their sustainability, in addition to factors such as stakeholder conflict and political negligence. The authors support the promotion of transdisciplinarity in addressing inland small-scale fisheries governance issues, which in this particular case, needs to go beyond looking at fisheries alone. They also offer suggestions for how to achieve transdisciplinarity, discussing the tradeoffs involved in achieving sustainability. Chapter 19, **Governing the governance: small-scale fisheries in Europe with focus on the Baltic Sea,** by Milena Arias-Schreiber, Sebastian Linke, Alyne E. Delaney, and Svein Jentoft, addresses the values and interests of governance institutions and how they align with small-scale fisheries in European fisheries in general, and in the Baltic in particular. They discuss how supportive the governance system is of the values and interests of the small-scale fisheries and what the scope is for advancing their governability. The chapter explores fishers' representation in the EU Baltic Sea Advisory Council for this purpose and discusses the consequences of a missing link between small-scale coastal fisheries' values and the overarching EU fisheries governance system, as regulated by the Common Fisheries Policy of the European Union. They conclude that correcting this would require transdisciplinary approaches.

Part 7: ***Towards transdisciplinarity in fisheries governance*** is about what this approach may involve in practice. If transdisciplinarity is seeking to bring different disciplines and knowledge types into both the research and governance of the multiple wicked problems that are associated with the resilience, empowerment, and, ultimately, the sustainability of small-scale fisheries, some methodological challenges need reflection. One should not assume that such an approach is straightforward to apply. Indeed one may argue that transdisciplinary research and governance are in themselves wicked problems. Single disciplines are too limited in their perspectives and methods to single-handedly define the problems and adequate solutions for small-scale fisheries. There are simply too many concerns that need to be recognized and sorted out, as they are sometimes not only inherently complex, but also in conflict. Bringing stakeholders, with their experience-based, but also interest-framed, knowledge into the process would hardly make the challenge any easier. Yet, that is what small-scale fisheries need to do. One of the first challenges is related to information about small-scale fisheries. In Chap. 20, titled **Beyond the basics: improving information about small-scale fisheries,** Melinda Agapito, Ratana Chuenpagdee, Rodolphe Devillers, Jennifer Gee, Andrew F. Johnson, Graham J. Pierce, and Brice Trouillet, argue that the multiple dimensions of small-scale fisheries and their great diversity globally, make data acquisition and integration challenging. So far, systematic data collection about the characteristics and contribution of small-scale fisheries is mostly lacking, and information systems drawing data from multiple sources, through onboard sampling and market studies, have been focused mostly on large-scale fisheries. The chapter discusses whether tools, approaches, and models used in assessing large-scale fisheries are applicable for small-scale fisheries. The authors present examples of TBTI initiatives to improve small-scale fisheries data provision, and suggest directions for how to proceed in the future. Chapter 21, by Mbachi Ruth Msomphora and Svein Jentoft, with the title **Transdisciplinary science for small-scale fisheries,** pinpoints what they believe are the essential principles of transdisciplinary research, applied to small-scale fisheries. They discuss a shift in science processes that has been going on for some time, often labelled as a move from Mode 1 to Mode 2, which has changed the ethos and practice of science, from a basically curiosity-driven approach embedded in academic disciplines to a more practical, externally driven, and funded process, where industries demand science-based solutions to their problems. This change has, they argue, watered down some of the traditional values and criteria of science. They point to the risk that the process may have gone too far, that traditional qualities of academic science are still worth defending, and that there is a need to find a more productive middle ground, which is what they think transdisciplinarity in small-scale fisheries should aim for, in order to enhance their governability. Chapter 22, **The principles of transdisciplinary research in small-scale fisheries,** by Alicia Said, Ratana Chuenpagdee, Alfonso Aguilar-Perera, Minerva Arce-Ibarra, Tek Bahadur Guru, Bonnie Bishop, Marc Léopold, Ana Isabel Márquez Pérez, Sérgio M. Gomes de Mattos, Graham J. Pierce, Prateep K. Nayak, and Svein Jentoft, argues that there is no tailor-made solutions for small-scale fisheries problems, which implies that solutions must be appropriate to context, but that the interactions between the natural and social systems require approaches that integrate different academic and stakeholder perspectives and knowledge, the idea underpinning the TBTI project. The chapter further elaborates what transdisciplinarity means and involves. The argument is also that small-scale fisheries governance requires institutional and academic innovation at local and national scales that facilitates transformative learning. Finally, in Chap. 23, titled **Transcending fisheries knowledge: from theory to integration,** Ratana Chuenpagdee and Svein Jentoft discuss what steps are required to go beyond transdisciplinary research to its implementation in real world situations of small-scale fisheries and their governance. They see this as something that requires more than the conventional dissemination of research findings, as something that should be accompanied by concrete actions. In their chapter, they are primarily interested in how transdisciplinary knowledge can be converted into programs of education for small-scale fisheries governors and stakeholders. They believe that new information technology, via the internet, can provide a useful platform for developing the on-the-ground learning processes and build local research and governance capacity that the implementation of the SSF Guidelines requires.

References

Bavinck M, Chuenpagdee R, Jentoft S et al (2013) Governability of fisheries and aquaculture: theory and applications. Bavinck, Chuenpagdee, Jentoft, Kooiman. MARE publication series, vol 7. Springer, Dordrecht

FAO (Food and Agriculture Organization of the United Nations) (2015) Voluntary guidelines for securing small-scale fisheries in the context of food security and poverty eradication. FAO, Rome

Flyvbjerg B (2001) Making social science matter. Why social inquiry fails and how it can succeed again. Cambridge University Press, Cambridge

Harris M (1976) History and significance of the emic/etic distinction. Annu Rev Anthropol 5:329–350

Jentoft S (2006) Beyond fisheries management: the phronetic dimension. Mar Policy 30:671–680

Jentoft S (2014) Walking the talk: implementing the international voluntary guidelines for small-scale fisheries. MAST. http://www.maritimestudiesjournal.com/content/13/1/16

Jentoft S, Chuenpagdee R (eds) (2015) Interactive governance for small-scale fisheries: global reflections, MARE publication series, vol 13. Springer, Dordrecht

Jentoft S, Chuenpagdee R, Barragán-Paladines MJ et al (eds) (2017) The small-scale fisheries guidelines: global implementation, MARE publication series, vol 14. Springer, Dordrecht

Kooiman J, Bavinck M, Jentoft S et al (eds) (2005) Fish for life: interactive governance for fisheries. Amsterdam University Press, Amsterdam

Rittel HWJ, Webber MM (1973) Dilemmas in a general theory of planning. Policy Sci 4:155–169

Chapter 2
Too Big To Ignore – A Transdisciplinary Journey

Ratana Chuenpagdee

Abstract While the concerns and challenges affecting fisheries today are applicable to both large- and small-scale fisheries, neither the impacts nor the solutions to address the problems are the same for both sectors. A large amount of research, particularly from social science studies, has helped enhance our understanding about small-scale fisheries and revealed several nuances that need to be considered in management and governance efforts. Compiling the existing information and bringing the researchers together to identify knowledge gaps in small-scale fisheries is the critical first step initiated by the Too Big To Ignore (TBTI) – Global Partnership for Small-Scale Fisheries Research. Elevating the profile of small-scale fisheries, addressing their marginalization, and promoting their viability and sustainability are major goals of TBTI that cannot be achieved only by researchers. A partnership approach is, therefore, taken to foster collaboration between researchers and fishers' associations, civil society organizations, environmental groups, and governments at all levels in the identification of the problems, as well as in the search for solutions and opportunities. This chapter presents an overview of the transdisciplinary journey that TBTI has taken.

Keywords TBTI · Global partnership · Transdisciplinarity · Big questions · Information system · Capacity building

2.1 A Global Look at Small–Scale Fisheries

More than 120 million people around the world depend directly on fishing and other fisheries-related activities, including processing and trading (HLPE 2014). Of these, small-scale fisheries represent about 90%, with the vast majority of them residing in developing countries, particularly Asia (FAO 2014). Small-scale fisheries contribute at least 25% of the world fisheries catches (Pauly and Zeller 2016), and unlike

R. Chuenpagdee (✉)
Department of Geography, Memorial University of Newfoundland, St. John's, NL, Canada
e-mail: ratanac@mun.ca

R. Chuenpagdee, S. Jentoft (eds.), *Transdisciplinarity for Small-Scale Fisheries Governance*, MARE Publication Series 21,
https://doi.org/10.1007/978-3-319-94938-3_2

large-scale fisheries, almost all of these catches are for human consumption (World Bank et al. 2012). The importance of small-scale fisheries has long been recognized, and recently heightened with the adoption of the Voluntary Guidelines for Securing Sustainable Small-Scale Fisheries in the context of Food Security and Poverty Eradication (hereafter SSF Guidelines; FAO 2015) by the member states of the Food and Agriculture Organization (FAO) of the United Nations in June 2014 at the 31st session of the Committee on Fisheries. With the SSF Guidelines in place, small-scale fishers and their organizations and governments, along with other stakeholders interested in supporting and promoting the sustainability of small-scale fisheries, including civil society organizations (CSOs), environmental groups, and the research community can work collaboratively and in concert. The numbers associated with small-scale fisheries, the diverse geography they operate in, and the importance of the contributions they make means that a great effort is required to achieve the goals stipulated in the SSF Guidelines, and elsewhere.

An example of the latter is the Sustainable Development Goals (SDGs), which has one target (14b) that specifically addresses small-scale fisheries access to resources and markets (United Nations 2015). The contribution of small-scale fisheries to jobs and livelihoods, local food security and poverty alleviation makes it imperative to link small-scale fisheries to other SDGs, particularly SDG 1 (no poverty), SDG 2 (zero hunger), SDG 5 (gender equality), SDG 8 (decent work and economic growth), and SDG 13 (climate action) (Diz et al., 2017; McConney et al. Chap. 10, this volume). Other SDGs that aim at improving quality of life of the people are highly relevant to small-scale fishing communities who often live in rural or remote areas with poor infrastructure and facilities; for example, SDG 3 on good health and wellbeing, SDG 4 on quality education, SDG 6 on clean water and sanitation, SDG 7 on affordable and clean energy, and SDG 10 on reduced inequalities. Finally, many small-scale fisheries are in close proximity of cities and urban areas, making it ideal for them to help achieve SDG 11 related to sustainable cities and communities and promote responsible consumption and production (SDG 12).

What are the real values and contributions of small-scale fisheries, one may ask? As an intergovernmental agency responsible for promoting and recommending actions on fisheries-related issues, FAO has convened several conferences, workshops, and expert consultations, with emphasis on small-scale fisheries. Most notably, an expert consultation on the 'Role of Small- Scale Fisheries in Poverty Alleviation and Food Security' was held in Rome, in July 2004, which expanded the Code of Conduct for Responsible Fisheries to focus on small-scale fisheries. These technical guidelines recognize the comprehensive characteristics of small-scale fisheries, as follows:

> Small-scale fisheries can be broadly characterized as a dynamic and evolving sector employing labour intensive harvesting, processing and distribution technologies to exploit marine and inland water fishery resources. The activities of this subsector, conducted full-time or part-time, or just seasonally, are often targeted on supplying fish and fishery products to local and domestic markets, and for subsistence consumption. Export-oriented production, however, has increased in many small-scale fisheries during the last one to two decades because of greater market integration and globalization. While typically men are

> engaged in fishing and women in fish processing and marketing, women are also known to engage in near shore harvesting activities and men are known to engage in fish marketing and distribution. Other ancillary activities such as net- making, boatbuilding, engine repair and maintenance, etc. can provide additional fishery-related employment and income opportunities in marine and inland fishing communities. Small-scale fisheries operate at widely differing organizational levels ranging from self-employed single operators through informal micro- enterprises to formal sector businesses. This subsector, therefore, is not homogenous within and across countries and regions and attention to this fact is warranted when formulating strategies and policies for enhancing its contribution to food security and poverty alleviation (FAO 2005, 4).

From the research community, early efforts offer some estimates that indicate the importance of small-scale fisheries in terms of catches and number of fishers involved (Chuenpagdee et al. 2006), also in comparison to large-scale sector (Pauly 2006), with better figures emerging later on (see, for instance, World Bank et al. 2012; Pauly and Zeller 2016). In addition to providing data and information that suggest the breadth of values and contributions of small-scale fisheries, some studies also point to how small-scale fisheries are more environmental friendly than their large-scale counterparts. For instance, they consume less fuel, have fewer discards, and make better use of the catch such that much less of the catch goes into reduction, when compared with large-scale fisheries (Pauly 2006; Teh and Pauly 2018).

Many studies argue that the values of small-scale fisheries go beyond those that are measurable, and that the richness and meaning of the sector is best portrayed through thick description and narratives. This also helps avoid the temptation to generalize fisheries as if all are the same, or the oversimplification in management policies that tend to treat small-scale fisheries as problems. The differences in fisheries are not only between large-scale and small-scale, but are also among small-scale fisheries in various locations and parts of the world. In fact, it is the diversity of small-scale fisheries, in the pre-harvest, harvest, and post-harvest activities that gives rise to their numerous values (see Johnson 2018; Johnson et al. Chap. 3, this volume). It is also for this reason that the SSF Guidelines do not provide a global definition for small-scale fisheries, but rather describe, generally, what they are and what they represent. The task of defining small-scale fisheries in a country is left to the member states, allowing contextualization to its social and political characteristics.

Despite the increasing interest in small-scale fisheries, many of them remain marginalized, ignored, or dismissed as relics of the past. In many countries, this marginalization is shown by inadequate financial, institutional, and scientific support for small-scale fisheries, and an under-representation of the concerns of people working in this sector in policy discussions (Béné 2003; Béné et al. 2007; Salas et al. 2007; Jentoft and Eide 2011). Schuhbauer et al. (2017) illustrate this with their studies on the distribution of subsidies, showing that small-scale fisheries receive only about 15% of the total subsidies given to fisheries globally. The under-appreciation of the importance of small-scale fisheries and the misplaced policies not only magnify the difficulty that the sector faces in maintaining viable livelihoods, but also increase their vulnerability to global change processes, such as urbanization, globalization, and climate change, which threaten their sustainability.

The local complexity and dynamic reality of small-scale fisheries poses major challenges to researchers, managers, policy-makers, environmental groups, fishers' organizations, and others interested in understanding and elevating the profile of this sector. Isolated work by individual researchers in many areas of the world is likely to effect change in those locations, but integration of knowledge generally does not occur due to the dissimilarity in the approaches and perspectives underlying the research. Thus, comparative analysis and global understanding of small-scale fisheries is unavailable. Consequently, issues and concerns important to small-scale fisheries continue to have low priority in national policies. In turn, management decisions are not adequately sensitive to the conditions and needs of this sector of the population, and do not fully acknowledge their roles and contributions to the society and global sustainability. The lack of fine-scale, comprehensive information about small-scale fisheries, and the lack of mechanisms to share and exchange knowledge are two key factors inhibiting the ability to properly address issues and concerns related to this sector. They also hinder the potential of small-scale fisheries to contribute to global food security and environmental stewardship.

This is the premise for the creation of the Too Big To Ignore – Global Partnership for Small-Scale Fisheries Research (TBTI; toobigtoignore.net). Funding for TBTI came from the Social Sciences and Humanities Research Council of Canada (SSHRC) through its inaugural partnership program in 2012. SSHRC's vision for partnership grants is to support formal partnerships, either between academic institutions or between academic institutions and non-academic partner organizations, to work collaboratively to achieve shared goals. The vision for a partnership approach to research fits well with the multiplicity of issues, concerns, and challenges affecting small-scale fisheries. The following describes how TBTI came to be, what it aims to do and why a partnership approach is needed. Next, it presents the big questions, highlighting key outputs and reflection on the lessons learned. The chapter concludes with the role of TBTI in promoting transdisciplinary research and capacity building for small-scale fisheries governance.

2.2 The Need for a Large–Scale Research Network

Small-scale fisheries occur in all types of waters, large or small, in coastal areas and open seas, rivers and streams, lakes and lagoons, and in urban and rural settings. The multi-nature of small-scale fisheries allows them to be part of a formal and informal economy, supporting livelihoods and food security of millions of people in all corners of the world, including the most remote areas. Despite the geographical spread, a large portion of small-scale fisheries is globally connected through mobility and migration, markets and trade, and financial institutions and governance. The 'global' nature of small-scale fisheries calls for concerted research, capacity building, and governance efforts from key fisheries stakeholders and supporting organizations around the world, which is what TBTI aims to offer.

The goals of TBTI are to elevate the profile of small-scale fisheries, to argue against their marginalization in national and international policies, and to develop research to address global food security and sustainability challenges in fisheries policy. The partnership has five specific objectives: (1) to provide evidence to promote recognition and understanding of the importance of small-scale fisheries to livelihoods, wellbeing, poverty alleviation, and food security; (2) to explore their potential contributions to economic growth and development, environmental sustainability, stewardship, and community resilience; (3) to assess their vulnerability to anthropogenic global change processes, such as the growth of large-scale fishing operations, climate change, aquaculture development, tourism, marine protected areas, the private enclosure of coastal spaces, urbanization, and migration; (4) to encourage policy discussions and contribute information for improving decision making about small-scale fisheries; and, (5) to advance knowledge and build local and global capacity in research and governance for the future of small-scale fisheries. Clearly, a single organization or a small group of researchers cannot achieve these big and ambitious goals on their own. When considering financial constraints and competing policy priorities around the world today, bringing together community groups, non-governmental organizations, and governments to work collaboratively with researchers makes sense. In addition to being cost-effective, such partnerships are also critical in sustaining local capacity, research, and governance efforts that are being developed through TBTI and other initiatives.

The long-term funding support from SSHRC (7-year program) makes a major difference in how the partnership has developed and is evolving. The SSHRC partnership program also requires partner organizations and research institutions involved in the research to contribute at least 35% of total funding as additional resources over the life of the project. The stipulation of in-kind or matching contributions to pursue collaborative research encourages mutual ownership and shared responsibility of the project, which, in turn, can help project deliverables, extend the network, and also sustain some of the activities after the funding ends. At the start of the project, TBTI worked with 62 researchers from 27 countries, and 15 intergovernmental organizations, environmental organizations, and non-governmental organizations based in Canada, Italy, Mexico, Namibia, Norway, South Africa, Sri Lanka, Thailand, and the USA (Fig. 2.1). Today, the network has expanded to more than 500 members and many more organizations working in at least 50 countries around the world.

Building a global partnership is not without challenges, however. Today's information and communication technologies, and the multiple networks that already exist among researchers, governments, and non-governmental organizations help make it possible. Within TBTI, significant effort has been allocated to foster interactions and build connection between members and partners, through tools such as the Internet, web conferencing, social media, online searchable databases, and file sharing systems. The TBTI website and the monthly digest circulated to members through Listserv and as postings on the website are effective means for outreach and

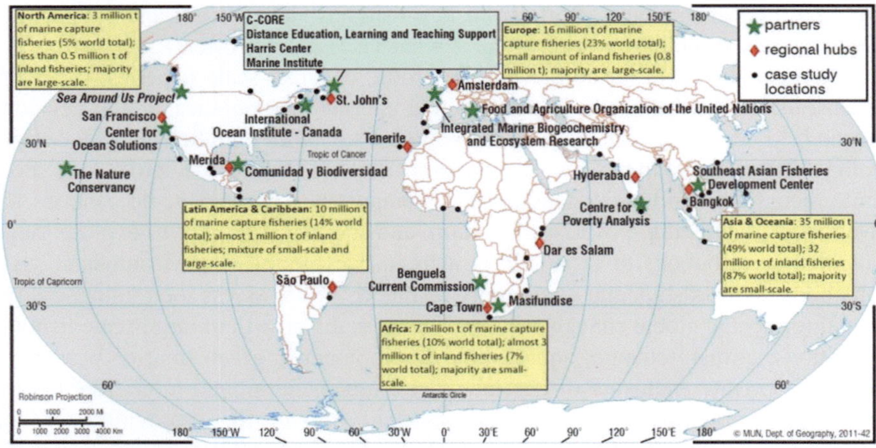

Fig. 2.1 The original structure of TBTI at the start of the project in 2012, showing partners, regional hubs, and case studies

knowledge dissemination about small-scale fisheries. TBTI is not only a virtual network, however. Many of its members are part of academic and professional organizations with regular meetings and/or participation in scientific conferences related to fisheries. Opportunities for TBTI members to meet occur at various venues, such as the MARE conference series, People and the Sea, held biennially in Amsterdam, and the World Small-Scale Fisheries Congress (WSFC), held every 4 years in different locations. Not only is the WSFC a flagship conference for small-scale fisheries, the first congress held in Bangkok in Thailand in 2010 also gave rise to the idea to create a global research network like TBTI.

Global partnership is not possible without regional networks. TBTI draws heavily on the existing networking and coordinated activities within regions and subregions in research and knowledge mobilization efforts. TBTI is coordinated into five regions to facilitate some of the regional work and outreach in Africa, Asia and Oceania, Europe, Latin America and the Caribbean and North America. These areas have different capacity, mechanisms, and institutional arrangements to promote collaboration and concerted actions. Within Latin America and the Caribbean, for instance, the three Regional Fisheries Bodies, namely the Caribbean Regional Fisheries Mechanism (CRFM), the Organisation of the Central American Fisheries and Aquaculture Sector (OSPESCA) and the FAO-Western Central Atlantic Fisheries Commission, were already working together, and strengthened their collaboration early in 2016 with a Memorandum of Understanding for further coordination for fisheries sustainability. Further, the region has strong and well-organized CSOs operating in many countries, along with several universities and research institutes. It also has a long-running annual conference series organized by the Gulf and Caribbean Fisheries Institute, entering its 71st conference in 2018. These regional networks and other coordinated work in the region are imperative to sustaining efforts to promote the wellbeing and viability of small-scale fisheries.

2.3 Identification of the Big Questions

As complex and dynamic social-ecological systems, small-scale fisheries pose major research and governance challenges. Social and natural scientists have done an enormous amount of research on this sector based on their disciplinary perspectives, and more recently from multi- and interdisciplinary approaches. Yet, the majority of research is small in scope and frequently takes place in remote places, thus inhibiting comprehensive understanding of the overall issues and challenges broadly affecting small-scale fisheries. Fishing people themselves are not empowered, often living in poor conditions, with little political connection. These conditions require innovative and transformative approaches to research, starting with the framing of the research questions, to how the research is conducted, information shared, and knowledge disseminated. We argue in TBTI that small-scale fisheries research requires an analytical framework that draws on multiple disciplinary foundations but also moves beyond individual disciplines towards a new transdisciplinary approach.

An integrative perspective underpinning a lot of research in TBTI is interactive governance theory (Kooiman 2003; Kooiman et al. 2005), which has been applied to investigate the specific and complex problems of governing fisheries in various settings around the world, such as Canada (Khan and Chuenpagdee 2014), India (Bavinck and Vivekanandan 2011), Indonesia (Triyanti et al. 2017), Malawi (Song and Chuenpagdee 2010), Nicaragua (Gonzalez and Jentoft 2011), Tanzania (Onyango and Jentoft 2010), Spain (Pascual-Fernandez and De la Cruz Modino 2011), and Thailand (Satumanatpan and Chuenpagdee 2015). The interactive governance perspective focuses on understanding the characteristics of the social and ecological systems that are being governed, the governing systems, and their interactions. It also offers an analytical framework using the concept of 'governability', which measures the relative ease or difficulty of governing a given fishery (Bavinck et al. 2013). This holistic and systemic approach, along with its analytical lens, offers a powerful tool to examine the complex small-scale fisheries problems. As suggested by Rittel and Webber (1973), however, identifying what these problems are is indeed a problem, whose definition requires an interactive governance process to determine.

Following a transdisciplinary perspective, TBTI research questions were identified through a participatory and bottom-up process. It began with the 'Top 100 Questions' online survey inviting small-scale fisheries stakeholders to contribute questions that they believed should be researched. Contributors to the surveys were mainly researchers and practitioners. In order to further elicit inputs from stakeholders, particularly fishers, CSOs, and policy-makers, regional multi-stakeholder workshops were organized in Thailand, Mexico, and South Africa to discuss these potential questions, which eventually led to consensus around big five research questions for TBTI (Box 2.1; see also Jentoft and Chuenpagdee Chap. 1, this volume).

Box 2.1 The Big Five Questions Guiding TBTI Research

Strengthening the base: What options exist for improving economic viability of small-scale fisheries and increasing their resilience to large-scale processes of change?

Broadening the scope: What aspects of small-scale fisheries need to be accounted for and emphasized in order to increase awareness of their actual and potential social contributions and their overall societal importance?

Enhancing the stewardship: What alternatives are available for minimizing environmental impacts and fostering stewardship within small-scale fisheries?

Defending the beach: What mechanisms are required to secure livelihoods, physical space, and rights for small-scale fishing people?

Governing the governance: What institutions and principles are suitable for the governance of small-scale fisheries?

While these questions can be independently examined, they are interdisciplinary in nature. The interconnectedness of these questions invites cross-fertilization of ideas and sharing of tools and approaches, resulting in the formation of multiple research groups, with numerous team members from a wide range of disciplinary backgrounds, training, and experience. More than 100 case studies guided by these questions have been conducted around the world by students and post-doctoral researchers funded through TBTI and other sources, as well as TBTI researchers and colleagues (see Appendix 2.1 for publication titles and volumes). Several contributors are practitioners and non-academic researchers working with CSOs. In addition to five research working groups addressing the big questions, TBTI structures two working groups to integrate and mobilize knowledge: the Global Synthesis and the Transdisciplinary Learning (see details in the next sections).

The release of the SSF Guidelines in 2014 coincided with the 2nd WSFC, which took place in Merida, Mexico in September of that year. Unlike other scientific gatherings, WSFC provides a platform for fishers, practitioners, policy-makers, and academics to interact, exchange information and knowledge, and discuss research and policy directions for sustainable small-scale fisheries. Of about 400 participants from 50 countries participated in the 2nd WSFC, nearly 50 people were small-scale fishers from all regions of the world, and close to 100 were students, early career

Fig. 2.2 TBTI research clusters showing original working groups in dark blue and new ones in lighter colour

scientists, and members of non-governmental organizations. The congress was an important milestone for TBTI as it prompted the network to evaluate the work and make necessary adjustments. Considering the issues raised in the SSF Guidelines and what they try to achieve, and assessing what the network had done up to that point, TBTI restructured its research focus into 15 clusters, retaining the original big questions, but also adding new topics that require more research attention (Fig. 2.2). By focusing on topics such as fish as food, market opportunities, global change responses, transboundary fisheries, inland fisheries, indigenous fisheries, women and gender in fisheries, and the SSF Guidelines, along with small-scale fisheries rights, values and wellbeing, economic viability, stewardship, and governing the governance, and through multiple case studies, TBTI research clusters are poised to provide evidence and directions to support small-scale fisheries sustainability worldwide. The research cluster topics correspond with the key aspects promoted in the SSF Guidelines, including governance of tenure, social development, value chain and trade, gender equality and equity, and climate change. The contributed chapters in this book present research findings and recommendations on the cluster topics (see Jentoft and Chuenpagdee Chap. 1, this volume and Chuenpagdee and Jentoft Chap. 23, this volume).

2.4 The Focus on Knowledge Sharing and Integration

As previously mentioned, information about small-scale fisheries around the world is scattered, and studies are generally local in scope. The number, diversity, complexity, and the geographical spread of small-scale fisheries around the world, including in remote places, also pose major challenges for data collection. Several groups and organizations have attempted to build capacity through cooperation, but mechanisms for broad-based synthesis of knowledge about small-scale fisheries worldwide are limited. Recognizing this need, and bearing in mind the specificity of small-scale fisheries, TBTI took an 'agile' approach to develop the 'Information System on Small-Scale Fisheries' (ISSF; issfcloud.toobigtoignore.net) as a web-based system, with a user-friendly interface and an iterative process to prioritizing key attributes to include in the database (Chuenpagdee et al. 2017). Through a 'crowd-sourced' platform, ISSF invites expert and non-expert users to contribute data and to explore information in map or table format. As such, ISSF offers a unique one-stop Web portal for data sharing, cross-fertilization of ideas, co-creation of knowledge, and collaborative building of research and governance capacity in small-scale fisheries. The number of records in ISSF continues to grow (more than 2500 records, contributed by 500 people from 60 countries), while the website garners about 1500 page views per month and continues to move closer to the top of Google search results for relevant terms.

The transdisciplinary nature in ISSF is reflected in the consultative approach and broad engagement taken to develop the platform, particularly related to the decisions about the interface, functionality, and content. Many considerations went into the technical aspects, like decisions between different technology options that could provide flexibility and good end-user experience. The use of drill down tools, map or table views, and aggregate views of data went through user-friendliness testing. In order to overcome some of the language barriers, a 'Google Translator Plugin' is embedded on the main page for automated translation into 90 languages. From the content perspective, since ISSF was the first data platform of its kind, it was tempting to make it be and do everything. Numerous discussions with the content experts and users took place throughout the two-year development phase in order to strike the balance between the 'need-to-know' and the 'nice-to-know.' Multi-stakeholder workshops were held, along with additional consultations using emails, web conferences, and online surveys.

As a knowledge platform, ISSF shares information about researchers and organizations working in small-scale fisheries, as well as presents key descriptions about research and activities that have taken place around the world. In the 'SSF profile', it offers a unique opportunity for researchers and practitioners to portray what small-scale fisheries look like using the same 20 questions to describe the ecological, economic, sociocultural, and governance characteristics associated with small-scale fisheries in a location. This approach enables easy comparison between

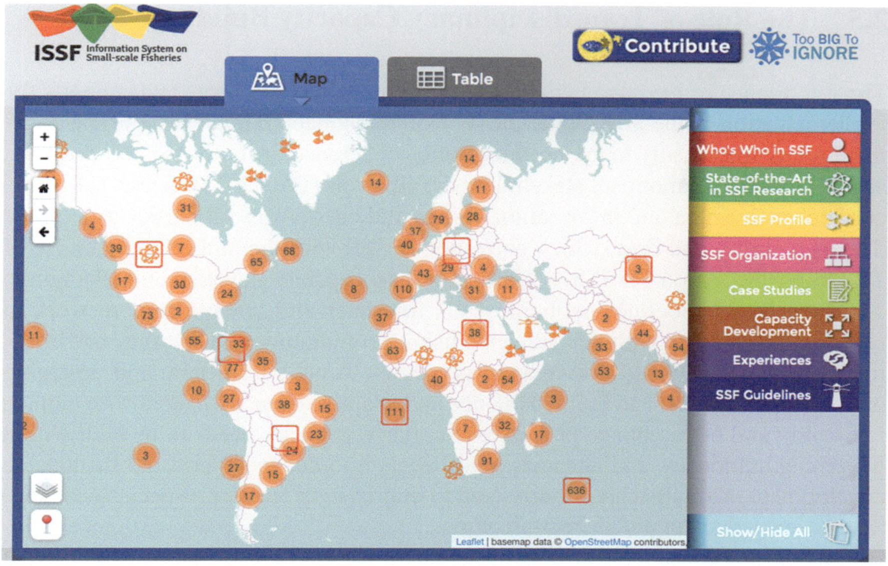

Fig. 2.3 ISSF web-interface showing the name of the eight data layers

fisheries and, with sufficient numbers of contributions, will allow data to be aggregated and synthesized to the national and regional levels. Another key aspect of ISSF is that it also caters to fishers and community groups through the 'case studies', 'capacity development', and 'experiences' layers (Fig. 2.3), which can accommodate textual data, photos, and videos. ISSF also contains a layer to help track activities related to the SSF Guidelines.

At the most basic level, ISSF helps put small-scale fisheries on the map, giving them visibility and raising their profile. With such an extensive and comprehensive information system, however, it is possible, for the first time, to learn more about small-scale fisheries in a location, compare them with others, conduct regional comparative 'state-of-the-art' analyses, to identify knowledge gaps and local capacity development needs, and, finally, to address key issues and concerns affecting small-scale fisheries sustainability. While a potentially powerful tool, ISSF is similar to other crowdsourced efforts that can only realize its potential when it is well utilized. As discussed in Chuenpagdee et al. (2017), challenges facing ISSF are in two major categories: data availability and quality, and accessibility issues due mostly to lack of or poor Internet connection or language barriers. The former can be dealt with through active promotion of the platform and vigilance in monitoring and validating data, while the latter is more difficult to address, as it requires structural change and institutional intervention.

2.5 The Role of Transdisciplinary Capacity Building

The importance of having an enabling environment that facilitates and supports the implementation of the SSF Guidelines is well recognized. Thus, Part 3 of the document refers specifically to the need for policy coherence, institutional coordination, and collaboration among all stakeholders at all levels (FAO 2015). It also calls for a comprehensive data and information system and encourages all parties to improve data availability and flow, which corresponds well with the underlying premise of TBTI and the vision for ISSF. With respect to capacity development, Paragraph 12 in the SSF Guidelines explicitly calls for states and other parties to work to enhance the capacity within small-scale fishing communities so that they can effectively participate in decision-making processes, benefit from new and emerging market opportunities, and build resilience and adaptive capacity to deal with climate variability and other change processes. Similarly, capacity needs to be built within governing authorities and agencies at all levels in order to support the sustainable development of small-scale fisheries and to improve governance, especially to facilitate co-management and decentralization.

Capacity building is an important mandate for TBTI. Like its research, TBTI takes the transdisciplinary perspective to both the training approach and to the development of the course contents. One of the key deliverables of the network is to develop an online transdisciplinary training program to build research and governance capacity at the local and global levels to deal with the pressing and emerging challenges facing small-scale fisheries, as well as be proactive in actions and interventions in the face of various large-scale change processes. As the first step, a team of researchers and practitioners from different disciplinary backgrounds and experiences participated in an online collaborative platform to discuss the various elements of the training program. After a couple of months of online interaction, the group of 17 researchers and practitioners met in Merida, Mexico, for an in-person discussion and testing of the case study approach for transdisciplinary training. The deliberation and outputs from the whole process, including the philosophy and the principles in a transdisciplinary approach to research and governance, as well as personal traits and skill sets to be developed through the training, reported in Aguilar-Perera et al. (2017) (see also Said et al. Chap. 22, this volume), provide the basis for the actual development of the 'Open Transdisciplinary Learning Platform' (or TD platform) that TBTI is launching. As described in Chuenpagdee and Jentoft (Chap. 23, this volume), the TD platform will be built as an open, online course, freely accessible to anyone interested in learning about transdisciplinarity in fisheries. We will also offer the course for in-situ training as a way to overcome any technological barriers, and to increase the outreach to certain end-users, like fishers and organizations with limited access to Internet.

The content of the training program is guided by interactive governance theory (Kooiman et al. 2005), particularly the governability assessment framework

(Chuenpagdee and Jentoft 2009), making use of the case studies produced within TBTI and beyond. The utility of the theoretical grounding and the functionality of the approach embedded in the training, especially the use of case studies to drive the discussion about the problem and as exercises for the system description, is validated in three trial runs in Vietnam, Thailand, and India. While all the tests were conducted in Asia, they varied in terms of settings and participants. The first test in Saigon, organized and sponsored by the Economy and Environment Partnership for Southeast Asia (EEPSEA), involved 30 environmental economists, working on different resource systems, including forestry, soil, water, and fisheries. In that 4-day training workshop, Saigon River served as an impromptu case study, where the participants worked in small groups to analyze the problems during the two-hour river cruise. The local experts participating in the workshop made the case study excursion a success and the workshop participants were able to describe the problems and the systems following the governability assessment template without much difficulty. The Saigon workshop serves as a model for the subsequent training exercises. Satumanatpan et al. (Chap. 12, this volume) describe the one in Thailand, which took place in Trat Province on the east coast, as a successful multi-stakeholder workshop that helped deepen the understanding of the problems in the estuarine ecosystem, as well as broaden the thinking about possible solutions. A similar experience occurred in the final test in Chilika Lagoon in Odisha State, India, where coastal practitioners and governments discussed issues and explored solutions to address the multiple problems in Chilika Lagoon. The next test of whether the online training will facilitate learning and help build transdisciplinary research and governance capacity will take place in the summer with the soft launch of the TD platform for the 'Training-of-the-Trainers' program, with an official launch scheduled for January 2019. As per the spirit of a transdisciplinary teaching and learning (Lang et al. 2012), the content of the TD platform will be modular in form, with living libraries of tools and resources, including regular uploads of new case studies, which will foster sharing of knowledge and experiences about small-scale fisheries among interested individuals, groups and organizations.

2.6 Looking Forward

Long-term funding for large-scale, transdisciplinary research on fisheries is a rarity. Through TBTI and with conscious efforts to engage broadly and inclusively, we have been able to set up research programs to address key concerns and big questions in small-scale fisheries. The timing for the establishment of such a global research network could not be better with the overall increasing interest in small-scale fisheries sustainability, particularly with the adoption and implementation of the SSF Guidelines. The transdisciplinary and network approach has been fruitful in achieving a large number of book volumes (printed and e-publications), as well as special issues (Appendix 2.1). These outputs also reveal interesting data, with about 40% of the total publications focusing on Latin America and the Caribbean, followed by about 25% in Asia and Oceania. The smaller proportion in Europe (about 15%) will receive a major boost once the edited volume about European small-scale fisheries is published. Thus, not counting Canada and the USA, the region with the smallest case study contribution is Africa. For a region rich with biodiversity and resources, and with fisheries contributing to food security of at least 200 million people (Sowman and Cardoso 2010), more research and knowledge mobilization on small-scale fisheries are warranted. As experienced in TBTI, however, there is a geographical limit and financial constraint that restricts the outreach and networking in large areas. Efforts at a sub-regional level may be more efficient.

TBTI headquarters is situated in St. John's, Newfoundland, on the east coast of Canada, where small-scale fisheries have a long tradition, but the Northern cod moratorium in 1992 is a well-known story. This setting has offered TBTI an excellent case study for transdisciplinary research. Along with other case studies in Alaska and many more in Europe, TBTI coverage of small-scale fisheries in the north is quite impressive. Importantly, it sends an important message that, like their counterpart in the south, small-scale fisheries in the north require attention and institutional support from their governments, and the SSF Guidelines are also applicable to them (Jentoft 2014).

With the theme 'Trandisciplinarity and Transformation for the Future of Small-Scale Fisheries', the 3rd WSFC, to be held in Chiang Mai in October 2018, is another opportunity for all stakeholders to get together and share information, exchange knowledge, discuss collaboration, and set directions and actions to support small-scale fisheries worldwide. The congress is an opportunity to celebrate what can be achieved when many actors and stakeholders come together in major transdisciplinary research partnerships like TBTI, which, in this case, not only contributes to addressing important concerns in small-scale fisheries but also reveals challenges that lie ahead.

Appendix 2.1

List of TBTI Book Volumes, E-book Publications and Journal Special Issues

Title	Editors	Publisher	Year	# of case study countries by region				
				ASO	AFR	EUR	LAC	NAM
Enhancing Stewardship in Small-Scale Fisheries: Practices and Perspectives	Patrick McConney, Rodrigo Pereira Medeiros, Maria Pena	E-publication: TBTI and CERMES	2014	1	1	1	7	1
Interactive Governance for Small-Scale Fisheries: Global Reflections	Svein Jentoft, Ratana Chuenpagdee	Springer MARE Publication Series	2015	9	5	6	11	2
Inter-Sectoral Governance of Inland Fisheries	Andrew M. Song, S.D. Bower, Paul Onyango, S.J. Cooke, Ratana Chuenpagdee	TBTI Publication Series (E-book)	2016	5	2	2	0	0
The Small-Scale Fisheries Guidelines: Global Implementation	Svein Jentoft, Ratana Chuenpagdee, María José Barragán-Paladines, Nicole Franz	Springer MARE Publication Series	2017	9	7	7	9	1
Social wellbeing and the values of small-scale fisheries	Derek Johnson, Tim Acott, Natasha Stacey, Julie Urquhart	Springer MARE Publication Series	2018	5	0	1	1	0
Viability and Sustainability of Small-Scale Fisheries in Latin America and Caribbean	Silvia Salas, María José Barragán-Paladines, Ratana Chuenpagdee	Springer MARE Publication Series	2018	0	0	0	14	0
The Meaning of Small: Diverse Values of Small-Scale Fisheries	Mirella de Oliveira Leis, Ratana Chuenpagdee	TBTI Publication Series (E-book)	2018	7	4	4	4	2
TOTAL			**138**	**36**	**19**	**21**	**56**	**6**

Notes: ASO (Asia-Pacific); AFR (Africa); EUR (Europe); LAC (Latin America and the Caribbean); NAM (North America, i.e. Canada and US); CERMES (Centre for Resource Management and Environmental Studies)

Special Issues

Medeiros RP, McConney P, Serafini TZ (eds) (2014) Enhancing ecosystem stewardship in small-scale fisheries. Desenvolvimento e Meio Ambiente 32 (10 papers)
Pinkerton E, Davis R (eds) (2015) Neoliberalism and global small-scale fisheries. Marine Policy 80:1–176 (18 papers)
Gasalla MA, de Castro F (eds) (2016) Enhancing stewardship in Latin America and Caribbean small-scale fisheries. Maritime Studies Thematic Series 15(15):1–119 (8 papers)

Forthcoming Book

Pascual Fernández JJ, Pita CB, Bavinck JM (eds) Small-scale fisheries in Europe: status, resilience and governance. Springer MARE Publication Series (26 country studies)

References

Aguilar-Perera A, Arce-Ibarra AM, Bishop D et al (2017) Towards sustainable small-scale fisheries: key considerations for transdisciplinary teaching and training. Too Big To Ignore Research Report R-04/2017, St. John's, NL, Canada, 40 pp
Bavinck M, Vivekanandan V (2011) Conservation, conflict and the governance of fisher wellbeing: analysis of the establishment of the Gulf of Mannar National Park and Biosphere Reserve. Environ Manag 47(4):593–602
Bavinck M, Chuenpagdee R, Jentoft S et al (2013) In: Bavinck M, Chuenpagdee R, Jentoft S et al (eds) Governability of fisheries and aquaculture: theory and applications, MARE publication series, vol 7. Springer, Dordrecht
Béné C (2003) When fishery rhymes with poverty: a first step beyond the old paradigm on poverty in small-scale fisheries. World Dev 31(6):949–975
Béné C, Macfadyen G, Allison EH (2007) Increasing the contribution of small-scale fisheries to poverty alleviation and food security, Fisheries Technical Paper 481. FAO, Rome, p xii
Chuenpagdee R, Jentoft S (2009) Governability assessment for fisheries and coastal systems: a reality check. Hum Ecol 37:109–120
Chuenpagdee R, Liguori L, Palomares MLD et al (2006) Bottom-up, global estimates of small-scale fisheries catches. Fisheries Centre Research Report 14(8). University of British Columbia, Vancouver
Chuenpagdee R, Rocklin D, Bishop D et al (2017) The global information system on small-scale fisheries (ISSF): a crowdsourced knowledge platform. Mar Policy. https://doi.org/10.1016/J.MARPOL.2017.06.018
Diz D, Morgera E, Wilson M (2017) Marine policy special issue: SDG synergies for sustainable fisheries and poverty alleviation. Mar Policy. https://doi.org/10.1016/j.marpol.2017.08.020
FAO (2014) The state of world fisheries and aquaculture 2014. FAO, Rome
FAO (2015) Voluntary guidelines for securing sustainable small-scale fisheries in the context of food security and poverty eradication. FAO, Rome
FAO (Food and Agricultural Organization of the United Nations) (2005) Increasing the contribution of small-scale fisheries to poverty alleviation and food security, FAO technical guidelines for responsible fisheries. No. 10. FAO, Rome
Gonzalez C, Jentoft S (2011) MPA in labor: securing the Pearl Cays of Nicaragua. Environ Manag 47(4):617–629

HLPE (High Level Panel of Experts) (2014) Sustainable fisheries and aquaculture for food security and nutrition. A Report by the High Level Panel of Experts on Food Security and Nutrition, Committee on World Food Security (CFS), Rome, Italy
Jentoft S (2014) Walking the talk: implementing the International Voluntary Guidelines for Small-Scale Fisheries. MAST http://www.maritimestudiesjournal.com/content/13/1/16
Jentoft S, Eide AH (eds) (2011) Poverty mosaics: realities and prospects in small-scale fisheries. Springer, Dordrecht
Johnson D (2018) The values of small-scale fisheries. In: Johnson D, Acott TG, Stacey N et al (eds) Social wellbeing and the values of small-scale fisheries, Mare publication series. Springer, Dordrecht
Khan AS, Chuenpagdee R (2014) An interactive governance and fish chain approach to fisheries rebuilding: a case study of the Northern Gulf cod in eastern Canada. Ambio 43:600–613
Kooiman J (2003) Governing as governance. SAGE Publications, London
Kooiman J, Bavinck M, Jentoft S et al (eds) (2005) Fish for life – interactive governance for fisheries. Amsterdam University Press, Amsterdam, pp 25–37
Lang DJ, Wiek A, Bergmann M et al (2012) Transdisciplinary research in sustainability science: practice, principles, and challenges. Sustain Sci 7(SUPPL. 1):25–43. https://doi.org/10.1007/s11625-011-0149-x
Onyango P, Jentoft S (2010) Assessing poverty in small-scale fisheries in Lake Victoria, Tanzania. Fish Fish 11(3):250–263
Pascual-Fernandez JJ, De la Cruz Modino R (2011) Conflicting gears, contested territories: MPAs as a solution? In: Chuenpagdee R (ed) World small-scale fisheries contemporary visions. Eburon, Delft, pp 205–220
Pauly D (2006) Major trends in small-scale fisheries, with emphasis on developing countries and some implications for the social sciences. MAST 4:7–22
Pauly D, Zeller D (2016) Catch reconstructions reveal that global marine fisheries catches are higher than reported and declining. Nat Commun 7:10244
Rittel HWJ, Webber MM (1973) Dilemmas in a general theory of planning. Policy Sci 4:155–169
Salas S, Chuenpagdee R, Seijo JC et al (2007) Challenges in the assessment and management of small-scale fisheries in Latin America and the Caribbean. Fish Res 87(1):5–16
Satumanatpan S, Chuenpagdee R (2015) Assessing governability of environmental protected areas in Phetchaburi and Prachuap Kirikhan, Thailand. MAST 14:17
Schuhbauer A, Chuenpagdee R, Cheung W et al (2017) How subsidies affect economic viability of small-scale fisheries. Mar Policy. https://doi.org/10.1016/j.marpol.2017.05.013
Song AM, Chuenpagdee R (2010) Operationalizing governability: a case study of a Lake Malawi fishery. Fish Fish 11(3):235–249
Sowman M, Cardoso P (2010) Small-scale fisheries and food security strategies in countries in the Benguela Current Large Marine Ecosystem (BCLME) region: Angola, Namibia and South Africa. Mar Policy 34(6):1163–1170
Teh LCL, Pauly D (2018) Who brings in the fish? The relative contribution of small-scale and industrial fisheries to food security in Southeast Asia. Front Mar Sci 5:44. https://doi.org/10.3389/fmars.2018.00044
Triyanti A, Bavinck M, Gupta J et al (2017) Social capital, interactive governance and coastal protection: the effectiveness of mangrove ecosystem-based strategies in promoting inclusive development in Demak, Indonesia. Ocean Coast Manag 150:3–11
United Nations (2015) Rethinking poverty: report on the world social situation 2010. Social Perspective on Development, New York City
World Bank, FAO, WorldFish Center (2012) Hidden harvest: the global contribution of capture fisheries. Report No. 66469-GLB. World Bank, Washington, DC

Part II
Broadening the Scope

Chapter 3
The Value of Values for Understanding Transdisciplinary Approaches to Small-Scale Fisheries

Derek S. Johnson, Annie Lalancette, Mimi E. Lam, Marta Leite, and Sölmundur K. Pálsson

Abstract The key lesson values bring to transdisciplinarity is that the latter should not be confused with synthesis. Rather, a careful theoretically-informed consideration of values suggests that engagements across difference – disciplinary, fisher and non-fisher, policy-maker and activist, male and female, wealthy and poor, etcetera – are necessarily extremely difficult. They involve challenging, and often failed, efforts to translate between varying perceptions, paradigms, and priorities. Communications among those who are different from one another are tricky and subject to accumulating histories that may bring actors together, but may also build suspicion, distrust, and conflict. In this chapter, we do not seek to propose an answer to the process questions of value-sensitive transdisciplinary engagement. Rather, we contrast implicit and explicit approaches to value through six approaches to studying small-scale fisheries: economic valuation; ecosystem services; political economy; social wellbeing; interactive governance; and, post-normal science. This comparative analysis shows not only the benefits, but also the challenges, that are at play in constituting a value-sensitive transdisciplinary approach to small-scale fisheries.

Keywords Held values · Assigned values · Objective values · Relational values · Translation · Knowledge · Power · Governance

D. S. Johnson (✉) · S. K. Pálsson
Department of Anthropology, University of Manitoba, Winnipeg, MB, Canada
e-mail: derek.johnson@umanitoba.ca; palssons@myumanitoba.ca

A. Lalancette
St. Mary's University, Halifax, NS, Canada
e-mail: annie.lalancette@smu.ca

M. E. Lam
Centre for the Study of the Sciences and the Humanities, University of Bergen, Bergen, Norway
e-mail: mimi.lam@uib.no

M. Leite
Natural Resources Institute, University of Manitoba, Winnipeg, MB, Canada

R. Chuenpagdee, S. Jentoft (eds.), *Transdisciplinarity for Small-Scale Fisheries Governance*, MARE Publication Series 21,
https://doi.org/10.1007/978-3-319-94938-3_3

3.1 Introduction

The project *Too Big to Ignore: Global Partnership for Small-Scale Fisheries Research* (TBTI) that stimulated this volume was conceived on a transdisciplinary foundation. TBTI recognized from its inception that small-scale fisheries are diverse, complex, and dynamic human enterprises that should be understood at the intersection of different academic and non-academic perspectives. TBTI's motivation is more than simply fostering holistic methods for describing small-scale fisheries, however. It also holds that transdisciplinary modes of understanding small-scale fisheries are necessary for governance that is more attuned to the particularities of small-scale fisheries. In TBTI's language, transdisciplinarity is thus a basis for better small-scale fisheries governability (e.g., Bavinck et al. 2013). The intent of this chapter is to argue that transdisciplinarity also has the benefit of bringing out the centrality of values to discussions of small-scale fisheries when values are so often left implicit.

The advantage of transdisciplinarity in relation to values is that it is premised on the idea that different ways of knowing coexist. As no one perspective is complete, and each has its own strengths, blind spots, and interests, transdisciplinarity advocates for bringing into dialogue different approaches to knowledge and ways of being in the world, with all the difficulties of translation that doing so poses (Brosius 2006). Implicit within different approaches to knowledge, both disciplinary and cultural, are often different understandings of what value is and what is of value (Brosch and Sander 2016b). Given this plurality of perspectives on value, we begin with a simple overarching definition of value as reflecting the importance something holds for us (Brosch and Sander 2016a). Values are reference points for evaluating something as positive or negative, as desirable or objectionable (Kaiser 2012). They are rationally and emotionally binding, giving long-term orientations and motivations for behaviour and action (Kaiser 2012). Not only does a plurality of value perspectives exist, but also, and more fundamentally, a plurality of values exists in modern society and among individuals and cultural groups. Complicating analysis, values, like culture, are not static, but dynamic. Transdisciplinarity does not offer a path to synthesize perspectives, but rather, by bringing attention to the variations and contradictions in how small-scale fisheries are valued or not valued, it offers an approach that brings different ways of knowing and valuing in relation to one another.

The plurality of values manifest in fisheries governance (Jentoft and Chuenpagdee 2009), and small-scale fisheries, is at the heart of wicked social policy problems (Rittel and Webber 1973). Such problems have no straightforward technical solution, but require the hard work of any human collaboration: patience, tolerance, and acceptance of mistakes and miscommunications; an acknowledgement of accumulating histories that may bring parties together, but may also build suspicion, distrust, and conflict; and the recognition that different parties in negotiations of value will have different power to affect discourses of value. Wicked problems tend to reoccur with new and challenging variations. The engagements across difference that transdisciplinarity fosters – beyond discipline, profession, gender, socio-economic class, etcetera – necessarily leads to translation problems across differences in perception, paradigm, and priority. Seeking consensus on the values of small-scale fisheries through transdisciplinarity is thus a wicked problem. Indeed, consensus may not be possible or even desirable for post-normal science problems at the science-policy interface, where facts are uncertain, values are in dispute, stakes are high, and decisions are urgent (Funtowicz and Ravetz 1993). Value plurality forces us to accept difference and embrace tolerance if we are to forge solutions together.

This chapter sketches the potential of linking transdisciplinarity to the study of values and valuing in small-scale fisheries. Our transdisciplinary approach is predicated upon relating contextually-specific ways of knowing. These may be academic disciplinary, interdisciplinary and multidisciplinary, as well as non-academic indigenous and local experiential ways of knowing (Berkes 1999; Lam 2014). Transdisciplinary knowledge is relational; it emerges through ongoing exchanges among differently positioned individuals and groups and thereby holds the promise of transcending boundaries and particular contexts.

Thus, the core of our chapter juxtaposes value perspectives in six approaches to studying small-scale fisheries: economic valuation; ecosystem services; political economy; social wellbeing; interactive governance; and, post-normal science. We reflect on the lessons that emerge from four case studies by researchers working from different disciplinary foundations in three countries: two cases in Canada and one in Australia and in Brazil. Given space limitations, we cannot be exhaustive in terms of approaches (we leave out commons theory, for example) or representative in terms of the areas of the world that we cover. We do not attempt to synthesize the approaches, but rather explore how they complement and contrast with each other. Our intention is to stimulate others to extend the analysis to other approaches and places relevant to small-scale fisheries.

3.2 Transdisciplinary Approaches to Values in Small-Scale Fisheries

As values permeate all aspects of human existence, their study has led to diverse value conceptions. Consequently, no coherent theory of values exists, but rather multiple taxonomies and understandings of values have emerged from different disciplinary lenses. Our objective here is not to review or to summarize those perspectives, as this has been done thoroughly (Brosch and Sander 2016b), but rather, we wish to highlight perspectives that have influenced the fisheries literature. We do this to contrast the perspectives and to open up new space for inquiry into values in small-scale fisheries.

Two value distinctions have been particularly influential: values as objects or relations and values as held or assigned (Brown 1984; Chan et al. 2016; Rabinowicz and Rønnow-Rasmussen 2016). The relation-object distinction is concerned with what values are (Chan et al. 2016). The relational perspective emphasizes how values emerge from social interchange and histories, while the objective points to the possibility of freezing value, so that perceptions of value can be translated among actors with a single currency, such as money (Graeber 2001). Objective values are collectively agreed and transacted upon. While those value 'objects' are simplifications, they are also necessary referents through which to share information and negotiate meaning. Held values, in contrast, are ideas that guide how humans navigate their worlds (Brown 1984). They are products of experience and upbringing that draw on social and cultural influences, including ideology and religious beliefs. Assigned values point to the human propensity to give greater or lesser importance to diverse aspects of life and the world, including material objects, ideas, subjective states, or affective relations with others (Brown 1984). From the perspective of economic theory, one's stated preferences are related to held values, while exhibited preferences in behaviours are closely connected with assigned values. A contrasting, more relational social science view sees held and assigned values as reciprocally connected, in the sense that they inform each other through one's life experiences (Johnson 2018; Song 2018), though in a complex, non-linear fashion. We argue that a rich understanding of values, appropriate to transdisciplinarity, has to encompass each of these distinctions, and the productive tensions among them, as facets of a broader notion of values.

The case of small-scale inland fishing on Lake Winnipeg in Manitoba, Canada, as summarized in Box 3.1, illustrates the power of held values to shape behaviour: commitment to fishing on Lake Winnipeg is driven in part by the sense that fishing realizes the held value of freedom. However, freedom is not a settled, universal value; rather, it is a value that is given meaning in opposing ways, with contradictory implications for fishery policy in Manitoba.

Box 3.1 Freedom and Fishing on Lake Winnipeg, Manitoba, Canada (Sölmundur Pálsson)

At 5 AM, a fisher on Lake Winnipeg looks outside his kitchen window to check the weather, even though he knows it really doesn't matter what the weather is: he still will go out to fish. Whether in 20 °C on open water or in −30 °C on the frozen lake, the fisher will go out to catch Walleye or Whitefish. Values drive his actions: the value of the potential catch, of course, by which he contributes to supporting his family, but also the value of the great satisfaction that his job provides. In this, the fisher acts consistently with the observation that fishers often have high levels of job satisfaction (Pollnac and Poggie 2006). Yet, this driving value of satisfaction from fishing as an occupation is not as straightforward as it seems, as fishers on Lake Winnipeg (and likely elsewhere) conceive of that satisfaction in fundamentally opposing ways. To understand this point, it is necessary to understand a little about the history and contemporary politics of fishing on Lake Winnipeg.

In August 2016, the newly elected conservative government in the province of Manitoba announced that they would opt out of the Freshwater Fish Marketing Act (1985) or, in other words, they would withdraw from their participation in the Freshwater Fish Marketing Corporation (FFMC). This had been a promise of the Progressive Conservative Party during the 2016 election

(continued)

Box 3.1 (continued)

campaign, reflecting the party's longstanding criticism of the FFMC. During the 2016 campaign, the Progressive Conservatives received support from groups of fishers from Lake Winnipeg and Lake Manitoba who were disgruntled with FFMC's operations. It was a controversial decision, for two main reasons: (1) the lack of broad consultation with fishers (especially First Nations' fishers with whom consultations are required by law), and (2) only a minority of fishers on Lake Winnipeg, much the largest population of commercial fishers in Manitoba, supported the decision.

The FFMC was established in 1969 to break the exploitative relation that existed between fishers and fish buyers by nationalizing the right to market and export fish from Manitoba. The mandate of the FFMC was simple: to stabilize the market for fish by buying all the legally caught fish that fishers offered. For more than 40 years, state intervention increased fisher incomes and autonomy and, with other state-imposed regulations, reduced resource overexploitation (Johnson and Pálsson 2015). In recent years, however, divisions have arisen among fishers over the FFMC. The debate amongst fishers concerns the political values that frame the satisfaction derived from fishing for different fishers.

Maritime anthropologists have identified the independence, or freedom of fishing, as a central component of how fishing conveys satisfaction (McGoodwin 1990). The debate over the FFMC shows that freedom is itself a contested value. One group of fishers sees freedom through a neoliberal lens, with freedom achieved only when the state and state bureaucracy exercise minimal control over fisher economic activity. Their rhetoric is of the "free market" that enables them to sell their fish to whomever they want. They argue that without the slow bureaucratic practices and the lack of incentives of the FFMC, they will receive higher prices and therefore enhance the financial security of their families. A second group of fishers, however, sees the state, acting through the FFMC, as a vital part of achieving their freedom. This group's rationale is tied closely to the collective memory of fishers as indentured to fish buyers in the years prior to 1969. They see the FFMC and, by extension, the state, as a buffer against the global market and larger companies that might seek to dominate the Lake Winnipeg fishery. This 'freedom' from competition with much more powerful external competitors is what allows fishers to sustain their own operations. Their ability to control their own operations allows them to be their own bosses and gives them certain freedoms: control over their own time and control over how much time they can choose to spend with their families (Johnson and Pálsson 2015).

The Lake Winnipeg case shows that the definition of freedom as a value is often taken for granted as the neo-liberal view of freedom from the state. The contrasting views held by different Lake Winnipeg fishers illustrates, however, that freedom is highly contested, as is the value basis by which fishers derive satisfaction from their professions.

3.2.1 *Economic Valuation*

Economic valuation encompasses approaches in fisheries economics that commonly inform fisheries policy, where neo-classical economic methods inform judgments of appropriate interventions to achieve societal benefits for, or through, fisheries. Economic valuation approaches are oriented towards objective and assigned values. This is evident in the classic contribution of fisheries economics to management that aspires to achieve bio-economic equilibria for individual species of fish over time (Gordon 1954; Clark and Munro 1975) and in debates over whether there are limits to the assumption that privatization will result in sustainable fisheries outcomes (Clark et al. 2010; Pitcher and Lam 2010). The economic valuation approach is present in arguments for the diverse material benefits of small-scale fisheries (Thomson 1980; Pauly and Zeller 2016). In all fisheries, efforts are made to identify optimal economic arrangements for realizing objective social benefits, such as maximum sustainable yield, economic return, or employment. Values are assigned, with focus on the benefits derived from particular fisheries or economic arrangements. Commonly, such benefits are reduced to, or commodified using, the standard metric of money.

An illustration of the economic valuation approach is given in Box 3.2, which focuses on a debate between indigenous fishers and the government of Australia. While indigenous fishers are keenly aware of the values that underpin their position in the debate, governmental fisheries management actors do not appear to recognize that their approach to economic valuation is anchored in commodity values, even with regard to conservation.

Box 3.2 Values as a Window to Understanding Divergent Perspectives in Fisheries Management (Annie Lalancette)

The case of the tropical rock lobster fishery in Torres Strait, northern Australia, illustrates how a discourse analysis focused on values can help understand why management perspectives between actors may differ – even when some objectives seem compatible (Lalancette 2017b). In the last decade, this fishery has been rocked by access and resource ownership disputes between indigenous Torres Strait Islanders and non-indigenous Australian fishers. Torres Strait Islanders assign a high value to tropical rock lobster or *kaiar* – it is the commercial fishery where they are the most active and Islanders view it as a key element of achieving self-determination. Although Islanders' practices support their cultural resilience and *kaiar* sustainability (Lalancette 2017b), they are under considerable pressure from decision-makers to change their approaches to fishing in order to increase their catch (Lalancette 2017a).

The main management objectives defined in the Torres Strait Fisheries Act (1984) are protecting the environment, protecting the Torres Strait Islander traditional way of life (focused on subsistence fishing), optimal utilization of

(continued)

Box 3.2 (continued)

fisheries, and economic development. These objectives promote values of conservation and utilitarianism framed by the maximum sustainable yield; i.e., tropical rock lobster should be caught up to a pre-determined target point and only conserved beyond that specific catch level. They also reflect neoliberal values such as individual freedom, wealth accumulation and commodity production. Such a perspective supports fisheries management that is single-species, technical and centralized. It favours measures such as limited access, output controls with a preference for individual transferable quotas, and incentives that encourage profit-maximizing behaviour (Lalancette 2017a).

While Torres Strait Islanders share some objectives with those outlined in Australian legislation, what they consider to be good management significantly differs from conventional fisheries management. Islanders' overarching aspiration for self-determination is expressed through objectives of 100% ownership of territories and resources. For Islanders, good fisheries management starts with them being in control and respecting *Ailan Kastom*, rather than a limited focus on subsistence fishing. This means that management should be decentralized and founded on customary marine tenure, in which traditional owners control access to resources based on local abundance and needs, kinship affiliations, personal relationships and reciprocity.

Torres Strait Islanders' concerns regarding the sustainability of *kaiar* are motivated by stewardship responsibilities and inter-generational equity. The *kaiar* fishery is important economically as a primary or supplementary income, for its welfare functions, and as one of the few employment alternatives in the region. *Kaiar* also supports subsistence and food security, learning, knowledge transmission, cultural practices and the maintenance of social and kinship networks and relations, ultimately strengthening Torres Strait Islander identity (Lalancette 2017b; Mulrennan and Scott 2002). For Islanders, a sustainable fishery is thus one that can provide for both the environment and people in the future. The general view is that what remains in the sea is a long-term investment, making the tenet "take only what you need" a central component of Torres Strait Islander management.

The Australian government sees tropical rock lobster as a cornerstone of economic development for the region, as do Torres Strait Islanders. However, Islanders emphasize community wellbeing and generally reproach wealth accumulation. Their desire to improve their economy does not supersede other values, such as spending time with family and in the community and participating in cultural and ceremonial life (compare to *Vida Simples*, Box 3.3). Islanders tend to prefer small boats with low overhead costs because it gives them the freedom to choose when and how much they will fish. Good management is thus one that can support the economic viability of their small-scale operations. It should accommodate various and flexible levels of efforts and participation so that fishers can adapt to changing responsibilities, oppor-

(continued)

Box 3.2 (continued)

tunities, and ecological conditions, as well as respond to different economic needs, capabilities and personal preferences.

Equity in benefit distribution is an important value for Torres Strait Islanders. Most Islanders are adamant that the fishery should provide equal opportunities to everyone in terms of access, catch and revenue – regardless of one's level of effort and participation. Islanders tend to disapprove of regulations or technologies that confer an advantage to some fishers over others. For example, there is a strong awareness and concern for free-divers who must compete with hookah divers. To level the playing field, Islanders have devised informal rules, such as prohibiting the use of hookah on top of reefs and hookah bans by some communities in their waters.

According to Torres Strait Islanders, good management is best achieved by a combination of restraint, careful adoption of technology so that fishing is not "too easy", respect of Torres Strait Islander institutions and norms, and adjusting effort to environmental conditions. As opposed to the conventional fisheries management view, these constraints are not considered "inefficiencies", but viewed as positive effort controls.

A comparison of the discourses of Torres Strait Islanders and Australian government managers shows that they share similar objectives in terms of sustainability: protecting the traditional way of life of Torres Strait Islanders and economic development. These sustainability objectives are informed, however, by different values. A nuanced understanding of values might be able to explain users' resistance to fisheries measures by highlighting potentially adverse consequences and their varied impacts on different groups.

3.2.2 Ecosystem Services

Ecosystems services' orientation to value is similar to economic valuation, as a result of its methodological debt to neo-classical valuation approaches. As pointed out by many (Agarwala et al. 2014; Hicks et al. 2016; Acott and Urquhart 2018), ecosystem services approaches to values were initially premised on a linear understanding of the services generated by ecosystems for human users. These were understood in assigned value terms, and as reducible to objective monetary measurements (Millennium Ecosystem Assessment 2005; Chan et al. 2012). This approach was able to estimate values for provisioning, regulating, and supporting ecosystem services, but was accompanied by a fourth category of cultural ecosystem services that fit awkwardly with the other three and required an understanding of culture as an object produced by natural processes. This is a reductionist understanding of culture that eviscerates it of its core interpretive and social relational quality, in contrast to recognizing that cultural values are both derived from and embedded within community and ecosystem relationships (Lam and Borch 2011; Lam and Pitcher 2012b).

In recent years, efforts have been made to reconfigure the ecosystem services approach to better understand how culture serves as a context through which ecosystem services are recognized and given value, rather than reducing culture to an ecosystem product. These efforts have sought to integrate attention to held values and a relational approach to values (Chan et al. 2016). Authors working in small-scale fisheries have played a significant role to reorient the conceptualization of value in ecosystem services. Hicks et al. (2016), for example, apply the idea of elasticity from economics to look at the social and cultural factors that shape the value given to ecosystem services. Acott and Urquhart go further and use a co-constructionist approach to human-environment relations to argue that ecosystem services are products of particular histories of human-environment engagements (Acott and Urquhart 2014, 2015; C. White 2018). Both cases demonstrate that economic valuation methods have to be complemented, and qualified, by qualitative research to reveal how definitions and valuations of ecosystem services vary by social context.

3.2.3 Political Economy

Political economy, in the contemporary Marxian sense, can be understood as a theory of value creation where value is the product of human labour. This situates political economy as focused on assigned and objective values: humans create objects that acquire (are assigned) value from the work that went into producing them. Political economy's analytical richness, however, comes from the additional crucial observation that all labour is social in two senses. First, labour is social in that it is a collective product, with products of labour made by people acting directly or indirectly in concert with each other. Second, what we choose to labour for is a social and cultural question, such that held values influence what is meaningful labour and what are meaningful products of this labour. Value, in the political economy sense, can be crystallized in socially produced objects (that may or may not be material, e.g., an informal fishing regulation), but which only take value because of the social relations that literally and symbolically make them. Political economy thus also focuses attention on relational and held values.

This theoretical starting point can lead to a profound analysis of the hidden dimensions and dynamics of power in the production of commodities within capitalism (see Graeber 2001; Taussig 2010 [1980]). It is important to acknowledge political economy's fruitful conceptualization of power in relation to value. Power is made manifest through the production of values or, in other words, material advantage can confer benefit in shaping the discursive ground out of which values emerge. Thus, in fisheries, held values such as the need for development, heightened commodity-based production, privatization, or environmental protection are not simply objects that can be rationally debated on their merits in public forums (as interactive governance holds), but are social constructs whose ambiguities are honed and manipulated in efforts to control both people and resources (Pálsson 2006; Pinkerton and Davis 2015). For example, fishery resources may be valued as commodities giving profit to commercial fishers or as living resources with ecosystem and social relationships for local and indigenous communities (Lam and Pitcher 2012b). These contrasting valuations of fish, as either alienable property that can be valued in a market or inalienable property that cannot, can favour industrial-scale fisheries over small-scale fisheries or fisheries over conservation or vice versa. The process of commoditization, analyzed as a societal selection pressure that favours an economy of things over relationships (Manno 2000), can be seen holding or assigning value to objects over relations.

3.2.4 Social Wellbeing

The distinction between relations and objects is central to the social wellbeing approach. Social wellbeing emerged from the field of international development built on the crucial insight that poverty is not just an economic matter of income, but a multi-dimensional phenomenon (Agarwala et al. 2014). Poverty involves not only material deprivation, but also relational incapacity, and subjective stress. These material, relational, and subjective dimensions constitute the analytical core of social wellbeing's multidimensional framing of ways in which people seek to live well in terms that make sense to them. Values are relevant to social wellbeing as guides to aspiration; they frame that which is necessary or preferable to live well (Fischer 2014; Johnson 2018). Social wellbeing thus has a relational sensitivity to values, including the recognition that material objects and ways of living are perceived through lenses informed by context, history, and practice (S. White and Ellison 2007).

For small-scale fisheries research, the three dimensions of social wellbeing (Coulthard 2012b) provide a way to structure the analysis of how meaning and value emerge in particular social-ecological contexts. How people, in relation to their environments, co-construct histories and influence behaviours with and of each other have valued material, subjective, and relational aspects that can lead to more or less sustainable fisheries. The insights that a social wellbeing approach may bring to understanding how particular values affect conditions for fisheries governance is illustrated in Box 3.3. Leite shows how a cultural orientation and particular frustrating histories of engagement help explain why one group of Brazilian small-scale fishers are leery of state efforts to involve them in governance.

Box 3.3 Small-Scale Fishers' Values as a Barrier to Participation in Fisheries Governance (Marta Leite)

> I love my life, this 'simple life'! It is the best life, you know? People from the city pay to have a little taste of our life, they want to rent our house, go fishing in our boat, buy our fish, isn't that funny? – Caiçara Fisher, Ubatuba, Brazil.

Understanding fishers' motivations and behaviours is fundamental to managing small-scale fisheries' long-term sustainability. Nevertheless, fishers' behaviours and choices are typically complex and comprise a set of economic, social and cultural considerations (Coulthard 2012a). The Social Wellbeing Approach offers a powerful tool to expand and deepen our understanding of how values shape small-scale fishers' behaviours. In the case below, it explains their engagement (or not) with local fisheries governance institutions.

Indeed, in Ubatuba, a Municipality on the Southeastern Coast of Brazil, small-scale *Caiçara* fishers' cultural value of "*Vida Simples*" (the Simple Life) has played a significant role in defining behaviours in times of change posed by rapid economic development, including fishers' disengagement with the fisheries governance process.

Vida Simples is an expression that speaks to the embrace of life that is both modest and humble. Faith in God, family 'togetherness', freedom to fish, autonomy and ties with friends and neighbours compose some of the central elements of this way of life. Moreover, humility, the *Caiçaras*' unique and ubiquitous sense of humour, and an aptitude for living life in the present were all described as important components of what it meant for participants to live the *Caiçara Vida Simples*. The embodiment of this lifestyle captures a major piece of *Caiçaras'* cultural identity, and, we argue, has important consequences for local fishers' lack of interest in taking part in meetings with government. This is occurring even as the discourse of inclusion features prominently in recent policy initiatives within Brazilian fisheries management institutions.

For *Caiçara* people, in large part due to an extensive history of loss of land and access to natural resources, a clear and persistent distrust in government institutions remains, creating significant challenges for fishers' participation in government initiatives for inclusive governance. Another aspect is the prevalence of policies representing conservation interests, both on land and at sea. The consequence is a tendency of fishers to resist engaging with fisheries governance, as observed in this research and demonstrated by other studies in the area (Bockstael et al. 2016; Trimble et al. 2014; Trimble and Berkes 2013).

We offer a complementary explanation, however, based on the concept of *Vida Simples*, from the perspective of the fishing community studied. We make the argument that *Caiçara* fishers also mediate their potential participation in government-driven initiatives based on deeply-rooted cultural tendencies.

(continued)

Box 3.3 (continued)

For example, some of the *Vida Simples* related values that act as barriers to engaging with governance are avoiding conflict, the embrace of a 'laidback' persona, appreciation for autonomy and independence, freedom from regimentation, living in the moment, and communicating through humour, as opposed to through formal and technical languages. These values conflict with the governmental model of engagement, i.e., the 'meeting', and discourage fishers' participation, resulting in low representation of fishers in government-organized functions.

Yet, while *Caiçara* fishers from Ubatuba generally choose not to engage in meetings with the government, they do find alternative and more "silent" ways of responding to their lack of voice in defining their own freedoms and wellbeing priorities. This is achieved, simply, by electing not to cooperate with imposed restrictions and to resist the power imbalances to which they are subject. While this takes several forms, one of the most prominent is through the practice of illegal fishing, which in turn only heightens the barriers between local fishers and government management agencies. Disobeying fishing policies and fishing illegally represent strategies to maintain their lifestyles in the face of change, but with severe consequences for fishing households, including marginalization, arrest and loss of fishing gear. These consequences have direct impacts on the overall wellbeing of the fisher and his or her family. We argue that only through a deeper cultural analysis can we better understand the challenges of encouraging fisher engagement and buy-in for fisheries management. A governance model that is culturally sensitive may offer a better chance of success.

3.2.5 Interactive Governance

Interactive governance theory provides a systems-based set of analytical tools for analyzing the characteristics and relations of governance in particular settings. It is not a formal theory of how governance works, but rather a set of tools for analyzing particular governance arrangements and their performance (pers. comm. Svein Jentoft). The premise of interactive governance is normative: enhancing understanding of the different components of particular governance arrangements and contexts, while also enhancing capacities for engagement among actors and groups in governing, will result in better governance outcomes. Interactive governance theory has been applied widely to fisheries in the past 15 years (Kooiman et al. 2005; Bavinck et al. 2013; Jentoft and Chuenpagdee 2015).

In its analysis of fisheries as systems, interactive governance identifies three different levels or orders of governance, addressing: (1) practical, day-to-day, operational questions; (2) institutional organization and maintenance concerns; and (3) attention to the ideal or philosophical underpinnings of governance systems. Values are integral to this last, third-order level of governance. In interactive governance theory, values are held, underpinning how governing is approached, but are also assigned, depending on the foundational held values of a given governance system. The interactive governance approach advocates reflexive understanding of value motivations as a basis for more transparent negotiations of interest in governance, but it is not fully relational, as it also sees values as objects that can be categorized and associated with particular individuals or groups (Song et al. 2013). Unlike the ecosystem services approach, where value is a benefit from the environments in which we live, or political economy, where value is the product of human labour, interactive governance does not posit a theory of how value is created, but rather just assumes that values shape social interactions.

3.2.6 Post-normal Science

Values are central in post-normal science approaches to problems at the science-policy interface, where facts are uncertain, values are in dispute, stakes are high, and decisions are urgent (Funtowicz and Ravetz 1993). Post-normal science warns against arbitrary separations between facts and values, embraces complexity, and focuses on the quality of the scientific process, which is seen as recursive (e.g., participatory and iterative) and reflexive (i.e., the analyst is part of the analysis). It stresses recognition of the plurality of publics – and hence the plurality of values – that typify wicked problems (Rittel and Webber 1973) and advocates compromise, rather than consensus. Issue resolution is via an 'extended peer community', involving scientific disciplinary experts, concerned citizens, and diverse stakeholders, engaged in deliberative dialogue to frame the problem, assess the options, and control the quality of knowledge (Van der Sluijs et al. 2008). Post-normal science is transforming science (Gluckman 2014) and governance (Pitcher et al. 2017), particularly in areas of conflict, where the use of evidence is contested due to different norms and values.

In fisheries, characterized by high system complexity and uncertainty (Ludwig et al. 1993; E. K. Pikitch et al. 2004; Dankel et al. 2012), fishery openings, quotas, harvest allocations, and other contested decisions that affect multiple participants and groups must be made prior to each fishing season. Implementing a transdisciplinary post-normal science approach, the Canadian Pacific herring fishery case study in Box 3.4 presents an innovative value- and ecosystem-based management approach to resolve resource conflicts. With an extended peer community of scientists, indigenous communities, artists, industry, government, and civil society, collaborative governance solutions Lam and Pauly (2010), Lam and Pitcher (2012a) were explored through a participatory stakeholder process to explicate their values and preferences for alternative fishery management strategies. The ecological impacts and risks of these strategies were also modelled. By thus making values more transparent in resource conflicts, such as between indigenous communities and industry or small- and large-scale fisheries, it is possible to foster sustainable and just fisheries policies and legitimate governance interventions.

Box 3.4 Explicating Values to Resolve Conflicts in Fisheries (Mimi E. Lam)

> The herring is the most important fish for all the islands, all the area. We're hoping that all the Haida Gwaii herring stocks will get better. I'd like to see it closed 'til all these areas get herring. – Haida elder and retired commercial fisherman, Haida Gwaii, Canada

Missing in traditional fisheries management and governance, and vitally important to recognize for small-scale fisheries, is the explication of values to aid decision-makers in resolving the inherent policy trade-offs that emerge when diverse stakeholders and communities have competing interests. Post-normal science offers a lens that explicitly recognizes and addresses the plurality of values that often exists in resource management contexts and gives rise to highly contested policy arenas. In these conflicted situations affecting diverse stakeholders, it is important to extend the peer community beyond scientists and policy-makers to involve holders of local experiential and indigenous knowledge, incorporating their beliefs and values in the decision-making process. We have developed an innovative, participatory value- and ecosystem-based management approach that combines practical ethics with ecological modelling (Lam et al. 2017; Pitcher et al. 2017; Surma et al. 2018). This approach can facilitate conflict resolution and decision-making at the science-policy nexus by explicating values often at the source of conflicts, but masked as stakeholders advocate for specific management strategies. Ultimately, by promoting more inclusive, transparent, and accountable governance, our approach can enhance ecological sustainability and societal welfare.

In western Canada, the Pacific herring fishery conflict between local and indigenous communities and the herring industry (see 'Herring Quest' video: http://perf.oceans.ubc.ca/projects/current-projects/herring-people/herring-quest-stakeholders-values/) reflects diverse ecological, economic, and cultural values of herring (Lam et al. 2017; Pitcher et al. 2017; and Surma et al. 2018). Pacific herring (*Clupea pallasii*) are small pelagic forage fish that play a pivotal ecosystem-provisioning role within marine food webs, linking predatory fish, marine mammals, and seabirds to its planktonic invertebrate prey (E. Pikitch et al. 2012; Ellen K. Pikitch et al. 2014; Surma et al. 2018). Pacific herring supports several commercial fisheries, including roe herring, spawn-on-kelp, and food and bait (Haas et al. 2016), and provides seasonal export revenue and income to local fishers and processors. Herring spawn-on-kelp is also a traditional food for many coastal indigenous peoples, including the Haida, the indigenous inhabitants of Haida Gwaii, a remote northern archipelago with a rich marine ecosystem and cultural heritage (von der Porten et al. 2016; Jones et al. 2017). The scientific basis for management of the fishery is uncertain due to complex ecosystem interactions that link herring to its predators and prey (Surma et al. 2018). Meanwhile, harvest quotas for herring are set annually (DFO 2016), which makes decisions urgent and often contested.

(continued)

Box 3.4 (continued)

In community-based research designed to facilitate collaborative governance solutions for the Pacific herring fishery conflict, Haida Gwaii community and herring industry members ranked a set of values and identified their preferred scenarios for fisheries management (Lam et al. 2017; Pitcher et al. 2017). While value rankings were similar for both groups, their scenario preferences differed markedly, suggesting that the link between values and preferences is nuanced and complex. Values may run counter to an individual's or group's interests in some settings, particularly in highly politicized contexts, such as the resource management conflict here. Also, individuals can have similar overarching values, but interpret or apply them differently, depending on their local circumstances and the differential benefits and impacts accrued from the fishery (illustrated by the value freedom, Box 3.1). However, focusing on common value priorities and scenario preferences, and excluding the extreme management options preferred by one group but not the other, reveals a compromise strategy to manage the roe herring and spawn-on-kelp fisheries separately, with distinct harvest control rules. This takes into account the differential values and impacts of the two fisheries. Explicating values thus opens up room for dialogue and compromise among diverse stakeholders and highlights the benefits of embedding a value- and ecosystem-based management approach within a post-normal science framework to resolve conflicts and policy trade-offs in fisheries.

The two central themes of this book, transdisciplinarity and small-scale fisheries, now need to be made explicit in relation to these various approaches to values. First, with regard to transdisciplinarity, all but the first approach above are more than disciplinary, but come to their multi-dimensional approaches in different ways. Among them, only post-normal science can strongly make the claim to be fully transdisciplinary in terms of integrating non-academic knowledge, but the clear attention of all except economic valuation to relational ways of understanding values suggests openness to inclusion of diverse ways of knowing. The held-assigned and relational-objective distinctions suggest a guide to how all six approaches relate to each other, while recognizing contradictions and incompatibilities. Second, as the references in each section show, each approach has generated literatures that indirectly or directly reflect on values in small-scale fisheries. Some of these contributions are reductionist, emphasizing the material assigned values of small-scale fisheries, while others are much more evocative, focusing on relational values, including meaning, sense of place, and spiritual or religious values. In arguments for the general societal value of small-scale fisheries (Johnson 2018), both are necessary, perhaps with the assigned-values group serving the tactical purpose of speaking the language of policy and the relational-values group truer to the fullness of values experienced by many small-scale fishers.

3.3 Conclusion

Transdisciplinarity is about the recognition that ways of knowing are plural and that, to address human problems, it is necessary to engage in ongoing efforts to create and sustain bridges between different perspectives on the world. Complicating that task are the held values and processes of valuing aspects of the world that guide human engagement with each other and with our environments. Power is implicit within values seen relationally, as values so often are the basis for ranking groups, objects, and ways of knowing against each other. Transdisciplinarity, therefore, needs to integrate attention to values in its translation efforts and to recognize that such efforts always need to be attuned to the relations of power within which they operate.

These transdisciplinary considerations are as relevant to questions of governance in small-scale fisheries as they are to any other human endeavour. In this chapter, we have reviewed some of the ways in which values emerge in different approaches to small-scale fisheries and we have provided four illustrations of the importance of taking values into account in the sector. Each of the cases shows in particular how attention to values forces consideration of different ways of knowing, and especially the importance of fishers' knowledge. But the attention to values in the cases shows that transdisciplinarity as an effort to communicate across different ways of knowing is far from straightforward. Lam shows, in the herring fishery case study, that even when we assume that a commonly understood set of values can be defined, different individuals and groups may interpret and act upon seemingly shared value priorities in divergent ways. Transdisciplinarity thus should be seen as a space for negotiation, hard choices, and trade-offs (Coulthard 2012a), rather than a path to finding a clean common ground. History is important as well, as Leite's case of *Vida Simples* makes clear. Efforts to build transdisciplinary understanding can face serious challenges due to the build-up of past interventions by the state or outsiders whose actions made little effort to take into account fisher values, such as those associated with the idea of *Vida Simples*. In the Lake Winnipeg case of Pálsson, different understandings of freedom, as a value, align with different political positions. Any effort to build transparent transdisciplinary space has to recognize that the very terms themselves are not settled and that some interpretations of fundamental value terms have greater political credibility than others. Lalancette shows that Torres Strait Islanders' knowledge systems rest on culturally important values that conflict with more powerful neoliberal values that are the basis for the Australian government's approach to assigning value to fisheries. All four cases suggest that transdisciplinary engagement has to recognize differences of power, which complicate the assumption that just heightening transparency with regard to values can foster trust.

To conclude, values complicate the transdisciplinary ideal in productive ways. They put a check on assumptions that transdisciplinarity is a silver bullet to resolve problems in small-scale fisheries. They suggest, rather, that transdisciplinarity is like any other human endeavour, in that it suffers translation problems, duplicity, frustrations, and conflict, all of which require hard work and patience to address.

Nonetheless, the principle of inclusion that underpins transdisciplinarity, and its foregrounding of the coexistence of different knowledge systems, makes it an important addition to the governance toolkit.

References

Acott TG, Urquhart J (2014) Sense of place and socio-cultural values in fishing communities along the English Channel. In: Urquhart J, Acott TG, Symes D et al (eds) Social issues in sustainable fisheries management. Springer, pp 257–278

Acott TG, Urquhart J (2015) People, place and fish: exploring the cultural meanings of inshore fishing through photography. In: Warren S, Jones P (eds) Creative economies, creative communities: rethinking place, policy and practice. Ashgate, London, pp 43–63

Acott TG, Urquhart J (2018) Co-constructing cultural ecosystem services and wellbeing through a place-based approach. In: Johnson D, Acott TG, Stacey N et al (eds) Social wellbeing and the values of small-scale fisheries. Springer, Cham

Agarwala M et al (2014) Assessing the relationship between human well-being and ecosystem services: a review of frameworks. Conserv Soc 12:437–449

Bavinck JM, Chuenpagdee R, Jentoft S et al (eds) (2013) Governability of fisheries and aquaculture: theory and applications, MARE publication series. Springer

Berkes F (1999) Sacred ecology. Taylor and Francis, Philadelphia

Bockstael E, Bahia NCF, Seixas CS et al (2016) Participation in protected area management planning in coastal Brazil. Environ Sci Pol 60:1–10

Brosch T, Sander D (2016a) From values to valuation: an interdisciplinary approach to the study of values. In: Brosch T, Sander D (eds) Handbook of value: perspectives from economics, neuroscience, philosophy, psychology, and sociology. Oxford University Press, Oxford, pp 397–404

Brosch T, Sander D (2016b) Handbook of value: perspectives from economics, neuroscience, philosophy, psychology, and sociology. Oxford University Press, Oxford

Brosius JP (2006) Between politics and poetics: narratives of dispossession in Sarawak, east Malaysia. In: Biersack A, Greenberg JB (eds) Reimagining political ecology. Duke University Press, Durham and London, pp 281–324

Brown TC (1984) The concept of value in resource allocation. Land Econ 60:231–246

Chan KMA, Satterfield T, Goldstein J (2012) Rethinking ecosystem services to better address and navigate cultural values. Ecol Econ 74:8–18

Chan KMA et al (2016) Opinion: why protect nature? rethinking values and the environment. Proc Natl Acad Sci 113:1462–1465

Clark CW, Munro GR (1975) The economics of fishing and modern capital theory: a simplified approach. J Environ Econ Manage 2:92–106

Clark CW, Munro GR, Sumaila UR (2010) Limits to the privatization of fishery resources. Land Econ 86:209–218

Coulthard S (2012a) Can we be both resilient and well, and what choices do people have? incorporating agency into the resilience debate from a fisheries perspective. Ecol Soc 17. https://doi.org/10.5751/ES-04483-170104

Coulthard S (2012b) What does the debate around social wellbeing have to offer sustainable fisheries? Curr Opin Environ Sustain 4:358–363

Dankel D, Aps R, Padda G, Rockmann C, Van der Sluijs J, Wilson D, Degnbol P (2012) Advice under uncertainty in the marine system. ICES J Mar Sci, Journal du Conseil 69:3–7

DFO (Fisheries and Oceans Canada) (2016) Stock assessment and management advice for British Columbia pacific herring: 2016 status and 2017 forecast vol DFO Canadian science advisory secretariat science response 2016/052. Nanaimo

Fischer EF (2014) The good life: aspiration, dignity, and the anthropology of wellbeing. Stanford University Press, Stanford

Funtowicz SO, Ravetz JR (1993) The emergence of post-normal science. In: Von Schomberg R (ed) Science, politics and morality: scientific uncertainty and decision making. SpringerE, Dordrecht, pp 85–123

Gluckman P (2014) Policy: the art of science advice to government. Nature 507:63–165

Gordon HS (1954) The economic theory of the common property resource: the fishery. J Polit Econ 62:124–142

Graeber D (2001) Towards an anthropological theory of value: the false coin of our own dreams. Palgrave, New York

Haas AR, Edwards DN, Sumaila UR (2016) Corporate concentration and processor control: insights from the salmon and herring fisheries in British Columbia. Mar Policy 68:83–90

Hicks CC et al (2016) Engage key social concepts for sustainability. Science 352:38–40

Jentoft S, Chuenpagdee R (2009) Fisheries and coastal governance as a wicked problem. Mar Policy 33:553–560

Jentoft S, Chuenpagdee R (eds) (2015) Interactive governance for small-scale fisheries: global reflections. Springer

Johnson D (2018) The values of small-scale fisheries. In: Johnson D, Acott TG, Stacey N et al (eds) Social wellbeing and the values of small-scale fisheries. Springer, Cham

Johnson D, Pálsson SK (2015) Governability and its discontents in the fishery of Lake Winnipeg since the late 1960s: the view from Gimli. In: Jentoft S, Chuenpagdee R (eds) Interactive governance for small-scale fisheries: global reflections. Springer, Dordrecht

Jones R, Rigg C, Pinkerton E (2017) Strategies for assertion of conservation and local management rights: a Haida Gwaii herring story. Mar Policy 80:154–167

Kaiser M (2012) Value isobars: the landscape and isobars of European values in relation to science and new technology. Final Report to the European Commission Research and Innovation DG. Project No. 230557

Kooiman J, Bavinck M, Jentoft S et al (2005) Fish for life: interactive governance for fisheries, MARE publication series. Amsterdam University Press, Amsterdam

Lalancette A (2017a) Creeping in? neoliberalism, indigenous realities and tropical rock lobster (kaiar) management in Torres Strait, Australia. Mar Policy 80:47–59

Lalancette A (2017b) Navigating the tides: indigenous perspectives and conventional fisheries Management in the Tropical Rock Lobster Fishery, Torres Strait, Australia. Dissertation, Concordia University

Lam ME (2014) Building ecoliteracy with traditional ecological knowledge: do, listen, and learn. Front Ecol Environ 12:250–251

Lam ME, Borch T (2011) Cultural valuing of fishery resources by the Norwegian Saami. In: Westra L, Bosselmann K, Soskolne C (eds) Globalisation and ecological integrity in science and international law. Cambridge Scholars Publishing, Cambridge, pp 361–376

Lam ME, Pauly D (2010) Who is right to fish? evolving a social contract for ethical fisheries. Ecol Soc 15:16

Lam ME, Pitcher TJ (2012a) The ethical dimensions of fisheries. Curr Opin Environ Sustain 4:364–373

Lam ME, Pitcher TJ (2012b) Fish commoditization: sustainability strategies to protect living fish. Bull Sci Technol Soc 32:31–40

Lam ME, Pitcher TJ, Kaiser M, Scott J, Surma S, Pakhomov EA, Millar K, Ward L, White A (2017) Values- and ecosystem-based management approach to the Pacific Herring fishery conflict in Haida Gwaii, Canada. Presentation at the international symposium on drivers of dynamics of small Pelagic Fish Resources, Victoria, BC, Canada, March 6–11, 2017

Ludwig D, Hilborn R, Walters C (1993) Uncertainty, resource exploitation and conservation: lessons from history. Science 260:17–36

Manno JP (2000) Privileged goods: commoditization and its impact on environment and society, Ecological economics series. Lewis Publishers, CRC Press LLC, Boca Raton

McGoodwin JR (1990) Crisis in the world's fisheries: people, problems, and policies. Stanford University Press, Stanford

Millennium Ecosystem Assessment (2005) Ecosystems and human well-being: synthesis. Island Press, Washington, DC

Mulrennan M, Scott C (2002) Mare nullius: indigenous rights in saltwater environments. Dev Chang 31:681–708

Pálsson G (2006) Nature and society in the age of postmodernity. In: Biersack A, Greenberg JB (eds) Reimagining political ecology. Duke University Press, Durham, pp 70–93

Pauly D, Zeller D (2016) Toward a comprehensive estimate of global marine fisheries catches. In: Pauly D, Zeller D (eds) Global atlas of marine fisheries: a critical appraisal of catches and ecosystem impacts. Island Press, Washington, DC, pp 171–181

Pikitch EK et al (2004) ECOLOGY: ecosystem-based fishery management. Science 305:346–347

Pikitch E et al (2012) Little fish, big impact: managing a crucial link in ocean food webs. Lenfest Ocean Program, Washington, DC

Pikitch EK et al (2014) The global contribution of forage fish to marine fisheries and ecosystems. Fish Fish 15:43–64

Pinkerton E, Davis R (2015) Neoliberalism and the politics of enclosure in North American small-scale fisheries. Mar Policy 61:303–312

Pitcher TJ, Lam ME (2010) Fishful thinking: rhetoric, reality, and the sea before us. Ecol Soc 15:12

Pitcher TJ, Lam ME, Kaiser M et al (2017) Hard of herring. In: Tortell P, Young M, Nemetz P (eds) Reflections of Canada: illuminating our biggest possibilities and challenges at 150+ years. Peter Wall Institute for Advanced Studies, University of British Columbia, Vancouver, BC, pp 112–119

Pollnac RB, Poggie JJ (2006) Job satisfaction in the fishery in two Southeast Alaskan towns. Hum Organ 65:329–339

Rabinowicz W, Rønnow-Rasmussen T (2016) Value taxonomy. In: Brosch T, Sander D (eds) Handbook of value: perspectives from economics, neuroscience, philosophy, psychology, and sociology. Oxford University Press, Oxford, pp 23–42

Raman S, Hobson-West P, Lam ME et al (2018) Science matters and the public interest: the role of minority engagement. In: Nerlich B, Hartley S, Raman S, Smith A (eds) Science and the politics of openness: here be monsters. Manchester University Press, Manchester, pp 230–250

Rittel HWJ, Webber MM (1973) Dilemmas in a general theory of planning. Policy Sci 4:155–169

Song AM (2018) How to capture small-scale fisheries' many contributions to society? – introducing the 'value-contribution matrix' and applying it to the case of a swimming crab fishery in South Korea. In: Johnson D, Acott TG, Stacey N, Urquhart J (eds) Social wellbeing and the values of small-scale fisheries. Springer, Dordrecht

Song AM, Chuenpagdee R, Jentoft S (2013) Values, images, and principles: what they represent and how they may improve fisheries governance. Mar Policy 40:167–175 8

Surma S, Pitcher TJ, Kumar R, Varkey D, Pakhomov EA, Lam ME, Belgrano A (2018) Herring supports Northeast Pacific predators and fisheries: insights from ecosystem modelling and management strategy evaluation. PLOS One 13 (7):e0196307

Taussig M (2010 [1980]) The devil and commodity fetishism in South America. University of North Carolina Press, Chapel Hill

Thomson D (1980) Conflict within the fishing industry. ICLARM Newslett 3:3–4

Trimble M, Berkes F (2013) Participatory research towards co-management: lessons from artisanal fisheries in coastal Uruguay. J Environ Manag 128:768–778

Trimble M, Araujo LG, Seixas CS (2014) One party does not tango! Fishers' non-participation as a barrier to co-management in Paraty. Brazil Ocean Coast Manag 92:9–18

Van der Sluijs J, Petersen A, Janssen P, JS R, JR R (2008) Exploring the quality of evidence for complex and contested policy decisions. Environ Res Lett 3:024008 von der Porten S, Lepofsky D, McGregor D, Silver J (2016) Recommendations for marine herring policy change in Canada: aligning with indigenous legal and inherent rights. Mar Policy 74:68–76

White C (2018) Symbols of resilience and contested place identity in the coastal fishing town of Cromer, Norfolk, UK: implications for social wellbeing. In: Johnson D, Acott TG, Urquhart J, Stacey N (eds) Social wellbeing and the values of small-scale fisheries. Springer, Dordrecht

White S, Ellison M (2007) Wellbeing, livelihoods and resources in social practice. In: Gough I, JA MG (eds) Wellbeing in developing countries: from theory to research. Cambridge University Press, Cambridge, pp 157–175

Chapter 4
Fish and Food Security in Small-Scale Fisheries

Philip A. Loring, David V. Fazzino II, Melinda Agapito, Ratana Chuenpagdee, Glenna Gannon, and Moenieba Isaacs

Abstract Fish is among the most eaten foods and traded commodities in the world, and small-scale fisheries provide food, jobs, and life satisfaction to billions of people worldwide. Yet, they are rarely recognized for these facts in global-level discussions about food systems and security. In this chapter, we argue that any discussion of food security, global or local, is incomplete if fisheries, and small-scale fisheries specifically, are not included. In this chapter, we discuss the many ways that small-scale fisheries contribute to local and global food security and to sustainable livelihoods in coastal communities. These include fish as an object of exchange and a marker of culture identity, and fisheries as a context in which people can connect their own health and well-being to the health of marine and freshwater ecosystems. The chapter begins with an introduction to the concepts of food systems and food security, the latter entailing more than just whether food is available, but also whether people have access to foods that are nutritious and culturally preferred. We conclude by discussing how a rights-based approach to food systems effectively brings these various ways that people engage with fisheries to the fore.

P. A. Loring (✉)
Department of Geography, Environment, and Geomatics and the Arrell Food Institute, University of Guelph, Guelph, Canada
e-mail: phil.loring@uoguelph.ca

D. V. Fazzino II
Department of Anthropology, Bloomsburg University, Bloomsburg, PA, USA
e-mail: dfazzino@bloomu.edu

M. Agapito · R. Chuenpagdee
Department of Geography, Memorial University of Newfoundland, St. John's, NL, Canada
e-mail: magapito@mun.ca; ratanac@mun.ca

G. Gannon
School of Environment and Sustainability, University of Saskatchewan, Saskatoon, SK, Canada
e-mail: glg875@mail.usask.ca

M. Isaacs
Institute for Poverty, Land and Agrarian Studies, University of the Western Cape, Cape Town, South Africa
e-mail: misaacs@plaas.org.za

R. Chuenpagdee, S. Jentoft (eds.), *Transdisciplinarity for Small-Scale Fisheries Governance*, MARE Publication Series 21,
https://doi.org/10.1007/978-3-319-94938-3_4

Keywords Food security · Food systems · Nutritional security · Fisheries sustainability · Small-scale fisheries

4.1 Introduction

It may seem difficult to imagine, given that fish is generally harvested for consumption, that the majority of fisheries research in the past 50 years or more has disregarded the role that fisheries play in local, regional, and global food systems (Loring et al. 2013; Olson et al. 2014). From the perspective of environmental conservation, fisheries have been treated primarily as populations of species of concern, with the act of fishing constituting a 'pressure' that managers must limit accordingly. From the perspective of economics and state-level policy, fisheries are largely treated as production systems for marketable commodities or as livelihood systems for coastal peoples. Yet, these segmented views of fisheries miss the important connections between them, connections made by fish and the roles that it plays in people's lives as food—not merely for consumption but also celebration and cultural heritage (Lyons et al. 2016; Bennett et al. 2018a, b). This oversight, and perhaps lack of perspective on fisheries from the context within which they are most meaningful, has impacted the ability of governing bodies worldwide to achieve outcomes in fisheries management that are both environmentally sustainable and socially just.

Our aim with this chapter is to help address this oversight, to draw attention to the numerous ways that communities and landscapes and seascapes are linked through fish as food. Fish as food is arguably the most important lens for thinking about the sustainability of fisheries and fishing communities, given how many people live next to or near bodies of water, and how many rely on fish for some or most of their daily protein consumption. Understanding fish as food, and fisheries systems as food systems, enables a transdisciplinary perspective—meaning that it enables us to look beyond any specific disciplinary framing of fisheries in an attempt to capture the diverse and nuanced ways that people interact with fish, and through fish, with each other.

The role that fish plays in food and food security is possibly most prominent in small-scale fishing communities (Bennett et al. 2018a, b; Teh and Pauly 2018). According to the Food and Agriculture Organization of the United Nations (FAO), the majority of the world fisheries are small-scale, providing income, livelihoods and food to the global population, over 4.5 billion people rely on fish for 15% or more of their protein (Béné et al. 2015). In remote places like the North American Arctic, as much as 80% of people's protein comes from subsistence fisheries (e.g., Fall 2014). Globally, demand for fish and other aquatic food products is growing, and for multiple reasons. As we will discuss below, fisheries provide an excellent, if not unmatched, source of nutrition. Fisheries are increasingly being thought of as 'local' alternatives to the global food system, and can also have a much lower carbon footprint than agricultural protein production, making them desirable from a climate change and global sustainability perspective (Béné et al. 2015). Finally, small-scale fisheries are also emerging as a venue through which many communities are asserting or trying to assert their rights and reclaim sovereignty, including food sovereignty, through initiatives that take control over how fisheries are managed, prosecuted, and marketed (e.g., Jones et al. 2016; Levkoe et al. 2017). These different production systems vary in terms of their contribution to local and global food security and have various consequences and implications that need to be taken into consideration when discussing fisheries sustainability.

A 'food systems' approach can be useful for understanding the myriad important roles that fish, as food, plays in local communities, and how fish links communities to regional and global systems. The concept of a food system is an analytical one, which attempts to capture all aspects of how people produce, transform, distribute, and consume food (Ericksen et al. 2010). At the local level, this can include a complex assemblage of local practices for agriculture, hunting, or fishing, as well as how foods are processed, shared, marketed, and eventually, consumed and celebrated, whether at home or in a restaurant, with family or alone. These local systems likewise are embedded within equally complex and complicated regional and global systems, connected through lines of trade, commodity and fuel markets, and global networks for transportation and shipping.

Here, the concept of a food system helps us to understand how fish, as food, moves and is transformed as it changes hands, from person to person, and plate to plate, taking on new meanings and significance at each step. The language of food systems is an open and conciliatory language, capable of facilitating engagement among fisheries professionals and researchers in different disciplines and sectors: economists working on global trade, political scientists interested in rights and dispossession, managers interested in ecosystem-based effects of fishing behaviors, and policy-makers interested in securing trade; all of these perspectives have a home in a food systems approach. Hence, any discussion of food systems inherently crosses disciplinary and political boundaries and provides a broad framework to effectively integrate a holistic consideration of the very essence of life.

Below, we bring into focus a small handful of the diverse ways that fisheries interact with global and local food systems. While this does not aim to paint a comprehensive picture of small-scale fisheries in countries around the world, because this would be exceedingly ambitious, it is an attempt to illustrate why thinking about fish as food is such an important and powerful lens for addressing numerous fisheries- and fishing community-related concerns. An overarching premise of this book is that small-scale fisheries matter, also in a normative sense, both locally and globally: locally in terms of the ways that they contribute to health, well-being, and sustainable development for impoverished and marginalized communities, and globally as a venue for transforming the global, unsustainable, and unjust food system into something that works better for people. The focus on fish as food in this chapter exemplifies the need for a broad and holistic perspective when thinking about the connections among small-scale fishers, fish and fishers, beyond workers converting capital for global markets. The sections below offer some of the much-needed justification for this premise, as well as guidance regarding how to put these ideas and values into practice.

4.2 Fisheries and Food Security

Fisheries around the world play essential roles in providing food security for many local communities, and they also contribute a fair share to food security globally. According to published articles in the 'Information System on Small-scale Fisheries' (ISSF[1]) of the Too Big To Ignore Global Partnership for Small-Scale Fisheries Research (TBTI[2]), small-scale fisheries are essential in providing access to fish as food for global population especially in Africa and Asia and Oceania regions. Notably, in these two regions, small-scale fisheries are predominantly discussed at a sub-national and national scales, which suggests that the relevance of subsistence fishing extends beyond local, community level where small-scale fisheries are more visible (Fig. 4.1). In Asia alone, small-scale fisheries contribute 25 million tons annually, which is more than half of the world's small-scale fisheries marine production, and 8 million tons or 70% of inland global production (Mills et al. 2011; FAO 2016). In Africa, 85% of fish harvesters are involved in small-scale fisheries and contribute 47% of the landed value (Sea Around Us 2016). About 90% of inland and marine fisheries catches in these two regions go directly to human food consumption. Additionally, fisheries provide jobs to 120 million people involved directly and indirectly in fisheries and 90% of these are related to small-scale fisheries, which take place mostly in developing countries (Mills et al. 2011). Hence, at a global level, increased access to small-scale fisheries could be an essential way to increase people's food security, as a large portion of income in developing nations goes to obtaining food needs (Banerjee and Duflo 2007). Based on the FAO (2016)'s most recently available data, fish accounts for a significant amount of the total protein available in many nations, as much as 21% in China, 23% in Japan, and 14% in Norway. The number is lower in the United States and Canada, but often much higher than these averages in North America's remote and rural areas. Fish's contribution to food systems is also much higher in small and developing coastal or island nations; for example, Maldives and Kiribati have a consumption rate of 180 and 72 kg per person per year respectively (FAO 2016).

Demand for seafood globally is also on the rise and expected to continue to rise for a number of reasons: fish are highly nutritious as we will discuss below, and dietary advisories in developed nations are increasingly recommending that people eat multiple large portions of seafood each week. Globally, seafood markets generally move the highest quality seafood away from smaller coastal communities and nations, to wealthy, developed nations (Watson et al. 2017). As such, lower-income fisheries dependent states rely on lower-quality seafood products, whether locally harvested or imported, to meet their own food needs (Watson et al. 2017).

The relationship among fisheries and food security often takes on dramatically different forms in the academic literature, depending on the scale and discipline of analysis. At a very high level, analyses might focus simply on the amount of fish

[1] https://issfcloud.toobigtoignore.net

[2] toobigtoignore.net

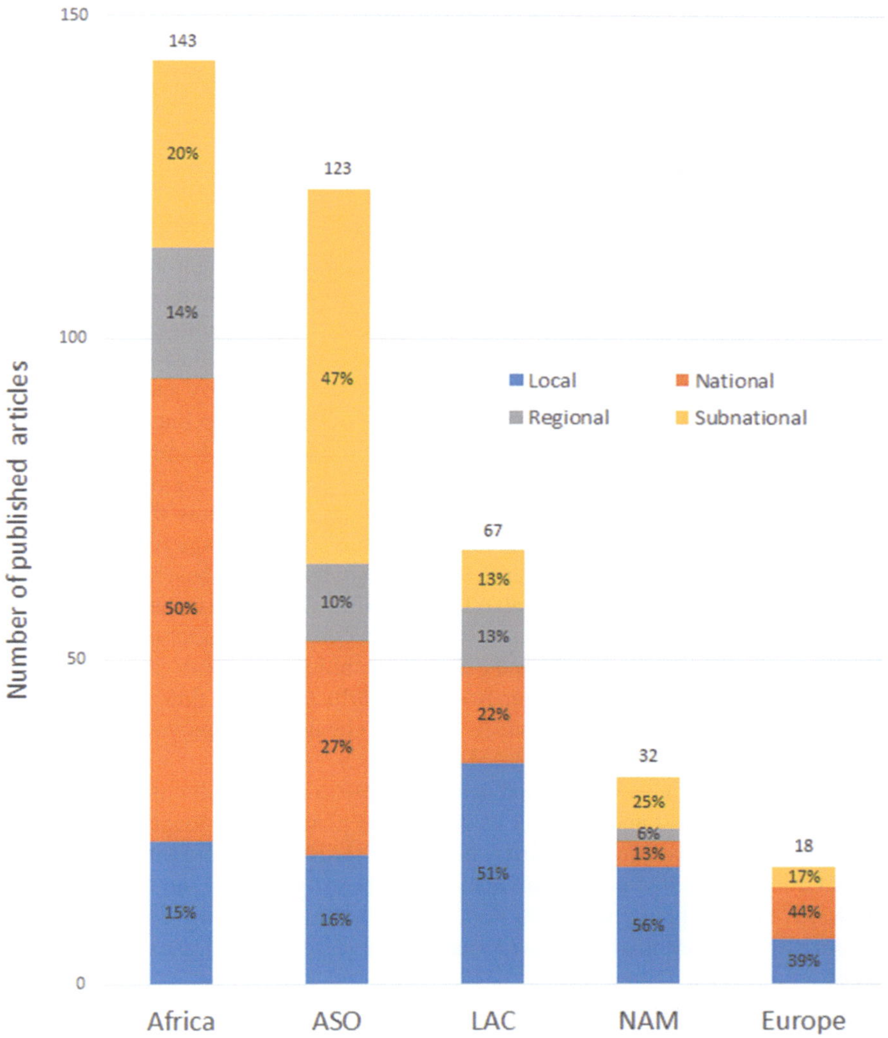

Fig. 4.1 Published articles discussing the importance of subsistence type of small-scale fisheries to food security. ASO = Asia and Oceania; LAC = Latin America and the Caribbean; NAM = North America (*Source: ISSF* see also Jentoft et al. 2017)

that could be harvested sustainably from a fishery, and compare this to demand (Merino et al. 2012). But this is a rather superficial approach to food security, which as we discuss below, conflates production, or the availability of food, with food security. The latter generally has more to do with whether people have access to foods, whether they are safe, and whether they are the foods that people prefer, than it is simply a matter of how much food exists. By comparison, household- and community-level approaches to food security are often far more multifaceted. In other words, it is not only a measure of how much food is being harvested or produced, food security at these levels is also understood to be an emergent property of a food system – the totality of activities, social institutions, material inputs and outputs, and cultural beliefs within a social group that are involved in the production, distribution and consumption of food.

Like conservation, food security can be discussed only in general terms at the global level, for example in terms of the total number of people living with food insecurity or hunger. At the local level, food security takes on a much more complex meaning and application, with multiple interacting factors—such as income, climate and climate change, socioeconomics, community infrastructure, and systems of land tenure—determining whether or not people can put healthy and culturally preferred food on their tables. Research in such places as Volta Basin in Africa and Mekong Basin in Southeast Asia shows, for instance, that poverty and food insecurity may not only be a consequence of resource overexploitation but is related also to other factors such as geographical and political isolation (Béné and Friend 2009).

A common framework used for household and community food security proposes that food security is a function of four interacting features (Ericksen et al. 2010): (1) availability, meaning whether or not sufficient food is produced and harvested; (2) access, meaning whether those foods are being distributed and marketed to those who need it, and if so, whether they can afford to purchase them; (3) utilization, meaning whether the food is safe and meets consumers' biophysical, psychological, and social and cultural needs; and (4) stability, a temporal dimension that recognizes that each of the first three features will fluctuate over time because of the influence and interaction of seasonality, weather, changes in employment, and other socioeconomic, cultural and ecological factors.

Food security can be undermined in any of these dimensions by multiple causes, and as such, it is best achieved via strategies that plan for and adapt to changes in food availability over time. Security, in this sense, is clearly more than just food production alone. Food production does have a bearing on food security, but so do the foods that are being produced, how, where, when, and by whom they are produced, how they are distributed and marketed, and whether all people have equal access.

More than simple commodities, it is essential to think about the benefits of fisheries as central to fishing people's rights—specifically, the right to food. The United Nations defines the right to food as,

> Regular, permanent and unrestricted access, either directly or by means of financial purchases, to quantitatively and qualitatively adequate and sufficient food corresponding to the cultural traditions of the people to which the consumer belongs, and which ensure a physical and mental, individual and collective, fulfilling and dignified life free of fear. (Ziegler 2008, 2).

This right, which is codified, at least most notably, by Article 25 of the Universal Declaration of Human Rights (United Nations 1948), underpins what has become known as a 'rights-based' approach to food systems (Chilton and Rose 2009). While parallel in many ways to the concept of food security, shifting the discourse from security to rights changes how people and governments must engage with food and food security. Specifically, it requires three kinds of action: respect for people's right to acquire food, protection of that right, and ensuring that people, through their own agency, have the opportunity fulfill that right in the ways that they see fit (Chilton and Rose 2009). A rights-based approach to food systems, and as such, to the fisheries within them, contends that multiple human rights, including the right to individual health, to food security, to pursue safe and fulfilling livelihoods, to gender equity, and to healthy and thriving ecosystems, can all be pursued through improvements to local food systems (Anderson 2008).

In rural Alaska, for example, Alaska Natives rely heavily on wild fish for a large component of their food system, and these are precisely the culturally preferred foods that contribute to a food secure situation (See Box 4.1) (Fall 2014). Although still abundant in many parts of Alaska, fish and game resources in some regions appear to be in decline, such as King salmon in rivers across the state (Lewis et al. 2015). Historically, Alaska Natives have been able to respond effectively to variability and change in the availability of fish and game through mobility and a flexible and seasonally specific strategy that includes multiple food options. More recently, this flexibility has been constrained by the increased cost of fuel necessary

Box 4.1 Country Food and Food Security in the Rural North

Communities in rural Alaska and northern Canada face much higher rates of food security than do those in other parts of the two developed nations. As of 2016 for the US and 2011 for Canada (the most recent years for which data are available), 12.3% and 8.4% of households are food insecure; by comparison, food insecurity rates are as high as 25% in Western Alaska and 69% in Nunavut, Canada. (Council of Canadian Academies 2014; Feeding America 2013). These stark inequities are arguably rooted in histories of exploitation and disempowerment, and systematically disadvantage historically colonized and exploited peoples in North America (P. Allen 2010). That being said, circumstances would be far worse for the indigenous residents of these places had they not maintained access to fish as a part of their traditional 'foodways' (Loring and Gerlach 2009; Kofinas et al. 2016).

Traditionally, the cultural, social and economic autonomy of these rural communities was centered on a flexible subsistence food system that, depending on region, included sea mammals, terrestrial mammals, birds, fish, and plant resources from formal or informal gardens (referred to in the North, generally, as 'country foods', see e.g., Gerlach et al., 2011). The portfolios of species harvested different from region to region, culture to culture (Wolfe and Walker 1987), yet, fish such as salmon, herring, and grayling, and arctic char, were and are a ubiquitous component (Nuttall et al. 2004). Many fish, like salmon and herring, are also cultural keystone species that are essential to the cultural health and survival of northern peoples (Thornton 1998; Loring and Gerlach 2009). Although northern foodways have been altered by multiple forces and stressors over the years, country foods today are still (albeit to

(continued)

Box 4.1 (continued)

various degrees) governed by traditional and localized patterns, and it is through subsistence hunting and fishing that many Indigenous people keep their cultures and traditions alive.

Country foods fit in to what scholars call a "mixed subsistence market economy": one characterized by a reinforcing combination of subsistence activities and cash generating paid employment (i.e. tourism, guided hunting, service sector). Typically, families invest a small portion of their household incomes (which are usually low) towards technologies and supplies for harvesting country foods, such as equipment fuel. These reinforcing systems are increasingly disrupted, however, by high costs of fuel for heating and transportation and of low quality store-bought foods (Gerlach et al. 2011; Council of Canadian Academies 2014). Nevertheless, country food harvests are extensive: in Alaska, for example, most rural residents (which form 17% of the state's population as of 2012) take part in the harvesting of wild game (60%) and fish (83%), and a still-higher portion uses these subsistence harvests (86% and 95% respectively) (see Table 4.1). This is due to the common practice of sharing harvests with extended family and the community, especially with households unable to fish or hunt, such as elders, the disabled, as well as single parents with young children (Fall 2014; Kofinas et al. 2016). Multiple scholars have estimated that only 30% of a community's households are responsible for 70% of its wild foods supply, specifically to have the ability to provide for those that are unable to do so (Wolfe and Walker 1987; Kofinas et al. 2016).

to power riverboats, all-terrain vehicles, and snowmobiles, by permanent settlement in fixed communities rather than seasonally shifting residential patterns, by a complicated patchwork of state, federal, and private land tenure regimes, and by policies for fish and game management that are out of sync with the land and seascape changes caused by climate change (Gerlach et al. 2011). More and more rural people are eating the high cost, nutritionally bankrupt foods available at small village stores, and while the availability of these foods does indeed provide a measure of protection against hunger in the strictest sense, they nevertheless provide little in the way of supporting individual and community health (Loring and Gerlach 2009; see Sect. 4.6 below). We use Alaska as the example here, but these circumstances mirror those being experienced for remote, natural resource dependent communities across the high latitude North and elsewhere around the globe.

Table 4.1 Percentage of households relying on country foods in rural Alaska. From Fall (2014)

Area	Harvesting game (%)	Using game (%)	Harvesting fish (%)	Using fish (%)
Arctic	63	92	78	96
Interior	69	88	75	92
South-central	55	79	80	94
Southeast	48	79	80	95
Southwest	65	90	86	94
Western	70	90	98	100
Total rural	**60%**	**86%**	**83%**	**95%**

4.3 Linking Fish Nutrition to Food Security

Increasingly, the high nutritional value of fish is recognized as an important aspect of how to build and maintain food security at local and national levels (Kawarazuka and Béné 2010; High Level Panel of Experts [HLPE] 2014; Isaacs 2016;). Fisheries NGOs such as the World Forum of Fisher People (WFFP), the World Forum on Fish harvesters and Fisher workers (WFF), the International Collective in Support of Fish workers (ICSF), La Via Campesina (LVC), and Food First Information and Action Network (FIAN) have been instrumental in advocating the importance of acknowledging fish nutrition as an essential component of food security and social justice. This is reflected in a number of international policy documents adopted by the Committee on Fisheries of the Food and Agriculture Organization of the United Nations (FAO), including the *Voluntary Guidelines on Responsible Governance of Tenure of Land, Fisheries and Forests in the context of National Food Security (Tenure Guidelines)*, the *Voluntary Guidelines on the Progressive Realization of the Right to Adequate Food in the Context of National Food Security (Right to Food Guidelines)*, and *Voluntary Guidelines for Securing Sustainable Small-Scale Fisheries in the Context of Food Security and Poverty (Small-Scale Fisheries Guidelines)*. The Committee on Food Security (CFS/FAO)'s HLPE report on *Sustainable Fisheries and Aquaculture for Food Security and Nutrition* likewise underscores the importance of small-scale fisheries in contributing to food security and nutrition (HLPE 2014). International instruments such as this help to create space for fish to play a dominant role in the contributing to many poor fishing communities' in marine, inland, riparian dependent on the rich nutrition provided by small-fish.

Consumption patterns of poor people show that people in underdeveloped and impoverished regions tend to eat the same foods every day, as dietary diversity is constrained due to the food availability, food access, and food affordability (Khor 2008). As Amartya Sen and others have shown, it is not that existing food systems do not produce enough food to feed the world's population, but that many people simply lack access, for a variety of social and political reasons, to healthful foods (Sen 1983; George 1986). Food insecurity, in the form of malnutrition, is often linked to consuming foods lacking in essential minerals and vitamins, and can manifest both in undernutrition (hunger) and overnutrition (obesity) (Popkin and Gordon-Larsen 2004; Khor 2008; Fazzino 2012). More and more, poor people are nutrient poor, even they live in food-rich areas. Market foods are often highly processed, high in sugar, salt, carbohydrates, and fats (Price 1939; Popkin and Gordon-Larsen 2004); they are easy accessible, cheap, often nutrient poor, made to taste good often contribute to the nutrient poor in vulnerable populations.

Increased access to fish could address several diet-related health problems for the poor. Fish, including small pelagic fish species such as sardines, and larger species such as salmon and herring, are rich in macro- and micronutrients, easily-digestible protein that contains all the essential amino acids, lipids with essential omega-3 fatty acids, key minerals and vitamins (Isaacs 2016). The lipid composition of these fish, especially small pelagics, is different from that of other proteins due to the long-chain, polyunsaturated fatty acids have a number of potential health benefits (Ramsden et al. 2009; Loring et al. 2010; Isaacs 2016). Only 150 mg of small pelagic fish per day contains the optimal amount of omega-3 docosahexaenoic (DHA) believed to be important during pregnancy, breastfeeding and infancy for neurodevelopment (Mozaffarian and Rimm 2006). Similarly, non-communicable diseases like heart diseases are common amongst urban and rural poor due to consuming unhealthy, cheap and processed foods (Seedat 2007); the omega-3 oils found in fish have also been found to reduce unhealthy cholesterol and triglycerides as well as increase healthy cholesterol (Ramsden et al. 2009; HLPE 2014).

Small pelagic fish also contain significant amounts of essential micronutrients, including lysine, methionine, Vitamins A, D, B1 and B2, as well as iron, phosphorous and calcium (High Level Panel of Experts 2014). Vitamin A is present in fish as retinol and anhydroretinol, which is more easily preserved through the cooking process, and more readily absorbed by humans than the form of Vitamin A found in vegetables (Thilsted et al. 1997). In addition, when fish is digested with bones and organs it can be slightly more effective source of calcium than milk (Titchenal and Dobbs 2007). Fish is likewise a rich source of iron, and also aids the bioavailability of iron of other foods in a meal, such as rice (Thilsted et al. 1997). Over 30% of people in the world are reported to be iron deficient, including up to 50% of pregnant women and children in developing countries, with approximately 2 billion being anaemic (HLPE 2014). Lack of iron impairs cognitive development, growth, and immune function, very often resulting in sub-optimal performance at school and in adulthood (Thompson 2011).

Fish, added to the traditional blend of staple diets, will therefore increase the utilization of micronutrients, indirectly boosting people's immune systems; this could be especially important for vulnerable populations such as young children, the aged and immune-compromised persons living with HIV and Aids (Thompson 2011). Finally, while fish in the food security literature is often discussed separately from the consumption of land-based foods, there may be important reasons to explore these together. Rice, maize and grains, for example, can be fortified with fish oil as a way to bring necessary nutrients to the poor (Isaacs 2016).

One noteworthy caveat to the many nutritional benefits of fish is that many fishes are now potentially contaminated with such toxicants as persistent organic pollutants (POPs), dioxins, and methylmercury (Corsolini et al. 2005; FDA 2006; Sunderland 2007). Mercury levels are high, for example, in fish and other marine animals in northern waters (Jewett and Duffy 2007), in part because of climate change and the northward atmospheric transport of mercury from coal burning at lower latitudes. However, it is difficult to develop straightforward or uniform guidance for whether the risks outweigh the benefits of consuming seafood, given regional variation in contamination levels and also the multiple benefits that fish can have to health (Egeland and Middaugh 1997; Loring and Duffy 2011). In one study focusing on methylmercury in food fish in Alaska, for example, Loring and colleagues (Loring et al. 2010) showed that benefits of high levels of omega-3 fatty acids in fish can outweigh the negative impacts of mercury with respect to both cardiovascular health and infant neurological development.

4.4 Fish, Food, and Culture

Nutrition, of course, is just one of many ways that food contributes to people's health and well-being. In many parts of the world, fish are extremely important aspects of local cultural identity and traditions. Salmon in the US Pacific Northwest and Alaska is considered to be a 'cultural keystone species' for its importance to local life and culture (Wolf and Zuckerman 2003; Amberson et al. 2016). Fishing, and sharing fish, are a way that many Indigenous peoples maintain social ties, both within and among families and with elders and other community leaders. Fish and seafood in general is ubiquitous in cultural iconography and art all around the word, from the Totem poles in Haida Gwaii, British Columbia to the carving of the Mao'ri in Ao'tea'roa (New Zealand). Several countries indicate the importance of fish by depicting them in their bank notes, coins and stamps. Fish and fishing people are also major themes in stories, folklore and songs. Fish may be the most commonly used non-human form represented in all culture, including in contemporary arts and design. Thailand, for instance, has an old saying, which has been turned into songs and dances, to signify the traditional role that fish plays in food security and the nation's prosperity, as a key accompaniment to rice, 'there are fish in the waters, there is rice in the fields.'

In many coastal and rural inland communities, where fishing and fish intertwine with food and culture, fishing activities contribute to maintaining social solidarity in the community (Freeman 2005; Stewart 2005; Kittinger et al. 2015). In the Hawaiian Islands, 20% of seafood obtained by small-scale fishers goes to socio-cultural activities as a way to keep social cohesion (Kittinger et al. 2015). In some of the Pacific Islands, about half of the small-scale fishers catch goes to household consumption, and cultural events and ceremonies, as well as to give away to friends and relatives.

Shared fish can be in the form of contribution or service to one's family (or *tautua* commonly practiced by Samoans), mutual kin assistance, or general support (Severance et al. 2013). Hence, fishing for food is a practice in which people find meaning, through which they learn from their elders, and ultimately establish their place in their community. In places where colonialism and development has displaced people from their traditional fishing roles, numerous negative health outcomes have emerged, including declines in physical health associated with a western diet—patterns of depression, substance abuse, family violence, and suicide in colonized places have all been attributed in various ways to the psychological impacts of not being able to fish or hunt (Kral 2012; Allen et al. 2014).

There can also be important gendered aspects of the various roles that fish, as food, play in supporting individual and community well-being, though these are typically overlooked in assessments of how small-scale fisheries contribute to food security (Harper et al. 2013). Women are often involved in pre- and post-harvest activities such as mending gears, packaging, processing, and selling the fish. This type of involvement of women can be done in their homes or in the community where they live, which means that these activities do not necessarily sacrifice their family-related responsibilities *(R. Biswal, personal communication)*. At the same time, it can contribute to the family's income which enhances the ability of the household to purchase other food needs. In many parts of the world, fishing is a way that women contribute to household and community food consumption, for example fisherwomen in Melanesia supply about 80% of the subsistence requirement of their communities (Kronen and Vunisea 2009).

Lastly, it is important to note that food preparation and value-adding play important roles in food security, especially in the context of accessing nutritious, culturally preferred, and affordable food. Various cultures have unique ways to prepare food, typically, influenced by resources available to them (see Box 4.2). Traditional preparation of fish and fish products uses fish that are available locally, which is in most cases, are from small-scale fisheries. Drying, salting, fermenting, smoking, and canning are useful methods to prolong the shelf life of fish and fish products. Also, dried, fermented, and canned fish are transported with ease to nearby towns or other parts of the country, thereby, increasing access to fish by communities away from the coasts and urban centers (Isaacs 2016). In some parts of the world, a powder form of fish increases distribution and consumption, and has been especially beneficial in addressing malnutrition in rural communities in Cambodia.

Box 4.2 Food and Culture: Appreciating Fermented Fish Sauce in the Philippines

In Southeast and East Asia, fermented fish sauce is deeply integrated into people's daily cuisines (Beddows 1998). Fermented fish sauce, also called *patis* in the Philippines, is regularly used in cooking or as condiments. *Patis* can add flavor to virtually all dishes and are also used in place of salt to enhance the flavor of many Philippine dishes. Hence, it is commonly observed in restaurants to keep fish sauce in the dining table, a reflection of how fish has become embedded in appreciating food in general (Rattagool 1985). The use of fermented fish in the Philippines has originated from the ways Filipinos have broadly depended on fish as food. Fish is relatively more affordable than other sources of proteins; hence about 70% of protein sources in the Philippines are from fish caught in small-scale fisheries (Muallil et al. 2013).

Although industries have become the major suppliers of fish sauce in urban areas, in the country, the practice of making fermented fish sauce has been, and still is, culturally significant in many coastal areas in the Philippines. An example is a town of Luna in La Union, where about 80% of fermented fish sauce, made mostly by families directly engaged in fishing (20% of the working population) goes to household consumption and the remaining 20% are shared with relatives and friends. The traditional process of fermentation is typically carried out by older women who have learned the know-how passed down to them by their parents and grandparents. In general, the homemade fermented fish sauce acquires exceptional taste and smell, thus considered high quality and the locals or connoisseur would know the difference *(M. Lim, personal communication)*.

Another important aspect of fermented fish sauce is that this typically makes good use of abundant small pelagic fish (e.g., varieties of mackerel, anchovies), which are relatively low-value compared with big fish species and are available year round. The preparation of fermented fish also makes use of locally available materials such as an earthen jar called Burnaj jars. The jar containing the fresh fish are sealed and kept buried, at least three-fourths of the jar, in the soil for 3 months to 1 year. The longer the fermentation is, the better it becomes *(M. Lim, personal communication)*. Although this may take a long time, it guarantees that this naturally induced fermentation process does not require additives or harmful chemicals that may be present in industrially made fermented fish sauce (Lopetcharat et al. 2001). As such, locally made fish sauce represents cultural and nutritional values of fish, as well as supports local economy and sustainability of the fisheries ecosystem.

4.5 Fish as a Linkage Between Ecosystem Health and Human Well–Being

In addition to the many ways that small-scale fisheries support community food security, taking a food systems approach to fisheries can also enhance efforts to conserve natural resources such as rangelands, fisheries, and other wild food resources. This is the case especially when explicitly linked to the precept that healthy ecosystems and sustainable harvests of natural resources are both part and parcel to sustainable food production systems, including relevant regulatory, management, cultural and social components of the system. However, often, this connection between sustainable harvest and food security are not explicitly defined (Foale et al. 2013; Loring 2013).

Therefore, as a start, food security goals should be explicitly defined, in the context of fish as food and its connection between ecosystem health and human well-being. This can be accomplished in part by tightening the connections between people and their ecosystems through their food choices (Sundkvist et al. 2005), i.e., by empowering people to be better stewards of the resources (Bennett et al. 2018a, b), by conducting social analysis and inter-sectoral policy and governance (Foale et al. 2013), and by providing a context within which people can respond to variability and change in the availability of resources by with a diverse portfolio of food options (Loring and Gerlach 2010; Kofinas et al. 2010).

An example is the case of the Pacific Island countries showing that fish is essential to health and well-being of people. With the changes in the diet of the Pacific Islanders, as a result of the influx of processed imported foods, the population has suffered from high rates of diabetes and obesity problems. However, if the Islanders would increase their fish consumption, these health issues would be minimized. Estimates show that for the Pacific Island population to be food secure, one avenue is to increase the supply of tuna by 12% in 2020 and 25% by 2025. Relatively, this required increase in domestic supply and is equivalent only to 2.1 and 5.9% respectively of the current industrial catch. However, to make the Pacific Islands population food secure is not straightforward. It would require an explicit goal ensuring that fish especially tuna is readily available for coastal communities. In the context of the Pacific Islands, this will require intervention specific to small-scale fishers to catch more tuna such as assisting them with the efficient use of the fish aggregating device (FADs). Another is by creating policy support at the regional level (e.g., ban on discarding small tuna and land this fish at regional ports) and country-level (e.g., increase distribution and access to affordable canned tuna for the inland population) (Bell et al. 2009).

An important step to making food security goals explicit is by examining the two-way relationship between resource governance and food security (Foale et al. 2013). There is a need to recognize that the process of achieving food security dimensions will have an impact on the ecosystem and may result in tradeoff and challenges. Box 4.3 shows examples where food security may align or conflict with sustainable fisheries harvests. A valuable lesson is that when the goals of resource management gear toward making people food secure, fish is seen not only a source of nutrition to support human health but also a provision of a healthy and well-managed environment.

Box 4.3 Food Security vs. Sustainable Fishery

Juan Fernandez Lobster fishery in Chile The Juan Fernandez lobster fishery has supported a small community but strongly dependent on the small-scale lobster fishery for more than 100 years. The users and boats have remained relatively stable from 41 to 57 boats from 1947 to date. A traditional tenure system called *marcas* has managed the fishery. The use of small boats and ownership of fishing spots are governed by unwritten rules under the *marcas* tenure system and have provided fishers with perceived equal access to the resource. The lobster fishery has remained sustainable to date for these additional reasons: (1) lobster stock is productive, (2) high compliance among fishers of the unwritten and formal rules, (3) the *marcas* tenure system is able to implement simple rules to protect reproductive female lobster, and (4) local engagement of fishers. Fishers' engagement led to positive responses and outcomes such as local fishers being actively involved and trained in collecting and maintaining fisheries datasets. Since fishers take an active role in management decisions, fishers cooperate with monitoring and implementing harvesting size regulations. Some fishers have also established marketing cooperative that allows them to participate in ecolabeling and get the support of Slow-Food movement which makes local fish consumed locally (Ernst et al. 2010, 2013).

Small-scale fishery in Solomon Islands: Food security vs. conservation The Solomon Islands population has been dependent on the small-scale fishery for food. However, the fishery is facing an increasing number of people reliant on fish as a major source of protein causing mounting pressure on the fisheries. Based on current trends of exploitation, the sea-cucumber and reef fisheries will not have the capacity to elevate the population above its local poverty situation while the fisheries will also be significantly degraded. Hence it will be a real challenge to both keep the community food secure and maintain environmental sustainability. However, a responsive governance system can help alleviate this problem by explicitly recognizing this tradeoff. A strategy may entail reducing the pressure on reef fisheries by developing the less exploited inshore tuna fishery to provide the required protein for the Island's population. This approach, however, would need policy support and behavioral and cultural adjustment among the communities actively engaged in fishing (Hardy et al. 2013).

4.6 Conclusion

The goal of this chapter has been to provide a sense of the many ways people can and do benefit from small-scale fisheries. Far more than just systems economic production, small-scale fisheries are essential to people's food and livelihood security and quality of life. They are a locus of cultural value and meaning, a setting for the transmission and sharing of knowledge and heritage, and a platform through which people connect with and steward their local ecosystems. The concept of the food system provides a unifying framework for these various interdependencies; in so doing, it also brings the discussion of fisheries into contact with other important aspects of people's livelihoods and food security. A successfully managed fishery, when recognized as part of a food system, involves more than just the prudent management of how many fish are taken, and gives equal credence to the multiple ways that fisheries enrich people's lives.

Food sovereignty, just as it is transforming how people view small-scale agriculture, can be just as transformative for how policy attends to small-scale fisheries. As already noted, food sovereignty means more than just food security—it means that people are in control of their food systems. In the case of small-scale fisheries, food sovereignty aligns directly with the three tenets of rights-based approaches to food systems noted above: respect, protection, and fulfilment of people's many human rights. Moving forward, we argue that fisheries management and governance discourses need to bring these issues of human rights and sovereignty to the fore, if fisheries, whether small or large scale, are to be both sustainable and just.

References

Allen P (2010) Realizing justice in local food systems. Camb J Reg Econ Soc 3:295–308. https://doi.org/10.1093/cjres/rsq015

Allen J, Mohatt GV, Fok CCT et al (2014) A protective factors model for alcohol abuse and suicide prevention among Alaska native youth. Am J Community Psychol 54:125–139. https://doi.org/10.1007/s10464-014-9661-3

Amberson S, Biedenweg K, James J et al (2016) "The heartbeat of our people": identifying and measuring how Salmon influences Quinault tribal well-being. Soc Nat Resour 29:1389–1404. https://doi.org/10.1080/08941920.2016.1180727

Anderson MD (2008) Rights-based food systems and the goals of food systems reform. Agric Hum Values 25:593–608. https://doi.org/10.1007/s10460-008-9151-z

Banerjee AV, Duflo E (2007) The economic lives of the poor. J Econ Perspect J Am Econ Assoc 21:141–167. https://doi.org/10.1257/jep.21.1.141

Beddows CG (1998) Fermented fish and fish products. In: Wood BJB (ed) Microbiology of fermented foods. Springer, New York, pp 416–440

Bell JD, Kronen M, Vunisea A et al (2009) Planning the use of fish for food security in the Pacific. Mar Policy 33:64–76. https://doi.org/10.1016/j.marpol.2008.04.002

Béné C, Friend RM (2009) Water, poverty and inland fisheries: lessons from Africa and Asia. Water Int 34:47–61

Béné C, Barange M, Subasinghe R et al (2015) Feeding 9 billion by 2050 – putting fish back on the menu. Food Secur 7:1–14. https://doi.org/10.1007/s12571-015-0427-z
Bennett NJ, Kaplan-Hallam M, Augustine G et al (2018a) Coastal and indigenous community access to marine resources and the ocean: a policy imperative for Canada. Mar Policy 87:186–193. https://doi.org/10.1016/j.marpol.2017.10.023
Bennett NJ, Whitty TS, Finkbeiner E et al (2018b) Environmental stewardship: a conceptual review and analytical framework. Environ Manag 61:1–18. https://doi.org/10.1007/s00267-017-0993-2
Chilton M, Rose D (2009) A rights-based approach to food insecurity in the United States. Am J Public Health 99:1203–1211. https://doi.org/10.2105/AJPH.2007.130229
Corsolini S, Ademollo N, Romeo T et al (2005) Persistent organic pollutants in edible fish: a human and environmental health problem. Microchem J 79:115–123. https://doi.org/10.1016/j.microc.2004.10.006
Council of Canadian Academies (2014) Aboriginal food security in Northern Canada: an assessment of the state of knowledge. Council of Canadian Academies, Ottawa
Egeland GM, Middaugh JP (1997) Balancing fish consumption benefits with mercury exposure. Science 278:1904–1905. https://doi.org/10.1126/science.278.5345.1904
Ericksen PJ, Stewart B, Dixon J et al (2010) The value of a food system approach. In: Food security and global environmental change. Earthscan, London, pp 25–45
Ernst B, Manríquez P, Orensanz JM et al (2010) Strengthening of a traditional territorial tenure system through protagonism in monitoring activities by lobster fishermen from the Juan Fernández Islands. Chile Bull Mar Sci 86:315–338
Ernst B, Chamorro J, Manríquez P et al (2013) Sustainability of the Juan Fernández lobster fishery (Chile) and the perils of generic science-based prescriptions. Glob Environ Change 23:1381–1392. https://doi.org/10.1016/j.gloenvcha.2013.08.002
Fall J (2014) Subsustence in Alaska: a 2012 update. Alaska Department of Fish and Game, Juneau
FAO (Food and Agriculture Organization of the United Nations) (2016) The state of world fisheries and aquaculture: contributing to food security and nutrition for all. FAO, Rome
Fazzino D (2012) The will to end hunger in the age of security. In: Coulter K, Schumann WR (eds) Governing cultures. Palgrave Macmillan, New York, pp 183–208
FDA (Food and Drug Administration) (2006) Mercury levels in commercial fish and shellfish. US Food and Drug Administration, Silver Spring
Feeding America (2013) Map the Meal Gap. In: Map the Meal Gap, Food Insecurity Estimates at the County Level. http://feedingamerica.org/hunger-in-america/hunger-studies/map-the-meal-gap.aspx. Accessed 3 Dec 2013
Foale S, Adhuri D, Aliño P et al (2013) Food security and the coral triangle initiative. Mar Policy 38:174–183. https://doi.org/10.1016/j.marpol.2012.05.033
Freeman MM (2005) Just one more time before I die: securing the relationship between Inuit and whales in the arctic regions. Indig Use Manag Mar Resour 67:59–76
George S (1986) How the other half dies: the real reasons for world hunger. Penguin Books, Harmondworth
Gerlach SC, Loring PA, Turner AM et al (2011) Food systems, climate change, and community needs. In: Lovecraft AL, Eicken H (eds) North by 2020. University of Alaska Press, Fairbanks, pp 111–134
Hardy P-Y, Béné C, Doyen L et al (2013) Food security versus environment conservation: a case study of Solomon Islands' small-scale fisheries. Environ Dev 8:38–56. https://doi.org/10.1016/j.envdev.2013.04.009
Harper S, Zeller D, Hauzer M et al (2013) Women and fisheries: contribution to food security and local economies. Mar Policy 39:56–63. https://doi.org/10.1016/j.marpol.2012.10.018
HLPE (High Level Panel of Experts) (2014) Sustainable fisheries and aquaculture for food security and nutrition. Committee on World Food Security. Food and Agriculture Organization of the United Nations, Rome

Isaacs M (2016) The humble sardine (small pelagics): fish as food or fodder. Agric Food Secur 5:27. https://doi.org/10.1186/s40066-016-0073-5
Jentoft S, Chuenpagdee R, Barragán-Paladines MJ et al (2017) The small-scale fisheries guidelines: global implementation. Springer, Cham
Jewett SC, Duffy LK (2007) Mercury in fishes of Alaska, with emphasis on subsistence species. Sci Total Environ 387:3–27
Jones R, Rigg C, Pinkerton E (2016) Strategies for assertion of conservation and local management rights: a Haida Gwaii herring story. Mar Policy. https://doi.org/10.1016/j.marpol.2016.09.031
Kawarazuka N, Béné C (2010) Linking small-scale fisheries and aquaculture to household nutritional security: an overview. Food Secur 2:343–357. https://doi.org/10.1007/s12571-010-0079-y
Khor GL (2008) Food-based approaches to combat the double burden among the poor: challenges in the Asian context. Asia Pac J Clin Nutr 17:111–115
Kittinger JN, Teneva LT, Koike H et al (2015) From reef to table: social and ecological factors affecting coral reef fisheries, artisanal seafood supply chains, and seafood security. PLoS One 10:e0123856. https://doi.org/10.1371/journal.pone.0123856
Kofinas GP, Chapin FS, BurnSilver S et al (2010) Resilience of Athabascan subsistence systems to interior Alaska's changing climate. Can J For Res 40:1347–1359. https://doi.org/10.1139/X10-108
Kofinas GP, BurnSilver S, Magdanz J et al (2016) Subsistence sharing networks and cooperation: Kaktovik, Wainwright, and Venetie, Alaska. School of Natural Resources and Extension, University of Alaska Fairbanks, Fairbanks
Kral MJ (2012) Postcolonial suicide among Inuit in Arctic Canada. Cult Med Psychiatry 36:306–325. https://doi.org/10.1007/s11013-012-9253-3
Kronen M, Vunisea A (2009) Fishing impact and food security–gender differences in finfisheries across Pacific Island countries and cultural groups. SPC Women Fish Inf Bull 19:3–10
Levkoe CZ, Lowitt K, Nelson C (2017) "Fish as food": exploring a food sovereignty approach to small-scale fisheries. Mar Policy 85:65–70. https://doi.org/10.1016/j.marpol.2017.08.018
Lewis B, Grant WS, Brenner RE, Hamazaki T (2015) Changes in size and age of Chinook Salmon Oncorhynchus tshawytscha returning to Alaska. PLoS One 10:e0130184. https://doi.org/10.1371/journal.pone.0130184
Lopetcharat K, Choi YJ, Park JW, Daeschel MA (2001) Fish sauce products and manufacturing: a review. Food Rev Int 17:65–88
Loring PA (2013) Alternative perspectives on the sustainability of Alaska's commercial fisheries. Conserv Biol 27:55–63. https://doi.org/10.1111/j.1523-1739.2012.01938.x
Loring PA, Duffy LK (2011) Managing environmental risks: the benefits of a place-based approach. Remote Rural Health 11:1800
Loring PA, Gerlach SC (2009) Food, culture, and human health in Alaska: an integrative health approach to food security. Environ Sci Pol 12:466–478
Loring PA, Gerlach SC (2010) Food security and conservation of Yukon River Salmon: are we asking too much of the Yukon River? Sustainability 2:2965–2987. https://doi.org/10.3390/su2092965
Loring PA, Duffy LK, Murray MS (2010) A risk-benefit analysis of wild fish consumption for various species in Alaska reveals shortcomings in data and monitoring needs. Sci Total Environ 408:4532–4541. https://doi.org/10.1016/j.scitotenv.2010.07.013
Loring PA, Gerlach SC, Harrison H (2013) Seafood as local food: food security and locally caught seafood on Alaska's Kenai Peninsula. J Agric Food Syst Community Dev 3:13–30. https://doi.org/10.5304/jafscd.2013.033.006
Lyons C, Blount B, Carothers C et al (2016) Considering communities in fisheries management. Mar Policy 74:288–291. https://doi.org/10.1016/j.marpol.2016.05.006
Merino G, Barange M, Blanchard JL et al (2012) Can marine fisheries and aquaculture meet fish demand from a growing human population in a changing climate? Glob Environ Change 22:795–806. https://doi.org/10.1016/j.gloenvcha.2012.03.003
Mills DJ, Westlund L, de Graaf G (2011) Under-reported and undervalued: small-scale fisheries in the developing world. Small-Scale Fish Manag Framew Approaches Dev World:1–15

Mozaffarian D, Rimm EB (2006) Fish intake, contaminants, and human health: evaluating the risks and the benefits. JAMA 296:1885–1899. https://doi.org/10.1001/jama.296.15.1885

Muallil RN, Cleland D, Aliño PM (2013) Socioeconomic factors associated with fishing pressure in small-scale fisheries along the West Philippine Sea biogeographic region. Ocean Coast Manag 82:27–33

Nuttall M, Berkes F, Forbes BC et al (2004) Hunting, herding, fishing, and gathering: indigenous peoples and renewable resource use in the Arctic. In: Arctic climate impact assessment (ACIA). Cambridge University Press, Cambridge, pp 649–690

Olson J, Clay PM, Pinto da Silva P (2014) Putting the seafood in sustainable food systems. Mar Policy 43:104–111. https://doi.org/10.1016/j.marpol.2013.05.001

Popkin BM, Gordon-Larsen P (2004) The nutrition transition: worldwide obesity dynamics and their determinants. Int J Obes 28:S2–S9

Price W (1939) Nutrition and physical degeneration: on the problems of the western diet and the obsession with nutrients. McGraw-Hill Book Company, Inc, New York

Ramsden C, Faurot K, Carrera-Bastos P et al (2009) Dietary fat quality and coronary heart disease prevention: a unified theory based on evolutionary, historical, global, and modern perspectives. Curr Treat Options Cardiovasc Med 11:289–301. https://doi.org/10.1007/s11936-009-0030-8

Rattagool P (1985) Fermented fish products of South East Asia. Trop Sci 25:61–73

Sea Around Us (2016) Sea Around Us|Fisheries, Ecosystems and Biodiversity. In: Sea Us Fish. Ecosyst. Biodivers. http://www.seaaroundus.org/. Accessed 23 Jan 2018

Seedat Y (2007) Impact of poverty on hypertension and cardiovascular disease in sub-Saharan Africa. Cardiovasc J Afr 18:316–320

Sen A (1983) Poverty and famines: an essay on entitlement and deprivation. Oxford University Press, Oxford

Severance C, Franco R, Hamnett M et al (2013) Effort triggers, fish flow, and customary exchange in American Samoa and the Northern Marianas: critical human dimensions of Western Pacific fisheries. Pac Sci 67:383–393. https://doi.org/10.2984/67.3.6

Stewart H (2005) The fish tale that is never told: a reconsideration of the importance of fishing in Inuit societies. Senri Ethnol Stud 67:345–361

Sunderland EM (2007) Mercury exposure from domestic and imported estuarine and marine fish in the U.S. seafood market. Environ Health Perspect 115:235–242

Sundkvist A, Milestad R, Jansson A (2005) On the importance of tightening feedback loops for sustainable development of food systems. Food Policy 30:224–239

Teh LCL, Pauly D (2018) Who brings in the fish? The relative contribution of small-scale and industrial fisheries to food security in Southeast Asia. Front Mar Sci 5. https://doi.org/10.3389/fmars.2018.00044

Thilsted SH, Roos N, Hassan N (1997) The role of small indigenous fish species in food and nutrition security in Bangladesh. Naga ICLARM Q 20:82–84;102

Thompson B (2011) 15 Combating Iron deficiency: food-based approaches. Combating micronutrient deficiencies food-based approaches 268. CABI, Wallingford

Thornton TF (1998) Alaska native subsistence: a matter of cultural survival. Cult Surviv Q 22:29–34

Titchenal CA, Dobbs J (2007) A system to assess the quality of food sources of calcium. J Food Compos Anal 20:717–724. https://doi.org/10.1016/j.jfca.2006.04.013

United Nations (1948) Universal declaration of human rights. Chelsea House Publishers, New York

Watson RA, Nichols R, Lam VWY et al (2017) Global seafood trade flows and developing economies: insights from linking trade and production. Mar Policy 82:41–49. https://doi.org/10.1016/j.marpol.2017.04.017

Wolf EC, Zuckerman S (2003) Salmon Nation. Ecotrust, Portland

Wolfe RJ, Walker RJ (1987) Subsistence economies in Alaska: productivity, geography, and development impacts. Arct Anthropol 24:56–81

Ziegler J (2008) Report of the special rapporteur on the right to food. UN Hum Rights Counc GE:08–10098

Chapter 5
Broadening the Knowledge Base of Small-Scale Fisheries through a Food Systems Framework: A Case Study of the Lake Superior Region

Kristen Lowitt, Charles Z. Levkoe, Andrew M. Song, Gordon M. Hickey, and Connie Nelson

Abstract Lake Superior is the largest and northernmost of the Great Lakes of North America. It supports a diversity of wildlife and fish species, along with commercial, recreational, and Indigenous fisheries that make vital contributions to nutrition, livelihoods, cultures, and food systems. However, this diversity of social and cultural values is not fully reflected in management practices that tend towards a 'resourcist' approach. This chapter seeks to 'broaden the scope', proposing a food systems framework as a way of grappling with the wicked problem of Lake Superior fisheries governance. Using a food systems framework, we look at the different values associated with fisheries, including the objective, subjective, and relational

K. Lowitt (✉)
Department of Geography, Brandon University, Brandon, MB, Canada
e-mail: lowittk@brandonu.ca

C. Z. Levkoe
Canada Research Chair in Sustainable Food Systems, Department of Health Sciences, Lakehead University, Thunder Bay, Canada
e-mail: clevkoe@lakeheadu.ca

A. M. Song
Australian Research Council Centre of Excellence for Coral Reef Studies, James Cook University, Townsville, Australia

WorldFish, Honiara, Solomon Islands
e-mail: andrew.song@jcu.edu.au

G. M. Hickey
Department of Natural Resource Sciences, McGill University, Sainte-Anne-de-Bellevue, QC, Canada
e-mail: gordon.hickey@mcgill.ca

C. Nelson
Food Security Research Network, Lakehead University, Thunder Bay, ON, Canada
e-mail: cnelson@lakeheadu.ca

R. Chuenpagdee, S. Jentoft (eds.), *Transdisciplinarity for Small-Scale Fisheries Governance*, MARE Publication Series 21,
https://doi.org/10.1007/978-3-319-94938-3_5

contributions they make to Lake Superior food systems. We explore these food-related values attached to fisheries by presenting three illustrative examples: The fisheries of Batchewana First Nation; Eat the Fish, a small business marketing local fish through alternative food networks in Northwestern Ontario; and Bodin's Fisheries in Wisconsin, a regional fish processor and retail outlet. We conclude by identifying ways of strengthening fisheries contributions to regional food systems and offer a set of transdisciplinary questions on fishery-food system linkages that may assist others in 'broadening the scope' of fisheries governance.

Keywords Food systems · Small-scale fisheries · Coastal communities · Great Lakes · Lake Superior

5.1 The Lake Superior Fisheries Governance System

Lake Superior (also known as *Anishnaabe Gichgamiing* in Anishinabek) is the largest of the five Great Lakes of North America and is a vital part of the social, ecological, and economic life for millions of people living in Canada, the United States, and across a number of Indigenous territories. Its expansive coastline and over 2500 islands support a diversity of wildlife and are home to rare and endangered species of birds, amphibians, and mammals, along with over 30 native species of fish (Minnesota Sea Grant 2014; The Lake Superior Partnership 2016). People living in the region rely on the lake for food, drinking water, economic activity, transportation, and recreation. For many Indigenous and settler communities, the lake is an important part of their identity and cultural heritage.

Lake Superior has a rich fishing history. Indigenous peoples relied on Great Lakes fisheries for sustenance and trade long before contact with European settlers four centuries ago. Increasingly, they are asserting their Aboriginal and treaty rights to the land and water and to fish, attesting to the importance of the fisheries to their communities, identities, cultures, and food systems. In addition to subsistence and commercial Indigenous fisheries, there are also non-Indigenous commercial fisheries and a recreational sport fishery on the lake. The social and ecological resilience of the Lake Superior watershed and its native fish populations have, however, been under considerable stress from pollution, over-fishing, invasive species, the stocking of non-native salmonines to support a sport fishing industry, and climate change, among other factors, for some time (McLaughlin and Krantzberg 2012; Brenden et al. 2013).

Jurisdictionally, the lake's borders are split between the Canadian and U.S. federal governments, the Canadian province of Ontario, three U.S. states (Minnesota, Wisconsin, and Michigan), as well as the traditional territories of numerous Canadian First Nations and Indian tribes. In 1954, driven primarily by the urgent need for transboundary coordination to address the sea lamprey (*Petromyzon marinus*) invasion, Canadian and U.S. governments signed the Convention on Great Lakes Fisheries which created the bi-national Great Lakes Fishery Commission (GLFC) (Gaden et al. 2013; Song et al. 2017). However, the mandate of the GLFC was largely limited to controlling invasive species and conducting related research in order to advise state, provincial, tribal, and federal agencies, which retained their jurisdictional authority over the management of fisheries (Song et al. 2017).[1]

Fisheries management in Lake Superior has been primarily organized through centrally coordinated, state-led structures designed to acquire more information on fish as a resource, construct more sophisticated prediction models, and refine and enforce command and control systems of regulation (McLaughlin and Krantzberg 2012). Such a 'resourcist' approach (see Berkes 2010) is based on a particular view of how humans and nature operate and interact. However, this view has been increasingly criticized as artificially separating human and natural systems, relying too heavily on positivist science, being too economic centric, and artificially excluding other ways of knowing and traditional knowledge systems. In the context of Lake Superior, examples of 'resourcist' thinking include the shift towards Individual Transferable Quotas (ITQs) in Ontario's commercial fisheries in the 1980s, the pursuit of modernization of fish harvesting and drives for economic efficiency, a lack of consideration of fishers' experiential ecological knowledge among resource managers, and a general failure to honour treaty obligations through Nation-to-Nation relationships with Indigenous communities (Norman 2015). This resourcist approach to Lake Superior fisheries management raises questions concerning the efficacy and equity of existing institutions to address the social-ecological complexity of Lake Superior fisheries (Henquinet and Dobson 2006; Krantzberg and Manno 2010; Langston 2017). Recognizing that Lake Superior fisheries have long been governed by multiple entities, individuals and, at times, contradicting epistemologies, we suggest that a shift away from the existing management ideology and towards broader governance approaches can better understand and account for the diversity of values and relationships important to sustainable fisheries. Fisheries governance in this context is a 'wicked problem', as it demands questioning the underlying assumptions that drive existing management relationships, with no universal solution for what more inclusive approaches to governance in support of sustainable fisheries should look like (Jentoft and Chuenpagdee 2009). Recognizing that values and sustainability are highly ontological questions, working towards more inclusive governance approaches will require deliberative and collaborative engagement among communities, policy-makers, and interdisciplinary researchers, as well as a consideration of the multiple interrelated systems that impact and are impacted by fisheries in the region.

It is within this context that we engage with a need to 'broaden the scope' of Lake Superior fisheries, proposing a food systems framework as a way of grappling with this wicked problem. This approach offers an opportunity to (re)value relationships, while considering a more complex set of social, cultural, and ecological questions that have direct implications for the future of the region's fisheries (Olson et al. 2014). In so doing, a food systems framework can help to broaden the knowledge

---

[1] Fisheries management responsibilities generally include fish population assessments, fish stocking, habitat restoration and design, and enforcement of harvesting regulations, including licensing (Song et al. 2017).

base vital to collaborative fisheries governance. We illustrate this food systems framework by presenting three cases: the fisheries of Batchewana First Nation; Eat the Fish, a small business marketing local fish through alternative food networks in Northwestern Ontario; and Bodin's Fisheries in Wisconsin, a regional fish processor and retail outlet.

5.2 A Food Systems Framework

A food systems framework enables us to broaden the scope on Lake Superior fisheries by looking within and beyond the fishery itself (e.g. the fish harvesters and fish stocks) to consider the interdependent social, economic, and ecological relationships within which fisheries are embedded (Charles 2005; Nelson et al. 2013; Levkoe et al. 2017). Food systems can be described as the overlapping processes and infrastructures involved in feeding a population (Tansey and Worsley 1995). This begins with the chain of activities such as growing and harvesting food along with the processing, distribution, marketing, wholesaling, retailing, consumption, and waste management. The food system is, thus, an analytical concept that shows how foods are transformed as they move from field or water to plate and beyond; it can be applied at different scales from the local to the global. A food systems framework is also integrative, as it demands consideration of a range of interdependent dimensions, including the geographical, political, economic, ecological, and social values and relationships that all come to interact around the production and consumption of food (Feenstra 2002).

Applying a food systems framework to fisheries is especially significant given the increasing awareness about the important contributions that fisheries make to food systems in fisheries-dependent regions (Lowitt 2013; Loring et al. 2013; Bell et al. 2015; Dey et al. 2016), and the recognized need for further research to understand fishery-food system linkages in particular settings (Lowitt 2013; Loring et al. 2014). Previous research by Levkoe et al. (2017) identified that a food systems perspective provides an innovative way of analyzing fisheries through the interdependencies between aquatic and terrestrial ecosystems, including outcomes for ecological and human health, food security, economically and culturally viable livelihoods, and community wellbeing based in principles of justice and democracy (Levkoe et al. 2017). A food systems framework is a useful tool for addressing wicked problems, because it considers multiple interrelated systems beyond only fisheries, as well as both scientific and traditional ways of knowing that, if embraced, may bring complementary perspectives.

In addition, a food systems framework entails a consideration of values. The concept of values has been addressed from a wide range of interdisciplinary interpretations including environmental philosophy, human geography, social psychology, anthropology, as well as ecological and 'mainstream' economics. Implicit in these different interpretations is an acknowledgment that values are ultimately a relativistic expression of what is desirable and are often contextual in nature (Song

2018). Therefore, what different individuals or groups assign value to, or the degree to which they desire certain values, are likely to differ depending on subjective interests, circumstances, and ideology. There has been an increasing interest in the field of fisheries to examine and apply the values concept to inform the field of resource governance and broaden its knowledge base (see, for example, Johnson et al. 2018).

In this chapter, we draw on the idea of the value-contribution matrix proposed by Song (2018) to explore the contributions fisheries make to local food systems around Lake Superior. Our approach emphasizes an understanding of value as contribution to a broad notion of human wellbeing, whether tangible or intangible in respect to three interrelated dimensions: objective, subjective, and relational. Objective values include the material features that are necessary for wellbeing, such as access to food and livelihoods; subjective values pertain to the perceptions, meanings, and emotions tied to wellbeing, such as cultural identity and sense of place; relational values emphasize how wellbeing is tied to social and cultural relationships and networks with other groups, whether these be households, communities, or organizations. In what follows, we use these categories as an open organizational framework (see Stacey et al. 2018) for understanding several key dimensions of fisheries from a food systems perspective, presenting three illustrative case studies in the Lake Superior region.

5.3 Three Illustrative Case Studies

Our three case studies were selected to represent a range of food and fisheries examples from across the region. They are based on 27 interviews undertaken in the summer of 2017 in the Lake Superior region with Indigenous and non-Indigenous governments, private, and civil society actors. The cases include the fisheries of Batchewana First Nation; Eat the Fish, a small business marketing local fish through an alternative food network in Northwestern Ontario; and Bodin's Fisheries in Wisconsin, a regional fish processor and retail outlet. In what follows, we describe the three cases and use them to identify the different values and potential ways of strengthening fisheries contributions to food systems in the Lake Superior region with an aim to broaden the scope on fisheries.

5.3.1 Batchewana First Nation Fisheries

The Batchewana First Nation (BFN) of Ojibways[2] is located on the northeastern shore of Lake Superior. The First Nation has a total population of 2400, approximately 72% of which live off-reserve. In Canada, First Nations is the term used to

[2] http://batchewana.ca

designate descendants of Canada's inhabitants prior to the arrival of Europe settlers (this group is distinct from Inuit and Métis peoples).

Prior to European contact, the Batchewana people were a hunter gatherer society. In the summer months, during the whitefish season, they would meet at an area known as Bawatung (Gathering Place) in what are now the cities of Sault Ste. Marie, Michigan and Sault Ste. Marie, Ontario (Tobias 2015). However, with the arrival of European settlers in the seventeenth century, Indigenous fishing activities were forcibly disrupted. The result of a broken treaty process, reserve system, and imposed legislation was that Indigenous peoples' control of fish, land, and water systems was significantly curtailed (Blair 1997; McCrimmon 2002; Tobias 2015). Despite charges laid against band members at various times by Crown government agencies, BFN members have continued to fish and assert their inherent Aboriginal and treaty rights (Tobias 2015).[3] BFN asserts their jurisdiction relying upon the Covenant Chain of mutual respect and agreements with the Crown, including the Two Row Wampum Belt, an early treaty that First Nations and colonial settlers would each travel in their own vessels, side by side with each under their own laws, customs, and ways (Batchewana First Nation 2017).

Today, BFN operates a subsistence fishery in Lake Superior and one of the largest commercial fisheries on the Canadian side of the Lake (see Fig. 5.1). Their harvests, including fresh fish and smoked products, are sold by several families at various local outlets, including at the farmer's market in Sault Ste. Marie, Ontario. Selling fish remains a vital component of household livelihoods, in addition to being an important source of food. They also operate under their own fisheries management system, predicated on traditional ecological knowledge and developed in alignment with oral teachings and responsibilities (Batchewana First Nation 2018). Fishing is, thus, closely tied to cultural identity and the intergenerational transfer of knowledge and relationships with the community and natural environment (Tobias 2015). As explained by the Chief:

> We've always maintained and managed the fishery based on what was happening with the environment, and the elders. Even today I watch how the fishermen fish. They fish that way. They fish with the weather, they fish with the seasons, they fish with what's happening with the environment. They move with whatever is happening.

Despite these rich historical connections and BFN's persistence in asserting their inherent rights to the land and fisheries, their jurisdiction and traditional management plan has not been formally recognized by the Province of Ontario, including the Ontario Ministry of Natural Resources and Forestry (OMNRF), which asserts they have primary management authority for the province's fisheries.

---

[3] Since the recognition of Aboriginal and treaty rights in section 35(1) of the Constitution Act, 1982, the inherent right of Indigenous peoples to govern themselves has been an accepted component of Canadian constitutional law (McNeil 2007). However, there remain different views about the content, scope, and nature of this inherent right (Indigenous and Northern Affairs Canada 2010; McNeil 2007).

Fig. 5.1 Batchewana First Nation fish harvesters unloading fish. (Photo Credit: Charles Z. Levkoe)

5.3.2 Eat the Fish

Eat the Fish[4] is a small independently owned business located in Thunder Bay, on the shores of Lake Superior, marketing locally caught and wild fish in Northwestern Ontario. Established in 2016, the two cofounders recognized that despite the active fishery, there were no processing options and few opportunities to purchase local fish in the region. Lake Superior has a long history of both Indigenous and commercial fisheries; however, the vast majority of fish are sold to the United States through wholesalers. Furthermore, the low price of fish and the aging population of fishers have put the future of the region's fisheries into question. One of the founders and owners commented that prior to starting Eat the Fish, most people "*had no idea that there was even a [commercial] fishing industry here.*" They decided, "*let's generate some interest in this, let's keep some of the fish here and let's see where we can go from there.*"

Each week, Eat the Fish sets up a fish stand at the busy Thunder Bay Country Market (see Fig. 5.2). They are one of the few options for buying wild-caught and native fish in the region, which they purchase from local fishers, many of whom have been fishing on the lake for generations. This direct-to-customer model offers fishers a better per pound price than they would receive from the wholesalers, while keeping fish caught in the region in Northwestern Ontario. It also provides eaters with a closer connection to fish harvesters and to the lake. As one of the cofounders

[4] http://eatthefish.com

Fig. 5.2 Bodin's Fisheries in Bayfield, Wisconsin. (Photo Credit: Charles Z. Levkoe)

explained, "*we're trying to share their [fish harvesters] stories… because we think it is important for people to take some pride in what we have here.*"

Eat the Fish has also helped to open local markets to underutilized native fish species such as burbot—typically a bycatch for fish harvesters due to low market demand. By offering education to marketgoers about native fish species that have typically been less popular as an eating choice, they have created another revenue stream for fish harvesters, while also reducing bycatch. Fish leftover each week after the market gets sold to restaurants specializing in local cuisine and another small business marketing local foods in the region.

While Eat the Fish has taken important steps in raising awareness and has increased local demand for fish and linked harvesters and eaters, the owners noted that they are concerned about the future of the region's fisheries. Looking ahead there is a need to better support the industry and encourage young people to enter the fishery and/or take over family owned enterprises in order for a sustainable commercial fishing industry to continue.

5.3.3 Bodin's Fisheries

Located in Bayfield, Wisconsin, Bodin's Fisheries[5] is a multigenerational family fish processing business and one of the few remaining processors located on the shores of Lake Superior (see Fig. 5.3). With the majority of fish harvested in Lake

[5] http://www.bodinfisheries.com

Fig. 5.3 Eat the Fish at the Thunder Bay Country Market. (Photo Credit: Charles Z. Levkoe)

Superior sent to larger markets in Chicago, Boston and New York, Bodin's is an important part of the local food infrastructure by allowing some locally caught fish to remain in the region. Bodin's specializes in processing fresh, wild-caught Lake Superior whitefish, herring, and lake trout, selling primarily to restaurants, grocers, and distributors within Bayfield, the state of Wisconsin, and nearby states of the American mid-west. They also produce their own in-house smoked and sugar-cured products. Value added processing is a key part of their business, enabling them to differentiate their fish from imported whitefish products and receive a higher price. As the operator of the business explained, "*instead of shipping to New York or Chicago, 90% of the fish we buy is being used locally now…I can put a knife to it, fillet and process it, smoke it, and sell it. I'm doing value added here because why would I want to sell it at a quarter of a pound and ship it.*"

Located adjacent to the fish processing plant, they operate a year-round retail storefront, where the public can purchase their fresh, frozen, and smoked products. It has become a destination for locals and tourists alike. In addition to whole and filleted fish, they also offer livers and cheeks, valued as a local delicacy. Over time, they have also built up a market for fish species that have traditionally been a less popular eating choice, such as burbot. Often discarded by fish harvesters because of a lack of market, Bodin's Fisheries started giving samples to restaurants along with recipes for how to prepare the fish, resulting in growing interest and demand.

The Bodin family has one fishing license that supplies some fish to the plant; yet, the majority of their fish is sourced from commercial harvesters, including many tribal commercial harvesters. Over the years, Bodin's has established relationships with these harvesters based on trust, offering them important infrastructural supports, such as access to dock space and ice in exchange for purchasing and processing their catch. The operator of the facility described the business in terms of "*creating a facility that markets the fish and creates a value, and they [fish harvesters] know that they're going to get paid every Friday as opposed to an unknown of when someone's going to pay them for their fish.*" Bodin's also recognizes that building cooperative relationships with other stakeholders is important to the sustainability of the Lake and their business engaging, for example, in dialogue with recreational fishers to raise awareness about commercial fish harvesting practices.

5.4 Discussion

These three examples highlight a range of values embedded in Lake Superior's fisheries as a part of the region's food systems, including those pertaining to health, livelihoods, identity, and culture. Analyzing our three cases using a food systems framework, we draw on the categories of objective, subjective, and relational to illustrate the benefits that communities derive from food systems. Table 5.1 presents a summary of these key values in relation to each of the cases. Broadening the scope of small-scale fisheries using a food systems framework is a wicked problem, as it has no universal technical solution, but rather putting it into practice depends on the collective judgement and interaction of stakeholders (Jentoft and Chuenpagdee 2009; Nelson and Stroink 2014).

The first category, objective values, refers to the diversity in and the extent to which material contributions are deemed important. Drawing from our three cases, a common objective value pertains to the ways that fisheries support the individual and community livelihoods of fish harvesters and processers, as well as those of retailers and restaurateurs. For BFN and Bodin's Fisheries, this is illustrated by their respective histories of reliance on the lake's fisheries for livelihoods, employment, and sustenance, which collectively contribute to the survival of their communities and families. From a food systems perspective, this has meant that fishing work is integrated into broader community life, as it extends tangible benefits (both direct and indirect) to others in the value chain. For example, Bodin's Fisheries, as a business, provides a range of infrastructural support to local fishers, such as ice and dock space, as well as strategic advice, as needed. For Eat the Fish, it creates options for commercial and Indigenous fishers to sell their catch locally as a way of maintaining livelihoods. For all three cases, there is long term material benefit that results from engaging in fishing practices that better consider ecological, social, and regional economic resilience. While large buyers who prioritize economic efficiency tend to have little connection to the region, local businesses and communities hold deeper interests that go beyond the monetary gains generated from catching and selling of fish.

Table 5.1 Summary of values in three cases of Lake Superior food systems

	Values		
	Objective	Subjective	Relational
Batchewana first nation fisheries	Access to fish for sustenance; main employer in the community; supports household and community livelihoods through harvesting and processing.	Connections to cultural identity, ceremony, and traditional food practices; asserting inherent rights to land/water/livelihood; continuing and developing traditional practices and teachings, and teaching these to the younger generation.	Reinforcing relationships with nature through maintaining traditional ecological knowledge; an important means to resisting settler colonialism through exercising jurisdiction over fisheries; nation-to-nation relationships not in place but are key to sustainability.
Eat the fish	Support for commercial fish harvester livelihood; provides sustainable fish options for the local market and enhances the local food systems; reduces waste through using bycatch.	Telling the fish to plate story; enhancing community perceptions of traditionally under-valued species for consumption; pride in the lake/region and local food system.	Building renewed social relationships between eaters and harvesters; contributing to/enhancing local food system.
Bodin's fisheries	Provides fresh fish to the regional market; supports commercial fish harvester livelihoods, including providing infrastructural supports, such as ice and dock space; economic benefit to small community (e.g. processing jobs, tourist attraction, local restaurants).	Telling the fish to plate story; part of the social identity of the family and town; creating market demand for traditionally unpopular fish species.	Informal relationships based on trust with commercial fish harvesters; building cooperative relationships with other users, including sports fishers, in the lake.

Second, subjective values embody perceptions, meanings, and emotions tied to individual and community wellbeing. All three cases illustrate that fishing can build the pride people have towards their region as well as their sense of identity. Bodin's Fisheries plays an important role in the town of Bayfield adding significantly to the social identity of the community; this social identity is also increasingly of interest to tourists seeking out culinary tourism attractions. For BFN, the maintenance and expansion of their fisheries has meant building closer connections to identity, ceremony, and traditional food practices. These values contribute to defining who they are, and their revitalization, which, as in the case of BFN, can offer a foundation for asserting their inherent rights to land, water, and livelihood as part of an ongoing resistance of settler colonialism. In addition, the cases show that fish can be used to tell a different kind of narrative about resource use when rooted in a food systems framework—that is, a comprehensive mosaic of activities and events that take place from lake to plate. For instance, by focusing on native species and marketing that entails a sense of place in the region, each initiative supports educational activities about the fisheries (both experiential and informational) that have enhanced public knowledge and interest in undervalued species that are part of the lake ecosystem.

Finally, relational values emphasize how wellbeing is influenced by social and cultural relationships formed within particular societal contexts. From a food systems perspective, all three cases embrace fish as much more than a resource for economic gain or nutritional benefit (although both of these aspects are also identified as important). Of particular interest was the role that fish played in respect to human relationships. Participants of Eat the Fish and Bodin's Fisheries, for instance, expressed a deep emotional and historical connection of fishers to the lake and the region. Arising in this shared context, the relationships that developed around the fisheries were in many cases based on mutual understandings and trust. Eat the Fish and Bodin's Fisheries both rely on relationships of trust established with fish harvesters to maintain their businesses, often working with them to come up with an equitable pricing scheme depending upon fishing conditions and prices in the market. For BFN, there was also a direct reference to fish as part of intergenerational teachings and inherent rights, highlighting the long term benefits sought to be sustained from the fishery. The fishery also provided a foundational meaning for BFN leaders in their political struggle to assert sovereignty and demand improved Nation-to-Nation relationships with the government.

While the specific values we identified vary, they collectively illustrate how a food systems framework can help to broaden the scope on small-scale fisheries by illuminating broader perspectives, ways of knowing, and relationships. These dynamic interrelationships, and the nested nature of different interrelated systems are a key feature of wicked problems (Nelson and Stroink 2014). For example, eaters are important actors from a food systems perspective that are rarely directly considered in fisheries governance. Here, we do not mean rational consumers that exert choice through markets, but eaters that enjoy interrelated objective, subjective, and relational benefits (such as access to food, ties to a fishing community, and knowing the fish harvesters) through more local food systems. While our examples highlight these place-based innovations, they are not the norm around Lake Superior, as fishery policy that is insufficiently adapted to local contexts has tended not to favour the development of local and regional markets that can support these contributions.

Values and relationships tied to reconciliation, justice, and knowledge systems are also central food systems issues that are a part of fisheries, as evident in many Indigenous communities around the lake. BFN continues to fish, for both sustenance and livelihood needs, and assert their jurisdiction to manage their fishery according to traditional teachings; however, there remains significant tension around the nature and scope of their jurisdiction and inherent rights, particularly with settler governments. This is another example of the nature of fisheries governance as a wicked problem, as the complex issues that play out around fisheries can be rooted in larger structural inequalities. From a food systems perspective, fisheries policy is inextricably linked to issues of food sovereignty and self-determination, although these principles remain to be adequately considered and implemented in Lake Superior fisheries (see Levkoe et al. 2017). Hence, there is an opportunity for more inclusive fisheries governance to play a more constructive role in reconciliation efforts by recognizing and engaging with a larger scope of values, including Indigenous worldviews; this may be an important first step towards redefining existing institutions of power and authority (Turner et al. 2013; Suzuki 2016; Lowitt et al. 2018).

Looking at fisheries from a food systems perspective can also help us identify other specific gaps that should be addressed in policy, including insufficient support for small-scale processors and lack of succession planning for fishing enterprises. For example, regional processors, such as Bodin's, play an important role in supporting the livelihoods of local harvesters with access to ice and dock space, which buyers located outside of the region and shipping to larger markets cannot offer. However, the role of smaller processors tends to be undermined by policy regulations that enable the leasing of quota (see Ontario Commercial Fisheries Association n.d.), and may thereby encourage the flow of fish out of the region and into the hands of larger, distant buyers. This policy likely undervalues the objective contributions that local processors can make to livelihoods as well as the subjective and relational values tied to the development of trust and cooperation among processors and harvesters. Additionally, there appears to be a lack of formal programs available to support succession planning for fishing enterprises around Lake Superior, creating uncertainty about what will happen to family-based enterprises and continuity in terms of intergenerational involvement in fisheries.

5.5 Conclusions

In this chapter, we highlight how the different elements that make up Lake Superior fisheries can be (re)envisioned through a food systems framework. For the different stakeholders and communities represented in our examples, fisheries are embedded in systems of social wellbeing that are tied to contributions to foods and livelihoods, the maintenance of social and cultural identities, and the functioning of different types of knowledge systems, among other contributions. Further research done in partnership with communities around Lake Superior will be an important way to critically analyze and validate these results and for painting a more complete picture of how these values are viewed and negotiated from within and between different communities.

Recognizing that values are broadly about what is accepted as desirable, there is a need to 'broaden the scope' of prevailing management approaches to better account for the range of individual and collective aspirations embodied by different stakeholders around the lake. However, how to understand, integrate, and address potential tensions among the diversity of values illuminated in this chapter remains a key challenge for fisheries management around Lake Superior. As a wicked problem, there is no single perspective on what is most desirable, or agreement on a specific path forward. In line with our thinking on how to address wicked problems, encouraging interaction between diverse actors and different ways of knowing will be key to further negotiating the range of values that our study illuminates as a way to broaden the scope on small-scale fisheries.

Building on the exploratory results presented in this chapter, we conclude by raising some transdisciplinary questions related to fisheries governance to assist others in using a food systems framework as a way of continuing to broaden the scope:

- How may a food systems framework contribute to the social transformations needed for a sustainable future for small-scale fisheries?
- What might a fisheries harvesting policy look like if it were developed with food systems rather than an economic efficiency framework?
- What are the key linkages between reconciliation with Indigenous peoples and fisheries policy, and how may reconciliation be put into action through a food systems framework to fisheries?
- How does a food systems framework support consideration of fisheries governance as a wicked problem?

Acknowledgements We are very grateful to all the research participants who contributed their time to be part of this project. We also acknowledge conceptual input from the Food: Locally Embedded, Globally Engaged (FLEdGE) research team. This research was supported by the Social Sciences and Humanities Research Council of Canada.

References

Batchewana First Nation (2017) Notice of assertions: Sovereignty. http://batchewana.ca/about/sovereignty/. Accessed 5 Apr 2018

Batchewana First Nation (2018) Natural resources. http://batchewana.ca/departments/natural-resources. Accessed 5 Apr 2018

Bell J et al (2015) Optimising the use of nearshore fish aggregating devices for food security in the Pacific Islands. Mar Policy 56:98–105. https://doi.org/10.1016/j.marpol.2015.02.010

Berkes F (2010) Shifting perspectives on resource management: resilience and the reconceptualization of 'natural resources' and 'management.'. MAST 9(1):13–40

Blair P (1996–1997) Solemn promises and solum rights: the Saugeen Ojibway fishing grounds and R. V. Jones and Nadjiwon. Ottawa Law Rev 28:125–143

Brenden TO, Brown RW, Ebener MP et al (2013) Great Lakes commercial fisheries: historical overview and prognoses for the future. In: Taylor W, Lynch AJ, Leonard NJ (eds) Great Lakes fisheries policy and management. Michigan State University Press, East Lansing, pp 339–397

Charles A (2005) Toward sustainable and resilient fisheries: a fishery-system approach to overcoming the factors of unsustainability. http://www.fao.org/docrep/009/a0312e/A0312E13.htm. Accessed 5 Apr 2018

Dey M, Gosh K, Valmonte-Santos et al (2016) Economic impact of climate change and climate change adaptation strategies for fisheries sector in Solomon Islands: implication for food security. Mar Policy 67:171–178. https://doi.org/10.1016/j.marpol.2016.01.004

Feenstra G (2002) Creating space for sustainable food systems: lessons from the field. Agric Hum Values 19(2):99–106. http://www.glfc.org/research/humandimensions.pdf. Accessed 5 Apr 2018

Gaden M, Goddard C, Read J (2013) Multi-jurisdictional management of the shared Great Lakes fishery: transcending conflict and diffuse political authority. In: Taylor W, Lynch A, Leonard N (eds) Great Lakes fisheries management and policy: a bi-national perspective. Michigan State University Press, Michigan, pp 305–338

Henquinet J, Dobson T (2006) The public trust doctrine and sustainable ecosystems: a Great Lakes fisheries case study. N Y Univ Environ Law J 14(2):322–373

Indigenous and Northern Affairs Canada (2010) The Government of Canada's approach to implementation of the inherent right and the negotiation of aboriginal self-government. http://www.aadnc-aandc.gc.ca/eng/1100100031843/1100100031844. Accessed 16 Apr 2018

Jentoft S, Chuenpagdee R (2009) Fisheries and coastal governance as a wicked problem. Mar Policy 33:553–560. https://doi.org/10.1016/j.marpol.2008.12.002

Johnson D, Acott T, Stacey N, Urquhart J (eds) (2018) Social well-being and the values of small-scale fisheries. Springer, Cham

Krantzberg G, Manno JP (2010) Renovation and innovation: it's time for the Great Lakes regime to respond. Water Resour Manag 24:4273–4285. https://doi.org/10.1007/s11269-010-9658-0

Langston N (2017) Sustainable Lake superior: an extraordinary lake in a changing world. Yale University Press, New Haven

Levkoe CZ, Lowitt K, Nelson C (2017) "Fish as food": exploring a food sovereignty approach to small-scale fisheries. Mar Policy 85:65–70. https://doi.org/10.1016/j.marpol.2017.08.018

Loring PA, Gerlach SC, Harrison H (2013) Seafood as local food: food security and locally caught seafood on Alaska's Kenai Peninsula. J Agric Food Syst Community Dev 3(3):13–30. https://doi.org/10.5304/jafscd.2013.033.006

Loring PA, Harrison HL, Gerlach SC (2014) Local perceptions of the sustainability of Alaska's Cook Inlet salmon fisheries. Soc Nat Resour 27(2):185–199

Lowitt K (2013) Examining fisheries contributions to community food security: findings from a household seafood consumption survey on the west coast of Newfoundland. J Hunger Environ Nutr 8(2):221–241. https://doi.org/10.1080/19320248.2013.786668

Lowitt K, Johnston-Weiser D, Lauzon R et al (2018) On food security and access to fish in the Saugeen Ojibway nation, Lake Huron, Canada. Great Lakes Res 44(1):174–183. https://doi.org/10.1016/j.jglr.2017.10.009

McCrimmon D (2002) Sustainable fisheries management in the great lakes: scientific and operational challenges. Lakes Reserv Res Manag 7:241–254

McLaughlin C, Krantzberg G (2012) An appraisal of management pathologies in the Great Lakes. Sci Total Environ 416:40–47. https://doi.org/10.1016/j.scitotenv.2011.12.015

McNeil K (2007) The jurisdiction of inherent right aboriginal governments. All papers. 261. http://digitalcommons.osgoode.yorku.ca/all_papers/261. Accessed 18 Apr 2018

Minnesota Sea Grant (2014, October 23) Lake Superior's fish species. http://www.seagrant.umn.edu/fisheries/superior_fish_species. Accessed 5 Apr 2018

Nelson C, Stroink M (2014) Accessibility and viability: a complex adaptive systems approach to a wicked problem for the local food movement. J Agric Food Syst Community Dev 4(4):191–206. https://doi.org/10.5304/jafscd.2014.044.016

Nelson C, Lowitt K, Nagy M et al (2013) Future research approaches to encourage small-scale fisheries in the local food movement. J Agric Food Syst Community Dev 3(4):177–181. https://doi.org/10.5304/jafscd.2013.034.020

Norman E (2015) Governing transboundary waters: Canada, the United States, and indigenous communities. Routledge Press, New York

Olson J, Clay PM, da Silva P (2014) Putting the seafood in sustainable food systems. Mar Policy 43:104–111. https://doi.org/10.1016/j.marpol.2013.05.001

Ontario Commercial Fisheries Association (n.d.) Quotas. http://www.ocfa.ca/fisheries-industry/quotas. Accessed 5 Apr 2018

Song A (2018) How to capture small-scale fisheries' many contributions to society? – introducing the 'value-contribution matrix' and applying it to the case of a swimming crab fishery in South Korea. In: Johnson D, Acott T, Stacy N, Urquhart J (eds) Social well-being and the values of small-scale fisheries. Springer, Cham, pp 125–246

Song AM, Temby O, Krantzberg G et al (2017) Institutional features of U.S.-Canadian transboundary fisheries governance: organizations and networks, formal and informal. In: Temby O, Stoett P (eds) Towards continental environmental policy? North American transnational networks and governance. SUNY Press, Albany, pp 156–179

Stacey N, Steenbergen DJ, Clifton J et al (2018) Understanding social wellbeing and values of small-scale fisheries amongst the Sama-Bajau of archipelagic Southeast Asia. In: Johnson D, Acott T, Stacy N, Urquhart J (eds) Social well-being and the values of small-scale fisheries, Springer, Cham, pp 97–123

Suzuki D (2016, December 1) Reconciliation requires recognizing rights-based fishing. [Blog post]. http://www.davidsuzuki.org/blogs/science-matters/2016/12/reconciliation-requires-recognizing-rights-based-fishing/. Accessed 5 Apr 2018

Tansey G, Worsley T (1995) The food system: a guide. Routledge Press, New York

The Lake Superior Partnership (2016) Lake Superior lakewide action and management plan 2015–2019. https://www.epa.gov/sites/production/files/2016-10/documents/lake_superior_lamp_2015-2019.pdf. Accessed 5 Apr 2018

Tobias J (2015) We are the land: researching environmental repossession with Anishinaabe elders dissertation. Western University, London

Turner N, Berkes F, Stephenson J et al (2013) Blundering intruders: extraneous impacts on two indigenous food systems. Hum Ecol 41:563–574. https://doi.org/10.1007/s10745-013-9591-y

Part III
Strengthening the Base

Chapter 6
Economic Viability of Small-Scale Fisheries: A Transdisciplinary Evaluation Approach

Anna Schuhbauer, Andrés M. Cisneros-Montemayor, and U. Rashid Sumaila

Abstract Economically viable small-scale fisheries are likely much better prepared to face global changes in climate, technology, and markets. 'Economic viability' is often mistakenly equated with financial viability, with profitability as the sole goal. However, in complex and dynamic systems, such as small-scale fisheries, social, ecological, and governance aspects also play essential roles. Here, we use a more inclusive definition of economic viability as the achievement of nonnegative net benefits to society over time. This chapter develops a methodological framework derived from a transdisciplinary process involving stakeholders from policy, science, and fishing communities. The study first presents the concept of basic economic viability and how it differs from financial viability, followed by an extended version of economic viability, including a much broader set of attributes drawn from economic, social, ecological, and institutional aspects of fishing. The applications and implications of both assessments are then illustrated using Mexican fisheries as a case-study. Results indicate that economic viability can be improved when managers and policy makers ensure that the distribution of total fisheries subsidies between small- and large-scale fisheries are not creating disadvantages, and rather contribute to controlling fishing pressure and supporting the use of non-destructive fishing techniques. Our analysis suggests that the assessment of economic viability can be used as a tool to better understand the underlying dynamics of fisheries. This understanding can help identify their vulnerabilities, and inform sound policies to

A. Schuhbauer (✉)
Fisheries Economic Research Unit, Institute for the Oceans and Fisheries, The University of British Columbia, Vancouver, BC, Canada

A. M. Cisneros-Montemayor
Nereus, Institute for the Oceans and Fisheries, The University of British Columbia, Vancouver, BC, Canada

U. R. Sumaila
Fisheries Economic Research Unit, Institute for the Oceans and Fisheries, The University of British Columbia, Vancouver, BC, Canada

The Liu Institute for Global Issues, The University of British Columbia, Vancouver, BC, Canada

R. Chuenpagdee, S. Jentoft (eds.), *Transdisciplinarity for Small-Scale Fisheries Governance*, MARE Publication Series 21,
https://doi.org/10.1007/978-3-319-94938-3_6

achieve balanced management regarding the social, economic, ecological, and institutional dimensions of small-scale fishing communities.

Keywords Small-scale fisheries · Economic viability · Profit maximization · Stakeholder engagement · Transdisciplinary approach · Mexican fisheries

6.1 Introduction

Small-scale fisheries face many challenges worldwide, including climate change, overfishing, pollution, and high-level social and economic forces, which jeopardize biodiversity, food security, and the wellbeing of coastal communities. Adaptive capacity in fisheries is key to overcoming these threats and is influenced by contextual factors from different dimensions (ecological, economic, social, and institutional). With adaptation strategies that are co-developed, new information, management tools, and analytical frameworks are required to aid fishers, managers, and policy-makers in making sound short- and long-term decisions. In this context, financial stability is known to be a key facilitator of success (Blasiak et al. 2017; De la Cruz-González et al. 2018), and a primary concern for managers and public administrators. Unfortunately, the open-access conditions within which most small-scale fisheries operate are characterized by low financial performance, directly increasing fishers' vulnerability. However, many small-scale fisheries are tied to cultural and traditional values and benefits, where profitability is not always a primary goal (Kronen 2004; Béné et al. 2010; Kraan 2011; Cisneros-Montemayor et al. 2016a). Furthermore, in complex and dynamic systems such as small-scale fisheries, social wellbeing and the maintenance of livelihoods play essential roles.

We argue that a fishery that is economically viable will be able to face threats such as global market shifts, industrialization, and climate change much better than a fishery that is not considered economically viable (Schuhbauer and Sumaila 2016) (Box 6.1). Economic viability, in the context of this study, is defined as the net economic contribution of small-scale fisheries to society. Understanding the economic viability of small-scale fisheries is an essential step in characterizing the long-term outlook for fishing communities, as well as providing fishers and policy-makers with important data to inform sustainable and equitable public policies. In the following sections, we summarize the definition of economic viability, explain the importance of a transdisciplinary approach, and introduce a novel methodology on how to assess the economic viability of small-scale fisheries.

Box 6.1
Understanding and improving economic viability will help prepare small-scale fisheries withstand threats they face.

6.2 A Transdisciplinary Approach to Economic Viability

Financial viability, economic performance, and ecological-economic viability are being used interchangeably to describe economic viability, even though these terms have different meanings. For example, the viability of a fishery has been assessed using simple economic indicators, such as return on investment, revenue over investment, cash flow, or the ratio of net cash flow over total earnings (Lery et al. 1999; Tietze et al. 2001; Adeogun et al. 2009; Ünal and Franquesa 2010). Models to assess a fishery's viability have been created and applied to add the biological and ecological dimensions of fishing to the economic dimension, mainly based on bio-economic and ecological-economic theory (Péreau et al. 2012; Gourguet et al. 2013; Maynou 2014). Increasing in complexity to integrate ecological, economic, and social aspects of a fishery, models based on viability kernel analysis have been created and applied, for example, to assess the viability of small-scale fisheries (Hardy et al. 2013; Cissé et al. 2015). Viability kernel analysis is based on viability theory, a complex mathematical framework that uses algorithms and constraints to capture and model the viability of dynamical systems (Aubin et al. 2011). The strength of viability theory is that it involves interdisciplinary investigations with constraints (restrictions) that can be of biological, social, or economic nature (Aubin et al. 2011).

As mentioned in the Introduction, small-scale fisheries are very complex and dynamic systems and it is important to consider social, economic, ecological, and governance aspects when finding ways to support their sustainability. While viability assessments based on viability kernel analysis have successfully integrated social, ecological, and economic indicators (Hardy et al. 2013; Cissé et al. 2015), these analyses need time, expertise in the field of mathematical modelling, and large data sets, all often very limited resources in small-scale fisheries (Charles 1991). Furthermore, while the interdisciplinary nature of new methods—such as the viability kernel analysis—is a step in the right direction, we argue that a transdisciplinary approach to the assessment of economic viability would be even more inclusive. A clear framework to assess the economic viability of small-scale fisheries is needed, guided by a key principle of transdisciplinary research, which is the co-production of knowledge (Mauser et al. 2013; Roux et al. 2017). To co-produce knowledge, actors from policy, science, and practice need to identify societally relevant problems and formulate research questions together (Hirsch Hadorn et al. 2006; Mauser et al. 2013).

Our definition of economic viability (Box 6.2), which forms the basis of the approach introduced and explained in this chapter, takes a first transdisciplinary principle, the complexity and limitations of small-scale fisheries into account. Here, we refer not only to the financial profit of a sole private firm (including individuals) when assessing economic viability, but also to the whole society or the public sector. Economic viability is accomplished when the net benefit of an activity to society is non-negative.

Box 6.2 Definition of economic viability
An active fishery is economically viable when net benefits to society $\geq$ zero.

In the following sections, we describe the development of the introduced framework and use a case study to illustrate its application. This methodology will help scientists, fisheries practitioners, managers, fishers, and other interested community members to better understand the underlying dynamics of the fisheries they study to support their sustainable and equitable management.

6.3 Methodology

6.3.1 *Attribute–Based Approach*

We used an attribute-based approach as the foundation of our economic viability assessment of small-scale fisheries. A literature review was carried out to identify a list of attributes for determining the economic viability of fisheries (Schuhbauer and Sumaila 2016). We engaged stakeholders through a participatory workshop aiming to propose, discuss, and establish attributes related to the viability of fisheries (Too Big To Ignore 2014). The workshop brought together fisheries managers, practitioners, fishers, community members, and scientists whose expertise lies in the fields of economic, social, governance, and ecological aspects of small-scale fisheries (Too Big To Ignore 2014). A framework was developed based on a comprehensive list of attributes derived from the workshop; this list, including definitions, provided the foundation for the assessment of economic viability presented here (Table 6.1). The framework, which we will introduce in the following sections, can be applied at various scales (local, regional, national, global), with selection criteria for the attributes including relevance, availability, measurability, and objectivity, in other words, whether the same result is obtained when the attributes are measured by different scientists at different times (Boyd and Charles 2006).

6.3.2 *Assessment of Economic Viability*

We distinguish here between financial viability, basic economic viability, and extended economic viability (Box 6.3). Financial viability, is defined as net benefits to a private entity, which includes the subsidies it receives from the government. Basic economic viability is defined as the net benefit to society that is based on the real undistorted cost of inputs and value of outputs (i.e. without subsidies). The extended economic viability broadens this basic approach by further including key attributes into the assessment.

The elements of basic economic viability are i) total revenue generated from fishing; ii) total cost of fishing; and iii) fisheries subsidies. Total revenue (landed value) is the product of ex-vessel prices and tonnes of landed catch. Ex-vessel prices are defined as the price that a fisher receives at the dock per unit weight of fish sold (Sumaila et al. 2007). Total costs of fishing include fixed and variable costs of fish-

Table 6.1 Attributes of economic viability, the geographical scale and timeframe of the attributes depend on the study's objectives

N°	Economic attributes (unit)	Definition
1	Landings (t)	Amount of fish in weight landed in all ports by a given fishing sector during the study period.
2	Ex-vessel price ($)	Price received by fishers at the dock or landing site per unit weight of fish (Sumaila et al. 2007).
3	Total cost of fishing ($)	Total cost represents the value of inputs at the next alternative best use. Cost is split up into fixed costs, which do not change with production (e.g. capital investment) and variable costs, which can vary based on inputs and outputs (e.g. fuel, crew, maintenance, refrigeration) and opportunity cost.
4	Subsidies ($)	Subsidies are defined here as financial transfers, direct or indirect, from public entities to the fishing sector.
5	Proportion of small-scale fisheries to total landed value (%)	The ratio of small-scale fisheries revenue to total revenue of the whole fishery.
6	Cost structure (ratio)	Cost structure is the ratio of fixed costs (e.g. capital investment) to variable costs (e.g. fuel, labor).
7	Cost per tonne of catch ($)	Cost divided per tonne of catch, useful when comparing different fisheries and countries.
8	Multiplier (factor)	Describes indirect income (income multiplier) and induced effects on society (economic multiplier) through fisheries. A multiplier is a factor reflecting the contribution of total contribution of an economic activity's final demand value to total economic output, including activities directly and indirectly dependent on it.
9	Employment (#)	Number of jobs highlights the contribution of fisheries employment including both commercial and subsistence marine small-scale fisheries.
10	Subsidy intensity (ratio)	Using the amount of subsidies provided per fisher involved in a specific fishing sector is used here as a measure of equity revealing how subsidies are distributed between different fishing sectors.
11	Fisheries discards (%)	How much of total catch is being discarded at sea compared to that landed at the ports.
12	Catch not used for direct human consumption (%)	The amount of landings that are used directly for human consumption compared to indirect human consumption (e.g. fodder for livestock) and industrial use (e.g. beauty products).

Box 6.3

Financial viability:

Net Benefits to the Private sector = Total Revenue – Total Costs + Subsidies

Basic Economic viability:

Net Benefits to Society = Total Revenue – Total Costs (– Subsidies)

Extended Economic Viability:

Integration of attributes beyond the economic dimension.

ing. It is important to note that economic costs, in contrast to accounting costs, are captured by the opportunity costs of input to fishing. Fisheries subsidies are defined as direct or indirect financial transfers from public entities to the fishing sector (Sumaila et al. 2010). Subsidies, therefore, help the fishing sector to make more money than it would have otherwise. See Appendices for more detailed descriptions about the terms used (Glossary – Appendix 6.1), the equations (Appendix 6.2), and the attributes and data sources (Appendix 6.3).

To estimate basic economic viability as net benefits from fishing we use a set of simple equations found in Appendix 6.2 (Eqs. 6.6, 6.7, 6.8, and 6.9). Net benefits to society are equal to total costs subtracted from total revenue, which is calculated for both small- and large-scale fisheries for the years available. Financial viability, as in net benefits to the private sector, on the other hand, is equal to total revenue minus the total costs plus the total subsidies provided to the assessed fishery. The key difference between net benefits to society versus the private sector are that the latter is distorted by subsidies. For society, the amount of subsidy paid to fishers is a transfer from tax payers. From the point of view of the private sector, on the other hand, subsidies are added, because each fishing unit benefits from the subsidy provided to it.

To better validate how economic and socioeconomic drivers of small-scale fisheries contribute to their economic viability, the following attributes are integrated defining the extended version of economic viability: proportion of small-scale fisheries landed value to total landed value, cost structure, economic impact of fishing (multiplier), employment, fisheries discards, and the use of fisheries products (Appendix 6.3). The list of attributes is dynamic, and individual attributes can be adjusted to the information available for the fishery in question.

6.4 Relating Attributes to Economic Viability

To analyze small-scale fisheries in light of basic and extended economic viability, the results from the attributes described above are brought together and compared to a large-scale fishery counterpart. To address how each assessed attribute is likely to affect the results of a basic economic viability assessment, we designed a simple point system, which scores each attribute of small-scale fisheries with a minus or a plus depending on whether the attribute that is being analysed is expected to impact the results of the basic economic viability analysis negatively or positively in direct comparison to their large-scale fishery. If there are known context-specific aims for a particular fishery, it would be possible to incorporate a weighting system to the calculation, though we have not done so here. Depending on the attribute, the resulting amount (number) is either considered better if higher, as for example number of jobs, or better if lower, as for example costs per catch, in comparison to the large-scale fishery. It should be noted that the scores are given under the assumption that all else remains the same (*ceteris paribus*).

Results from both basic and extended economic viability approaches are expected to enhance the understanding of the complex dynamics of small-scale fisheries and to compare them to their large-scale fisheries counterparts. Furthermore, they can be used to recommend policy and management strategies to improve the economic viability of these fisheries, which, we argue, supports ecological sustainability and social wellbeing of small-scale fisheries communities. In the following section, we use a case study to exemplify how the here introduced and described methodological framework to assess basic and extended economic viability can be used.

6.5 Case Study: Mexico

6.5.1 Case Study Context

To demonstrate the workings of our approach, we use available national-level fisheries data for a developing country, Mexico. We chose Mexico because its marine small-scale fisheries is among the largest globally, based on catch and employment, but remains largely unregulated (Teh and Sumaila 2013; Cisneros-Montemayor et al. 2013). Available data and policies have tended to emphasize large-scale fisheries, such as those for sardine, tuna, and shrimp (e.g. Garcia-Caudillo et al. 2000; Lluch-Cota et al. 2007; Ishimura et al. 2013; Punt et al. 2016). The small-scale fisheries sector includes around 70,000 small vessels (*pangas*), catching around 900 thousand tons of fish and invertebrates per year (Cisneros-Montemayor et al. 2013). Some regional and local studies, as well as efforts by national government institutions, have illustrated the social, economic, political, and ecological importance of small-scale fisheries in Mexico (Smith et al. 2009; Cisneros-Mata 2010; Salas et al. 2011; Espinoza-Tenorio et al. 2011; Conapesca 2013). The Mexican small-scale fisheries sector is embedded in a wide cultural context providing food and employment for hundreds of thousands of people and contributing to coastal social and economic development (OECD 2006a; Lluch-Cota et al. 2007).

6.6 Defining Mexican Small- and Large-Scale Fisheries

Based on the definition of the Mexican National Commission of Aquaculture and Fishing, the marine small-scale fishing sector includes artisanal commercial fisheries, who sell most of their catch at local markets and often keep a portion of it for household consumption. Almost all small-scale fisheries use *pangas*, open-deck fiberglass boats around seven meters in length, usually with 50–115 horsepower outboard engines. The most common fishing gears used are gillnets, hook-and-line, hookas (a regulator and on-board air-compressor), traps, and a range of small bottom-trawl nets. Large-scale (or industrial) fisheries, on the other hand, include vessels with a covered deck, inboard engine (almost exclusively diesel), mechanical winches, and fishing gear including otter trawls, purse-seiners, and longlines. There is an offshore fleet targeting tunas and billfishes, and a large coastal fleet targeting shrimp and small pelagic fishes (e.g. sardines). Although recreational fisheries in Mexico are a significant industry for some regional economies and interact strongly with marine ecosystems and other fisheries (Cisneros-Montemayor et al. 2012), they are not included in the assessment carried out in this study.

6.6.1 Basic Economic Viability

6.6.1.1 Data Sources of Basic Economic Viability

Fisheries specific studies and government reports are the sources of information used to compute revenues, fishing costs, and subsidies of Mexican fisheries and form an important cornerstone for our economic viability assessments (OECD 2006a; Conapesca 2013; Cisneros-Montemayor et al. 2013; Ramírez-Rodríguez and Almendárez-Hernández 2013). However, not much is known of these main economic indicators (costs, revenues, and subsidies) and their impact on small- compared to large-scale fisheries over time. Here, these will be provided and analyzed for the period from 2000 to 2012 at a national level to clearly illustrate how the assessment of basic economic viability works using the approach introduced above. Data sources are displayed in Table 6.2.

6.6.1.2 Results of Basic Economic Viability

Despite potential data uncertainty and limitations, results from the basic economic viability analysis suggest that small-scale fisheries are more economically viable than large-scale fisheries looking at national averages over the 13 year study period (Table 6.3).

As expected, net benefits to society (basic economic viability) are lower compared to the net benefits to the private sector (financial viability) for both small- and large-scale fisheries (Table 6.3 and Fig. 6.1). The difference between basic economic and financial viability is much more visible in large-scale fisheries as they receive a much greater share of the subsidies compared to the small-scale sector. This difference, furthermore, has been increasing over time with the increase in subsidies provided to large-scale fisheries (Fig. 6.1).

Table 6.2 Main economic indicators of Mexican fisheries and its data sources essential for basic economic viability, attributes 1–4

N°	Economic attributes (unit)	Data sources
1	Landings (t)	Sea around us (Cisneros-Montemayor et al. 2013) for 2000–2012 (www.seaaroundus.org last accessed January 2017)
2	Ex-vessel price ($)	Sea around us and fisheries economic research unit databases (www.seaaroundus.org last accessed January 2017) and the National Commission of Aquaculture and fishing annual fisheries reports (CONAPESCA) (Conapesca 2013; Swartz et al. 2013; http://www.conapesca.sagarpa.gob.mx/wb/, last accessed January 2017).
3	Total cost of fishing ($)	Calculation based on data from cost structure (fixed and variable costs) of individual fishing units, scaled up to fishing fleets and used changes over time of cost of fuel and vessel numbers to create a timeline (2000–2012) for both small- and large-scale fisheries. Large-scale fisheries: Data was used from shrimp, sardine, and tuna fisheries and scaled up using number of boats to the whole fishing fleet (Gillet 2008; Agroprospecta 2010; Lam et al. 2011). Fixed costs was estimated based on (Lasch 2005). Small-scale fisheries: Information was used from (OECD 2006a; Lam et al. 2011; Ramírez-Rodríguez and Almendárez-Hernández 2013).
4	Subsidies ($)	Data was gathered from fisheries reports, CONAPESCA annual reports, peer-reviewed articles, OECD (organisation for economic co-operation and development) reports and gray literature (OECD 2006b; Lara and Guevara-Sangines 2012; Ramírez-Rodríguez and Almendárez-Hernández 2013; Sumaila et al. 2016). Once we gathered information on total subsidies for each year (2000–2012), data from CONAPESCA annual reports were used to split subsidy into small- and large-scale fisheries for each year.

Table 6.3 Basic economic viability (BEV) assessment of small- and large-scale fisheries averaged over the years 2000–2012 (constant 2015 USD million)

	Large-scale fisheries (USD)	Small-scale fisheries (USD)
Total revenue	711	2011
Total cost of fishing	678	1969
BEV	33	42
Subsidies	132	51
Financial viability	165	93

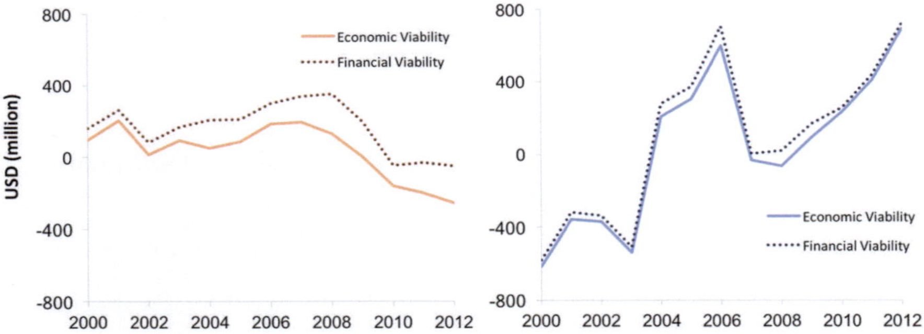

Fig. 6.1 Net benefits (NB) presented by sector before subsidies (financial viability) and after subsidies (economic viability) in constant 2015 USD million, large-scale fisheries on the left (red) and small-scale fisheries on the right (blue)

6.6.2 *Extended Economic Viability*

6.6.2.1 Data Sources for Extended Economic Viability

Attributes (5–12 in Table 6.4) were analyzed at a national level, using annual data from 2012 or the closest year as available. All amounts are converted into constant 2015 USD.

6.6.2.2 Results of Extended Economic Viability Assessment

Based on extended economic viability total points shown in Table 6.5, small-scale fisheries have a positive score of 5 (out of a maximum of 8). This means that, based on the extended economic viability assessment, prospects for small-scale fisheries are better compared to large-scale fisheries. It is important to keep in mind that the analysis is based on the ceteris paribus concept (i.e. where all else remains the same) (Table 6.5).

For a more detailed description of the results of the economic viability assessments of Mexican small-scale fisheries and large-scale fisheries see Schuhbauer (2017).

Table 6.4 Extended Economic Viability Attributes 5–12, its data sources and notes for Mexican fisheries

N°	Attributes	Data sources	Notes
5	Proportion of small-scale fisheries to total landed value (%)	Total revenue data mainly based on catch and price data from Cisneros-Montemayor et al. (2013), FAO statistics and CONAPESCA annual reports (2012) to calculate the proportions of total revenue to the revenue generated by small-scale fisheries for each region.	There are known issues and uncertainties associated with catch and revenue data in Mexico, but these data are increasingly available for more regions and species groups.
6	Cost structure (ratio)	See methods of basic economic viability for a detailed description of the cost and cost structure estimates for Mexican small- and large-scale fisheries. In summary, data for variable costs (fuel, labor, and running costs) and fixed costs (depreciation and interest) were calculated using various data of either a single fishery and/or individual fishing units. The results were then scaled up to national level using fishing effort data (number of boats and fishers) per fishing sector and/or fishery for the year 2012 (Agroprospecta 2010; Conapesca 2013; Ramírez-Rodríguez and Almendárez-Hernández 2013; Greer 2014).	The total cost of fishing and the cost structure may vary depending on the fishing gear and vessel type.
7	Cost per tonne of catch ($)	See data sources of cost data of basic economic viability, of both small- and large-scale fisheries, divided by tonnes of catch using data from Cisneros-Montemayor et al. (2013) for the year 2012.	
8	Multiplier (factor)	The National Institute of statistics and geography (INEGI) provided data for the year 2008, which include total revenue from fishing and the total added value, together these produce the total economic impact from fishing. INEGI data are available by municipality, however, not by fishing sector. As Mexican large-scale fisheries focus on a few species whose landings are concentrated at a few ports (FAO 2003; Erisman et al. 2011), we used the data of municipalities where the large-scale fishing ports* are located to estimate the economic multiplier for large-scale fisheries. Data from all other coastal municipalities was used to estimate the economic multiplier for small-scale fisheries, as small-scale fisheries are found spread along the coastlines with various landing sites. The multiplier was then used to estimate the total economic impact based on total revenue.	* The ports include: Ensenada, Comondú, Manzanillo, Tapachula, Mazatlán, Guaymas, and Progreso. Most of the processing plants of the large-scale fisheries sector are located in these municipalities with about 80% of all plants of the Pacific found in the Gulf of California, where on the Atlantic side almost 50% of the processing plants are found in Progreso, Yucatan (FAO 2003).

(continued)

Table 6.4 (continued)

N°	Attributes	Data sources	Notes
9	Employment (#)	Information from annual fisheries reports (Conapesca 2013) show the number of fishers combined in aquaculture and fisheries as a total national value, illegal and informal fishers are probably not captured here. To disaggregate the number by fishing sector, we used the number of active fishing vessels and average number of crew per type of fishery from official statistics to estimate the reported number of fishers active in each fishery. Then we used the fraction of unreported to reported catch data from Cisneros-Montemayor et al. (2013) to reconstruct the number of total fishers per sector aimed to cover numbers of both licensed and unlicensed fishers.	Unfortunately, these numbers probably still do not capture the whole sector, which includes people collecting seafood from the shore.
10	Subsidy intensity (ratio)	Subsidy and total revenue data was used from (Schuhbauer 2017) for the years 2012 for both Mexican large- and small-scale fisheries. Total number of fishers was used from employment data.	Finding detailed information on subsidies has been a challenge and results are estimates based on the best data available.
11	Fisheries discards (%)	This analysis was possible due to catch reconstruction data that indicate whether the catch (both reported and unreported) was landed or discarded disaggregated into fishing sector (Cisneros-Montemayor et al. 2013).	
12	Catch not used for direct human consumption (%)	Official Mexican fisheries reports indicate what percentage of landings is used for either direct human consumption, indirect human consumption (e.g. animal fodder), and industrial use, also shown separated by sector (Conapesca 2013).	Using these data, we compared the actual landings that are directly used for human consumption in comparison to animal fodder (Agri- and aquaculture) as well as for industrial use (e.g. beauty products).

6.6.3 Policy Recommendations and Conclusions Based on the Economic Viability Assessment of Mexican Fisheries

Following our analytical framework, and based on this study's results, we derive specific recommendations to improve basic and extended economic viability of Mexican fisheries. The inequity of subsidy distribution has increased over time, with significant benefits to large-scale fisheries. As shown in many other studies at global and national levels (e.g. Schorr 2005; Sumaila et al. 2010; Charles 2011; Cisneros-Montemayor et al. 2016b; Schuhbauer et al. 2017), subsidies are in need

Table 6.5 Results from the extended economic viability assessment for Mexican small- and large-scale fisheries

Extended economic viability attributes	Results		Score	
	Small-scale fisheries	Large-scale fisheries	Small-scale fisheries	Large-scale fisheries
5. Proportion of small-scale fisheries to total revenue (%)	70	30	+	–
6. Fuel costs / costs of labor (ratio)	1.5	1.6	–	+
7. Cost per tonne of catch (USD)	1755	771	–	+
8. Multiplier (factor)	1.54	1.49	+	–
9. Employment (# x 1000)	308	57	+	–
10. Subsidy intensity (subsidies per fisher USD) and by total revenue (ratio)	180	8710	–	+
	0.01	0.37		
11. Fisheries discards (% of total catch)	9	22	+	–
12. Catch not for direct human consumption (%)	0	58	+	–
EEV total score			+ 5	+ 3
BEV 2012 (USD in million)			690	–258

Scores of extended economic viability (EEV) of small-scale fisheries, positive (+) and negative (–) impact on basic economic viability (BEV), the higher the EEV score the stronger the impact on BEV. The score is given in direct comparison to the large-scale fisheries sector, in other words, if the score of small-scale is better than the score of large-scale the attribute will receive a (+) and if the contrary is true it receives a (–)

of reform and capacity-enhancing types urgently need to be reduced (to zero, if possible). The monitoring of active boats and fishers needs to be improved and updated regularly, for both small- and large-scale fisheries, to inform stock and economic performance assessments, more sustainable fishing operations, and reductions in bycatch and discard rates. Discards are often comprised of juvenile fish, so their protection will improve ecosystem health in already overfished areas. Together with reductions in costs from stock recoveries, better and more selective fishing technology, and regulated fishing effort, this will improve economic viability (Box 6.4).

We conclude that the analysis of attributes that influence economic viability (e.g. subsidies, discard rates, and what fish is used for), provides us with key information not only to understand profitability and efficiency of fisheries but also to see what can be done to improve ecological sustainability and tackle social issues such as food security. Understanding how changes in fisheries management and policy could improve these attributes from economic, ecosystem, and social dimensions and with them the economic viability of small-scale fisheries, will help these fisheries be much better prepared to face the threats and challenges they currently face. These results are important to help bridge the current knowledge gap that has been identified for small-scale fisheries, not only in Mexico, but worldwide. This is essential to policy making and management that would not only improve economic viability, but also the sustainability of the fish stocks upon which they rely.

Box 6.4
To improve economic viability of small-scale fisheries we need to:

1. Use fisheries subsidies to improve monitoring and enforcement;
2. Regulate fishing effort and improve fishing technologies;
3. Focus on food security, employment and, ecological factors;
4. Include small-scale fishers in decision-making processes.

There are advantages and disadvantages to both basic and extended economic viability approaches. If time and data availability allow, it is ideal to carry out both assessments, contributing to the best possible understanding of the overall economic state of the fisheries under study. However, if time and data are limited, which is unfortunately the case for most small-scale fisheries, information from the basic economic viability assessment is very valuable on its own. While focusing on numbers alone can have its drawbacks, numbers are invaluable when it comes to standardization and comparison of results. Indeed, results from a straightforward calculation of basic economic viability (as defined in this study) are comparable not only over time and across sectors, but also across regions and countries. Extended economic viability, on the other hand, is a more comprehensive approach. While it, therefore, requires more data and input from stakeholders, it can be more flexible and adapt to the information that is available for the fisheries under study. It also has the benefit of encouraging various participants in fishery systems to openly highlight and discuss their benchmarks for success, potentially leading to increased collaborative solutions.

6.7 Discussion and Conclusion

This study is the first attempt to clearly define and assess economic viability of small-scale fisheries. The literature has shown incoherence surrounding the definition of the term economic viability, not only for fisheries specific studies but also in general. Economic viability has often been used interchangeably with economic performance and financial viability. While both economic performance and financial viability are important for fisheries to be able to survive in the long term, these definitions fail to integrate other key attributes (e.g. subsidies, employment, and food security) that make a fishery economically viable. Complex models based on bio-economics and viability theory have successfully been applied to analyze the viability of small-scale fisheries (Gourguet et al. 2013; Hardy et al. 2013; Cissé et al. 2015). The application of these models, however, has only been possible where expertise and enough data are accessible, often resources and information that are scarce especially in small-scale fisheries.

Here, we have shown that incorporating the first stage of a transdisciplinary approach—the co-design of the methodological framework—provides a solid foundation for an economic viability assessment, as illustrated using the Mexican case study. Using a transdisciplinary approach that encompasses not only the co-design of a project but also co-produces knowledge and facilitates the co-dissemination of the produced knowledge is an essential next step. Once actors from policy, science, and practice are involved throughout the entire study, problems such as the uneven distribution of subsidies and the lack of representative data can be tackled in a collaborative manner (Box 6.5).

The methodology described in this chapter and its application show how important it is to analyze small- and large-scale fisheries separately in order to properly address needs specific to each fishing sector. As mentioned earlier, the lack of data in many countries is a huge challenge to this kind of research. Data transparency and accessibility are key factors that are in need of improvement. We hope this study not only brings attention to the data quality problem of small-scale fisheries (see Mexico as an example) but will also help to correct it.

An important caveat that needs to be acknowledged in this study is that the importance of cultural values of small-scale fisheries in society has been mentioned but not analyzed or discussed. This is beyond the scope of this research project; however, it would be useful for future research to incorporate cultural values into the economic viability assessment framework. As this is the first study to analyze economic viability of small-scale fisheries at national and regional levels, it would be interesting to repeat this assessment in other countries as well as at a global scale.

Marine fisheries are threatened globally, and small-scale fisheries are the ones that suffer the most as a result of these threats. Fortunately, the current fisheries crisis has already been recognized and solutions have been proposed. These include the reduction of harmful fisheries subsidies (Milazzo 1998; Sumaila et al. 2016), the implementation of marine protected areas, and more generally, reduction of fishing efforts (Pauly et al. 2002; McClanahan et al. 2006; Edgar et al. 2014; Sumaila et al. 2015; Boonzaier and Pauly 2016). In addition, fishing effort management through ecosystem-based approaches and the allocation of access to groups within this management context are needed (Mora et al. 2009; Sumaila 2010; Fulton et al. 2014; Edgar et al. 2014; Christensen et al. 2015). International agreements have been quite positive, especially in relation to small-scale fisheries, during the last few years. For example, the United Nations' Sustainable Development Goals include targets such as reducing illegal fishing, protecting marine biodiversity and providing small-scale fisheries access to natural resources and markets (Targets 14.4 and 14.b in United Nations (2016)).

Box 6.5
While complex models exist to determine a fishery's viability, we have introduced a simple approach to assess the economic viability of small-scale fisheries showing how a transdisciplinary approach can be applied.

Another important step towards improving the current state of small-scale fisheries has been the development of voluntary guidelines for securing sustainable small-scale fisheries in the context of food security and poverty eradication (FAO 2015). These guidelines, while voluntary, have been developed through a human rights-based approach—recognizing human rather than market needs as central to resource management—and are a sign that the importance of small-scale fisheries is increasingly being recognized across the world as they play major roles in food security and nutrition (Jentoft et al. 2017; Singleton et al. 2017).

These policy advancements over the last years geared towards achieving sustainable small-scale fisheries are important both for international and national guidance. However, considering the large knowledge gap that exists in small-scale fisheries research, in order to achieve agreed targets and effectively implement set policies, it is clear that more research needs to be carried out with a transdisciplinary focus. The application of the methodology introduced and described herein would contribute to filling some of this knowledge gap at global, national and local levels.

Appendices

Appendix 6.1: Glossary

Total Revenue
Total revenue is calculated as the ex-vessel price (for each fish species) times the total landings of the fishery. The total revenue can then be calculated for each year. Total revenue is calculated for each unit, which represents a fishing vessel or a company that owns more than one vessel, which is then summed up to produce the total revenue for the entire fishery.

Total Costs of Fishing Total costs of fishing are calculated by adding variable costs to fixed costs. Variable costs comprise of fuel, maintenance, running (e.g. docking fees), labor and opportunity costs. In the case of small-scale fisheries where vessel owners operate the vessel, labor costs are assumed to be equal to opportunity cost. Fixed costs consist of vessel and gear depreciation and interest paid on capital costs. Total costs can be calculated for each year and each unit and then summed up where useful.

Fisheries Subsidies Fisheries subsidies are direct or indirect financial transfers from public entities (i.e. tax revenues) to private firms (here, the fishing sector) (Milazzo 1998; OECD 2006b; Sumaila et al. 2010). Subsidies have been categorized, depending on their expected impact on fish stocks over time, into beneficial, capacity-enhancing and ambiguous (Sumaila and Pauly 2006). Many different forms such as financial assistance, tax breaks, fisheries management, and research, as well as direct capital infusion are considered fisheries subsidies (see Abdallah and Sumaila 2007). The amount of total subsidies per fishing sector received from the government can be calculated for each year and for each category.

Appendix 6.2: Equations

Total Revenue
Total Revenue (TR) is calculated as follows:

$$TR = P \cdot TL \tag{6.1}$$

Where P denotes ex-vessel prices (prices are used by species) and TL is the total fisheries landings of each species. TR is calculated for each year t:

$$TR_t = \sum_{U}^{u=1} (TRt_t)_u \tag{6.2}$$

Where u (unit) represents a fishing vessel or a fishing company that owns more than one vessel and U denotes the total number of units for each year t.

Total Costs of Fishing
Total Costs (TC) are calculated as follows:

$$TC = VC + FC \tag{6.3}$$

Where VC represent variable costs, which comprise of fuel, maintenance, running costs (e.g. docking fees), and labour. In the case of small-scale fisheries, where vessel owners operate the vessel, labour costs are assumed to be equal to opportunity cost. FC denotes fixed costs, which consist of vessel and gear depreciation and interest paid on capital costs.

TC is calculated for each year t as:

$$TC_t = \sum_{U}^{u=1} (TC_t)_u \tag{6.4}$$

Subsidies
The amount of total subsidies per fishing sector received from the government can be calculated for each year.

$$TS_t = \sum_{U}^{u=1} (TS_t)_u \tag{6.5}$$

Basic Economic Viability

To estimate basic economic viability (BEV) as Net Benefits (NB) from fishing, the following calculations are carried out for both small- and large-scale fisheries for the years available, based on the estimates from Eqs. 6.1, 6.2, 6.3, 6.4 and 6.5:

$$NB_t^S = TR_t^S - TC_t^S \tag{6.6}$$

where the superscript, S, denotes society.

Equation 6.7 is used to calculate the average basic economic viability (net benefits to society: NB^S) over the years under study here (n):

$$\overline{NB^S} = \frac{1}{n}\sum_{n}^{t=1} NB_t^S \tag{6.7}$$

Financial viability, as in net benefits to the private sector, on the other hand, are calculated as expressed in Eq. 6.8. The key difference between Eqs. 6.6 and 6.8 is that the economic viability is distorted by subsidies (Eq. 6.5).

$$NB_t^p = TR_t^p - TC_t^p + TS_t \tag{6.8}$$

where the superscript, p, denotes the private sector. Average net benefits to the private sector:

$$\overline{NB^p} = \frac{1}{n}\sum_{n}^{t=1}\left(NB_t^p\right) \tag{6.9}$$

For society, the amount of subsidy paid to fishers is a transfer from tax payers (Eq. 6.6). It should be noted that in formulating the equations above, we assume that the only distortion is the subsidy received by the fishing unit. From the point of view of the private sector, on the other hand, subsidies are added, because each fishing unit benefits from the subsidy provided to it (Eq. 6.8).

It should be noted that the scores are given under the assumption that all else remains the same (ceteris paribus), we therefore assume the following:

$$\frac{\partial BEV}{\partial A_i} > 0 \text{ for } i = 5,8,9,10 \tag{6.10}$$

where BEV = Basic Economic Viability and A_i = the result of each attribute.

$$\frac{\partial BEV}{\partial A_i} < 0 \text{ for } i = 6,7,11,12 \tag{6.11}$$

Equations 6.10 and 6.11 show how the result of each attribute impact BEV, the higher the attribute in Eq. 6.10 and the lower each attribute in Eq. 6.11 the better it will be for BEV.

Appendix 6.3: Tables Showing Data Sources for Each Attribute

Sources of information on the different attributes at a global scale can be found in Lam et al. (2011), Teh and Sumaila, (2013), (Dyck and Sumaila 2010), FAO website and reports, Sea Around Us data bases (http://www.seaaroundus.org/). For case studies we recommend to check different literature, for example, government reports, peer reviewed articles, and FAO country profiles, as well as, if possible, to conduct surveys.

Table A Data sources for elements of basic economic viability (attributes 1–4 from Table 6.1), the geographical scale and time frame of the attributes depend on the study's objectives

N°	Economic attributes (unit)	Sources, measures, and notes
1	Landings (t)	For national numbers see FAO and SAUP (sea around us project, specifically catch reconstruction data) database.
		For case studies check literature, for example, government reports, conduct surveys, and monitor the landing.
2	Ex-vessel price ($)	For national numbers see fisheries economic research unit (FERU) and SAUP database (Sumaila et al. 2007; Swartz et al. 2013)
		For case studies check literature, for example, government reports, conduct surveys, log book, and buyer record.
3	Total cost of fishing ($)	For national numbers see FERU database (Lam et al. 2011).
		For case studies check literature, for example, government reports and/or conduct surveys.
		Total cost includes opportunity costs, here represented as labour costs, which makes it different from accounting cost (Lam et al. 2011).
4	Subsidies ($)	For national numbers see Sumaila et al. (2010, 2016)
		For case studies check literature, for example, government reports, conduct surveys, and interview key informants.
		Subsidies help the sector become more profitable than it would otherwise (Sumaila et al. 2010).

Table B List of attributes (5–12 from Table 6.1) included in the extended economic viability assessment of small- and large-scale fisheries, their sources, and measures

N°	Attributes	Notes and sources
5	Proportion of small-scale fisheries to total landed value (%)	To highlight differences between the small- and large-scale fisheries sectors representing total revenue from fishing.
6	Cost structure (ratio)	To simplify the cost structure, we calculate the ratio of fuel to labor costs, using fuel as an indicator for efficiency and labor as a social indicator. As small-scale fisheries are often owner-operated and fishing crew are paid with a share of the catch revenue rather than a set wage, opportunity cost is used. We argue that the lower the ratio of fuel to labor cost, the better its impact on basic economic viability
7	Cost per tonne of catch ($)	This facilitates comparisons across the sectors and regions, the lower the costs the better for economic viability.
8	Multiplier (factor)	A multiplier, which has a social component, was included as part of the economic dimension, as it captures the value of demand in relation to an activity. This multiplier can be interpreted as an output and a degree of opportunities for employment that exist for a particular society. To measure the total economic impact, it is important to look beyond the landed value that each fishery produces and estimate the direct and indirect impact from fishing on the economy. Direct impact is considered all money spent by fishers to carry out fishing (e.g. boats, motors, paint, fuel, etc.). Indirect impact refers to the impact arising from purchased inputs, spent by the sellers of goods to fishers (e.g. equipment needs to be bought to sell fuel, produce nets, catch bait, etc.). When these impacts are added together it reflects the total impact of fishing on a country's economy. We argue the higher the multiplier the better it is for the fishery's economic viability.
9	Employment (#)	Employment type describes how a worker is employed. For example, the international labor organization (ILO) classifications are: Employees, employers, own-account workers, members or producer cooperatives, contributing family workers, workers not classifiable by status. These can be adjusted to fisheries (e.g. boat owner, paid by catch share, employed by a company/cooperative). Employment is considered part of the socioeconomic dimension, showing how many people depend on fisheries for their work and livelihoods. The number of people employed in a fishery and the number of dependents accounted for here can be direct and indirect. Direct employment includes all people who fish, this includes men, women, and children, as well as encompasses all fishing activities, whether they happen from the shore, for example, collecting invertebrates, or fishing from a boat. Indirect employment includes people engaged in the post-harvest sector, any processing and selling of the catch. The more people employed in the sector the better it is for economic viability.

(continued)

N°	Attributes	Notes and sources
10	Subsidy intensity (ratio)	We consider subsidy-intensity part of socioeconomics, as it can highlight inequities in subsidy distribution among sectors, regions, and countries. The term subsidy intensity is used to express the ratio of subsidies to total revenue. Furthermore, subsidies per fisher were estimated to better understand equity in the distribution of subsidies to each fishing sector and the different regions, thus reflecting the impact on people involved in fishing in comparison to number of fishing firms or vessels. It needs to be clarified, especially for fishers employed in large-scale fisheries, that the resulting amounts do not reflect what each fisher actually receives, but instead what the company receives or saves in costs as subsidies. The main goal is to compare the 'subsidies per capita' of small-scale and industrial fisheries. If this number is high, we expect it to have a positive impact on economic viability.
11	Fisheries discards (%)	Understanding discards is important as part of the ecosystem component because discards are known to harm the environment and exacerbate overfishing. Furthermore, the ratio of discards to landings from fisheries is an essential aspect in relation to food security and poverty, especially in developing countries where millions of people suffer from undernourishment (example Mexico: (Olaiz-Fernandez et al. 2006).
12	Catch not used for direct human consumption (%)	The end destination for fisheries landings is a key aspect that impacts ex-vessel prices and food security, and therefore, represents economic and social components of fisheries. The demand for fishmeal production globally is growing and with it the use of fish not for direct human consumption (the World Bank 2013; Cashion et al. 2017). This has become a challenge for food security, especially in regards to animal protein and micronutrients at a global level (Naylor et al. 2000; Béné et al. 2016; Majluf et al. 2017). The higher the percentage of the landings is destined for direct human consumption the better, we assume, the outcome for economic viability.

References

Abdallah PR, Sumaila UR (2007) An historical account of Brazilian public policy on fisheries subsidies. Mar Policy 31:444–450. https://doi.org/10.1016/j.marpol.2007.01.002

Adeogun OA, Abohweyere PO, Ogunbadejo HK et al (2009) Economic viability of small-scale marine capture fisheries in the bonny area, rivers state, Nigeria. Mar Resour Econ 24:195–203

Agroprospecta (2010) Unidades representativas de producción acuícola y pesquera. Panorama económico 2008–2018. Universidad Autónoma Chapingo, Chapingo

Aubin J-P, Bayen AM, Saint-Pierre P (2011) Viability theory – new directions. Springer, Berlin/Heidelberg

Béné C, Hersoug B, Allison EH (2010) Not by rent alone: Analysing the pro-poor functions of small-scale fisheries in developing countries. Dev Policy Rev 28:325–358. https://doi.org/10.1111/j.1467-7679.2010.00486.x

Béné C, Arthur R, Norbury H et al (2016) Contribution of fisheries and aquaculture to food security and poverty reduction: assessing the current evidence. World Dev 79:177–196. https://doi.org/10.1016/j.worlddev.2015.11.007

Blasiak R, Spijkers J, Tokunaga K et al (2017) Climate change and marine fisheries: least developed countries top global index of vulnerability. PLoS One 12:e0179632. https://doi.org/10.1371/journal.pone.0179632

Boonzaier L, Pauly D (2016) Marine protection targets: an updated assessment of global progress. Oryx 50:27–35. https://doi.org/10.1017/S0030605315000848
Boyd H, Charles A (2006) Creating community-based indicators to monitor sustainability of local fisheries. Ocean Coast Manag 49:237–258. https://doi.org/10.1016/j.ocecoaman.2006.03.006
Cashion T, Le Manach F, Zeller D et al (2017) Most fish destined for fishmeal production are food-grade fish. Fish Fish 18:837–844. https://doi.org/10.1111/faf.12209
Charles AT (1991) Bio-socio-economic dynamics and multidisciplinary models in small-scale fisheries research. In: Durand J-R, Lemoalle J, Weber J, Symposium International ORSTOM-IFREMER, Montpellier (FRA), 1989/07/3–7 (eds) La recherche scientifique face à la pêche artisanale = research and small-scale fisheries. ORSTOM, Paris, pp 603–608
Charles A (2011) Small-scale fisheries: on rights, trade and subsidies. MAST 10:85–94
Christensen V, Coll M, Buszowski J et al (2015) The global ocean is an ecosystem: simulating marine life and fisheries. Glob Ecol Biogeogr 24:507–517. https://doi.org/10.1111/geb.12281
Cisneros-Mata MA (2010) The importance of fisheries in the Gulf of California and ecosystem-based sustainable co-management for conservation. In: Brusca R, Brusca RC (eds) The Gulf of California: biodiversity and conservation. University of Arizona Press, Tucson
Cisneros-Montemayor AM, Christensen V, Arreguín-Sánchez F et al (2012) Ecosystem models for management advice: an analysis of recreational and commercial fisheries policies in Baja California Sur. Mexico Ecol Model 228:8–16. https://doi.org/10.1016/j.ecolmodel.2011.12.021
Cisneros-Montemayor AM, Cisneros-Mata MA, Harper S et al (2013) Extent and implications of IUU catch in Mexico's marine fisheries. Mar Policy 39:283–288. https://doi.org/10.1016/j.marpol.2012.12.003
Cisneros-Montemayor AM, Pauly D, Weatherdon LV et al (2016a) A global estimate of seafood consumption by coastal indigenous peoples. PLoS One 11:e0166681. https://doi.org/10.1371/journal.pone.0166681
Cisneros-Montemayor AM, Sanjurjo E, Munro GR et al (2016b) Strategies and rationale for fishery subsidy reform. Mar Policy 69:229–236. https://doi.org/10.1016/j.marpol.2015.10.001
Cissé AA, Doyen L, Blanchard F et al (2015) Ecoviability for small-scale fisheries in the context of food security constraints. Ecol Econ 119:39–52. https://doi.org/10.1016/j.ecolecon.2015.02.005
Conapesca (Comisioon Nacional de Acuacultura y Pesca) (2013) Anuario Estadistico de Acuacultura y Pesca. SAGARPA/CONAPESCA, Mexico
De la Cruz-González FJ, Patiño-Valencia JL, Luna-Raya MC et al (2018) Self-empowerment and successful co-management in an artisanal fishing community: Santa Cruz de Miramar, Mexico. Ocean Coast Manag 154:96–102. https://doi.org/10.1016/j.ocecoaman.2018.01.008
Dyck AJ, Sumaila UR (2010) Economic impact of ocean fish populations in the global fishery. J Bioecon 12:227–243. https://doi.org/10.1007/s10818-010-9088-3
Edgar GJ, Stuart-Smith RD, Willis TJ et al (2014) Global conservation outcomes depend on marine protected areas with five key features. Nature advance online publication 506:216–220. https://doi.org/10.1038/nature13022
Erisman BE, Paredes GA, Plomozo-Lugo T et al (2011) Spatial structure of commercial marine fisheries in Northwest Mexico. ICES J Mar Sci J Cons 68:564–571. https://doi.org/10.1093/icesjms/fsq179
Espinoza-Tenorio A, Espejel I, Wolff M (2011) Capacity building to achieve sustainable fisheries management in Mexico. Ocean Coast Manag 54:731–741. https://doi.org/10.1016/j.ocecoaman.2011.07.001
FAO (2015) Voluntary guidelines for securing sustainable small-scale fisheries in the context of food security and poverty eradication. Food and Agriculture Organization of the United Nations, Rome
FAO (Food and Agriculture Organization of the United Nations) (2003) Information on fisheries management in the United Mexican States. In: Last Accessed Oct 2016. http://www.fao.org/fi/oldsite/FCP/en/mex/body.htm. Accessed 19 Oct 2016
Fulton EA, Smith ADM, Smith DC et al (2014) An integrated approach is needed for ecosystem based fisheries management: insights from ecosystem-level management strategy evaluation. PLoS One 9:e84242. https://doi.org/10.1371/journal.pone.0084242

Garcia-Caudillo JM, Cisneros-Mata MA, Balmori-Ramirez A (2000) Performance of a bycatch reduction device in the shrimp fishery of the Gulf of California, México. Biol Conserv 92:199–205. https://doi.org/10.1016/S0006-3207(99)00053-1

Gillet R (2008) Global study of shrimp fisheries, Part 2: Shrimp fishing in Mexico. FAO, Rome

Gourguet S, Macher C, Doyen L et al (2013) Managing mixed fisheries for bio-economic viability. Fish Res 140:46–62. https://doi.org/10.1016/j.fishres.2012.12.005

Greer K (2014) Considering the 'effort factor' in fisheries: a methodology for reconstructing global fishing effort and CO2 emissions, 1950–2010. Master of Science, University of British Columbia

Hardy P-Y, Béné C, Doyen L, Schwarz A-M (2013) Food security versus environment conservation: a case study of Solomon Islands' small-scale fisheries. Environ Dev 8:38–56. https://doi.org/10.1016/j.envdev.2013.04.009

Hirsch Hadorn G, Bradley D, Pohl C et al (2006) Implications of transdisciplinarity for sustainability research. Ecol Econ 60:119–128. https://doi.org/10.1016/j.ecolecon.2005.12.002

Ishimura G, Herrick S, Sumaila UR (2013) Stability of cooperative management of the Pacific sardine fishery under climate variability. Mar Policy 39:333–340. https://doi.org/10.1016/j.marpol.2012.12.008

Jentoft S, Chuenpagdee R, Barragán-Paladines M-J (2017) The small-scale fisheries guidelines – global implementation. Springer, Amsterdam

Kraan M (2011) More than income alone: the anlo-ewe beach seine fishery in Ghana. In: Jentoft S, Eide A (eds) Poverty mosaics: realities and prospects in small-scale fisheries. Springer, Dordrecht, pp 147–172

Kronen M (2004) Fishing for fortunes?: a socio-economic assessment of Tonga's artisanal fisheries. Fish Res 70:121–134

Lam VWY, Sumaila UR, Dyck A et al (2011) Construction and first applications of a global cost of fishing database. ICES J Mar Sci J Cons 68:1996–2004. https://doi.org/10.1093/icesjms/fsr121

Lara J, Guevara-Sangines A (2012) Apoyos al sector pesqueroo en Mexico 2005–2011. WWF, Mexico

Lasch R (2005) Technical, operational and economic characteristics of selected fishing vessels by country. Econ perform fish Effic mar capture fish 53–76

Lery J-M, Prado J, Tietze U (1999) Economic viability of marine capture fisheries: findings of a global study and an interregional workshop. FAO, Rome

Lluch-Cota SE, Aragón-Noriega EA, Arreguín-Sánchez F et al (2007) The Gulf of California: review of ecosystem status and sustainability challenges. Prog Oceanogr 73:1–26. https://doi.org/10.1016/j.pocean.2007.01.013

Majluf P, De la Puente S, Christensen V (2017) The little fish that can feed the world. Fish Fish 18:772–777. https://doi.org/10.1111/faf.12206

Mauser W, Klepper G, Rice M et al (2013) Transdisciplinary global change research: the co-creation of knowledge for sustainability. Curr Opin Environ Sustain 5:420–431. https://doi.org/10.1016/j.cosust.2013.07.001

Maynou F (2014) Coviability analysis of Western Mediterranean fisheries under MSY scenarios for 2020. ICES J Mar Sci J Cons 71:1563–1571. https://doi.org/10.1093/icesjms/fsu061

McClanahan TR, Marnane MJ, Cinner JE et al (2006) A comparison of marine protected areas and alternative approaches to coral-reef management. Curr Biol 16:1408–1413. https://doi.org/10.1016/j.cub.2006.05.062

Milazzo M (1998) Subsidies in world fisheries a reexamination. World Bank, Washington, DC

Mora C, Myers RA, Coll M et al (2009) Management effectiveness of the world's marine fisheries. PLoS Biol 7:e1000131. https://doi.org/10.1371/journal.pbio.1000131

Naylor RL, Goldburg RJ, Primavera JH et al (2000) Effect of aquaculture on world fish supplies. Nature 405:1017–1024. https://doi.org/10.1038/35016500

OECD (2006a) Agricultural and fisheries policies in Mexico. Organisation for Economic Co-operation and Development, Paris

OECD (2006b) Financial support to fisheries: implications for sustainable development. Organization for Economic Co-operation and Development, Paris
Olaiz-Fernandez G, Rivera-Dommarco J, Shamah-levy T et al (2006) Encuesta Nacional de Salud y Nutricion. Instituto Nacional de Salud Publica, Cuernavaca
Pauly D, Christensen V, Guénette S et al (2002) Towards sustainability in world fisheries. Nature 418:689–695. https://doi.org/10.1038/nature01017
Péreau J-C, Doyen L, Little LR, Thébaud O (2012) The triple bottom line: meeting ecological, economic and social goals with individual transferable quotas. J Environ Econ Manag 63:419–434. https://doi.org/10.1016/j.jeem.2012.01.001
Punt AE, MacCall AD, Essington TE et al (2016) Exploring the implications of the harvest control rule for Pacific sardine, accounting for predator dynamics: a MICE model. Ecol Model 337:79–95. https://doi.org/10.1016/j.ecolmodel.2016.06.004
Ramírez-Rodríguez M, Almendárez-Hernández LC (2013) Subsidies in the jumbo squid fishery in the Gulf of California, Mexico. Mar Policy 40:117–123. https://doi.org/10.1016/j.marpol.2013.01.005
Roux DJ, Nel JL, Cundill G et al (2017) Transdisciplinary research for systemic change: who to learn with, what to learn about and how to learn. Sustain Sci 12:711–726. https://doi.org/10.1007/s11625-017-0446-0
Salas S, Bjørkan M, Bobadilla F et al (2011) Addressing vulnerability: coping strategies of fishing communities in Yucatan, Mexico. In: Jentoft S, Eide A (eds) Poverty mosaics: realities and prospects in small-scale fisheries. Springer, Dordrecht, pp 195–220
Schorr DK (2005) Artisanal fishing: promoting poverty reduction and community development through new WTO rules on fisheries subsidies. UNEP, Geneva
Schuhbauer A (2017) The economic viability of small-scale fisheries. Dissertation, University of British Columbia
Schuhbauer A, Sumaila UR (2016) Economic viability and small-scale fisheries — a review. Ecol Econ 124:69–75. https://doi.org/10.1016/j.ecolecon.2016.01.018
Schuhbauer A, Chuenpagdee R, Cheung WWL et al (2017) How subsidies affect the economic viability of small-scale fisheries. Mar Policy 82:114–121. https://doi.org/10.1016/j.marpol.2017.05.013
Singleton RL, Allison EH, Le Billon P et al (2017) Conservation and the right to fish: international conservation NGOs and the implementation of the voluntary guidelines for securing sustainable small-scale fisheries. Mar Policy 84:22–32. https://doi.org/10.1016/j.marpol.2017.06.026
Smith W, Bizzarro J, Cailliet G (2009) The artisanal elasmobranch fishery on the east coast of Baja California, Mexico: characteristics and management considerations. Cienc Mar 35:209–236
Sumaila UR (2010) A cautionary note on individual transferable quotas. Ecol Soc 15:36
Sumaila UR, Pauly D (2006) Catching more bait: a bottom-up re-estimation of global fisheries subsidies. Fish Cent Res Rep 14
Sumaila UR, Marsden AD, Watson R et al (2007) A global ex-vessel fish price database: construction and applications. J Bioecon 9:39–51. https://doi.org/10.1007/s10818-007-9015-4
Sumaila UR, Khan AS, Dyck AJ et al (2010) A bottom-up re-estimation of global fisheries subsidies. J Bioecon 12:201–225. https://doi.org/10.1007/s10818-010-9091-8
Sumaila UR, Lam VWY, Miller DD et al (2015) Winners and losers in a world where the high seas is closed to fishing. Sci Rep 5:8481. https://doi.org/10.1038/srep08481
Sumaila UR, Lam V, Le Manach F et al (2016) Global fisheries subsidies: an updated estimate. Mar Policy 69:189–193. https://doi.org/10.1016/j.marpol.2015.12.026
Swartz W, Sumaila UR, Watson R (2013) Global ex-vessel fish price database revisited: a new approach for estimating 'missing' prices. Environ Resour Econ 56:467–480
Teh LCL, Sumaila UR (2013) Contribution of marine fisheries to worldwide employment. Fish Fish 14:77–88. https://doi.org/10.1111/j.1467-2979.2011.00450.x
The World Bank (2013) Fish to 2030: prospects for fisheries and aquaculture. The World Bank
Tietze U, Lery J-M, Lasch R (2001) Techno-economic performance of marine capture fisheries and the role of economic incentives, value addition and changes of fleet structure, Findings of a global study and an interregional workshop. FAO, Rome

Too Big To Ignore (2014) Workshop on economic viability of small-scale fisheries in Africa, Oceanic Bay hotel, Bagamoyo, Tanzania, 23–25 April 2014. http://toobigtoignore.net/wp-content/uploads/2014/03/EconomicViability_SSF_WorkshopReport_Tanzania_April2014.pdf. Accessed 3 Apr 2017

Ünal V, Franquesa R (2010) A comparative study on socio-economic indicators and viability in small-scale fisheries of six districts along the Turkish coast. J Appl Ichthyol 26:26–34. https://doi.org/10.1111/j.1439-0426.2009.01346.x

United Nations (2016) Progress towards the sustainable development goals

Chapter 7
Gender Perspective in Fisheries: Examples from the South and the North

Katia Frangoudes and Siri Gerrard

Abstract Women have an active role in fisheries and aquaculture all over the world where fisheries activities related to resources, like fish, shell, and seaweed, take place. Women's participation in fisheries is diverse as they are involved in different ways depending on the cultural, social, and material conditions. In Western areas, women's contribution has mostly been performed on land, while in Southern areas, more examples of women fishing or collecting shells are found. However, everywhere, women in fisheries have either fewer rights than men or completely lack formal rights and political attention. Knowledge about women's roles, gender relations, women's ways of life in fishing areas, and the changes women and men face, is scarce and varied but necessary in order to conduct transdisciplinary research. Studies within social and cultural fishery research, most often carried out by means of qualitative methods, try to get deeper into women's lives, their actions or practices, identities, their relations with men, and how women and men as categories are constructed, most often within a specific timespan and places. Constructing gender and gender relations differs from gender as a variable in the fields of fisheries and aquaculture. Both approaches are applied, but highlight different aspects and gender issues in fishery-related societies. An integrated gender dimension in transdisciplinary approaches is necessary to achieve sustainability of resources and society, by bringing together scientists working on different areas and different stakeholders to find responses to major problems.

Keywords Gender relations · Small-scale fisheries · Transdisciplinary approach · Feminist research · Gender equality

K. Frangoudes (✉)
Univ Brest, Ifremer, CNRS, UMR 6308, AMURE, IUEM, 29280 Plouzané, France
e-mail: katia.frangoudes@univ-brest.fr

S. Gerrard
The Artic University of Norway, KVINNFORSK, UiT-The Arctic University of Norway, Tromsø, Norway
e-mail: siri.gerrard@uit.no

R. Chuenpagdee, S. Jentoft (eds.), *Transdisciplinarity for Small-Scale Fisheries Governance*, MARE Publication Series 21,
https://doi.org/10.1007/978-3-319-94938-3_7

7.1 Introduction

This chapter brings the reader into the literature about gender in fisheries and fisheries communities. Its objective is to provide more insight into worldwide fishery knowledge about women's tasks and practices, their roles, their relations to men, their situations in different places or contexts, and some of the changes they have faced. Gender research is an important component for transdisciplinary research, which has as a principle the integration of all scientific disciplines to solve societal problems (Lang et al. 2012). Equal rights between women and men within fisheries industry, communities, and households is one such problem. Gender equality is also a part of sustainable fisheries.

The chapter starts by presenting general narratives on women related to fisheries, the international conventions concerning gender, and efforts applied by development agencies and civil society. Then it develops a few examples of perspectives, concepts, and methods applied in social and cultural fisheries research. It continues with examples of gender-based ethnographies from around the world. It exemplifies women's participation in household work related to fisheries, pre-harvesting, harvesting, selling and marketing, and processing, often taking place in specific communities or contexts, with the objective of highlighting the differences that women in fisheries experience compared to men. The conclusion develops the challenges that women and men in fisheries face today, and how the knowledge of gender studies should be integrated in transdisciplinary research.

7.2 Searching for Women's Visibility in Fisheries: Feminist Research and International Legal Framework

Women in fisheries have been marginalized mainly because they were invisible for the outside world. Women's work in small-scale fisheries was given little attention since they seldom own vessels or work on board fishing vessels, and thus only participate in harvesting to a small degree (e.g. Gerrard 1983; Nadel-Klein and Davis 1988; Porter 1993). This image of women's participation in fisheries dominated in the public perception for many years, represented by fisheries scientists, civil servants, fisheries managers, and others who focused on fish stocks and fish production conducted predominately by men (Williams 2008). They seldom paid attention to the cultural and social structure. This outsider view that made women in fisheries invisible still exists.

People living in fisheries and resource-dependent communities in Western and Southern countries know that women of fishing households contribute to different tasks in fisheries, for example preparing bait for longlines, repairing nets, harvesting, selling, marketing and processing fish, shellfish, or seaweed. Usually, the women themselves and the communities see this contribution as part of women's household tasks and not as a permanent paid job. (Gerrard 1983; Frangoudes et al.

2008, Frangoudes 2011; Hitomi 2009). The main contribution of women in small-scale fisheries has been carried out on land, except where they harvest fish, and gather shellfish and seaweed near the shore (e.g. Frangoudes et al. 2008). Therefore, women's activities on or close to the shore may contribute to their invisibility.

At the same time, social and cultural scientists (sociologists, anthropologists, historians, ethnologists) studying the sexual division of labor within fishing households and communities have shown the importance of women's contribution to fisheries activities. Many spoke about women's involvement in fish harvesting near the shore in the South Pacific islands (e.g. Bataille-Benguigui 1992; Kronen and Vunisea 2009), and in Europe (e.g. Bratrein 1976; Hoefnagel 1991; Meltzoff 1995; Cabrera 1997; Willson 2015). Others study the role of women in marketing and processing fish, especially in West Africa (e.g. Weigel 1987; Overå 2000; Thorpe et al. 2014). There is also documentation on women mending fishnets, helping cleaning and maintaining the boat (e.g. Gerrard 1975; Escallier 2014), renting out rooms, cooking, and washing the fishermen's work clothing (Balsvik 1989). Women's contributions were studied as part of the household activities or paid piecework. Nevertheless, such work was important for fishers' efficient operations and sustain livelihoods (Gerrard 1983). The main impression from Western countries is that men are onboard the vessel, while women are taking charge of the tasks on land (Nadel-Klein and Davis 1988). However, when the men were absent or dead, women were fishing and could even be skippers of bigger boats (Bratrein 1976).

Women in fisheries became a research theme thanks to feminist researchers who focused on gender equality and women empowerment within households, communities, employment, and participation in decision-making processes, and on the lack of women related issues in fishery politics (Gerrard 1975, 1983; Neis 1993; Porter 1993; Skaptadottir 2000). They focused on women's working roles and relations, often in challenging situations, using qualitative methods and analyzing such important components in fishing, processing, selling and marketing, often in the context of specific local communities. Concepts like *women as the ground crew* or shore crew, *mothers at the helm,* and *the veiled crew* were developed (Gerrard 1983; Porter 1993; Jentoft 1989; Thiessen et al. 1992). The masculine and patriarchal character of the Western fisheries was pointed out later (Munk-Madsen 1998, 2000; McGrath et al. 1995; Power 2005). Some of these studies reveal women's and men's situations, relations, and identities, and are excellent examples demonstrating how gender is constructed in different global contexts. Later on, following this early research, other researchers identified and studied women's roles and relations in fisheries and fisheries communities and expanded worldwide, based on qualitative and quantitative data (Neis et al. 2005; Frangoudes 2011; Harper et al. 2017; Kleiber et al. 2015).

7.2.1 International Legal and Institutional Frames Securing Women and Gender Equality

Research in developing countries (e.g. Boserup 1970), uncovered early a strong male bias in development aid. Slowly, international agreements and international institutions, like the United Nations or the World Bank, have increased the awareness of women and 'gender' equality. The UN Convention on Elimination of All Forms of Discrimination Against Women (UN-CEDAW), adopted by the UN Assembly in 1979, is one example of the promotion of equal rights between women and men. UN-CEDAW defines discrimination against women as

> any distinction, exclusion or restriction made on the basis of sex which has the effect or purpose of impairing or nullifying the recognition, enjoyment or exercise by women, irrespective of their marital status, on a basis of equality of men and women, of human rights and fundamental freedoms in the political, economic, social, cultural, civil or any other field (UN 1979).

UN-CEDAW is relevant for strengthening the contribution of women's participation in fisheries, especially after the Beijing conference and its action plan in 1995. The signatory countries are committed to promoting equal opportunities for women and men and acting against discrimination against women. Other agreements linked to fisheries and oceans ensure women's rights, for example the Convention of Biological Diversity (1992), reiterating the need for the participation of women at all levels of policy making, and the UN Fish Stocks Agreement, adopted in 1995, calling on states to ensure access to subsistence fisheries for artisanal and small-scale coastal fishers and woman fishworkers. Other international agreements secure women rights, for example, the UN Agenda 21 (Chap. 17), the International Labor Organization (ILO) that protects women's rights to safe working conditions at the workplace or at home, the Millennium Declaration, and the Millennium Development Goals signed in 2000, referring to the promotion of gender equality and the empowerment of women. The Voluntary Guidelines for Securing Sustainable Small-scale Fisheries (SSF Guidelines; FAO Overå 2014) is a concrete tool that can improve women's situations, if applied. This text does not only include gender equity and equality as one of its 13 guiding principles but also elaborates on gender in detail, throughout (Kleiber et al. 2017).

All these conventions demonstrate changes in the basic ideas behind development strategies, from Women in Development (WID) in the 1970s to Gender and Development (GAD) in the 1990s (Razavi and Miller 1995). In 1997, the United Nations Economic and Social Council introduced the idea of gender mainstreaming. In 2016, the 5th goal of Sustainable Development Goals is focusing on gender equality: "Achieve gender equality and empowerment for all women and girls" (UN n.d.)

Such international agreements call for more attention to women, gender, and gender equality in general, with agreements related to women in fisheries opening up new horizons for research. Gender equality is a principle enshrined in legislation and politics of many countries from the South (e.g. Mexico, Chile) and the North (e.g. European Union (EU), Canada, Norway). Gender equality today means equality between all genders and it is based on the principle that all humans are free to develop their personal abilities and make choices without being limited by stereotypes, prejudices, and rigid gender roles. There should be no discrimination on the grounds of gender in the allocation of resources or benefits, or in the provision of, access to, and use of services. It also means that the different behaviors, aspirations, and needs of men and women are considered, valued, and favored equally. Gender equality is measured in terms of equality of opportunities.

Now, a new generation of scholars interested in new dimensions of women and gender in fisheries comes from several academic disciplines, including the traditional fishery disciplines, like marine biology and economy, or other interdisciplinary academic traditions. Examples of this new research are the evaluation of unreported catches of women, the economic value of women's unrecognized contribution, women's place in the value chain, and topics on women and climate changes or disasters. (Harper et al. 2017; Kleiber et al. 2017). Other studies treat women and men as variables of major themes, like disasters, biodiversity, and aquaculture (e.g. Gopal et al. 2016, 2017). Many studies and projects using gender analysis have been conducted in different continents because international funders, development agencies, and non-governmental organizations (NGOs) often demand the inclusion of gender.

7.3 Conceptualizing and Constructing Gender in Contexts and Empowerment

Gender analysis is "the systematic gathering and examination of information on gender differences and social relations in order to identify, understand, and redress inequities based on gender" (Reeves and Baden 2000, 2) As such, gender is a descriptive and diagnostic tool that serves to introduce a gender perspective into development fisheries projects and activities.

Defining gender is the first step when we study women's practices, situations, and gender relations in fisheries. Gender is seen as a relational concept referring to socially constructed identities, roles, and relations of women and men, as opposed to sex, which refers to the biological characteristics of women and men. While sex is biologically determined and carries no implications for how a person is perceived or acts in society, gender is multidimensional, referring to a continuum where gender is related to specific, social, and cultural contexts, involving gender practices, roles, and identities, in other words revealing how gender is defined, perceived and lived.

Depending on the historical or political context, different ideas about social and economic behavior, expectations, and responsibilities are associated with women and men. These preconceptions shape power relations, patterns of decision-making power, social exclusion and inclusion, and the norms that underpin social, economic, and political inequalities. Gender reveals the status of women; gender practices, roles, and identities are constructed by hierarchies and power relations between women and men. Gender relations can give rise to social patterns where women or men can end up in a subordinate or adverse position (Munk-Madsen 1998). The constructed character of gender indicates that gender is neither static nor unchangeable. Conceptions of women and men are subject to change over time. What is supposed to be feminine or masculine is historically and contextually defined (Power 2005; Gerrard 2013).

Gender also has to be studied in relation to other social phenomena, since women and men are not homogeneous groups. For instance, social status, ethnic origin, or skin color can be a deciding factor together with gender. Intersectionality is a concept that emphasizes how power relations, rather than being based on additive principles, should be understood as dynamic interactions (Collins 1998; Crenshaw 1989, 1997, 2005). Rather than examining gender, race, class, and nation as distinctive hierarchies, intersectionality examines how they mutually construct one another (Collins 1998).

7.3.1 Gender Empowerment: What to Measure

Gender empowerment is the enhancement of assets and capabilities of diverse individuals and groups to engage, influence, and hold accountable the institutions that affect them. For many years, feminists claimed that women's empowerment was necessary to achieve gender equality. Empowerment as a basis for development programs first succeeded when the international institutions, for example the World Bank, understood that women's empowerment could be operationalized through some indicators measuring women's capabilities, like fertility and demographic transition, children's wellbeing, economic development, and poverty alleviation (Malhotra et al. 2002). Policies and development studies, often applying quantitative methods, emphasized impacts on specific interventions on women's empowerment and comparison between countries.

Feminists and development theorists have contested and questioned such an approach. For feminists, the value of the empowerment concept is based on its indistinctiveness. Any attempt to measure empowerment should be discussed and clarified from the beginning. At least, it is important to find out what is possible and impossible to measure. Naila Kabeer focused her work on the issue of empowerment and its measurement. For her, the concept of empowerment deals with the possibility, or being denied the possibility, to make choices. In her thinking, people who exercise a great deal of choices in their life may be powerful but not necessarily empowered because they never have been disempowered. Analysis of power requires the definition of the concept of *choice*. Choice implies possibility of alternatives and the ability to choose otherwise. The ability to exercise choices incorporates three interrelated dimensions: resources (access and future claims to material, human, and social resources); agency (decision making, negotiation); and achievements (wellbeing, outcomes) (Kabeer 2002).

7.4 Women in Fisheries – A Few Examples

For researchers, development agencies, NGOs, and others, gender analysis can be the first step to observe the tasks men and women perform within the fishing households and in fisheries pre-harvesting, harvesting, and post harvesting activities. It is important to describe women's situations in relation to men, as well as exploring and understanding inequalities and discrimination between men and women, and their causes. Identifying inequalities and their causes can help target objectives for resolving the gender differences and design actions that can contribute to eliminating or decreasing them. It is necessary to have disaggregated data by sex (if available), and show how assets and resources are distributed. If such data are lacking, we must gather them, perhaps in a more qualitative way, and try to explain the gender differences related to the specific fishery tasks we want to highlight. In relation to power, one can look at women's and men's participation in decision making, resources management, and policy making for everyday life in the households.

7.4.1 Pre-harvesting and Household Work

Historically, in Northern countries, putting bait longlines, and making and mending nets were carried out on land by women and old fishermen. These activities can be paid or unpaid, depending on location and the situation of the persons involved. For example, women who belong to a fishery household as a fisher's wife rarely get paid (Frangoudes et al. 2008; Hitomi 2009). For them, these tasks are part of household activities. However, women in Brittany mending nets for the family boat were able to tell how much time they devoted to this task because of the real difference between domestic tasks and others tasks (Frangoudes and Keromnes 2008; Frangoudes 2013). In other places, such as Galicia (Spain) and Peniche (Portugal), women mending nets are paid (Frangoudes 2011), but remuneration is low. In countries like Norway and France, this work is mechanized and often done in factories. In Hatiya Islands, Bangladesh, divorced or isolated women collectively construct nets for local boats. Making a net requires 1 month of work for ten women, and thus this job provides a small salary to assure their subsistence. In the Azores, the fabrication of longlines provides paid jobs to local women. The number of women employed by fishers depends on the number of longlines the fishers deploy (Neilson et al. 2014). The shift to hand lines observed recently may leave women jobless, with little alternatives. These examples demonstrate that women perform important pre-harvesting tasks. Payment can vary depending on local practices. Other common aspects are the low remuneration and the lack of other jobs alternative within fishing communities.

Women's contributions in fisheries are often combined with work for the household members. In the lake fisheries in Cameroon, women in Muslim households are responsible for preparing food, while men often bring the fish or the meat. Women are also responsible for taking care of the children. They often supply the household economy by growing important vegetables and produce that is sold outside the house by the children (Gerrard 1996). In this way, they support the household economy, which is especially important when the catches are bad. In the past, the same gender pattern was found in coastal areas of Norway, where the fishers' wives took care of the children, sowed, knitted the clothing, and were responsible for the animals and the little farm. So, women's work was crucial for the household (Gerrard 1975; Bratrein 1976).

7.4.2 Harvesting Activities, Property Rights, and Management of Territories and Resources

Harvesting with vessels is men's activity. This male pattern can be explained in many ways, either by religious believes, history, or culture taboos. In France, being a fisher was a privilege granted by the kings to men in the Navy. These soldiers on land were allowed to fish and, since the nineteenth century, to farm shellfish. This tradition lasted until the mid-twentieth century. Thus, women were excluded from both activities. Wives or widows of soldiers were allowed to gather fish, shellfish, and crustaceans from the shore to feed children when the husbands were absent. The distinct gender division between work is also associated with cultural taboos that prevent women from working at sea; for example, the idea that water should never reach the knees of women (Hoefnagel 1991). In Norway, France, and Greece, women's presence on a boat meant bad fishing luck. Getting a big halibut was, on the contrary, associated with good luck and related to having slept with the wife during the stay on shore. As cultural taboos vary from place to place, it is important to identify such taboos or explanations.

Nowadays, few women are boat owners or crew members onboard. Around Europe, like in Norway and France, there are examples of women shareholders and managers in fishing companies with large vessels, in some cases inherited from their fathers. In Cornwall, a woman is managing a fishing company with a number of small-scale fishing vessels. These women are often daughters of former fishers who expanded their fishing activity into a vessel owning company, often with many vessels (Nilsen 2008). The same tendencies are seen in some of the aquaculture companies in Greece, Norway, and France. The number of female fishers is unfortunately unknown as only few countries produce sex-disaggregate data (Chile, Brazil, Japan, and Norway). In others countries, if such data are available, they are difficult to find or are incomplete (Frangoudes et al. 2008; Frangoudes 2013). Thus, the number of women appearing in statistics is usually lower than in reality. (Harper et al. 2017).

Gendered information on quota ownership is interesting because the size and number of quotas tell something about wealth and power. Such information is difficult to get even when available. A gender analysis of quota allocation will proba-

Box 7.1 Gender Inequalities and Quotas: The Case of Norway

In Norway, quotas, initiated in 1990 for the coastal fishing fleet, are today organised in a closed and open group. Full-time fishers that had fished cod of the 62-degree north were allocated the quota in the closed group for free. Part-time fishers were allocated quotas in the open group. Women coastal fishers belong to the open group because they do not have the necessary capital required to buy a quota in the closed group (Gerrard 2016). Women and men listed in the national fishery register, owning boats listed in the national fishing vessel register, can become fishers in the open group. Because women often meet other obstacles related to their roles as women and fishers there are still hurtles to jump before we can speak about equality in fishing. In 2018, special rules were made for fishers in the open group stating that these fishers could not hire a skipper to run the boat. Exceptions were made in case of illnesses and political meetings. When a young mother in her maternity leave complained the rules were changed. However, maternity leave allowance does not provide enough income for many when it comes to paying bank loans. This example shows how inequalities between women and men are persisting in fisheries even in countries where gender equality is part of the constitution (Gerrard 2016).

bly show that fisherwomen have few and small quotas compared to men, probably because they enter into fishing later. Box 7.1 shows an example of gender inequalities in accessing such quota systems.

7.4.3 *Women in Harvesting*

Harvesting fish, and gathering shellfish and seaweed near the shore are activities that women dominate (Kleiber et al. 2015). In many cases the catch is for family subsistence, but in some contexts women may also sell their catch. Throughout the South Pacific women and children fish in the intertidal area (Chapman 1987). For example in Tonga, "gathering is a female activity" (Bataille-Benguigui 1992) and women only fish to satisfy family consumption needs In the Philippines, women also dominate shellfish gathering in inertial habitats (Kleiber et al. 2015). In Zanzibar, women gather seaweed on shore and engage in algae farming (Torre-Castro et al. 2017).

In Galicia, women gather shellfish on foot. These women were recognized as professional in the mid-1990s. Ninety percent of shellfish gatherers are women, while men harvest shellfish on boats. Women shellfish gatherers succeeded in regulating their activity through a licensing system that provided each permit-holder with a quota (Frangoudes et al. 2013). Resources are managed by a women's organization with the support of scientists within the territory allocated by the fisheries authorities. Each year, they produce a management plan and submit it for validation to the regional fisheries authorities. This became possible as soon as the women, through vocational training, understood the benefits attached to the status of professional gatherers (Frangoudes et al. 2013).

In Japan, women dive to harvest abalones or seaweed. They are not recognized as professional fishers, and consequently do not have the right to use boats to reach the fishing grounds. Fishers (husbands or otherwise) bring them to the harvesting area, and they pay these men with a percentage taken from the sale of abalones. Usually one boat can carry between four to eight women. As a result, men's income is often higher than women's income. Men are members of fisheries cooperatives (FCA) and discuss abalone management while women who are the harvesters cannot participate because they are not members of the FCA. Only one member from each family is allowed to be a member and it is always the man as explained in Box 7.5 (Miki 2009; Lim et al. 2012) (Box 7.2).

Box 7.2 Seaweed Harvesters Gulf of Mannar, India

Women and men living in fishing communities around the Gulf of Mannar were dependent on seaweed harvesting for their livelihoods. In the past, women were hiring a catamaran to go the islands where they harvest by diving. The establishment of The National Park in 1986 put an end to this activity because all harvesting under the Indian law was forbidden in the area they had used. Restrictions of seaweed harvesting started in 2000 because the authorities thought women's diving was destroying coral. Seaweed harvesting was the only source of income for these women and they resisted the decision by establishing new management measures, for example reducing the number of harvesting days and not using modern equipment, such googles and gloves. Despite all these efforts "women seaweed collectors face harassment on a regular basis as seaweed collection around the islands is officially forbidden" (Yemaya, n°24 March 2007). Since this first period, women's harvesting is restricted. Now, they harvest only 12 nights per month. Public authorities also imposed a ban for 45 more days.

In 2014, following training organized by the International Collective in Support of Fishworkers (ICSF), women formulated new claims including integrating the welfare scheme and being compensated for the 45-day ban imposed by the authorities. But, the fisheries department allowed only male heads of families to get such social benefits, since these benefits should cover the needs of the family. This thinking shows that women are dependent on men. After negotiation, the Tamil Nandu State decided to recognize women seaweed harvesters as fishers; the fisheries department provided them with an identity card, and provided protective equipment, such gloves and scissors to cut seaweed. This example shows how the implementation of Marine Protected Areas impact marginal groups, such as women, who lost their rights to the resource. It shows how women resist unfair decisions and how training and support can emancipate women (Yemaya, n°24 March 2007, ICSF).

Women in many coastal areas in the world are the main harvesters of resources located close to the shore. These activities can be covered by formal rights. This happens in some countries when women build up their capacity, build organizations, and manage the resources by following vocational training, if available. The Japanese example shows that in places that women are not integrated in resources management, building capacity to claim equal rights is still difficult.

7.4.4 Post-harvesting Activities

Most women are involved in post-harvest, selling, or processing fish or fish produce. This work can be paid or unpaid depending on the countries and how the work is carried out. From an individual or artisanal level, these two activities can become more industrial, especially when it concerns processing (Gerrard 1986; Frangoudes 2011). A few examples from Africa, Europe, and Asia will illustrate different types of work and show women's participation in the two activities, as well as their concerns and opportunities.

7.4.4.1 Selling and Artisanal Processing of Fish

From West to East Africa, South and East Asia, Mexico, and Europe, women are important agents in the marketing and selling of fish and sea products, as well as in artisanal processing. In some countries selling and processing are practiced together. Different forms of selling are observed: direct sale of raw fish to consumers; selling of dry fish in markets; and managing fish markets or big fish companies, where women are a central workforce. Artisanal processing can be an individual activity but as investment is required, often women join forces to find the capital to expand their activity (e.g. Côte d'Ivoire). In other places, policies or development projects targeting fisheries communities' development offer the possibility to women and men to develop individual or collective projects, like in Japan.

Women are the main fish sellers in Ghana and Togo, both located in West Africa. In Ghana, where pirogue fisheries constitute an exclusively male task, women are in charge of selling and processing fish and have a sizeable amount of capital, which is used for credit to fishers to buy boat equipment (Overå 2000). As a result, women have exclusive rights to the catches that are landed. Some of these women are wealthy with a high degree of social prestige. They belong to matrilineal ethnic groups. In the household, women alone are responsible for their children and the household economy. This could be one factor encouraging women to be efficient fish merchants (Overå 2000). Their capacity to manage and control money illustrates women's economic empowerment. In Togo, a group of women called 'nana Benz' also controls the selling of fish and provides credit for the fishers. Their position in society is recognized through their name, which comes from the French word nana (girls) and the Mercedes-Benz cars they drive (Weigel 1987). Nana Benz women are the main actors of the fisheries sector in Togo because they control the fish marketing, as well as the credit given to fishers.

Women, both from the South and the North who process and sell fish products, are functioning as micro-enterprises and they are self-employed. This form of organisation allows women to perform different roles, such as housekeeping, reproductive activities, and caregiving, maintaining social cohesion in the community, including supplementing the family's income by working inside or outside fisheries. The development of the fish trade and processing can be seen as a household survival strategy usually developed by women to secure a decent livelihood for their families. Women develop survival strategies to cope with environmental degradation or economic impoverishment. Different studies (Medard et al. 2001; Cruz-Torres and McEwee 2012; Peke 2013) underline the main challenges faced by women selling: access to resources (fish, credit to purchase fish, transport to market); protection and maintenance of vending space; and, the lack of organization. Without organizing women cannot claim their rights and participate in decision making. Box 7.3 shows an example of the market of Guadalajara (Mexico) in which women can be successful fish traders.

In artisanal processing women can be workers or business leaders according to the type of organizations in the activity: artisanal or industrial. The three following examples (Boxes 7.4, 7.5 and 7.6) show the reasons to develop this activity in coastal communities. In these examples, fish processing is the only job that women can find especially when they are migrants or immigrants, or to add value to the fish harvested by the community members, or because the survival of the community is at stake. The common pattern in the three examples is that all are collective initiatives requiring the organization of women and men.

Box 7.3 Guadalajara, Mexico

In the fish market of the city, women own 24% of the retails shops and 27% of wholesale companies; 80% of retails and wholesale companies are managed by women. While men negotiate fish prices with fishing cooperatives, women are managing the business and selling fish. Women domination within this fish market is explained by the following: some inherited family businesses; others showed better abilities than men to manage money and the family enterprise and gained their parents' confidence. Women's capacity to manage the family company allows them to lead the association that manages the market. Thanks to better education and parents' confidence, women have developed entrepreneurial capacities.

Author: C. Pedroza, Universidad Nacional Autónoma de México

Box 7.4 Federation of Women Processors and Fishmongers in Côte d'Ivoire

In Abidjan, fish drying is an activity conducted by migrant women who are heads of households. They have often accepted bad working conditions. After years of discussion, in 2010 migrant women established an association to fight for better working conditions and their inclusion in local society. Their main aim is to access fish resources during the low fishing season, including access to some pelagic fish that the European Union vessels under bilateral fishing agreement fish in the area, and also to have access to imported frozen small pelagic fish from Mauritania or Senegal. The access to the fish of EU vessels was negotiated by the national federation of women and now some of the fish is supplied at the local market.

Adapted from Yemaya n°47 2014, ICSF

Box 7.5 Fisherwomen Take Over Japanese Government Initiative to Empower Themselves

Following the fourth world conference on women held in Beijing in 1995, Japanese authorities encouraged the establishment of women groups in all economic sectors, including fisheries. The objective of these groups was the promotion of women entrepreneurship. By 1998, 1158 women's fisheries groups operated under the authority of Fisheries Cooperative Association (FCA), the main organization representing small-scale fisheries interests.

The members of these groups were mainly wives of small-scale fishermen who wished to increase their family income through fish processing and marketing, including delivery to restaurants and at-home meals. One obstacle faced by many of the groups was in marketing the new fish products they were producing. Working with academic researchers in 2003 these women established the Japanese forum of fisherwomen groups: UMI HITO KURACHI (Sea People Life). With the network women's capacity for entrepreneurship and marketing improved.

Since 2003, some women's groups have quit because they felt FCA was an obstacle to innovative functioning and to new markets development. The split from the FCA offered women the opportunity to work more effectively and to escape the paternalistic influence of male organizations. Now women are wishing to establish their own independently of the FCA because only men representing fishing enterprises can be full individual member of cooperatives.

Author: K. Soejima, National fisheries University, Shimonoseki.

Box 7.6 Processing Groups in Thailand
Inspired by 'One Village One Product' program in Japan, Thailand launched its national campaign '*One Tambon One Product*' (OTOP) in 2001 to promote local entrepreneurship and community development. The program has been running successfully with more than 7000 districts participating, each attempting to offer products that are unique to their areas, focusing particularly on improvement of quality and marketing. The Thai Department of Fisheries, through its Fishery Development Division, has also been supporting the OTOP program by promoting 82 fish processing groups in 45 provinces in coastal and upland areas. Women are the key actors and members of many of these groups and wives of fishers get together to develop products from freshwater and the sea. The description of these programs tells colourful stories about the importance of women in supporting the local economy and livelihoods of their households. The composition and objectives of the groups are evidenced in the Pak Nam Lang Suan Fishers' Group of Chumphon Province, southern Thailand: about 70 members are part of this group and the majority of them are fishers' wives. They work together to produce traditional products like roasted shrimp with chilli mix and mackerel cracker as leisure time activities.

Author: R. Chuenpagdee, TBTI

7.4.4.2 Women in Industrial Processing

In Iceland, Norway, France, and Portugal, women were important members of workforces to process fish harvested by local vessels. Filleting fish and canning sardines employed women from fishing communities. After the introduction of the ITQ system in Iceland, the processing of fish was centralised in urban areas. Many workers came from countries in Asia or from the Baltic area as Icelandic women with high education tried to avoid such jobs that are considered 'easy' or 'unskilled'. Men's jobs were defined as 'difficult' or 'skilled' (Juliusdottir et al. 2013). Based on this principle, women earn lower salaries compared to men. Many processing companies and owners of larger fishing vessel today sell their fish to Asia, Balticum, and North Africa, where it is processed by other women before the products are returned to European countries to be sold.

In Southern countries, women had to abandon traditional processing and moved to processing plants working mainly for export. This is the case in many African countries around Lake Victoria, like Uganda, and Tanzania, and countries in West Africa, like the Ivory Coast, and in Asia, like Sri Lanka, India, and Bangladesh. Since the 1990s, processing of Lake Victoria Nile-perch moved from the hands of local populations to big filleting factories. Women from rural areas and men were often hired as factory-workers. Working conditions were hard and women's wages lower than men's. In order to survive, workers were forced to increase their working capacity and working hours. The consequences of this industrial change were different for women workers with a regular salary compared to women practicing traditional processing, who lost their jobs or, for some time, processed the waste from the fish production line and in this way created new jobs (Nakato 2005).

7.5 Women Organizations and Participation in Decision Making

Despite their relative invisibility within the fisheries sector, fisherwomen formally organize themselves and claim their rights within the sector and outside. Women organizations can be present at all levels, local (e.g. Spain, France, Italy), national (e.g. Australia, Japan, Spain, France, Norway, Brazil), and also in transnational organizations, for example in the EU and in the Caribbean (Alonso-Poblacion and Siar 2018). In Europe, women's organizations started fighting for the defence of the fishing industry (France), communities and families, and safety at sea (Norway), moving later to claims related to women's rights and sustainable fisheries. In France, women advocated for a legal recognition of their invisible contribution, for training and education to satisfy their need to participate in enterprises, and resource management. These organizations brought their claims into the public arena and became one of the players in the decision-making process, not only in matters of women's rights but also in social issues concerning the fishing industry. European fisherwomen's organizations can be divided into three categories: those regrouping women involved in harvesting and ancillary activities; those involved in fisheries enterprise management; and, wives of crew members. Each of these entities defined its own functioning, usually based on the history of the fishing sector and on the law of each country. In some countries, women's organisations are part of male organizations; others are independent.

In Norway, women were organized in the Fisherwomen's Association linked to the Fishermen's Association. They worked to improve fishermen's safety and together with other women's associations tried to improve welfare at national and community levels. They did not engage in fishery policies, but many fishers' wives, especially in the northern Norway, were active in political networks and actions against the quota policy, concentration of ownership of the trawlers, and the outsourcing of tasks and jobs (Gerrard 2016). In 2018, a network for women fishers was established in order to improve their situation as women professional fishers. Some of them unionized in the established organizations (Gerrard 2018).

The Gender in Aquaculture and Fisheries is an international network regrouping scientists, managers, and practitioners related to gender in fisheries. It holds a regular international congress in gender in fisheries and updates its website (http://www.genderaquafish.org/). In later years, networks and organizations often recruit highly educated women working in aquaculture, sales, and consultancy to arrange meetings, conferences, and to produce information material, especially in Western countries. Women in Seafood Industry (https://wsi-asso.org/) and Women in Seafood (http://intrafishevents.com/womeninseafood/) are examples presented in Boxes 7.7 and 7.8.

At the European Union level, national and local women's organizations built the AKTEA network in 2006. Its objectives are similar to the Australia network, extending to the defense of the fisheries industry and communities. AKTEA achieved some of its objectives through lobbying the European institutions, such as the modification of the EU directive 86/613 establishing the collaborative spouse status by the directive 2010/41/ EU recognizing unpaid contribution, and their entrance to the

Box 7.7 Women's Industry Network: A Seafood Community in Australia
In 1998, women of South Australian created their organization as an inclusive body, not subject to the kind of inter-fishery antagonisms that were perceived to characterize the 'men's committees'. One of the key roles of the network was to recognize the social impact of the policies rarely taken into consideration at that moment. In 2000, the group became national, Women's Industry Network a Seafood Community, and is acting for the recognition and the promotion of the existing contribution of women to the fishing industry, to build the capacity of members and other women in the industry to play a more influential role in governance, and to contribute to positive community engagement and education.

Author: T. King (2018)

Box 7.8 European Union Recognizes Women Unpaid
In 1986, the European Union recognized the unique position of collaborating spouses: the role that a female usually plays in supporting her male partner in his efforts for self-employment. The Council Directive 86/613 called on member states, where a contributory social security system existed, to take measures to ensure that the spouses who assist their partners in their work and who are currently not protected under the self-employment scheme have the opportunity to benefit from its provisions. In 2010, after fisherwomen organizations voiced their claims, a new directive was voted (2010/41/UE) "on the application of the principle of equal treatment between men and women engaged in an activity in a self-employment capacity", with the objective of urging member states to give visibility to women's unpaid contribution. Assisting spouses or life partners of fishers are now able to contribute to social security, which in turn gives them access to social benefits, including pensions, maternity leave, training, and access to professional fishers' organizations. This legal recognition is not simply a matter of social rights; it is also about social recognition of women's contribution in the fisheries.

advisory councils of fisheries at regional seas. This is perhaps one of the most important political victories and gives hopes for others.

The SSF Guidelines contributed to the establishment of new forms of networks or groups promoting gender equality in fisheries. One of them is the Gender in Fisheries Team (GIFT), established in 2016, which includes people interested in gender in Caribbean small-scale fisheries: fisher organization leaders, scientists, NGO staff, inter-governmental organization officers, and others The objective is to introduce the gender dimension in national and regional fisheries policies (GIFT n.d.).

Women's organizations benefit from external support like the European organizations, including AKTEA. This support is given by scientists, NGOs, or national fisheries authorities. All are volunteer organizations and have limited financial resources (Frangoudes et al. 2014). Other organizations like GIFT have been supported by international agreements. Today, all of them act as lobbying organizations for more visibility and rights defense and protection. They have difficulties in accessing the same support as male organizations, even in countries with gender sensitive policies.

7.6 Conclusion

These examples demonstrate that fishery-related gender tasks worldwide are of the same kind, especially in pre-harvesting, harvesting, and processing, even though these tasks can be organized, recognized, and valued in different ways. However, most of the examples reveal that some women go fishing, often close to land, and most women work on or close to the shore, while men fish on their boats. This is no explanation for why women's work has been, and often still, is less valued and less visible. Women's work in the households and communities have been difficult, time-consuming, and taken for granted. Men's fishing and its association with the monetary economy seems to have benefited from a wider interest expressed by the fishers themselves, sometimes through their associations and unions and by external supporters like NGOs and fishery civil servants and politicians. This interest reaches from the local, national, and international level. Fishery development agencies have focused on women as fishers to a small degree. Regulations, laws, and politics seldom take women's pre-harvesting and harvesting tasks into consideration. Despite all that, there are current examples of women being vessel owners. In the North, men and some women in extraordinary situations have had better chances to accumulate capital while women, especially in southern countries, are often the sole breadwinner with little income.

The trading of fish has different consequences in different countries. West Africa provides examples of women being the main and best traders, while East African women's trading conditions have worsened because of the industrialization of factories and filleting production. Local traders lose their jobs while women from rural districts become factory workers, often with lower wages than men. Patterns we know in the Western world seem to be transferred to new settings in Africa and Asia. Further, the recruitment pattern of workers in the Western filleting factories sees women and men from abroad (e.g. from the states near the Baltic Sea, including Poland) replacing local women who have higher levels of education and decided to go into other jobs or migrate to urban areas. Women worldwide go into fishery jobs that require highly skilled people.

Fishery-related women's engagement in politics, organizations, and networking in many parts of the world gives new opportunities and improves women's working and living conditions. This engagement has been shaped by the challenges that women face. Sometimes the solutions are local, for example better working conditions for processers in the Ivory Coast. Sometimes the solutions have a national aim, for example community and capacity building for the members of the FCA groups in Japan, or visualizing women in fishery and governance in Australia. Sometimes the solutions are international, for example getting a gender perspective in fishery planning and increasing knowledge about gender in Caribbean Islands, and increasing the focus and the influence of women and their work at the EU level. This means that their marginal role can, little by little, be changed to acquire more influence and power.

Fishery women's political engagements are materialized in networks, women groups, and formal organizations. If these entities can be maintained and continue their work, and the fishers' associations and official politics include women's special needs and demands, there is hope that women's conditions can be improved and taken into consideration at all levels. Fisheries research must be more transdisciplinary and include women's situation and problems, often brought forward by gender researchers. However, without more gender focus by male researchers such a transformation will take more time. It took 25–30 years before the problems were stated by women and gender researchers, until FAO managed to develop her most forceful politics that hopefully may benefit women in small-scale fisheries (FAO 2014).

There are still many challenges, for example in food security, as women in the household are often the main providers. If women's conditions in fishing, shell collecting, seaweed and aquaculture production, marketing and distributions of such products can be improved, they could contribute to food security on a wider scale. However, this can only happen if the natural and social living conditions are stable and effectively improved.

The World Bank (2011, 2) stated that "the change in the patterns of climatic variability is likely to add to the already high vulnerability of poor households, which would exacerbate the incidence, severity, and persistence of poverty." We know from Bangladesh that women and children are usually the most helpless and disadvantaged. During flash flooding, food insecurity and sanitation are considered to be major issues for affected women. They have to travel long distances to access drinking water, for example. In such situations, traditional roles are often reinforced, girls' education suffers, and women's abilities to diversify their livelihoods (and therefore their capacity to access income-generating jobs) are diminished (Masika 2002; Denton and Parikh 2003).

In conclusion, new opportunities and new challenges have occurred, and women and men have to adapt to these changes. Nevertheless, the examples given in this chapter give us hope; women and men on equal terms can develop a sustainable fishing and fishery politics that lead to good conditions for both women and men. In order to succeed, women must gain more influence and have access to ownership. Applying a gender perspective in fisheries research, making this type of research more transdisciplinary and carried out by women and men can help us, not only to understand, but also to create means to a more equal, viable, and sustainable world.

References

Alonso-Población E, Siar S (2018) Women's participation and leadership in fisherfolk organizations and collective action in fisheries: a review of evidence on enablers, drivers and barriers, FAO fisheries and aquaculture circular no.1159. FAO, Rome

Balsvik RR (1989) Vardø: grensepost og fiskevær, 1850–1950. Vardø kommune, Vardø

Bataille-Benguigui MC (1992) Pêcheurs de mer, pêcheurs de terre: la mer dans la pensée tongienne. Études rurales No. 127/128, La terre et le Pacifique, pp 55–73. http://www.jstor.org/stable/20125318

Boserup E (1970) Women's role in economic development. St. Martin's Press, New York

Bratrein HD (1976) Drivandes kvinnfolk: om kvinner, lønn og arbeid. Universitetsforlaget, Tromsø

Cabrera Soccoro G (1997) Los hombres y las mujeres de la mar: Isla de La Graciosa, in Spanish

Chapman MD (1987) Women's fishing in Oceania. Hum Ecol 15(3):267–288

Collins PH (1998) It's all in the family: intersections of gender, race, and nation. Hypatia 13(3):62–82. https://doi.org/10.1111/j.1527-2001.1998.tb01370.x

Crenshaw K (1989) Demarginalizing the intersection of race and sex: a black feminist critique of antidiscrimination doctrine, feminist theory and antiracist politics. University of Chicago Legal Forum 1989 (1). http://chicagounbound.uchicago.edu/uclf/vol1989/iss1/8

Crenshaw K (1997) Intersectionality and identity politics: learning from violence against women of color. In: Shanley ML, Narayan U (eds) Reconstructing political theory: feminist perspectives. Pennsylvania State University Press, University Park, pp 178–193

Crenshaw K (2005) Mapping the margins: intersectionality, identity politics, and violence against women of color (1994). In: Bergen RK, Edleson JL, Renzetti CM (eds) Violence against women: classic papers. Pearson Education New Zealand, Auckland, pp 282–313

Cruz-Torres ML, McEwee P (eds) (2012) Gender and sustainability: lessons from Asia and Latin America. The University of Arizona Press, Tucson

Denton F, Parikh J (2003) Gender: a forgotten element. Tiempo 47:27–28. http://www.tiempocyberclimate.org/portal/archive/issue47/t47a6.htm

Escallier Chr (2014) Les pêcheurs de Nazaré (Portugal), editions l'Harmattan p 373

FAO (Food and Agriculture Organization of the United Nations) (2014) Small-scale fisheries – International Guidelines on Securing Sustainable Small-Scale Fisheries [SSF Guidelines]. http://www.fao.org/fishery/. Accessed 19 Feb 2018

Frangoudes K (2011) Women's contribution in small scale fisheries in the European Union. In: Chuenpagdee (ed) World Small-Scale Fisheries Contemporary visions. Eburon, Delft, pp 101–116

Frangoudes K (2013) Women in fisheries: a European perspective, note for the European Parliament, Policy Department B: Stuctural and Cohesion Policies. http://www.europarl.europa.eu/studies. Accessed 10 May 2018

Frangoudes K, Keromnes E (2008) Women in artisanal fisheries in Brittany, France. Development 51(2):265–270

Frangoudes K, Carroll M, Holmyard N et al (2008) The role of women in the sustainable development of European fisheries areas. Brussels: European Parliament, Directorate General for Internal Policies of the Union, Policy Department B: Structural and Cohesion Policies, Fisheries (IP/B/TRAN/IC/2007_033, PE 389.586)

Frangoudes K, Marugán Pintos B, Pascual-Fernández J (2013) Gender in Galician shell-fisheries: transforming for governability. In: Banvinck M, Chuenpagdee R, Jentoft S et al (eds) Governability of fisheries and aquaculture: theory and applications, Mare Publications series. Springer, Dordrecht

Frangoudes K, Marugan-Pintos B, Pascual-Fernández J (2014) Women's organisations in fisheries and aquaculture in Europe: history and future prospects. In: Urquhart J, Accott GT (eds) Social issues in sustainable fisheries management, MARE Publications series. Springer, Dordrecht

Gerrard S (1975) Arbeidsliv og lokalsamfunn. Mag. Art-thesis, Department of Social Sciences. University of Tromsø, Tromsø,

Gerrard S (1983) Kvinner i fiskeridistrikter – fiskerinæringens bakkemannskap. In: Hersoug B (ed) Kan fiskerinæringen styres? Novus Forlag, Oslo, pp 217–241

Gerrard S (1986) Kvinners makt og avmakt – et kjønnsrolleperspektiv på forvaltning av faglige interesser, Alta: FDH- rapport

Gerrard S (1996) Gender and ethnicity: the organization of fishery in Mbakaou, a village in Adamaoua Province, Cameroon. In: Ngaoundéré Anthropos. University of Ngauoundéré, Ngauoundéré

Gerrard S (2013) Mobilities, materialities, and masculinities: interconnected mobility practices in Norwegian coastal fisheries. Nor Geogr J 67(5):312–319

Gerrard S (2016) Kvinner i kystfiske – Gjester ved menns bord? Sigtona Halrynjo og Mari Teigen (red.):Ulik likestilling i arbeidslivet. Gyldendal Akademisk, Oslo

Gerrard S (2018) Then and now- women in Norway's fisheries, Yemaya, n° 568

GIFT (n.d.) Gender In Fisheries Team (GIFT) http://www.cavehill.uwi.edu/cermes/projects/gift/overview.aspx. Accessed 29 Mar 2018

Gopal N, Williams MJ et al (2016) Gender in aquaculture and fisheries: the long journey to equality. Asian Fish Sci (Special Issue) 29S

Gopal N, Williams MJ, Gerrard S et al (2017) Gender in aquaculture and fisheries: engendering security in fisheries and aquaculture. Asian Fish Sci (Special Issue) 30S

Harper S, Grubb C, Stiles M, Sumaila R (2017) Contributions by women to fisheries economies: insights from five maritime countries. Coast Manag 45(2):91–106

Hitomi N (ed) (2009) Women in Japanese fishing communities. Agriculture and Forestry Statistics Publishing Inc, Tokyo

Hoefnagel E (1991) Seaweed, status and the symbolic domain: seaweed harvesters in Portuguese coastal communities. MAST 4(2):40–61

Jentoft S (ed) (1989) Mor til rors. Report. Norges Fiskerihogskole (Norwegian Fishery College), Tromso

Júlíusdóttir M, Dís Skaptadóttir U, Karlsdóttir A (2013) Gendered migration in turbulent times in Iceland. Norsk geografisk tidsskrift (Nor Geogr J) 67(5):266–275. https://doi.org/10.1080/00291951.2013.847483

Kabeer N (2002) Reflections on the measurement of women's empowerment. In: Discussing women's empowerment-theory and practices, SIDA Studies n°3, Stockholm

Kleiber D, Harris ML, Vincent CJA (2015) Gender and small-scale fisheries: a case for counting women and beyond. Fish Fish 16:547–562

Kleiber D, Frangoudes K, Snyder H et al (2017) Promoting gender equity and equality through the small-scale fisheries guidelines: experiences from multiple case studies. In: Jentoft S, Chuenpagdee R, Barragán-Paladines MJ et al (eds) The small- scale fisheries guidelines: global implementation. Springer, Dordrecht

Kronen M, Vunisea (2009) Fishing impact and food security-Gender differences in finfisheries across Pacific Islands countries and culture groups in SPC Women in Fisheries Information Bulletin # 19. February 209

Lang JD, Wiek A, Bergmann M et al (2012) Transdisciplinarity research in sustainable science: practice, principles and challenges. Sustain Sci 7(Supplement 1):25–23. https://doi.org/10.1007/s11625-011-0149-x

Lim CP, Ito Y, Matsuda Y (2012) Braving the sea: the Amasan (women divers) of the Yahataura fishing community, Iki island Nagasaki Prefecture, Japan. Asian Fish Sci 255:29–45

Malhotra A, Schuler SR, Boender C (2002) Measuring women's empowerment as a variable in international development. Report. Gender and Development Group of the World Bank. http://siteresources.worldbank.org/INTEMPOWERMENT/Resources/486312-1095970750368/529763-1095970803335/malhotra.pdf. Accessed 15 Apr 2018

Masika R (ed) (2002) Gender, development, and climate change. Oxfam Great Britain, Oxford

McGrath C, Neis B, Porter M (eds) (1995) Their lives and times – women in Newfoundland and Labrador: a collage. Killick Press, St, John's

Medard M, Sobo F, Vgatunga T et al (2001) Women and gender participation in the fisheries sector in Lake Victoria, presented at the 6th Asian Forum, Kaohsiung, Taiwan, 29 November 2001

Meltzoff SK (1995) Marisquadoras of the shellfish revolution: the rise of women in co-management on Illa de Arousa, Galicia. J Polit Ecol 2:20–38

Miki N (2009) Conditions for women working at sea. In: Hitomi N (ed) Women in Japanese fishing communities. Agriculture and Forestry Statistics Publishing Inc

Munk-Madsen E (1998) The Norwegian fishing quota system: another patriarchal construction. Soc Nat Resour. Int J 11(3):229–240. https://doi.org/10.1080/08941929809381075

Munk-Madsen E (2000) Wife the deckhand, husband the skipper: authority and dignity among fishing couples. Women's Stud Int Forum 23(3):333–342. https://doi.org/10.1016/S0277-5395(00)00091-1

Nadel-Klein J, Davis DL (eds) (1988) To work and to weep: women in fishing economics. ISER, Memorial University of Newfoundland, St. John's

Nakato M (2005) Women and globalisation in Uganda fisheries. In: Frangoudes K, Pascual-Fernandez J (eds) Women in fisheries and aquaculture: lesssons from the past, current actions and ambitions for the future, Conference Proceedings, Santiago de Compostela (Spain) 10–13 November, 2004. Associacion Canaria de Anthropologia, La Laguna Tenerife

Neilson LA, Gabriel R, Moura Arroz A et al (2014) Perspectives about the sea in the Azores: respecting narratives that sustain inshore fishing communities. In: Urquhart et al (eds) Social issues in sustainable fisheries management, MARE Publications series. Springer, Dordrecht

Neis B (1993) From 'shipped girls' to 'brides of state': the transition from familial to social patriarchy in the Newfoundland fishing industry. Can J Reg Sci 16(2):185–211

Neis B, Binkley M, Gerrard S, Maneschey MC (eds) (2005) Changing tides: gender, fisheries and globalization. Fernwood Publishing, Black Point

Nilsen GB (2008) Bokført til to millioner kroner, men er verdt 200 To kvinnelige redere eier ei lita gullegruve. Fiskeribladet. https://fiskeribladet.no/nyheter/?artikkel=8129. Accessed 4 May 2018

Overå R (2000) Les femmes chefs d'entreprise dans la pêche piroguière au Ghana. Le cas de la ville de Fante de Morre. In: Cheveau JP, Jul-Larsen E, Chaboud C (eds) Les pêches piroguières en Afrique de l'Ouest. Dynamiques institutionnelles: pouvoirs, mobilités, marchés. Karthala, Paris, pp 299–324

Peke S (2013) Women fish vendors in Mumbaï: a study report, ICSF, 66p. www.ICSF.net. Accessed 10 May 2018

Porter M (1993) Place and persistence in the lives of Newfoundland women. Avebury, Aldershot

Power NG (2005) What do they call a fisherman? Men, gender and restructuring in the Newfoundland fishery. Institute of Social and Economic Research, Memorial University of Newfoundland, St. John's

Razavi S, Miller C (1995) From WID to GAD: conceptual shifts in the women and development discourse, Occasional Paper no 1. United Nations Research Institute for Social Development, United Nations Development Programme, Geneve

Reeves H, Baden S (2000) Gender and development: concepts and definitions, BRIDGE (development – gender). Institute of Development Studies, University of Sussex Report no 55. http://www.bridge.ids.ac.uk/sites/bridge.ids.ac.uk/files/reports/re55.pdf. Accessed 9 May 2018
Skaptadóttir UD (2000) Women coping with change in an Icelandic fishing community. Women's Stud Int Forum 23(3):311–321. https://doi.org/10.1016/S0277-5395(00)00089-3
Thiessen V, Jentoft S, Davis A (1992) The veiled crew: an exploratory study of wives reported and desired contributions to coastal fisheries enterprises in northern Norway and Nova Scotia. Hum Organ 51(4):342–352
Thorpe A, Pouw N, Baio A et al (2014) 'Fishing and everybody business': women's work and gender relations in Sierra Leone's fisheries. Fem Econ 20(3):53–77. https://doi.org/10.1080/13545701.2014.895403
Torre-Castro M, Fröklin S, Börjesson S et al (2017) Gender analysis for better coastal management- increasing our understanding of social ecological sea-scape. Mar Policy 83:62–74
UN (United Nations) (1979) The convention on the elimination of all forms of discrimination against women (CEDAW). http://www.un.org/womenwatch/daw/cedaw/cedaw.htm. Accessed 14 Apr 2018
UN (n.d.) Sustainable development goals. Goal 5: achieve gender equality and empower all women and girls. https://www.un.org/sustainabledevelopment/gender-equality/. Accessed 10 May 2018
Weigel JY (1987) Nana et pêcheurs de port de Lomé: une exploitation de l'homme par la femme? Polit Afr 27:37–46
Williams M (2008) Why look at fisheries through a gender Lens? Development 51:180–185. https://doi.org/10.1057/dev.2008.2
Willson M (2015) Seawomen of Iceland. Survival on the edge. University of Washington Press, Seattle
World Bank (2011) The poverty impacts of climate change: a review of the evidence. Policy Research Working Paper [online] WPS5622, 2. http://www-wds.worldbank.org/servlet/WDSContentServer/WDSP/IB/2011/04/04/000158349_20110404100435/Rendered/PDF/WPS5622.pdf. Accessed 17 Mar 2012

Chapter 8
Markets, Distribution and Value Chains in Small-Scale Fisheries: A Special Focus on Europe

José J. Pascual-Fernández, Cristina Pita, Helga Josupeit, Alicia Said, and João Garcia Rodrigues

Abstract Several factors affect the ability of small-scale fishers to secure their livelihoods. Particularly relevant is their capacity to sell their fish, receive remunerative prices, and add value to their catches. In general, catches from small-scale fisheries have a superior quality and freshness, but this does not always lead to better prices or higher demand. Too often, local fishing catches are not sufficiently differentiated in the market from those coming from imports, larger scale fisheries, aquaculture,

---

This chapter is based on research conducted under the project "Governance challenges for sustainable small-scale fisheries: creating synergies with marine conservation and tourism" (GOBAMP II, CSO2013-45773-R, financed by Ministry of Economy and Competitiveness of Spain), with the support of the Vice-rectorate of Research of the University of La Laguna, Tenerife. We also acknowledge support from the "Too Big To Ignore: Global Partnership for Small-Scale Fisheries Research". This chapter is also based on research related to the Interreg project *Macarofood (MAC/2.3d/015) "Valorización de productos marinos de la Macaronesia: turismo, gastronomía y capacitación profesional"*. The Fisheries Local Action Group of Tenerife contributed financially to the research with two projects and supported the process. Cristina Pita would like to acknowledge INTERREG project Cephs & Chefs (EAPA 282/2016) "Octopus, squid, cuttlefish, sustainable fisheries and chefs" and FCT/MEC national funds and FEDER co-funding, within PT2020 Partnership Agreement and Compete 2020, for the financial support to CESAM (UID/AMB/50017/2013). Our gratitude to the fruitful comments of two anonymous reviewers and the book editors.

J. J. Pascual-Fernández (✉)
Instituto Universitario de Investigación Social y Turismo, Universidad de La Laguna, Santa Cruz de Tenerife, Spain
e-mail: jpascual@ull.es

C. Pita
Department of Environment and Planning & Centre for Environmental and Marine Studies (CESAM), University of Aveiro, Aveiro, Portugal
e-mail: c.pita@ua.pt

H. Josupeit
INFOPESCA, Montevideo, Uruguay
e-mail: helga.josupeit@infopesca.org

R. Chuenpagdee, S. Jentoft (eds.), *Transdisciplinarity for Small-Scale Fisheries Governance*, MARE Publication Series 21,
https://doi.org/10.1007/978-3-319-94938-3_8

or even from furtive fishing. This chapter focuses on strategies adopted by small-scale fishers to add value and improve the market penetration of their catches. These strategies must embrace a wide range of actors and issues. There is also a need to know whether the fishing resources available can cope with demand, or in what ways consumers' preferences could be refocused on other marine species to make more efficient use of potential resources. This requires the contribution of a variety of scientists, ranging from the fields of natural sciences to economics and marketing, with social sciences also playing an important role. Consequently, transdisciplinary research in this area is needed in order to optimize the income of fishing families and make it easier for consumers to access locally caught fish. This chapter will also provide a practical example to illustrate transdisciplinary research to improve small-scale fisheries capacity for innovation in marketing being developed in Tenerife (Spain). These strategies involve collaborations between scientists, local government, and the industry to produce an innovative small-scale fisheries branding initiative.

Keywords Markets · Labels · Community-supported fisheries · Artisanal fisheries · European Union · Canary Islands

8.1 Introduction

Small-scale fisheries are extremely important around the world, as their activities and practices contribute significantly to marine stewardship. In Europe, they provide employment and welfare, while increasing food sovereignty and supporting communities in coastal areas. In the European Union (EU), about 80% of the fishing fleet (approx. 85,000 vessels) belong to small-scale fisheries.[1] They employ over 40% of EU fishers (90,000) and landings from EU small-scale fisheries are worth

[1] Definition of small-scale fisheries is a complex issue in Europe and beyond. EU Regulation n° 508/2014 (European Maritime Fund and Fisheries-EMFF) describe SSF as follows: "*small-scale coastal fishing means fishing carried out by fishing vessels of an overall length of less than 12 meters and not using towed fishing gear as listed in Table 3 of Annex I to Commission Regulation (EC) No 26/2004*". A similar definition was present in the previous European Fisheries Fund (Regulation 1198/2006) (Natale et al. 2015). In any case, each EU country may define these fleets in a different way, and consequently to homogenise a definition is very difficult, even in Europe. If we adopt an international perspective these inconsistencies only increase. There is a tendency to consider the EU definition in different realms, so it constitutes a general reference.

A. Said
Too Big To Ignore Global Partnership for Small-Scale Fisheries Research, Memorial University of Newfoundland, St. John's, NL, Canada
e-mail: alicia.said@mun.ca

J. Garcia Rodrigues
Campus Do*Mar – International Campus of Excellence Galicia-North Portugal, Faculty of Political and Social Sciences, University of Santiago de Compostela, Santiago de Compostela, Spain
e-mail: joao.rodrigues@rai.usc.es

around 2 billion euros annually (25% of the total revenue generated by EU fisheries) (Guyader et al. 2013; Macfadyen et al. 2011). The small-scale fisheries sector has been characterized historically by its diversity and capacity to adapt to a wide range of challenges, but with increasing difficulties after industrial fisheries and world markets development. Competition with larger fleets for resources and markets is one of these challenges, and the catches of both fleets are often undifferentiated in the market. Competition from recreational fisheries for resources is also an issue, as they target similar species and fishing areas. Furthermore, urban and tourism development affect many activities along the coast, on beaches or in harbour areas, potentially taking over areas previously used almost exclusively by small-scale fisheries. This results in displacement and a situation referred to as 'ocean and coastal grabbing' (Said et al. 2017).

The European fishing sector has always been deeply affected by public policies. National states have influenced the development of fisheries with a range of policies and measures that have included subsidies, laws promoting or restricting specific fishing techniques, and market interventions, amongst others. Food security concerns and industrial development priorities have also often contributed to shape these policies. More recently, sustainability considerations have taken centre stage in national policies. However, for many years, especially from the last decades of the nineteenth century onwards, European fishing policies have mainly focused on increasing productivity and facilitating the development of capital-intensive fisheries with larger and more productive vessels. By contrast, small-scale fisheries have rarely received the same degree of attention, and many policies appear to have been developed only to favour larger fleets. The lobbying capacity of the large-scale fisheries sector may explain this privileged focus, but perhaps we also need to take into account the lack of visibility of small-scale fisheries. These policies not only have consequences for fishing operations but also for the marketing of catches. For instance, increasing restrictions have recently been placed on the direct marketing of small-scale fisheries catches in many areas of Europe, making it increasingly difficult for traditional selling patterns that involve members of fishing households, especially women (Pascual Fernández 1991; Frangoudes et al. 2008a, b).

The voices of small-scale fishers have, traditionally, barely been heard at regional, national, or European levels; particularly, as in most countries, small-scale fisheries are not well organized at the national level. The reformed Common Fisheries Policy (CFP) and the European Maritime and Fisheries Fund (EMFF) encourage us all to reflect on how small-scale fisheries in Europe are being influenced by public policies at different levels (regional, national, or European). The FAO *Voluntary Guidelines for Securing Sustainable Small-Scale Fisheries in the Context of Food Security and Poverty Eradication* (FAO 2015), approved in June 2014, have finally put small-scale fisheries at centre stage, and it is necessary to analyze how EU or national policies can comply with these guidelines.

Markets constitute an added challenge for small-scale fisheries in Europe and around the world. Several factors affect the capacity of small-scale fishers to sell their fish, receive remunerative prices, and to add value to their catches. For instance, existing national and regional regulations as well as globalized marketing schemes may restrict market access. The supply chain, from catch to consumers, in general, has not been researched enough and there are few articles on this subject. For instance, it is necessary to determine how local small-scale fishery catches interact with those from large-scale fleets in different scenarios, and how world markets can affect local fishing strategies. In most countries, facilitating the entrance of catches from world markets has hampered the selling of local fish using traditional marketing strategies ill-adapted to new scenarios, and where supermarkets chains and large companies control most fish markets. In addition, there is increasing competition from aquaculture that has developed with the support of subsidies from national governments and the EU, as is evident in the EMFF (REGULATION (EU) No 508/2014) and especially in the document about *Innovation in the Blue Economy* (COM (2014) 254 final/2) that clearly favours aquaculture funding.

With this panorama, it is important to gather data on existing strategies for adding value to small-scale fishery catches, and under what circumstances these strategies could provide real alternatives for small-scale fisheries. In this context, a special focus must be placed on how small-scale fisher organizations can become involved in fish marketing. In general, we can say that local fresh fish obtained by small-scale fisheries with sustainable gears are often not adequately differentiated from the catches of industrial fleets or from refrigerated or frozen fish coming onto the market. In this chapter, we focus on why it is important to understand the challenges of small-scale fisheries, and discuss ways to make them more resilient through innovative marketing systems. Specifically, we show how transdisciplinary analysis, through the direct involvement of different stakeholders, including academics, fisher organizations, consumers, and other actors of the value chain, are necessary to identify trade-related concerns and strategies that could engender small-scale fisheries' sustainability.

Understanding the context around different marketing strategies for small-scale fisheries is important. In the next section of this chapter we describe the main initiatives and strategies in published literature to promote small-scale fisheries products and improve market opportunities around the world. In the third section a general perspective about challenges in the trade of small-scale fisheries products in Europe is presented, highlighting some elements of the Common Market Organization. In general, a transdisciplinary focus is not usual in the development of adaptive strategies to improve marketing. Frequently, specific social actors or organizations take the initiative to innovate in the market, with different degrees of success. In the fourth section we provide an example of a partnership between scientists and the industry, where a transdisciplinary approach is implemented to understand the various links and connections, such as restructuring of existing policies and consumer behaviour, which need to be considered when applying new market systems to improve the sustainability of small-scale fisheries. Research about small-scale fisheries markets innovation greatly benefit from such a focus, as the challenges faced by small-scale fisheries are not simply economic, nor biological, but also cultural and societal. In this context, a diversity of disciplines have something to say about what is happening with the innovation initiatives and may help to overcome the innovation challenges for small-scale fisheries in the new market scenarios.

8.2 Main Initiatives and Strategies to Promote Small-Scale Fisheries Products and Improve Market Opportunities Around the World

Small-scale fisheries face market competition from large-scale fishery products, foreign imports, and aquaculture products, and many small-scale fishing operations are pursuing new ways to market and sell their catch (Stoll et al. 2015). Over the last decade, a diversity of experiences to promote and differentiate small-scale fisheries products have been developed (Bolton et al. 2016; Godwin et al. 2017).

The scientific literature on marketing strategies to promote small-scale fisheries products is scarce. A review of the literature reveals that this topic has hardly been studied and the research has been developed only in a few countries, therefore, there is considerable scope for further research. Review of the literature also revealed that there are several strategies for adding value to small-scale fisheries catches, including certification of origin schemes, ecolabels, direct selling, alternative food networks, and collective labels. This section will identify and discuss the most common existing arrangements in literature for small-scale fisheries products and their supply chains, as potential solutions for their current situations. The strategies most commonly highlighted in the literature as employed by the small-scale fisheries sector to promote products and improve their market opportunities include: direct marketing strategies or short, local supply chains, and certification and labelling initiatives. In this section we will discuss each of these strategies individually by giving examples, and highlight how these systems can work in different contexts. We also acknowledge that different opportunities need to be adapted to the social and governance systems within which they operate.

8.2.1 Direct Marketing Arrangements

Direct marketing arrangements or short, local supply chain schemes have been a traditional characteristic of small-scale fisheries. A certain research and publication bias exist in this realm, with most recent publications focusing on schemes in North America.

The most common direct marketing arrangements are community-supported fisheries (CSFs). CSFs programs are arrangements between fishers and consumers, usually cutting out the middleperson, in which consumers provide upfront payments to fishers in exchange for scheduled seafood deliveries. CSFs are employed to increase profits for local fishers, provide high quality seafood to interested consumers, and directly engage consumers in using fishery products (Brinson et al. 2011).

CSFs programs are increasing, especially in the USA and Canada. These schemes have emerged as a way to reduce the environmental impacts associated with seafood production, distribution, and consumption (McClenachan et al. 2014). Literature

Table 8.1 Examples of strategies employed by the small-scale fisheries sector to promote their products and improve market opportunities

Initiative type	Name of the initiative	Relevant literature
Community-supported fishery	'Port Clyde Fresh Catch', 'Walking Fish', 'Revolutionary Fish', 'Off the Hook', 'Community Seafood'.	Brinson et al. (2011), Campbell et al. (2014), Stoll et al. (2015), Bolton et al. (2016).
Direct marketing	Pesca en linea	Van Holt and Weisman (2016).
MSC certification	Mexico Baja California pole and line yellowfin and skipjack tuna; Sian Ka'an and Banco Chinchorro lobster fishery; hanging fishery for Japanese scallops; Waterhen Lake walleye and Northern pike Commercial Gillnet fishery; Normandy and Jersey lobster.	Pérez-Ramírez et al. (2016), Wakamatsu and Wakamatsu (2017), Swartz et al. (2017).
Label of origin	'Bar de ligne de la pointe de Bretagne'; 'Pride Fish'.	Charles et al. (2003), Swartz et al. (2017).
Label of sustainability	Marine Eco-Label Japan	Swartz et al. (2017)

Note: MSC = Marine Stewardship Council

about CSFs tends to focus on discussing CSFs' main challenges, opportunities, market benefits, non-market benefits, sustainability considerations, and policy implications (Brinson et al. 2011; McClenachan et al. 2014). Campbell et al. (2014) conducted a survey of consumers involved in CSFs programs, and noted that these consumers were interested in accessing high-quality, fresh, local seafood, and in supporting fishing communities. Consuming seafood distributed by local CSFs reduces the average seafood carbon footprint when compared to consuming from large-scale fisheries (McClenachan et al. 2014). CSFs also distribute a subset of highly abundant species, not generally targeted by fishers due to their low market value, thus contributing to increasing local fisheries' sustainability (McClenachan et al. 2014). On the other hand, it is important to note that short local supply chains are restricted to niche markets (Oosterveer and Spaargaren 2011).

The most studied CSF initiative is 'Walking Fish', an initiative from Beaufort, in North Carolina (USA), implemented in 2009 (Bolton et al. 2016; Brinson et al. 2011; Campbell et al. 2014; Stoll and Johnson 2015). Examples of other CSFs in the literature include 'Off the Hook' (Canada), 'Port Clyde Fresh Catch' (USA), 'Community Fish' (USA) (Brinson et al. 2011) (Table 8.1).

Other interesting market strategies are locally-based collaborative markets in the south of Thailand, where fishers avoid selling their catch to individual middlepersons by organizing local markets. See Box 8.1 for a detailed description. This strategy can also be found in Europe and other areas of the world, even though the research and publications about these initiatives are scarce.

Box 8.1 Locally-Based Collaborative Markets, South of Thailand
Ratana Chuenpagdee and Kungwan Juntarashote.

Many small-scale fishers are economically dependent on fish traders and middle-persons, who set the prices for their catches. Some of the fishers also borrow money from these fish traders and intermediaries, through informal arrangements, and thus are obliged to sell at whatever price. The lack of negotiating power in the selling and marketing of their products makes small-scale fishers vulnerable and reduces viability. One of the ways for small-scale fishers to have better control of the value chain is through organizing collaborative marketing systems within communities. This is what the small-scale fishers in Phang-nga and Krabi Provinces, south of Thailand were able to do, partly due to their involvement in the CHARM project (Coastal Habitat and Resource Management; an EU-Government of Thailand collaborative project initiated in 2003). Rather than selling directly to individual buyers, small-scale fishers in these locations have agreed to bring their catches to market places that they organize themselves in their communities. This results in a higher volume of each product, which makes it possible for them to determine suitable prices. Buyers and traders then come to these central markets, and, in some instances, bid for the catches in an auction. This way of organizing the market basically shifts the power in the value chain from buyers to fishers, and has helped raised fishers' income. Such a system depends, however, on the collaboration of fishers, social cohesion within the communities and strong leadership. Since its establishment in 2006, the system in Phang-nga Province still functions, while it is no longer working in Krabi Province. With higher prices received, some of the fishers in Krabi started to sell to individual buyers, rather than bringing the catches to the central markets and sharing the costs of operating them. Running these locally-based collaborative markets takes extra administrative effort and commitment from fishers to bond and stay together. Thus, governments should provide financial support to such communities to help sustain these initiatives, along with other incentives to encourage new ideas for market development, including post-harvest value-added products that could be supplied to tourism and niche markets.

8.2.2 Certification and Labels

Concern over the collapse of many wild-caught fisheries has led to a focus on seafood-certification and consumer-driven support of sustainable seafood (Logan et al. 2008). Ecolabel initiatives, with assurances of transparency and sustainability, have been on the increase worldwide for over a decade now. However, these standards are not without controversy, with some authors perceiving certification to be an increasingly pervasive form of market governance through which retailers and

NGOs are able to exert control over producers of primary products in order to secure their commercial and institutional interests (Belton et al. 2011).

Foley and McCay (2014) argue that ecolabeling and certification privatizes fisheries governance by creating new institutions with property rights. Labels and certification organisations allocate rights and responsibilities to certificate holders, and thus provide a 'licence' to access new markets. These schemes may also create new conflicts in the contexts of common-pool resources (Foley 2012). Despite the negative side of these schemes, Foley and McCay (2014) also argue that label and certification providers (e.g. the Marine Stewardship Council) may promote cooperation and collective action, as they usually encourage different actors that target the same stock to cooperate and seek certification together.

Studies that have focused on certification standards and small-scale fisheries, especially those focused on global standards such as the Marine Stewardship Council (MSC) label, tend to be quite critical of these initiatives. For some time now, several authors have tried to raise our attention to the fact that MSC labels tend to marginalize small-scale operators. For instance, Ponte (2008) analysing MSC labels noticed that developing countries' fisheries, and small-scale fisheries in particular, have been marginalized by ecolabels. Belton et al. (2011) assessed the outcomes of emerging certification standards intended to govern production of a new global commodity. In the case of Pangasius catfish from Vietnam and Bangladesh, which specially focuses on the environment (one of the key areas which standards seek to regulate), the authors concluded that certification is likely to result in greater differentiation and polarisation between larger and smaller farm operators, increasingly excluding the latter from access to Western European and North American markets, with minor local environmental gains.

Moreover, small-scale fisheries have several constraints to access MSC certification and standards, including: lack of funding to access the assessment, especially in developing countries (Jacquet et al. 2010); high maintenance costs to maintain the label; lack of quantitative data to run the assessment process; being multi-gear and multi-species fisheries increases the complexity of certification; and, low price premiums (consumers may show low level of awareness towards sustainability undermining market-driven mechanisms) (Wakamatsu and Wakamatsu 2017). Issues related to food security in communities that depend heavily on fisheries as a source of protein and certified fisheries that are destined for fishmeal are also points to consider (Christian et al. 2013).

Despite the problems faced by small-scale fisheries to access MSC certification, there are several small-scale fisheries certified with the MSC standard (Table 8.1). The published literature focusing on analysing MSC certified small-scale fisheries includes works by Wakamatsu and Wakamatsu (2017), Swartz et al. (2017) and Pérez-Ramírez et al. (2016). They examined MSC certified small-scale fisheries and described the benefits and constraints for them to obtain MSC certification, noting that the maintenance or increase in market-share was the main motivation to pursue certification (Pérez-Ramírez et al. 2016). Besides that, certification allowed the products to enter new markets, increased publicity and improved management (Wakamatsu and Wakamatsu 2017). The main challenges faced by small-scale fisheries pursuing certification are related to the lack of scientific information, weaknesses in management, and the costs related to the certification process (Pérez-Ramírez et al. 2016).

Other certification schemes implemented in the market and focused on small-scale fisheries products include labels of origin; for example, in France and Japan. The 'Bar de ligne de la pointe de Bretagne' is a label of origin for seabass fished with lines from Brittany (France). This label was created with the objective of distinguishing high quality wild-caught seabass from that produced and sold by aquaculture companies (Charles et al. 2003). The 'Pride Fish' is a label of origin from Japan implemented in 2014, which supports several local small-scale fisheries by promoting increased consumption of local, seasonal seafood (Wakamatsu and Wakamatsu 2017). Another interesting example is a marine ecolabel from Japan. This Japanese label of sustainability has emerged as an alternative to MSC certification. It already includes 23 fisheries targeting different species. This certification places great emphasis on co-managed fisheries and is finding a lot of interest in local markets but lacks the support of major retail partners that the MSC has obtained (Swartz et al. 2017).

In Spain, there are a number of collective quality labels that have been developed with the support of small-scale fisher organizations, though scientific publications about these cases are currently non-existent. This is an example of how the scientific coverage of many initiatives related to small-scale fisheries is very limited. The fourth section of this chapter provides a detailed step-by-step description of a transdisciplinary approach to create a label in the Canary Islands (Spain).

8.3 Challenges in the Trade of Small-Scale Fisheries Products in the European Union

Small-scale fisheries play an important role in fish production in the world, but quantifying their role is very difficult. In general, the small-scale fisheries sector is greatly important in the EU as it contributes to the economies of coastal communities, especially in relatively isolated areas and on islands, where job opportunities are limited (Pita et al. 2010). However, it is difficult to quantify the amount of fishery products landed in the EU that come from small-scale fisheries, though some estimate one quarter of the overall EU fisheries catch value is from small-scale fisheries (Macfadyen et al. 2011).

However, some generalization is possible with regard to products and market channels. Products from small-scale fisheries generally supply the fresh fish market. They can be sold locally, close to the landing place, but later may travel long distances to major urban areas, as happens with octopus in Portugal or tuna in the Canary Islands. In this section, we will provide a discussion on the trade-related challenges that this sector face, and by giving examples from the EU context, we explain that problems can occur at different levels and can affect small-scale fisheries in many ways. We explain issues related to the social, natural, and governance systems within which small-scale fisheries operate, and highlight the need for the interconnections between these systems to be investigated when tackling small-scale fisheries trade problems. Our understanding of these problems could be improved through a multiple disciplinary approach in order to recognize the various links and connections between these issues, such as the problems emanating from fishing policies, challenges in the organization of the sector, and other issues concerning certification and marketing standards. All these issues can determine the sustainability of small-scale fisheries. From a marketing point of view, small-scale fisheries in many areas of the EU are in difficult situations at the moment. Their traditional market outlets are shrinking due to increased use of supermarkets, which buy aquaculture or imported products rather than local products, even though in some southern countries, the relevance of local fresh fish may be stronger in these sales channels. In some EU countries, small-scale fishers also face difficulties to sell directly to consumers for various reasons. For example, it may be compulsory to sell catches at local auctions, where consumers cannot participate directly. Fishers may also need to comply with complex regulations to sell directly, and selling on the street is usually banned. Moreover, even though European regulations may facilitate direct selling in small quantities, their implementation in national regulations is neither automatic nor easy, as has happened in Spain or Portugal, for instance. Thus, the shrinking numbers of fishmongers and reduction in traditional outlets for small-scale fisheries products are creating a major problem for the sales of these products. In addition, subsidies are generally going to the industrial sector, creating unfair competition for the small-scale fisheries sector (Schuhbauer et al. 2017).

The objective of the EU's Common Market Organization (CMO) is to strengthen the role of actors on the ground: producers are now responsible for ensuring the sustainable exploitation of natural resources and must be equipped with instruments to improve the marketing of their products. At the same time, consumers should receive better information on the products sold on the EU market, which, regardless of their origin, must comply with market rules. However, the CMO is not favouring small-scale fisheries producers; in fact, it is backing industrial producers with its information and promotional material. The five main areas covered in this section are: sector organization, marketing standards, consumer information, competition rules, and market intelligence.

8.3.1 Sector Organization

An individual fisher will not be able to easily access the market directly, apart from some occasional possibilities to sell to restaurants or, in some instances (when legal) to customers at the docks. Therefore, small-scale fisheries organization at different levels is vital. As already underlined, several types of organizations are linked to this sector across Europe, but tend to focus on fishing needs, rather than on the marketing needs. Although EU funding exists to provide the necessary mechanisms for fishers to organize themselves through Producer Organizations (POs) and to devise production and marketing plans (EC 508/2014 Art. 66), the process of establishing such POs involves a lot of administrative and financial burden, which small-scale fishers are not always able to take on due to insufficient capital and organizational capacity. A change of direction, taking into account examples from Spain, is exemplified by a number of cases where organizations not only manage the auction of the first sale of the catches, but also innovate in sales by differentiating the product with collective labels that are easily recognised in the market (*Pescado de Conil* in Andalusia). In this sense, the organizations can adapt themselves to the new scenario or new ones can be created to promote direct marketing. However, this is a process that clearly constitutes a major challenge and should be accompanied by capacity building of the main actors, including small-scale fishers and their families.

8.3.2 Marketing Standards

Any direct sales of small-scale fisheries products have to respect the quality and sanitary standards of the EU market. The freshness of the products sold has to be of the highest standards, and this may create challenges due to lack of infrastructure on small-scale vessels or in the harbours to ensure adequate quality conservation and fulfil sanitary requirements. Due to this, infrastructures devoted to ice production, to cold storage or to ultra-freezing are a frequent priority.

In addition, a logo for small-scale fisheries has been proposed and would include a reference to the fisheries co-management and represent a guarantee of a sustainable use of resources (Josupeit 2016). Management measures in place, such as minimum landing sizes and closed seasons and areas, will need to be respected by the fishers operating under this logo. Importantly, labelling of small-scale fisheries products represents a great opportunity to inform the consumer of the marketing standards of these products. Thus, in some countries, small-scale fisheries are developing collective and/or quality labels to differentiate their catches on the market, emphasizing their quality and freshness. In any case, the success of these initiatives in the market is not easy, with a number of failures for a variety of reasons. Time is needed for any labelling to be recognized in the market, together with promotion and assurance of quality standards. This usually requires at least strong fisher organizations backing the transformation and a long term commitment from them, not easy collective action requirements.

8.3.3 Consumer Information

Consumer behaviour with regard to food is changing in the EU. There is more interest in safe and environmentally-sound and sustainable products (Bailey et al. 2016; Oosterveer and Spaargaren 2011). The idea of buying products from local food producers is growing, especially for products coming from local agriculture, and more recently from fisheries. Initiatives to label local food products as 'local' or 'municipal' and KM Zero certifications (such as fishboxes) are mushrooming, though at a slower pace than for agricultural produce. The issue of carbon footprints has been in the media for seafood products, such as imports to the UK from as far away as New Zealand or Alaska, and whether this is an environmentally-sound practice, however, the recent economic crisis has wiped out many of these considerations. Although, in northern Europe, ecolabels for fishery products are increasingly demanded by consumers and supermarkets, as consumers show an interest in sustainable fishery management and fish production. In fact, the Marine Stewardship Council (MSC) is the most widely used label for fish products in north Europe and the world.

Consumer information should include various items that highlight the specific quality of small-scale fisheries products. Informed consumers could select the right fish product, based on the information of its origin, and are willing to pay a higher price for a local product, whose freshness and quality are guaranteed by a label or standard. Another issue to be studied further is what the consumer really wants to know about the fish product, whether the present labelling requirements (name of species, Latin name, capture or culture, country of origin (freshwater or culture fish) or FAO area of capture, best before, etc.) are really what consumers expect from a label for fishery products. Finally, an important area of concern is the issue of mislabelling or fraud, by which a product is sold under a different name, which is for a higher priced fish. Some studies already carried out state that in some cases even more than 30% of fish in restaurants is being mislabelled (i.e., EU Project Food Integrity, see: https://goo.gl/nLXLqu) (Pardo et al. 2018).

8.3.4 Competition Rules and Market Intelligence

Small-scale fisheries face major challenges due to the supply of imported fish stocks from third countries, and aquaculture products, which saturate local markets. By shortening the value chain and adequately differentiating their catches, small-scale fisheries products can become very competitive with other animal protein or imported fishery products, but this requires a level of organization that is not usual in fisher communities.

The European Market Observatory for Fishery and Aquaculture Products (EUMOFA) is concentrating on information on general market trends, without dedicating any specific interest to the small-scale fisheries sector. A specific request to better serve this sector should be made, which could help fishers understand more about the market environment in which they are operating. Producer Organizations could have a role in demanding and channelling market intelligence for small-scale fisheries, but many of these organizations do not have small-scale fisheries as their priorities. Price information should also be tailored to the needs of small-scale fishers. Further problems related to data and statistics that are affecting the viability of small-scale fisheries include information on the catches which are imported to the EU from third countries.

8.4 Transdisciplinarity in Action: An Example of Small-Scale Fishery Marketing in the Canary Islands

The Canary Islands is an archipelago of seven major islands divided in two provinces. A region of Spain, it is located close to the Saharan coast, near Morocco and Mauritania.

The Canary Islands fisheries have always played a significant role in the food security and identity of the islanders. For a long time, low volumes and limited varieties of canned, dry, or salted fish were the only imports of fish products to the islands. Most of the fish consumed on the Canary Islands used to originate either from the islands' coasts or from neighbouring African waters. In the latter case, they were salted in Canarian ships that had fished the area for centuries. Only in recent decades has consumption of fish from distant latitudes become commonplace. The increase in the islands' population has also favoured this process; population has more than doubled since 1960, with more than 2 million inhabitants plus 16 million tourists in 2017.

Fishing in Tenerife is mostly carried out by small scale vessels, less than 12 m in length, with limited engine power, capturing a great diversity of catches using low impact gears. Small-scale fisheries in the Canary Islands face challenges such as the management of increasingly scarce resources, especially demersal species, and challenges linked to the globalization of the fishing markets and to the growing importance of trade and transnational relations. Currently in the archipelago, it is possible to buy fish products from the most diverse origins, refrigerated or frozen, and there is often confusion in the market between products from local fisheries, with a remarkable quality and freshness, and others that have been on ice or refrigerated for a long time. An adequate differentiation in the market seems essential for the improvement of marketing local catches so as to preserve their value. We will provide a detailed example of how to carry out an examination of small-scale fishery markets and how to add value to their products taking a transdisciplinary approach.

In 2014, a transdisciplinary team of researchers from the University of La Laguna, including biologists, sociologist, anthropologist, economist, and tourism and marketing specialists developed a project, in collaboration with the Island Government (Cabildo de Tenerife) and some fisher organizations, such as Islatuna, on the market challenges for small-scale fisheries products on the island of Tenerife with funding from the Fisheries Local Action Group of the Island

(FLAG).[2] They worked together in order to achieve a holistic perspective about what was happening in small-scale fish markets in Tenerife and how to improve the penetration and value of these catches on the island. It was thought that without this holistic and transdisciplinary perspective it would be impossible to cope with the challenges of the project. The project aims were not related only to basic research analysing the market conditions and potential strategies. It was conceived as an action research initiative, devised to support innovation in the marketing of small-scale fisheries products, in alliance with key fisher organization and Island Government (*Cabildo de Tenerife*). This alliance with key actors exemplifies the applied and transdisciplinary focus of the project, intended to generate new knowledge about the market, and to facilitate and impulse strategies to transform it. The project was developed in two phases. First, an analysis of the main marketing channels of Tenerife's small-scale fisheries catches was conducted, detecting those elements that prevented adequate addition of value. Second, based on the data and information obtained, improved marketing strategies were proposed after a discussion with representatives of the sector, forging an alliance with the stronger fisher organization on the island.

The data gathering for the project aimed at involving as many of the actors related to small-scale fisheries markets as possible. Fishers and their organizations, intermediaries of different relevance, supermarket chains, fish shops, hotels, restaurants, and consumers were all involved. Island Government (Cabildo de Tenerife) was also involved as a promoter of the initiative from the beginning. Specific methodologies were devised to obtain information from each of these groups of actors. We are detailing in this section some elements of the methodology used in the process, later the results, proposals, and actions adopted are detailed.

8.4.1 Research Methods in Small-Scale Fisheries Markets in Tenerife

All *cofradías* (traditional fishers' organizations) were visited to establish an initial diagnosis of the situation about marketing channels, conducting semi-structured interviews with key actors in each organization (10 *cofradías*). Next, we performed semi-structured interviews with fishers and family members, who sell fish catches, contacting most of the island's fishers directly and obtaining indirect data from the

[2] Two reports were ellaborated from this project: Pascual-Fernández, J.; Dorta Morales, C.; Melgar Ramírez, S.; Chinea Mederos, I.; Baute Díaz, N.; Revilla Hernández, M.; Borges Perera, C.; Santana Talavera, A.; de la Cruz Modino, R.; Gutiérrez Taño, D. (2014). *Acciones para mejorar el valor añadido de los productos de la pesca artesanal de Tenerife*. La Laguna (Tenerife): ULL - Fundación General Univ. La Laguna (unpublished). Pascual-Fernández, J.; Santana Talavera, A.; de la Cruz Modino, R.; Dorta Morales, C.; Melgar Ramírez, S.; Chinea Mederos, I.; Baute Díaz, N.; Revilla Hernández, M.; Borges Perera, C. (2014). *Estudio para el análisis de viabilidad e implantación de una marca colectiva de los productos pesqueros de Tenerife*. La Laguna (Tenerife): ULL - Fundación General Univ. La Laguna (unpublished).

rest. The initial data gathered suggested it was necessary to define two different areas for data collection: one focused on white fish and shrimp, the other on small pelagic and especially tuna catches. For the latter, we collected information via a fruitful collaboration with a producer organization (ISLATUNA).

In order to gather data from companies related to the commercialization (distribution and importation) of fresh fish products, we collected data from official registries. By checking this information against the information obtained from previous surveys of other actors, we obtained a list of the most important intermediary companies in Tenerife. It was possible to interview all the relevant intermediaries detected, and the results from this research were very interesting for shaping the analysis and conclusions.

Tourism and catering are especially relevant on Tenerife due to its economic dependence on service-related activities. In 2014, there were 202 hotels on the island. These hotels were contacted by e-mail asking them if they served fresh fish of any kind, and then fieldwork was conducted to complete the results *in situ* to check whether fresh fish was offered or not and their origin. As for restaurants, a sample of 117 restaurants was selected and visited by researchers. In both hotels and restaurants, data were gathered through observations and semi-structured interviews, asking, for instance, what kinds of fish were sold, how many suppliers were used, who the suppliers were, etc.

To analyse fish marketing by fishmongers, a semi-structured interview was conducted with fishmongers identified in Tenerife. Data were also collected from the observations made during the visits to the fishmongers, for instance related to the way in which the fish were labelled and identified. Most of the fishmongers on Tenerife were interviewed (32 fishmongers out of 35).

We requested information and opinions from those responsible for purchasing fish for the food distribution groups that operate in Tenerife. Semi-structured interviews were designed to obtain information from five of the six supermarket chains working on the island. The interview had a similar design to that of the fishmongers, focusing on aspects like the distribution of fishery products among the group's establishments, the percentage of fresh fish sales in Tenerife and other islands, strategies for obtaining a supply of fresh fish, and questions related to the development of collective labelling in Tenerife.

In the final stage of our research, when most of the data had already been processed and some conclusions and potential improvements defined, a phone survey was conducted to confirm the viability of introducing a collective label on Tenerife, as well as the viability of strategies designed to improve tuna marketing. A total of 600 home purchasing managers were interviewed by telephone.

8.4.2 Some Results from Market Research in Tenerife

A mixed methods approach was employed in the research, but the most relevant information collected was qualitative. Quantitative data from a diversity of sources was also collected, but confidence in some of these data was low. Analysing all the information collected, it is evident that a large percentage of the fresh fish marketed in the Canary Islands and identified by consumers as local is actually imported, especially for demersal species. The lack of clear and easy identification of origin has made this possible; even if fish catches are associated with an identification label of first sale origin, it is not easily understood by consumers or not clearly visible at points of sale or consumption. In this sense, the system of commercialization in Tenerife is not very transparent, and consumers usually receive little information about what they are buying. Too frequently, especially in restaurants, fish from furtive fishing or African waters (Morocco, Mauritania, Senegal) are sold as local catches, thus deceiving consumers.

Regarding hotels and restaurants, we discovered that around 10% of hotels offered some kind of fresh fish in their restaurants (only 21 hotels out of 202), mainly 5-star hotels and a few 4-star ones, with very few (5, all five-star hotels) offering local fish. Several reasons explain this outcome, such as the focus on minimum prices by most of the hotels on the island, and also the lack of an organized offer to supply these establishments with the products to suit the price, volume, and tourist-preference requirements.

By contrast, for other species, like small pelagic and small shrimp, local catches can supply the local market adequately. However, there are some inefficiencies due to monopolistic behaviour of actors in the value chain that hampers access to these resources for some fishmongers. The price and availability of these catches make them attractive for most consumers, so they are widely present in some supermarkets but not so frequent in restaurants or hotels. The most striking fact found in the analysis of the market was how the vast majority of tuna catches were exported refrigerated or frozen, while the local market only received from 5% to 15% of such catches, depending on the species and circumstances. With information from our research, we detected that the potential local demand for tuna was much higher than current local supply to the islands, and it would only require some processing to improve the local market penetration (i.e. loins vacuum refrigerated). In general, tuna catches are relevant on the islands, but access to some tuna resources have become especially difficult since 2008, specifically in the case of bluefin tuna (*Thunnus thynnus*) as this species is under a strict quota management. The Spanish quota has mostly benefitted purse seiners (only six industrial boats for 2018) linked to the tuna ranching industry, predominantly focused on lucrative foreign markets such as Japan and the USA. At the same time, the Canarian fleet has had a very limited access to the quota, even given their historical rights and despite other legal criteria that should be taken into account for the quota distribution. Thus, the artisanal fleet of the Canary Islands (more than 240 boats) has less (255 MT) than the quota assigned to one (260 MT) of these six purse seiners in 2018.[3] This is not an exception in southern Europe, as the Maltese case exemplifies (Said et al. 2016),

[3] Resolution of 2nd March 2018, by the General Secretary of Fisheries, in which the quota allocation of red tuna and the specific census of the authorised fleet to catch red tuna was published «BOE» num. 64, 14 March, 2018, pages 29949 to 29,957.

and provides a clear idea of the challenges that small-scale fisheries may face in order to access some of the most valuable resources.

8.4.3 Some Proposals and Actions

After the analysis of the market, two proposals were placed on the table. First, the need to differentiate local small-scale fisheries products from fresh fish imports and poachers' catches was clear, and a collective label was one of the potential solutions. Most of local consumers are keen to know where and how they can access local quality fish from this sector, though it should be noted that the total volume of local catches cannot satisfy all the potential demand. Nonetheless, given this low-supply and high-demand reality, if local small-scale fisheries catches are traceable and are differentiated from imported species, they could get a higher price than they are currently obtaining. It is difficult to know how much more, and when this may happen. Taking into account similar experiences developed in the mainland (i.e. label *Pescado de Conil*), it took some time until the label got a premium in the market, but now they get higher demand and better prices. That is the objective for the Tenerife market, besides clearly differentiating small-scale fisheries fish in points of sale and consumption. The effects are not only local, as some of the catches like tuna are exported and sold refrigerated in the mainland, and when some of the fish arrived with the label stick inserted in their head or tail the positive response was instantaneous. Most or all of the fish exported refrigerated from Tenerife is labelled nowadays with a stick inserted in the head or the tail of the fish (Fig. 8.1).

Fig. 8.1 *Pesca artesanal* label stick inserted in a tuna

Tenerife's Island Government (*Cabildo*) promoted the idea of a collective label from the very beginning, and our research demonstrated that it was possible through the implementation of certain strategies to involve different actors and specific requirements to get the permit to use the label. A challenge detected for the label implementation was the collective action requirements to develop and implement the initiative, not easy to accomplish without having strong fisher organizations in place. In Tenerife the stronger fisher organizations (PO) have more easily assumed the new strategy, the weaker ones have had more difficulties. This initiative has gone ahead, but the change in market conditions has been slow. It has been adopted by some restaurants that now announce the serving of local fish and demonstrate their commitment to local fisheries. In total, as of March 2018, 20 entities use this collective label (*cofradías*, fishmongers, or restaurants) and another 26 are interested and arranging the paperwork. For further advancement in this area, increasing the strength of fisher organizations is very important in order to improve the control and labelling of all small-scale fisheries catches.

Taking into account our research, it is essential that small-scale fisher organizations take a stronger role that represents their interests and gives them a voice with respect to the market. Indeed, the role of collective organizations in this context should be clearly enhanced. However, achieving this goal will not be easy, and in some areas it would seem almost impossible because of the traditional lack of involvement of these organizations in fish marketing. There have, however, been some noteworthy examples of organizations on the Canary Islands that have developed successful marketing strategies that could assure the long-term viability of small-scale fisheries in certain areas. This role of fisher organizations needs to focus on clearly differentiating small-scale fisheries products in the market, while at the same time reducing the prevalence of furtive catches and undifferentiated imports from local fish.

The second initiative is related to tuna. This is a very different case as most of the catches in the Canary Islands (5000–13,000 tons/year) are under the control of a few organizations: two producer organizations (Islatuna in Tenerife and Optuna 42 in Lanzarote), one enterprise (Pescados Ramón e hijos S.L.) and some smaller cooperatives or organizations. In this area, we collaborated with Islatuna from the beginning of our research and some pilot projects have been developed to enhance the presence of tuna catches in local markets. One such project was developed with a wholesale distributor of food for restaurants and involved supplying tuna loins from local catches and distributing them vacuum refrigerated. This was a success because the quality of the loins was much better than the ones previously distributed (imported from the Pacific) on the islands by this wholesaler. The increase in tuna loins on markets has been significant from 2015 to now, but still far from achieving its full potential growth. A key factor has been the strong external demand of national and world markets that makes it easier to send many thousands of tons to distant markets by these enterprises or other organizations traditionally dedicated to export. This export tradition with tuna catches has historical roots dating back nearly 200 years. Therefore, reshaping organizations or enterprises to cope with local market demands and fish processing involves some transformations that may not be easy to accomplish. This is our current endeavour, involving producer organizations in the process and other actors in the value chain. For example, currently, a pilot project with schools is being developed for the introduction of skipjack and tuna loins in school meals in Tenerife, with the perspective to expand to all the islands, inspired in an Interreg MAC project, MACAROFOOD (http://www.macarofood.org/).

8.5 Conclusions

The research developed on positioning small-scale fisheries' catches in the market is currently very scarce. It is necessary to understand better how small-scale fisheries products are sold, the value chain involved and the challenges faced in order to receive the price deserved for their quality. Too frequently, traditional patterns of fish marketing are ill-adapted to new cultural consumption models, and the capacity of small-scale fishers to innovate and cope with the new circumstances is not usually up to the challenge. The situation may vary in northern and southern countries, but in general the available knowledge about small-scale fisheries markets is still very low.

In the European Union, small-scale fisheries have been overlooked in fish marketing projects and policies. The most prominent example of this is the implementation of the new Common Market Organization (CMO), which concentrates on industrial fisheries. Thus, there is an urgent need to redirect the CMO towards the small-scale fisheries sector, which is in extreme need of assistance and promotion due to changes in retailing in the EU. These changes have favoured a concentration in food marketing places. Interestingly, while farmers' markets are booming in all EU countries, and consumers are attracted to local food production, with the KM Zero concept, small-scale fisheries have been excluded from this movement in most countries.

In order to design alternative marketing strategies for small-scale fisheries a holistic knowledge of the natural and social sides of the market scenario is required. A transdisciplinary focus on the elements of the system to be governed is necessary (Kooiman et al. 2005), with inputs from researchers of diverse disciplines who can collaborate on the problem and search for solutions. No less important is the analysis of the governing system, the institutional dimension of the fishery, with a special focus on fisher organizations and their interactions with the institutional frameworks that facilitate some selling behaviours and preclude others. There is also the cultural dimension, that is to say, what people want to eat does not necessarily coincide with what is available. We can find examples of species that are highlighted as the *summum* of the local appreciation for quality fish (i.e. *Espada preto* in Madeira – *Aphanopus carbo*) and are completely ignored in other nearby markets (Canary Islands and Azores Islands), creating different scenarios that need to be taken into account. In this arena, only holistic transdisciplinary analysis can ask the right questions and provide sound answers to market challenges.

A good example of the capacity to intervene with a transdisciplinary focus in local markets is what is being implemented in Tenerife. It is not easy to change global markets, but it is more accessible to induce changes in local ones for the benefit of small-scale fisheries with the acceptance of consumers, which demand quality and local products. The labelling initiative is facilitating this market differentiation, even the implementation process is slow and complex, as producers, consumers, and other actors in the value chain need to find an advantage in this labelling. Compensating all the collective action efforts invested in the development and implementation of this strategy is not easy, and the support from the *Cabildo* to the initiative has been essential. It is not only a matter of quality and differentiation, also necessary is a better match of local production, circulation strategies for catches, and local demand. The findings about the high potential local demand for tuna and the definition of market strategies to accomplish this request constitute a good example. A holistic knowledge of resources availability, production, markets, and population preferences (actual or potential) is needed, and for this purpose a disciplinary perspective is usually too restrictive. Furthermore, the key role of fishers' organizations need to be highlighted. Without strong organizations, market innovation in small-scale fisheries is restricted to personal initiatives that may succeed, but have a limited role to transform local markets. Engaging fisher organizations in innovation processes is facilitated by transdisciplinary research and support from public authorities, but also needs leadership in these organizations and a focus on medium or long term strategies. Besides this, from the researcher's side, a key factor is creating a relationship of mutual confidence and collaboration with these organisations, which facilitate the continuous interchange of information and support in decision making, both ways.

References

Bailey M, Bush S, Oosterveer P et al (2016) Fishers, fair trade, and finding middle ground. Fish Res 182:59–68. https://doi.org/10.1016/j.fishres.2015.11.027

Belton B, Haque MM, Little DC et al (2011) Certifying catfish in Vietnam and Bangladesh: who will make the grade and will it matter? Food Policy 36:289–299. https://doi.org/10.1016/j.foodpol.2010.11.027

Bolton AE, Dubik BA, Stoll JS et al (2016) Describing the diversity of community supported fishery programs in North America. Mar Policy 66:21–29. https://doi.org/10.1016/j.marpol.2016.01.007

Brinson A, Lee M-Y, Rountree B (2011) Direct marketing strategies: the rise of community supported fishery programs. Mar Policy 35:542–548. https://doi.org/10.1016/j.marpol.2011.01.014

Campbell LM, Boucquey N, Stoll J et al (2014) From vegetable box to seafood cooler: applying the community-supported agriculture model to fisheries. Soc Nat Resour 27:88–106. https://doi.org/10.1080/08941920.2013.842276

Charles E, Boude J-P, Murray A, Paquotte P (2003) Coastal fishing: resource's enhancement and preservation. Ocean Coast Manag 46:421–437

Christian C, Ainley D, Bailey M et al (2013) A review of formal objections to Marine Stewardship Council fisheries certifications. Biol Conserv 161:10–17. https://doi.org/10.1016/j.biocon.2013.01.002

FAO (Food and Agriculture Organization of the United Nations) (2015) Voluntary guidelines for securing sustainable small-scale fisheries in the context of food security & poverty eradication. FAO, Rome. http://www.fao.org/3/a-i4356e.pdf. Accessed 15 Apr 2018

Foley P (2012) The political economy of Marine Stewardship Council certification: processors and access in Newfoundland and Labrador's inshore shrimp industry. J Agrar Chang 12:436–457. https://doi.org/10.1111/j.1471-0366.2011.00344.x

Foley P, McCay B (2014) Certifying the commons: eco-certification, privatization, and collective action. Ecol Soc 19(2):28. https://doi.org/10.5751/es-06459-190228

Frangoudes K, Carroll M, Holmyard N et al (2008a) The role of women in the sustainable development of European fisheries areas. European Parliament, Brussels. https://goo.gl/QGSqgb. Accessed 17 Feb 2018

Frangoudes K, Marugán-Pintos B, Pascual-Fernández JJ (2008b) From open access to co-governance and conservation: the case of women shellfish collectors in Galicia (Spain). Mar Policy 32:223–232. https://doi.org/10.1016/j.marpol.2007.09.007

Godwin SC, Francis FT, Howard BR et al (2017) Towards the economic viability of local seafood programs: key features for the financial performance of community supported fisheries. Mar Policy 81:375–380. https://doi.org/10.1016/j.marpol.2017.04.009

Guyader O, Berthou P, Koutsikopoulos C et al (2013) Small scale fisheries in Europe: a comparative analysis based on a selection of case studies. Fish Res 140:1–13. https://doi.org/10.1016/j.fishres.2012.11.008

Jacquet J, Hocevar J, Lai S et al (2010) Conserving wild fish in a sea of market-based efforts. Oryx 44:45–56. https://doi.org/10.1017/s0030605309990470

Josupeit H (2016) Small-scale fisheries markets: value chain, promotion and labelling. European Parliament, Brussels. https://goo.gl/C27CHD. Accessed 17 Feb 2018

Kooiman J, Bavinck M, Jentoft S et al (eds) (2005) Fish for life: interactive governance for fisheries. Amsterdam University Press, Amsterdam

Logan CA, Alter SE, Haupt AJ et al (2008) An impediment to consumer choice: overfished species are sold as Pacific red snapper. Biol Conserv 141:1591–1599. https://doi.org/10.1016/j.biocon.2008.04.007

Macfadyen G, Salz P, Cappell R (2011) Characteristics of small-scale coastal fisheries in Europe. European Parliament, Brussels. https://goo.gl/JaF6nJ. Accessed 17 Feb 2018

McClenachan L, Neal BP, Al-Abdulrazzak D et al (2014) Do community supported fisheries (CSFs) improve sustainability? Fish Res 157:62–69. https://doi.org/10.1016/j.fishres.2014.03.016

Natale F, Carvalho N, Paulrud A (2015) Defining small-scale fisheries in the EU on the basis of their operational range of activity: the Swedish fleet as a case study. Fish Res 164:286–292. https://doi.org/10.1016/j.fishres.2014.12.013

Oosterveer P, Spaargaren G (2011) Organising consumer involvement in the greening of global food flows: the role of environmental NGOs in the case of marine fish. Environ Polit 20:97–114. https://doi.org/10.1080/09644016.2011.538168

Pardo MÁ, Jiménez E, Viðarsson JR et al (2018) DNA barcoding revealing mislabeling of seafood in European mass caterings. Food Control. https://doi.org/10.1016/j.foodcont.2018.04.044

Pascual Fernández J (1991) Entre el mar y la tierra: los pescadores artesanales canarios. Ministerio de Cultura – Interinsular Canaria, Santa Cruz de Tenerife

Pérez-Ramírez M, Castrejón M, Gutiérrez NL et al (2016) The Marine Stewardship Council certification in Latin America and the Caribbean: a review of experiences, potentials and pitfalls. Fish Res 182:50–58. https://doi.org/10.1016/j.fishres.2015.11.007

Pita C, Dickey H, Pierce GJ et al (2010) Willingness for mobility amongst European fishermen. J Rural Stud 26:308–319. https://doi.org/10.1016/j.jrurstud.2010.02.004

Ponte S (2008) Greener than thou: the political economy of fish ecolabeling and its local manifestations in South Africa. World Dev 36:159–175. https://doi.org/10.1016/j.worlddev.2007.02.014

Said A, Tzanopoulos J, MacMillan D (2016) Bluefin tuna fishery policy in Malta: the plight of artisanal fishermen caught in the capitalist net. Mar Policy 73:27–34. https://doi.org/10.1016/j.marpol.2016.07.025

Said A, MacMillan D, Schembri M et al (2017) Fishing in a congested sea: what do marine protected areas imply for the future of the Maltese artisanal fleet? Appl Geogr 87:245–255. https://doi.org/10.1016/j.apgeog.2017.08.013

Schuhbauer A, Chuenpagdee R, Cheung WWL et al (2017) How subsidies affect the economic viability of small-scale fisheries. Mar Policy 82:114–121. https://doi.org/10.1016/j.marpol.2017.05.013

Stoll JS, Johnson TR (2015) Under the banner of sustainability: the politics and prose of an emerging US federal seafood certification. Mar Policy 51:415–422. https://doi.org/10.1016/j.marpol.2014.09.027

Stoll JS, Dubik BA, Campbell LM (2015) Local seafood: rethinking the direct marketing paradigm. Ecol Soc 20(2):40. https://doi.org/10.5751/ES-07686-200240

Swartz W, Schiller L, Rashid Sumaila U et al (2017) Searching for market-based sustainability pathways: challenges and opportunities for seafood certification programs in Japan. Mar Policy 76:185–191. https://doi.org/10.1016/j.marpol.2016.11.009

Van Holt T, Weisman W (2016) Global production network mapping for transforming socio-ecological systems. Curr Opin Environ Sustain 20:61–66. https://doi.org/10.1016/j.cosust.2016.06.003

Wakamatsu M, Wakamatsu H (2017) The certification of small-scale fisheries. Mar Policy 77:97–103. https://doi.org/10.1016/j.marpol.2016.12.016

Chapter 9
Governing for Viability: The Case of Velondriake Locally Managed Marine Area in Madagascar

Lovasoa Cédrique Augustave

Abstract Until recently, Andavadoaka was a small fishing village in the southwest of Madagascar with an economy based on selling dried fish, sea cucumbers, and octopus to collectors. The fisheries went through a major change with the arrival of an export company focusing on fresh octopus trade, while facing many issues, including overexploitation and degradation of the coral reef ecosystem. The concern about sustainability and viability of the fishing communities led to the establishment of a temporary no-take zone, as part of the locally managed marine area (LMMA), facilitated by a partnership between governments, research institutions, environmental organizations, and villagers. The conservation initiative received support from all stakeholders and the model had been transferred to other areas. While the no-take zone is considered successful in reducing the pressure on the octopus fishery, the LMMA faces many challenges, including lack of funding and human resources, resource use conflicts, and lack of compliance, especially by fishers from outside the areas. These problems are complex and interconnected, and cannot be solved without a proper understanding of the whole fisheries system, throughout the entire fish chain. In this chapter, the governability assessment framework is applied to examine the natural and social characteristics of the fishing communities and the institutional and policy requirements for the successful implementation of the LMMA. The chapter offers insights into the LMMA and discusses the role that a transdisciplinary perspective can play in promoting long-term sustainability and viability of fishing communities in the area.

Keywords Octopus fishery · LMMA · Conservation · Small-scale fisheries · Viability · Governability · Madagascar

L. C. Augustave (✉)
Norwegian College of Fishery Science, UiT- The Arctic University of Norway, Tromsø, Norway

R. Chuenpagdee, S. Jentoft (eds.), *Transdisciplinarity for Small-Scale Fisheries Governance*, MARE Publication Series 21,
https://doi.org/10.1007/978-3-319-94938-3_9

9.1 Introduction

Fisheries are important sources of nutrition, employment, and welfare (EC 2015). Under a good management system, they can bring benefits, with a strong ecological and socioeconomic basis for operation, so that they can continuously provide wealth and wellbeing to the local population. As complex and dynamic systems, fisheries need to be addressed, not only on the biological, oceanographic, and ecosystem issues that influence the performance of fish stocks, but also on the social, economic, and institutional considerations that may affect the outcome of the management systems. In short, fisheries represent dynamic systems with interacting components (Charles 2000), and it is these interactions that create both the problems and the opportunities for governance, according to Kooiman et al. (2005).

Using the governability concept, a fishery system can be more or less governable depending on: (i) the characteristics or properties (diversity, complexity, dynamics, and scale) of the natural and social systems that are being governed; (ii) the capability of the governing system to address challenges created within the systems-to-be-governed; and, (iii) the overall quality of their interactions (Jentoft and Chuenpagdee 2009).

In the context of small-scale fisheries, governance challenges are mostly related to the high level of diversity, complexity, and dynamics in fishing communities. The governing capacity tends to be low mainly because small-scale fisheries do not receive high priority in decision making and funding support. This does not mean that small-scale fisheries are less governable than large-scale fisheries, however. A critical aspect determining governability is whether the governing mode corresponds well with the characteristics of the fisheries (Kooiman 2008). Thus, while a top-down, centralized management system might work in industrialized fisheries, studies show that co-governance and community-based management models are more applicable to small-scale fisheries (Jentoft and Chuenpagdee 2015). Providing an environment and building capacity for local fishers to participate in the management of fisheries resources are imperative for strengthening the viability of the communities, as well as for improving the overall governance quality.

Under this premise, this chapter examines the role of fishers and community members in the management of local fisheries resources in a small fishing village, Andavadoaka, on the west coast of Madagascar, where the first Velondriake Locally Managed Marine Area (LMMA) was established in 2004 (United Nations 2012). The Halieutic Institute of Marine Sciences and the UK-based conservation organization, Blue Ventures, started to work together to monitor the region's marine environment with a goal to protect marine and coastal biodiversity whilst improving livelihood sustainability in the region. This led, eventually, to the creation of a temporary octopus no-take zone (NTZ) within the coral reefs areas. The fisheries closure was established based on the partnership between the villagers, the research institution, and non-governmental organizations (NGOs). The villagers took full control over the management of the NTZ by means of the Velondriake Association, a management committee for the LMMA. The Velondriake LMMA was deemed to a good model, and had been replicated in other locations on the southwest coast of Madagascar. The government granted financial support to the community, and implemented 50 more NTZs along the coast through the project, *Projet d'Appui aux Communautés des Pêcheurs*, funded by the African Development Bank, which aims to support small-scale fisheries through sustainable development of the traditional marine fisheries in South West Madagascar (Cripps and Harris 2009; Mayol 2013).

While the NTZs were favourably seen as a good management measure, the overall idea of the LMMA management model and the actual role of the NTZs in supporting livelihoods and promoting wellbeing need to be carefully examined. This is particularly important because not all the NTZs are successful, and some implementations face serious obstacles. From a governance perspective, it is important to recognize that the NTZs may not work in all contexts, similar to other area-based management tools, especially marine protected areas (see Jentoft et al. 2015). As such, the effectiveness of the LMMA in other coastal communities may need to be evaluated before further expansion.

This chapter aims to enhance understanding about the LMMA as a governing institution and the system boundary of the study. Specifically, the study applies the governability assessment framework to examine the small-scale fisheries system in the village of Andavadoaka, and analyses the conditions and factors that foster or inhibit the successful implementation of the Velondriake LMMA, particularly the NTZs. Beginning with the background and setting of Andavadoaka, the LMMA is introduced, followed by the results of the governability analysis. The chapter concludes with a discussion about the role that a transdisciplinary perspective can play in analysing the governance challenges and in improving fisheries governance for long-term sustainability and viability of the fishing communities in the area.

9.2 The Village and the Fisheries

Andavadoaka is a small fishing village located on the southwest coast of Madagascar, in the Morombe division of Toliara Province (Fig. 9.1). The village lies on the edge of a shallow lagoon protected from the open ocean by a series of reefs that support substantial coral growth, providing a vital resource base for local traditional and artisanal fisheries. The 1200 villagers of Andavadoaka are primarily Vezo, a semi-nomadic sea-faring people who rely on healthy ocean waters and these coral reefs to support their fishing livelihoods, and for food and income.

Not long ago, the village economy was based on selling of dried fish, sea cucumbers, and octopus to collectors. However, the arrival of the export company known as '*Compagnie de pêche frigorifique de Toliara*' (Copefrito) changed the target of their fisheries to focusing mainly on fresh octopus (*Octopus cyanea*), large reef fish, such as emperor (*Lethrinus nebulosus*, *L. harak*, *L. obsoletus*), snapper (*Lutjanus sebae*, *L. notatus*, *L. bohar*), and grouper (*Epinephelus fasciatus*, *E. hexagonatus*, *E. rivulatus*). Year after year, villagers experienced signs of overexploitation of the stocks, which led them to seek a sustainable fisheries approach.

Fig. 9.1 The Southwest coast of Madagascar showing Andavakoaka, Nosy Fasy and Nosy Hao. (Source: image satellite from google Earth 2018)

Octopus cyanea is the most exploited and abundant species in the Andavadoaka region, representing about 95% of local catches (Benbow et al. 2014). Known as big blue octopus, the organism is found on reefs and in shallow waters in the Indo-Pacific. Hunting by day, *O. cyanea* are dioeceous and females lay between 150,000 and 700,000 eggs in a single clutch (Caveriviere 2006; Van Heukelem 1973). The reproduction of the organism has been documented to occur throughout the year in both Tanzania and Madagascar, with reproductive peaks in June and December (Caveriviere 2006).

As the most economically important fishery in southwest Madagascar (Raberinary and Benbow 2012), heavy pressure has been put on the octopus stock and concerns were raised about the overexploitation of the fishery. A call for better management of the octopus fishery was made, in order to implement measures to protect and preserve the octopus. The original idea came in 2001 when Blue Ventures realized, through research in cooperation with Halieutic Institute of Marine Sciences, that detrimental fishing techniques and population pressures were causing the decline in the octopus catches. As such, Blue Ventures began its campaign in 2003 and invited other NGOs to join in the design of a research and conservation project that would be managed by local communities to save the threatened marine ecosystems (SEED 2018).

One of the main outcomes of these efforts was the partnership's promotion of environmental education and raising of awareness about the importance of marine resource management, with the intention of establishing Madagascar's first community-run marine protected area. The coastal fishing village of Andavadoaka was chosen as the pioneer for implementing the first temporary octopus fishery closures, in other words, the NTZ. The project won the SEED Award[1] in 2005 for its pilot project design (SEED 2018). The positive experience in Andavadoaka village led to the replication of the same octopus fisheries initiative along the coastline, and eventually to the formation of Velondriake Locally Managed Marine Area (LMMA) in 2006 (United Nations 2012). Nowadays, the periodic octopus fishery closure form the bedrock of broader marine resource management initiatives centred on Andavadoaka (Gardner et al. 2017).

9.3 The Governability of Velondriake LMMA

As an experimental temporary octopus closure, Velondriake LMMA was considered a good conservation model. Velondriake, the name of the conservation plan, means '*to live with the sea*'. It is a network of marine and coastal protected areas aimed at protecting biodiversity from unsustainable overexploitation and ensuring that these marine systems remain healthy and productive for future generations. Like many protected areas initiatives, the fishery closure represented a big barrier for some fishers who depend on fishing as the main livelihood. Many of them also felt that the octopus fishery closure was imposed on them. As such, Velondriake LMMA was never fully welcomed by all fishers, and opposition began. Poachers and free riders, like fishers from outside the village, complicated the issue by reaping the benefits of the increased fish stocks. In addition, the lack of basic infrastructure and the remote location of Velondriake villages made it challenging for effective monitoring, surveillance, and control. As a governing institution, the LMMA has yet to fully meet its purposes.

[1] SEED is a global partnership for action on sustainable development and the green economy, supported by United Nations Development Programme and other organizations. The SEED Award is an annual award given to innovative and locally led entrepreneurial projects in developing and emerging economies countries (https://www.seed.uno/awards/about.html).

There are many aspects related to the Velondriake LMMA experiences that are in accord with the nature of the 'wicked problems' described by Rittel and Webber (1973). There is no real consensus about what the problems are, for instance, whether the decline in the octopus stock in Andavadoaka and the deterioration of coral reefs are due to fishing pressure alone or combined with other factors. Further, the octopus NTZ in Velondriake LMMA was deemed successful by some governing actors, including those who initiated it, but not necessarily so for other stakeholders, particularly some fishers in the village. While it is not likely that a single natural or social variable can explain the situation, understanding the relationship and the interaction between the components of the natural and the social systems that are the targets of governance can help make necessary structural adjustments[2] to the governing system, thus increasing the overall quality and governability (Jentoft 2007). The application of the governability assessment framework on the Velondriake LMMA can help explain, in a detailed and comprehensive manner, the challenges that the LMMA is facing and what may be done to improve its governance. As posited by Kooiman et al. (2005), factors that foster or hinder governance are mostly related to the diversity, complexity, dynamics, and the scale of the natural and the social systems that are being governed, the governing system, as well as the governing interactions occurring in the area. The key findings of the governability analysis are presented below.

9.3.1 The Diverse, Complex and Dynamic Fisheries Systems

The Andavadoaka reef system is characterized by three types of reefs: inshore fringing reefs, seaward fore-reefs, and lagoonal patch reefs (Gillibrand et al. 2007), serving as habitats for the economically important octopus (Raberinary and Benbow 2012), finfish, and other marine species. When looking at the natural system-to-be-governed, the LMMA offers many exploitable species, such as the large reef and pelagic species of fish, that are living in the coral reefs, but the reef octopus, *O. cyanea*, is seen to be the most vital commodity for the community of Andavadoaka. Among the fish species, about 334 species from 58 families are found (Gillibrand et al. 2007). Besides the common invertebrates and coral reef fish, many rare, protected or endangered species have also been sighted on reefs in the region. Some of these are the napoleon wrasse (*Cheilinus undulatus*), the giant grouper (*Epinephelus lanceolatus*), as well as several species of sharks, rays, and sea turtles. All of these sightings occurred on the barrier and patch reefs (Nadon et al. 2005), suggesting important relationships between marine animals and the reef habitats. For these reasons, the reefs are the key feature in the Andavadoaka ecosystem requiring protection (Nadon et al. 2005). Prohibition of octopus fishing in the reef areas during certain a period of time, at least for 6 months long, was, therefore, considered important to implement, in order to protect octopus from destructive fishing methods (Blue Ventures 2004).

[2] Structural adjustments refer to any kind of program/workshop funded by proponents in light of improving governability.

The diverse and complex natural ecosystem enables lucrative fisheries in the area. A number of fisheries stakeholders are involved in the fisheries value chain, including large-scale and small-scale operators,[3] fish collectors, and export companies. Several NGOs operate in the area, namely Blue Ventures, Wildlife Conservation Society, and World Wide Fund for Nature (WWF). When adding researchers and research institutions, the diversity and complexity of the social system-to-be-governed are likely high, which may lead to lower governability, unless these actors have good relationships and interactions.

The governability of the LMMA is also affected by changes in both the natural and the social systems. The Malagasy coral reefs are undergoing threats from at least three sources: (i) bleaching caused by abnormal warm surface waters; (ii) excessive sedimentation; and, (iii) overfishing (Gabrié et al. 2000). The El Niño event in 1998 caused a rise of the sea temperature and, therefore, severely affected the reefs in the region. As for the sedimentation, it has been reported that the effects on the reefs are not too serious as the nearest rivers are at least 100 km away to the north (Mangoky river) and to the south (Manombo river) (Nadon et al. 2005). Overfishing seems to be an issue, however, for the community of Andavadoaka, as the population grows at a high rate of 4% a year, and fishing is the primary economic activity for 71% of the villagers (United Nations 2003). Besides, some fishing methods that are used to catch commercial species, like octopus and sea cucumbers, are destructive. For instance, studies show that the use of illegal beach seines in the area have destroyed the physical structure of the reefs, leading to major changes in the biodiversity (Blue Ventures 2004).

The changes in the post-harvest also have direct impacts on the governability of the LMMA. The commercial fisheries have long been associated with the traditional dried and salted fish. However, a dramatic dwindling (90% since the 1950s) of the fish stocks, in numbers and in diversity (United Nations 2003), changed the market demand to fresh products, such as octopus. Export to France was increased by 35%, for instance. Octopus catches are sold to the local collectors, who are employed by the export company, Copefrito. Because fishers rely entirely on private collectors to access markets, they have little negotiation power over prices (Gardner et al. 2017). Overall, fishers do not have high security over their fishing livelihoods. In addition to variability in the natural system due to seasonality, the physical extent of the coral reefs serves as a natural boundary for the octopus fisheries in the LMMA. Fishers also lack technology and gear, like proper diving equipment, that would allow them to operate in the deeper and remote fishing areas. The dynamics of the natural and social systems cause insecurity and unpredictability that affect the governability of the LMMA.

[3] Large-scale and small-scale operators refer respectively to fisheries' semi-industrial and artisanal fleets.

9.3.2 *The LMMA as a Governing Institution*

The LMMA governing system consists of the regional octopus committee, known as Velondriake Association, which oversees the management of the area and ensures timely decision making and execution of the management actions. Local government authorities are involved in the committee, which helps strengthen the institutional support for Velondriake. As the LMMA involves 23 villages, a sub-committee referred to as 'Vondrona' is also created in each village to facilitate the local management.

The work of the Velondriake Association and the Vondrona in the management of the LMMA, and in establishing the NTZs, is supported by an advisory group, comprising of researchers from Halieutic Institute of Marine Science and NGOs, such as Blue Ventures, Wildlife Conservation Society, and WWF. Research and stock assessments are, for instance, conducted by researchers from these organizations, which are then used to set the term for the conservation of the octopus fisheries (Fig. 9.2). Blue Ventures, in particular, works closely with the Velondriake Association, not to implement projects directly, but to provide the committee with materials and technical support to enable them to manage and regulate the fisheries in the area (Blue Ventures 2004). As part of this support, Blue Ventures also help the Velondriake Association hold meetings and discuss possibilities for local rules and management strategies. With inputs from the export companies and other researchers, the Velondriake Association implements the NTZs as temporary octopus fishery closures for 7 months of a year.

As suggested in Fig. 9.2, however, fishers are at the receiving end of these rules and regulations, which explains some of the violations observed in the area, thus lowering the governability of the LMMA. While some fishers may have been part of the early discussion about the LMMA and the NTZ, the diversity observed in the fishing groups means that getting all of them to agree with the conservation and the management plan would require time and effort, and a process that they consider legitimate.

An interesting aspect in the governing system is the establishment of local rules called 'Dina' to manage the octopus stock within the NTZ. While Dina was crafted based on inputs from the community, commercial trawlers regularly operating within the shallow coastal waters were unmindful of these fisheries restrictions that have been agreed by the local resource users (United Nations 2003). The same applies to fishers migrating annually from other villages, who are mostly unaware

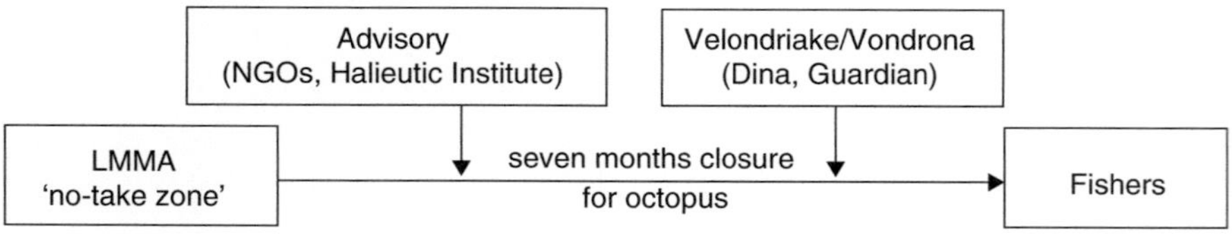

Fig. 9.2 LMMA management scheme

of the Velondriake Dina (Cripps and Harris 2009). These outside fishers often broke the local law by practicing illegal fishing in the villages. The application of the local law to individuals from outside of the community has proven to be a protracted and political process, with the decisions being biased as the exterior communal authority must look after their electorates (Cripps and Harris 2009). In addition, fishers using illegal gears in the area, like beach seines, also dismiss these local rules. The non-compliance on Dina affects the performance of the NTZs and the governability of the LMMA, as a management model in the area.

The interactions between the stakeholders and the management committees were not easy to facilitate. The application of the Dina has become particularly difficult in four villages: Ankindranoke, Ampasilava, Bevato, and Belavenoke. The greatest barrier to effective enforcement of Dina was that villagers were unwilling to apply Dina for fear of causing offence to their own family lineage (Cripps and Harris 2009). In addition, the rules and regulations applied to the Velondriake are found to be vulnerable to local politics, which often weaken the authority of the Velondriake Association in the villages. When Velondriake was created, it was appealing and supported by the communal mayor and its political party, in which two of them were member of the Velondriake management committee. The arrival of a new mayor who came into power with a different political regime that opposed to the functionality of Velondriake system had weakened the authority of Velondriake. As a result, the application of Dina was often disregarded in some villages where the mayor came from.

Managing the ecosystems and conflicting needs of resource users is a major challenge to environmental management decision makers in the region (United Nations 2003). As the conservation measures are locally based, there is a need for communication between fisheries authorities working at the state level government and the Velondriake Association. Besides, the implication of community associations' status as a non-profit entity (as stipulated by Malagasy law 60,133) was interpreted by the government in such a way that they cannot perform income-generating activities and, therefore, cannot fundraise for the project. Their only income is from the LMMA member fees (Mayol 2013). The lack of financial support presents a significant barrier to the long-term financial sustainability of the LMMA.

In June 2012, during the LMMA forum, the village leaders raised their voice on how the community should proceed to legally generate income in order to sustainably finance the LMMA (Mayol 2013). The forum was hosted by Blue Ventures and the Velondriake Association in Andavadoaka and involved 55 representatives from 18 LMMAs, comprising of 134 coastal villages. Other NGOs and LMMA partner institutions, such as Halieutic Institute of Marine Sciences, Cetamada, Conservation International, Honko, the Wildlife Conservation Society, and WWF were also present. The forum aimed to discuss the common challenges facing the communities with regards to marine resource sustainability and management, and to share possible solutions to meet these challenges. The second part of the forum was held in Toliara, south of Andavadoaka, and involved government representatives from all the relevant regional agencies, particularly the Ministry of Fisheries and one representative from Madagascar's protected areas system. Thus, round two offered an opportunity for communication and for providing government support to make the LMMA function as intended.

9.4 Lessons from the Velondriake LMMA

The Velondriake LMMA has several elements to make it a successful governance models. However, as revealed by the governability analysis, the LMMA faces many challenges, starting with the relationship between fishers and the NTZs. The importance of fisheries for the wellbeing of the Andavadoaka villagers was the primary reason for the establishment of the NTZs. As catches decreased, the village leaders sought a more sustainable method of fishing, which ended up in the first conception of the NTZ. The idea of creating the NTZ was not welcomed by all fishers. Their reaction ranged from suspicion to disapproval (United Nations 2003), which suspended the initial idea, until it was further debated and eventually approved. Fishers' reservations were related to the fear of losing the privilege of free access, thereby leading to a lack of income. In order to gain acceptance and support, intensive workshops were organized to raise community awareness about potential long-term benefits that the NTZ would provide. The involvement of the NGOs since the step zero, particularly Blue Ventures and the research institute, Halieutic Institute of Marine Sciences, was critical to the process.

Madagascar is ranked among the least food secure nations in the world (GFSI 2015; Gardner et al. 2017). The high dependence on small-scale fisheries for both food security and income (Le Manach et al. 2012; Barnes-Mauthe et al. 2013; Gardner et al. 2017) does not help facilitate the implementation of the LMMA. The inclusion of the local law, Dina, in the project management plan aimed to ensure that villagers respect the 6–7 month closure period of the NTZ in order to allow octopi to grow to a marketable size, as well as to avoid fishing the endangered species. However, the law has been violated many times, even after the enforcement. It has been reported that 70% of the fishers illegally dive at night with torches within the reserves (Cripps and Harris 2009), thus lowering the governability of the NTZ. Villagers took actions against these law-breakers by recruiting LMMA guardians. Unfortunately, socioeconomic issues, like poverty, persist and some fishers continue to violate the law in order to make a living. Deep conflicts between villagers over its application occur as the result. To address the problem, Blue Ventures along with Madagascar's wider marine conservation network, clarified the legal power of Dina. They were also able to secure the institutional support from the Ministry of Fisheries to sponsor an official resource management officer for the area (SEED 2018). The implementation of such solutions has reduced the problem to some extent, improving the governability of the LMMA.

Despite the challenges, there are positive lessons from the application of the NTZ in the Andavadoaka region. These include, but are not limited to, the economic benefits that the local communities earned from the conservation action plan, the satisfaction of being heard and to be part of the direct management process of the NTZ, and the major social behavioural change of the community to appreciate and strive for more NTZ to be implemented in the future. The Velondriake LMMA has done well in involving local and regional government representatives in its work since the inception, and has also been able to influence government approaches to community resource management through its demonstrable successes.

It should be noted, however, that despite its national importance, the research institute, Halieutic Institute of Marine Sciences, does not yet have the technological or human capacity to perform in-house analysis to independently inform coastal management decisions (Blue Ventures 2004). The future of the LMMA lies in the continuation of the partnership between these various stakeholders, but the financial support from the government and the NGOs is very limited and only lasts for a certain period of time.

The management of the NTZ takes place within a co-governance system between local institutions and stakeholders, through which a strong commitment and involvement of all stakeholders are key to the process. This type of governance at a local level allows fishers to tailor and blend traditional and modern management practices to best suit their needs with support from the government authorities. The incorporation of Dina by the Velondriake Association into the classification of its temporary and permanent reserves reflects a national process of institutionalizing these customary laws, and has rooted conservation regulations in both local traditions and decentralized government authority (United Nations 2012). Furthermore, stakeholders gather together and help to carry out research, observations, monitoring, conservation, and to establish management plan of the NTZ. The LMMA concept expands to thriving national and regional networks throughout the Indo-pacific, Africa, Western Indian Ocean, and Latin America regions, which is collectively known as the LMMA Network (Govan et al. 2008; Mayol 2013). The success of this governance mode, thus, has no boundary and can expand beyond the local scale.

9.5 Strengthening the Base with Transdisciplinary Perspective

Based on the governability assessment performed in the Andavadoaka LMMA, one can say that it is difficult to manage diverse, complex, and dynamic natural and social systems as well as to put in place a robust governing system that delivers good performance in governance. This is largely because any form of management depends, to a large extent, on good interactions within and between stakeholders in the governing system and in the system that is being governed. In addition to communication and agreements, stakeholders and institutions need to be committed to the process and to the agreed upon rules and regulations. As previously discussed, technical solutions related to the financial support for research and monitoring activities, the fundamental change in improving respect of local law and rule compliance, and enhancing the understanding of the overall systems are required. It is in this context that the transdisciplinary perspective can play a role in expanding options for the communities, both in terms of conservation and economic opportunities, which should then contribute to improving the governability of the LMMA.

According to Lang et al. (2012), a range of actors needs to work together to deal with real-world problems. Through interactive and participatory research, they need to reconcile values and preferences, and integrate the best available knowledge to understand the problems and explore possible options.

The concept of transdisciplinarity is already reflected through the Velondriake LMMA, with stakeholders working together to address a common problem and create new knowledge. However, the focus was very confined to achieving conservation and protection of the resources. The conservation problem in the area is a classic wicked problem that is difficult to solve. Even as the challenges related to the LMMA have been addressed through partnership between several fisheries stakeholders and involving the local communities, the problems still persist. Lawbreakers, conflicts between villagers, use of detrimental fishing gears, and local politics are all in constant motion and constitute day-to-day challenges, hampering the governability of the LMMA. Finding a common ground between the different stakeholders can be facilitated through a transdisciplinary process that emphasizes the importance of the sociocultural aspects of the villages. This involves asking questions, like why some villagers follow the local rules, while others do not. The relationship between poverty and illegal fishing activities and rule violation needs to be examined. Such knowledge is crucial for enhancing governability, and should not go unnoticed by researchers. The issue is linked to personal identity, brought to life by a continuous exposure of an individual in a set of social phenomena, and is, therefore, believed to make a great impact in human thoughts, feelings, and behaviours. Thus, the repetitive exposure to a particular social phenomenon creates a 'culture' that defines each individual and explains the way one works and responds to any situations he/she is in. Through this, new knowledge is gained and can contribute to improving governability through better alignment of the management measures in accordance with human sociocultural behaviours, and vice versa.

Research in new markets and technological developments that broaden economic opportunities and strengthen viability for the local communities can help reduce dependency on the fisheries. Villagers of Andavadoaka base their income not just on the octopus trade, but also on various alternatives, such as the development of mariculture and ecotourism. As such, communities should equitably benefit from these alternative income-generating opportunities and take an active role in the implementation of the Voluntary Guidelines for Securing Sustainable Small-Scale Fisheries (FAO 2015). These two activities allow the communities to maintain income and employment during the NTZ closure time. Mariculture provides the community with two possible farming activities: seaweed (*Euchema denticulatum*) and sea cucumber (*Holothuria scabra*). The partnership between Madagascar holothurie, Blue Venture and the Velondriake Association made it possible to implement sea cucumber farming using hatchery-reared juveniles produced in Toliara.

The project has found to be successful and has been expanded to two villages: Belavenoke and Ambolimoke. The seaweed farming has three sites that have been selected for an experimental growing lines using local seaweed: Tampolove, Lamboara, and near Ampasilava. *Euchema* has grown successfully, however, the growing rate is not sufficient to make the activity commercially viable (Cripps and Harris 2009). With respect to eco-tourism, several activities have been promoted, including a wide range of homemade and handmade activities, from cooking to embroidery, or accessories for sale to tourists by the Velondriake Women's

Associations. The latter is extremely valuable as it aligns well with the call in the SSF Guidelines to promote and support the work of women.

Lastly, the transdisciplinary perspective can help improve the overall situation by shifting focus from improving the LMMA to understanding the underlying values, images, and principles of the stakeholders involved (Song et al. 2013). Following interactive governance, the look at these meta-order elements can help demystify images, clarify the ideas about what is and what should be, and make explicit the different values held by stakeholders (Jentoft et al. 2010; Luomba et al. 2016). For instance, while some fishers participated in the discussion about the Velondriake LMMA, they may not share the same images about what the LMMA would represent to researchers or government officials. Transdisciplinary practices based on understanding the meta-order can help facilitate communication and improve interactions between stakeholders within and between communities, as well as with government authorities.

The commitment and synergy between the stakeholders in the Velondriake LMMA has led the project into its achievements, with replication and impacts on the government policy (United Nations 2012). Yet, as the governability assessment of the Velondriake LMMA reveals, a diversity of limitations and challenges exists in terms of project conception and implementation, and the project performance and management. Ways and mechanisms that can be used to address these problems beyond the traditional solutions can be explored through a transdisciplinary process, involving all stakeholders in a similar way to that of the Velondriake LMMA, but with due attention to the values, principles, and images that different stakeholders may hold. This also implies that the replication of the Velondriake LMMA model to other areas be re-examined and revisited before further expansion. As Rittel and Webber (1973) articulate about the wicked problems, there is a certain temptation to apply the same tool to a problem that seems familiar. Lessons from the governability assessment of the Velondriake LMMA reiterate the importance of context in the governance of small-scale fisheries, and the application of the transdisciplinary perspective to the problems experienced in the LMMA can help broaden the thinking around tools, approaches, and mechanisms that can be employed to address the problems.

References

Barnes-Mauthe M, Olesen KLL, Zafindrasilivonona B (2013) The total economic value of small-scale fisheries with a characterization of post-landing trends: an application in Madagascar with global relevance. Fish Res 147:175–185

Benbow S, Humber F, Oliver TA et al (2014) Lessons learnt from experimental temporary octopus fishing closures in south-west Madagascar: benefits of concurrent closures. Afr J Mar Sci 36(1):31–37. https://doi.org/10.2989/1814232X.2014.893256

Blue Ventures (2004) Marine protected area the context Andavadoaka: an opportunity for community based marine resource management. GlobalGiving, pp 1–6

Caveriviere A (2006) Principaux traits de vie du poulpe Octopus cyanea en zone tropicale. MADADOC, Antananarivo, pp 39–47
Charles AT (ed) (2000) Sustainable fishery systems. Blackwell Science Ltd., Oxford. https://doi.org/10.1002/9780470698785
Cripps G, Harris A (2009) Community creation and management of the Velondriake marine protected area. Blue Ventures Conserv 6(3):692–702. https://doi.org/10.1038/ismej.2011.124
EC (European Commission) (2015) Sustainable fisheries for sustainable development, fisheries. https://ec.europa.eu/fisheries/sustainable-fisheries-sustainable-development_en. Accessed 28 Sept 2017
FAO (Food and Agriculture Organization of the United Nations) (2015) Voluntary guidelines for securing sustainable small-scale fisheries in the context of food security and poverty eradication. FAO, Rome
Gabrié C, Vasseur P, Randriamiarana H et al (2000) The coral reefs of Madagascar. In: McClanahan T, Sheppard CRC, Obura DO (eds) Coral reefs of the Indian Ocean. Oxford University Press, New York
Gardner JC, Rocliffe S, Gough C et al (2017) Value chain challenges in two community-managed fisheries in western Madagascar: insights for the small-scale fisheries guidelines. In: Jentoft S, Chuenpagdee R, Barragán-Paladines MJ et al (eds) The small-scale fisheries guidelines: global implementation. Springer, Cham, pp 335–354
GFSI (2015) Global food security index 2015. The economist intelligence unit, London
Gillibrand CJ, Harris AR, Mara E (2007) Inventory and spatial assemblage study of reef fish in the Area of Andavadoaka, South- West Madagascar (Western Indian Ocean). West Indian Ocean J Mar Sci 6(2):183–197
Govan H, Aalbersberg W, Tawake A et al (2008) Locally-managed marine areas: a guide to supporting community-based adaptive management. Locally-Managed Marine Area (LMMA) Network, Suva
Jentoft S (2007) Limits of governability: institutional implications for fisheries and coastal governance. Mar Policy 31(4):360–370. https://doi.org/10.1016/j.marpol.2006.11.003
Jentoft S, Chuenpagdee R (2009) Fisheries and coastal governance as a wicked problem. Mar Policy Pergamon 33(4):553–560. https://doi.org/10.1016/J.MARPOL.2008.12.002
Jentoft S, Chuenpagdee R (eds) (2015) Interactive governance for small-scale fisheries: global reflections. Springer, Cham
Jentoft S, Chuenpagdee R, Bundy A et al (2010) Pyramids and roses: alternative images for the governance of fisheries systems. Mar Policy 34:1315–1321
Jentoft S, Pascual-Fernandez JJ, la Cruz Modino D et al (2015) What stakeholders think about marine protected areas: case studies from Spain. Hum Ecol 40:185–197. https://doi.org/10.1007/s10745-012-9459-6
Kooiman J (2008) Exploring the concept of governability. J Comp Policy Anal: Res Pract 10(2):171–190. https://doi.org/10.1080/13876980802028107
Kooiman J, Bavinck M, Jentoft S et al (eds) (2005) Fish for life: interactive governance for fisheries. Amsterdam University Press, Amsterdam. https://doi.org/10.5117/9789053566862
Lang DJ, Wiek A, Bergmann M et al (2012) Transdisciplinary research in sustainability science: practice, principles, and challenges. Sustain Sci 7(Suppl. 1):25–43. https://doi.org/10.1007/s11625-011-0149-x
Le Manach F, Gough C, Harris A et al (2012) Unreported fishing, hungry people and political turmoil: the recipe for a food security crisis in Madagascar? Mar Policy 36:218–225
Luomba J, Chuenpagdee R, Song MA (2016) A bottom-up understanding of illegal, unreported, and unregulated fishing in Lake Victoria. Sustainability 8:1062. https://doi.org/10.3390/su8101062
Mayol T (2013) Madagascar's nascent locally managed marine area network. Madag Conserv Dev 8(2):91–95. https://doi.org/10.4314/mcd.v8i2.8
Nadon MO, Griffiths D, Doherty E (2005) The coral reefs of Andavadoaka, southwest Madagascar. Blue Ventures, London, p 29

Raberinary D, Benbow S (2012) The reproductive cycle of Octopus cyanea in Southwest Madagascar and implications for fisheries management. Fish Res 125–126(126):190–197. https://doi.org/10.1016/j.fishres.2012.02.025

Rittel HWJ, Webber MM (1973) Dilemmas in a general theory of planning. Policy Sci 4:155–169

SEED (2018) 2005 Seed winner Madagascar's first community-run marine protected area. https://www.seed.uno/images/casestudies/SEED_Case_Study_Blue_Ventures_Madagascar.pdf. Accessed 10 Apr 2018

Song AM, Chuenpagdee R, Jentoft S (2013) Values, images, and principles: what they represent and how they may improve fisheries governance. Mar Policy 40:167–175

United Nations (2003) Village of Andavadoaka, Madagascar: marine reserves for octopus. In: Innovation for sustainable development: local case studies from Africa. United Nations, New York, pp 14–17

United Nations (2012) Equator initiative case studies local sustainable development solutions for people, nature, and resilient communities. Village of Andavakoaka, p 17

Van Heukelem WF (1973) Growth and life-span of *Octopus cyanea* (Mollusca: Cephalopoda). J Zool 169:299–315

Part IV
Enhancing the Stewardship

Chapter 10
Stewardship and Sustainable Practices in Small-Scale Fisheries

Patrick McConney, Rodrigo Pereira Medeiros, José J. Pascual-Fernández, and Maria Pena

Abstract Humans have harvested fish and other marine resources for millennia. So why be concerned today about stewardship and sustainable practices? The answers are not simple; they are changing over time, as climate changes, and perspectives vary. In this chapter, we examine the diverse facets and factors associated with stewardship and sustainable practices in small-scale fisheries, from a transdisciplinary perspective. We look at sustainability and stewardship through different lenses and examine some of the most prominent and promising initiatives applied on different scales, at multiple levels of governance, with diverse stakeholders. It is a tour that includes ecology, socioeconomics, and governance in combination rather than from one viewpoint. Critically important instruments include the Code of Conduct for Responsible Fisheries and the Small-scale Fisheries Guidelines. Topics touch on fishing gear and methods; marine protected areas; livelihoods; illegal, unreported, and unregulated fishing; postharvest and trade; organizations ranging from fisher cooperatives to the Marine Stewardship Council; and the sustainable practices associated with each. Our approach is one of agency in keeping with the central theme of enhancing stewardship to make a difference.

Keywords Stewardship · Sustainability · Small-scale · Fisheries

---

The research by Jose Pascual-Fernández in this chapter was conducted under the project 'Governance challenges for sustainable small-scale fisheries: creating synergies with marine conservation and tourism' (GOBAMP II, CSO2013-45773-R, financed by Ministry of Economy and Competitiveness of Spain. We also acknowledge support from the 'Too Big To Ignore: Global Partnership for Small-Scale Fisheries Research'.

P. McConney (✉) · M. Pena
University of the West Indies, Cave Hill, Barbados
e-mail: patrick.mcconney@cavehill.uwi.edu; maria.pena@cavehill.uwi.edu

R. Pereira Medeiros
Universidade Federal do Paraná, Apucarana, Brazil
e-mail: rodrigo.medeiros@ufpr.br

J. J. Pascual-Fernández
Instituto Universitario de Investigación Social y Turismo, Universidad de La Laguna, Santa Cruz de Tenerife, Spain
e-mail: jpascual@ull.edu.es

R. Chuenpagdee, S. Jentoft (eds.), *Transdisciplinarity for Small-Scale Fisheries Governance*, MARE Publication Series 21,
https://doi.org/10.1007/978-3-319-94938-3_10

10.1 Introduction

The notion of stewardship is fundamentally transdisciplinary. So are the sustainable practices that implement the very broad concept of stewardship in fisheries-related activities. In this chapter, we introduce both the conceptual and practical aspects of stewardship. First, we explain the terms, followed by examples to illustrate their application in small-scale fisheries. Given the breadth of how the concept has been interpreted and used by diverse disciplines and fisheries stakeholders, and has evolved over time, we cannot cover all facets of stewardship. We provide an overview of some of the more current and practical globally important topics.

The fisheries value chain—some say from hook to cook—provides a means of organizing our examples by fisheries activities and associated livelihoods. Activities and livelihoods cut across many other topics in this book. Examples of inter-sectoral approaches to stewardship and sustainable practices, such as marine protected areas (MPAs), integrated coastal zone management (ICZM), and the sustainable development goals (SDGs), are addressed in relation to the challenges experienced along the entire fisheries value chain. Challenges include habitat degradation, resource depletion, market failures, inequitable distribution of benefits and burdens, various deficiencies in governance, and the threat of all being exacerbated by impacts of climate change and variability.

In examining the concepts, we touch on initiatives that have redefined our thinking about stewardship and sustainability, such as the Marine Stewardship Council (MSC) and the rapidly growing Slow Food movement. Indeed, browsing topics is all that can be accomplished in a short chapter dealing with issues that have long histories, from before biblical times to the present. Because of this, we also provide you with links and recommendations for additional resources that can broaden your knowledge. Read on to be enlightened and make your own discoveries.

10.2 Key Concepts

Let us briefly examine the concepts of 'stewardship' and 'sustainable practices', linked by the notion of 'transdisciplinarity'. These three terms convey different meanings, some depending on discipline (Bennett et al. 2018). However, since our focus is on transdisciplinarity, our aim is to, primarily, achieve a shared understanding of the two concepts to allow us to use them consistently, for innovation and to practically solve problems. We use transdisciplinarity to mean "an applied practice, evolving from current problems of the world that need to be practically solved… throughout the continuous work and collaboration between researchers from different scientific disciplines, stakeholder participation, as well as communication and dissemination" (Angelstam et al. 2013, 256). There are, of course, other definitions including some that differently define research, teaching and learning, and practice.

The first part of our definition speaks mainly to sustainable practices, and the latter part mainly to stewardship.

10.2.1 Stewardship

For many, the concept of stewardship has its origins in theology, found in several religions. In this context, stewardship is often interpreted as humans having dominion over the Earth. The consequent belief is that humans are responsible for the world, and should take care of it (for God) (Worrell and Appleby 2000). Although many authors refer to ecosystem or environmental stewardship, we prefer to leave the term unqualified since it draws upon so many different perspectives, besides nature. Unpacking stewardship, there is the notion that since humans have inherited the world, they:

- Should have a sense of **ownership** over natural resources
- Need to exercise both individual and collective **responsibility**
- Will demonstrate **accountability** in stewardship within society
- May anticipate some sort of **reward** for being good stewards.

Combining theology with recent transdisciplinary approaches to natural and social sciences we see that stewardship includes property rights, collective action, and incentives related to species, habitats, and ecosystems (Medeiros et al. 2014). Stewardship is cross-scale and multi-level with regard to interactive governance (Jentoft and Chuenpagdee 2015). It incorporates the scales of time, space, institutions, jurisdictions, and more, (Cash et al. 2006) at multiple levels of governance and human organization, which range from the individual to planetary in scope (Galaz et al. 2012). The above constitutes our mosaic definition of the term, as it is used in this chapter. We must not only fully understand conceptual underpinnings, but also know how to practically enhance stewardship in small-scale fisheries. Transdisciplinarity assists fisheries stakeholders in real-life situations.

In the early days of Too Big To Ignore (TBTI), the global partnership for small-scale fisheries research, a group of collaborators examined the issue of enhancing stewardship (McConney et al. 2014a). They used a practical framework that said first, one had to understand the social-ecological impacts on, and of, fisheries. Then, armed with that shared understanding, stakeholders needed to engage in participatory monitoring and evaluation of the bidirectional impacts to also share deeper appreciation of trends and shocks in both the natural and social components of the small-scale fisheries social-ecological system. Finally, stewardship could be enhanced when the stakeholders took collective action to use the shared information to make decisions on responsible fisheries. Thus, stewardship is expressed, and can be observed, at the level of the individual (especially in leaders) and at the level of the collective (as whatever informal or formal structure it may take). This progression, along with feedback pathways (illustrated by grey counterflow arrows), is shown in Fig. 10.1.

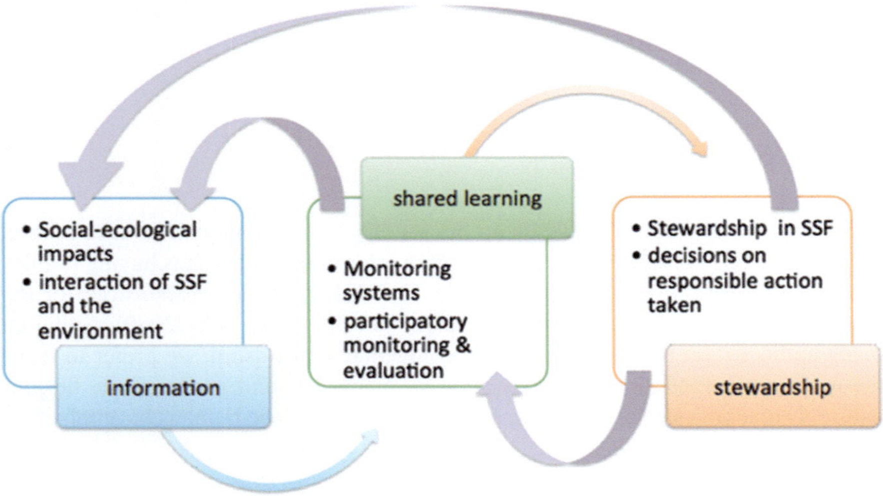

Fig. 10.1 Framework for "Enhancing the stewardship". (Source: McConney et al. 2014a)

The individual behavior and collective action for responsible fisheries, mentioned above, are primarily the 'sustainable practices' that we discuss next before getting to the transdisciplinary examples along the value chain and beyond.

10.2.2 Sustainable Practices

Similar to the concept of stewardship, sustainable practices also go back millennia, for example in the 'traditional' societies of Oceania and elsewhere (Johannes 1978). These practices were, and occasionally still are, deeply embedded in culture and governance at the local level, through the power of community leaders and a complex system of taboos and entitlements (Olomola 1993; Sanders et al. 2011). Sustainable practices in remote locations are still being 'discovered' by conservation interests. A distinctive feature is that most of these traditional practices are aimed at sustainable use by the adjacent community, or for some clear socioeconomic purpose, rather than conservation of biodiversity for its own sake, as is sometimes the aim of nature-focused environmental non-governmental organizations (NGOs) (McConney et al. 2014b). With modernization, however, such biocultural sustainable fisheries systems have become threatened even in isolated places (Sterling et al. 2017).

Globally, the Code of Conduct for Responsible Fisheries (FAO 1995) has been highly influential in promoting the use of sustainable practices all along the value chain, and use in supporting or related activities, such as research and coastal management. Such practices optimally utilize fisheries resources along the entire value chain in ways that do not harm ecological or social systems, and contribute to

Table 10.1 Code of Conduct for Responsible Fisheries and the SSF Guidelines establish a basis for stewardship and sustainability

Code of conduct for responsible fisheries and its series of technical guidelines	Voluntary guidelines for securing sustainable small-scale fisheries in the context of food security and poverty eradication (the SSF guidelines)
Nature and scope of the code	Objectives
Objectives of the code	Nature and scope
Relationship with other international instruments	Guiding principles
Implementation, monitoring, and updating	Relationship with other international instruments
Special requirements of developing countries	Governance of tenure in small-scale fisheries and resource management
General principles	Social development, employment, and decent work
Fisheries management	Value chains, post-harvest, and trade
Fishing operations	Gender equality
Aquaculture development	Disaster risks and climate change
Integration of fisheries into coastal area management	Policy coherence, institutional coordination, and collaboration
Post-harvest practices and trade	Information, research, and communication
Fisheries research	Capacity development
	Implementation support and monitoring

human wellbeing. More recently, the Voluntary Guidelines for Securing Sustainable Small-Scale Fisheries in the Context of Food Security and Poverty Eradication (SSF Guidelines) (FAO 2015) are becoming increasingly more influential, specifically for small-scale fisheries governance (Jentoft et al. 2017), and are referred to often in the chapter. The summary contents of these two key international instruments are shown in Table 10.1. Additionally, there are several other international fisheries instruments that operationalize a transdisciplinary approach to stewardship, incorporating guidance on biophysical, socioeconomic, and governance issues. We introduce some of these instruments later in the chapter.

As a global non-binding (voluntary) agreement espousing stewardship, the 1995 Code was revolutionary in terms of wide acceptance and adoption not only by states, but also the private sector and civil society. The 2014 SSF Guidelines (also voluntary) were revolutionary, mainly in terms of the process of their development through consultation with civil society from 2008, but especially for the engagement of fisher organizations at all levels of governance. This exhaustive consultative process was conducive to the early implementation of the draft instrument by some states and fisher stakeholders even before it was formally finalized by the nation-states. Since neither instrument is binding, there is no force of law with rights and obligations, but both instruments offer compelling arguments for stewardship and sustainable practices that have been accepted and acted upon. The language in the SSF Guidelines is necessarily aimed at nation-states, as required in the United Nations system, but the SSF Guidelines are also very much owned by fishing industry stakeholders seeking an international policy platform for supporting the many components of stewardship and sustainable practices. The scope of both is expanding. The remainder of this chapter provides examples of stewardship and sustainability in practice along the fisheries value chain.

10.3 Stewardship and Sustainability

Chapin et al. (2009) argue that the sustainability of ecosystems is the foundation of stewardship. They say that unless ecosystem-based management and ecosystem-based approaches are employed, habitat degradation and destruction will undermine all other efforts to be stewards of remaining coastal and marine resources and their environments. Eco-labeling and sustainability certification of fisheries, such as MSC, Friend of the Sea and similar initiatives, exemplify a trend towards market-based sustainable practices (Foley and McCay 2014). The three principles of the MSC fisheries standard use primarily ecological aspects of fisheries in assessing eligibility for the eco-label certification (see Box 10.1).

Eco-labels are often linked to the postharvest traceability of seafood through chains of custody, so that the cook can identify the hook (or at least the vessel) associated with the seafood item and assure the consumer of its compliance with

Box 10.1

Principle 1: Sustainable target fish stocks

A fishery must be conducted in a manner that does not lead to over-fishing or depletion of the exploited populations and, for those populations that are depleted, the fishery must be conducted in a manner that demonstrably leads to their recovery.

Principle 2: Environmental impact of fishing

Fishing operations should allow for the maintenance of the structure, productivity, function and diversity of the ecosystem (including habitat and associated dependent and ecologically related species) on which the fishery depends.

Principle 3: Effective management

The fishery is subject to an effective management system that respects local, national and international laws and standards and incorporates institutional and operational frameworks that require use of the resource to be responsible and sustainable.

Source: MSC 2014 (also see www.msc.org/about-us/standards/fisheries-standard)

conservation and management measures anywhere in the world. Global eco-labels are not the only response to this need for differentiation in the market of responsible fisheries. In some countries the development of collective labels or quality labels linked to small-scale fisheries are a response to this need, helping small-scale fishers to position their products in scenarios dominated by world markets and large scale fisheries (see Chap. 8 by Pascual-Fernandez et al. in this volume). Kittinger et al. (2017) argue that, in addition to ecological concerns, stewardship and sustainability are also about socially responsible seafood. Soliman (2014) reminds us that the religious or ethical values of society translate into norms that are codified into environmental and fisheries legislation, thereby transforming stewardship and sustainability practices into rules. Such formal and informal rule systems extend from the global to local level, and are expressed in very different ways in various locations depending on culture, societal norms and values, legal system, and diverse, situation-specific, dynamic institutions and interactions that require inter- and transdisciplinary approaches to understanding governance and governability fully (Jentoft and Chuenpagdee 2015).

The Convention on Biological Diversity is a prominent international instrument or set of rules that addresses conservation, going beyond fisheries to engage other economic sectors. The Convention's principled ecosystem perspective recognizes that biodiversity conservation is necessary, but not sufficient, for ecosystem stewardship (Secretariat of the CBD 2012). Other instruments that address land-based and marine sources of pollution and debris, ocean dumping of wastes, climate change and variability, and much more, contribute to the global governance framework of rights and obligations that underpin stewardship at the level of nation-states necessary for enabling policy.

Also at the global level, the chapters of the First World Ocean Assessment (UN 2016), as outputs of the United Nations *Regular Process for Global Reporting and Assessment of the State of the Marine Environment, including Socioeconomic Aspects*, provide a transboundary perspective on marine stewardship worldwide. Significantly, the process recognized the need to incorporate the human dimension. This is a growing trend that strengthens transdisciplinarity. It is consistent with the view that stewardship and sustainability are often best visualized through scenarios that combine perspectives and present alternative pathways to future states at multiple levels on various scales. Scenarios and nested scales featured in the Millennium Ecosystem Assessment (2005) were another major transdisciplinary undertaking that are still relevant to the environmental context of fisheries today. It heralded the notion of ecosystem services, which are also transdisciplinary.

Much of the above transdisciplinary thinking has recently been integrated into crafting the global, practical, and problem-solving SDGs previously mentioned. The SDGs are a worldwide transformative call to action to improve both our environment and human wellbeing by 2030. All of the 17 linked SDGs are relevant to small-scale fisheries treated as social-ecological systems, but most fisheries stakeholders will pay special attention to SDG 14 on Life Below Water (Fig. 10.2).

Fig. 10.2 The global Sustainable Development Goals (SDGs), highlighting SDG 14

SDG 14 aims to "Conserve and sustainably use the oceans, seas and marine resources", and it has a number of targets to be achieved and indicators to be monitored in so doing (see Box 10.2).

Although only target 14.b specifically mentions small-scale fisheries, all targets are relevant to these fisheries and require transdisciplinary approaches to be reached as planned. The 2017 United Nations Ocean Conference and its associated initiatives focused global attention on SDG 14 with pledges of stewardship. United Nations agencies, such as the FAO (2017), set out how their programs articulate with SDG 14. Fanning and Mahon (2017) examine how stakeholders at regional and sub-regional levels in the Wider Caribbean Region can implement SDG 14 initiatives. Like other scholars (e.g. Folke et al. 2005), they stress the need for policy that enables multi-level governance and, especially, self-organization for stewardship at the local level to support learning, decisions, and adaptation at higher levels.

Scaling down from transboundary to national and local levels, a number of technical-scientific guidebooks, individually and combined, outline practical transdisciplinary approaches to stewardship and sustainability. Several tackle biophysical, socioeconomic, and governance issues in an integrated fashion using linked indicators for measuring, monitoring, and evaluating, so as to learn and adapt (e.g. Bunce and Pomeroy 2003; Pomeroy et al. 2005). Such guidance is what the framework for enhancing stewardship calls for, and guidance on cross-scale interactions and networks for communication and collective action is also important to transdisciplinarity by engaging stakeholders. Kalikoski and Franz (2014) address these aspects of how fisher organizations articulate with the SSF Guidelines, especially at the local level.

Box 10.2 SDG 14 TARGETS

14.1 By 2025, prevent and significantly reduce marine pollution of all kinds, in particular from land-based activities, including marine debris and nutrient pollution

14.2 By 2020, sustainably manage and protect marine and coastal ecosystems to avoid significant adverse impacts, including by strengthening their resilience, and take action for their restoration in order to achieve healthy and productive oceans

14.3 Minimize and address the impacts of ocean acidification, including through enhanced scientific cooperation at all levels

14.4 By 2020, effectively regulate harvesting and end overfishing, illegal, unreported and unregulated fishing and destructive fishing practices and implement science-based management plans, in order to restore fish stocks in the shortest time feasible, at least to levels that can produce maximum sustainable yield as determined by their biological characteristics

14.5 By 2020, conserve at least 10 per cent of coastal and marine areas, consistent with national and international law and based on the best available scientific information

14.6 By 2020, prohibit certain forms of fisheries subsidies which contribute to overcapacity and overfishing, eliminate subsidies that contribute to illegal, unreported and unregulated fishing and refrain from introducing new such subsidies, recognizing that appropriate and effective special and differential treatment for developing and least developed countries should be an integral part of the World Trade Organization fisheries subsidies negotiation

14.7 By 2030, increase the economic benefits to Small Island developing States and least developed countries from the sustainable use of marine resources, including through sustainable management of fisheries, aquaculture and tourism

14.a Increase scientific knowledge, develop research capacity and transfer marine technology, taking into account the Intergovernmental Oceanographic Commission Criteria and Guidelines on the Transfer of Marine Technology, in order to improve ocean health and to enhance the contribution of marine biodiversity to the development of developing countries, in particular small island developing States and least developed countries

14.b Provide access for small-scale artisanal fishers to marine resources and markets

14.c Enhance the conservation and sustainable use of oceans and their resources by implementing international law as reflected in UNCLOS, which provides the legal framework for the conservation and sustainable use of oceans and their resources, as recalled in paragraph 158 of The Future We Want.

Source: United Nations Sustainable Development Goals (https://www.un.org/sustainabledevelopment/)

Stewardship and sustainability, especially at the community (of interest, geospatial or virtual) level involve more than technical-scientific solutions. Although not as prominent as in the so-called 'traditional' societies, the dimensions of culture, values, ethics, and social norms have to be taken into account. These are crucial, but often overlooked, aspects of governance (Bavinck et al. 2005). For example, at the local level Parrill and Vodden (2014) describe modern day fisheries for the endangered American eel in which First Nations (Indigenous) people utilize methods and equipment that are considered environmentally responsible and deeply embedded within their strong culture of stewardship. Applying external solutions in such cases undermines the aims of management and conservation. Yet, adopting without either adapting or assessing external interventions is customary in many developing countries that seek to benefit from donor assistance. Local stakeholders with adequate capacity are often in the best position to evaluate and shape the compatibility and likely success of interventions (de Mattos et al. 2017). This is one of the reasons why stewardship is also associated with subsidiarity, so decision making can occur at the lowest effective level (Béné et al. 2007). This applies all along the fishery value chain, as shown in the following sections.

10.4 Supporting Services, Pre-harvest, and Harvest

Stewardship does not start with putting a boat to sea. It starts with pre-harvest vessel and gear design and construction, provisioning, fueling, and other support services (Kent and Himes-Cornell 2016). These activities are all part of the social-ecological system, and responsible fisheries choices about them transcend disciplines. In the fishing industry, individuals, households, firms, fishing enterprises, fisher organizations, or other entities make these choices, which can both reflect and impact stewardship. For example, environmental awareness can increase the use of less polluting and more efficient engines and fuels, reduce the use of plastics and items that contribute to hazardous marine debris, and encourage the purchase of gear and equipment that adds to fishing safety and resource sustainability. Fisheries authorities, enforcement agencies, and other public service entities also have roles to play in supporting responsible practices and providing incentives or assistance (Singleton et al. 2017). Academics can use transdisciplinary approaches to investigate and encourage conservation ethics. They provide insight into what motivates people to collectively support stewardship and sustainability (Paterson et al. 2010; Thompson 2008), and what facilitates collective action (Gibson et al. 2000). In addition, conservation communication uses science to inform social marketing that changes attitudes and behavior (McConney et al. 2016).

More nation states are ratifying the Agreement on Port State Measures to Prevent, Deter and Eliminate Illegal, Unreported and Unregulated Fishing (FAO 2009). This is a major global instrument to combat illegal, unreported, and unregulated (IUU) fishing, which many consider the major threat at sea that undermines all efforts at stewardship. Article 2 of the Agreement states that its objective is to: "prevent, deter and eliminate IUU fishing through the implementation of effective port State measures, and thereby to ensure the long-term conservation and sustainable use of living marine resources and marine ecosystems". Implementing the Port State Agreement requires at least a combination of legal, biological, and vessel operation information. Port state measures, in addition to IUU vessel listing, trade-related measures, and sanctions, typically include a mixture of marine science and governance requirements related to:

- Prior notification of entry into designated ports
- Restrictions on port entry and port inspections
- Landing or transshipment of certain fish species
- Restrictions on supplies of goods and services
- Documentation on management compliance

Downscaling stewardship from a global to national level is often accomplished by international and national plans of action, such as those promoted by FAO. These address several topics in the Code of Conduct for Responsible Fisheries in a very practical scientific-technical manner. A suite of them, for example, address incidental seabird catches, shark conservation, and fishing capacity, with practical guidelines for reducing turtle mortality (FAO Fisheries and Aquaculture Department 2009). In the case of seabirds, there are specific technical and operational instructions for improving gear technology to reduce incidental seabird capture in ways that do not substantially reduce fish catch. Several types of scientists and technicians work with fishers to produce these transdisciplinary solutions and enhance sustainable practices.

In fishing operations the harmful impacts of some trawl and entangling net operations on the benthos, and the use of destructive and unselective fishing methods, such as explosives and poisons on coral reefs, have alarmed fisheries managers, scientists, and civil society. Habitat degradation, bycatch, and discards have been persistent problems in many fisheries resulting in millions of tonnes of seafood wastage that could otherwise have contributed to food security, nutrition, and poverty reduction especially in tropical countries (Eayrs and Bert 2007). Beyond scientific literature, there are now guidelines on how to address bycatch (FAO 2011). Projects with stewardship as a core value seek to build capacity through experiences from site-level case studies (e.g. Guanais et al. 2015) that help to disseminate information and encourage sustainable practices through responsible fisheries.

Actively engaging women and youth is crucial in such projects as harvest decisions are often made at the household level. Taking gender and age into account in stewardship is also part of the steps of understanding and monitoring the fisheries system to inform decisions (Neis et al. 2013). For example, natural hazards, climate change, and climate variability have different impacts on men versus women and adults versus youth (de la Torre-Castro et al. 2017). Consequently, adaptive response and roles in recovery also differ, and must be appreciated, as stewardship is inclusive and intergenerational. Learning from the harvest side is summarized in Box 10.3. These points are also relevant to the remainder of the fisheries value chain and apply to postharvest, distribution, and consumption in the next section.

Box 10.3 Key Transdisciplinary Learning from the Harvest Side

- Stewardship and sustainable practices requires responsible use of resources and habitats in human activities
- Marine science, technology, culture, economics, etc. combine to either control or create harmful impacts
- Community-based monitoring, fishing safety innovation, energy efficiency, etc. are good practices
- Overfishing, IUU fishing, habitat degradation, bycatch and discards, pollution, etc. are harmful practices
- Climate change and variability, including extreme weather, make adaptive capacity crucial in stewardship

10.5 Postharvest, Distribution, and Consumption

The marketing links from harvest to postharvest regarding seafood preferences (species, size, etc.) often determine the types, quantities, and quality of fishery products offered to the buyers from dockside to supermarket and restaurant (Hosch and Blaha 2017). It is a growing trend, especially in richer countries, for the consumer to be more discerning about seafood when cost is not the main factor determining purchase. The sustainability of fisheries was previously mentioned in the discussion of eco-labeling and the MSC fisheries standard for certification (Martin et al. 2012). The chain of custody for easy product traceability, from source to consumer, is another feature that supports stewardship. Science, management, technology, and business combine to make traceability systems successful (Hosch and Blaha 2017). The buyer or consumer can also be brought into stewardship through much less elaborate means, such as buying circles that can support small-scale and family fishing operations more directly in all countries (Brinson et al. 2011).

Sustainable seafood supply and preparation guides for chefs, and similar consumption guides for institutional buyers and consumers, often supplement and reinforce traceability systems and eco-labels. They are powerful market mechanisms for effecting stewardship and instruments of advocacy often accompanied by communication campaigns (Jacquet and Pauly 2007). Caution needs to be exercised, however, as seafood is not an undifferentiated commodity like other meat and vegetable items (Bellmann et al. 2016). A blanket campaign on tuna, for instance, can encompass a variety of types of fisheries, target species, bycatch, gears, scales, and methods of operation, markets and much more. Such a broad initiative could be inaccurate, misleading about many fisheries that are not intended to be included, and potentially harmful to poor small-scale fishery households. Mislabeling persists since most consumers are ignorant about seafood (Jacquet and Pauly 2008), and this ignorance facilitates their misinterpretation of information on sustainability even if it is very accurately and carefully crafted. Such collateral damage can seriously undermine efforts to enhance stewardship (Stratoudakis et al. 2016). Conservation

NGOs need to be responsible in this regard and small-scale fishers need to be better organized in order to be able to improve the differentiation of their products and their positioning in the markets (see Chap. 8 by Pascual-Fernández et al., this volume). More needs to be done to improve the marketing of small-scale fisheries catches. Better understanding of consumer preferences, and changes in these preferences are still needed at national and local levels for domestic and tourism markets. For instance, in some areas the consumer demand is focused on scarce species, while abundant and nutritious resources, highly appreciated in other places, are completely disregarded, creating unnecessary burden on the ecosystem. Culture, ecology, economics, and social marketing are of concern in this transdisciplinary scenario.

The international slow food movement that emphasizes good, clean, and fair food products, including seafood, is also closely aligned with stewardship and sustainability (see Box 10.4). In addition to ecosystem stewardship, food security, health, and nutrition, the slow food/slow fish initiative also pays attention to the working conditions and other livelihood features of those who produce seafood. It is well aligned with the value chain, employment, decent work, and gender content of the SSF Guidelines. The slow food movement aligns well with the SSF Guidelines and stewardship generally.

Maintaining our broad perspective on fisheries stewardship and sustainability as societal issues, we note that fishing communities, especially in remote rural areas, may have poor access to health, education, transport, utilities, and other expected public good social services. The public sector deficiencies constrain their contribution to poverty alleviation (Béné et al. 2007). Other limitations may be due to environmental drivers such as overfishing, pollution, and climate change. Fishers can remain in structural poverty due to state inaction, despite the effective leadership that is essential for fisher self-organization (Blackman and Almerigi 2017). All these factors make it difficult for small-scale fishers and fishery workers to make their voices heard, defend their human rights and tenure rights, strive for gender equality, and secure the sustainable use of the fishery resources on which they depend, as addressed in the SSF Guidelines (FAO 2015). The SSF Guidelines aim for stewardship and

Box 10.4
Slow Food Manifesto for Quality (www.slowfood.com/about-us/our-philosophy/) summarized:

Slow Food envisions a world in which all people can access and enjoy food that is good for them, good for those who grow it, and good for the planet. Our approach is based on a concept of food that is defined by three interconnected principles: good, clean and fair.

- GOOD: quality, flavorsome and healthy food
- CLEAN: production that does not harm the environment
- FAIR: accessible prices for consumers and fair conditions and pay for producers

Box 10.5 Some key transdisciplinary learning from the postharvest side

- Stewardship flourishes when ecosystem-based human wellbeing is addressed via social-ecological systems
- Eco-labels, traceability, national level quality assurance, and marketing programs assist sustainable practices
- Entire societies can seek to be stewards of sustainable futures through their seafood consumer preferences
- Market-based or other equitable distributions of tangible benefits and incentives from stewardship are key
- Institutional support to fisher organizations is a key factor in achieving collective action for stewardship

sustainability to be fostered, and sectoral marginalization prevented, by enabling comprehensive human resource development. This includes promoting fisher stewardship to obtain higher returns from seafood markets. From this perspective, small-scale fisheries stewardship flourishes when human wellbeing is prioritized, and entire societies seek to be stewards of sustainable futures. Integrated approaches are needed for this to happen, and we turn to some of these in the next section. Some key learning from the postharvest side is summarized in Box 10.5.

10.6 Inter-sectoral Coordination Mechanisms Support Stewardship

The ecosystem approach to fisheries (EAF) requires that other economic sectors and their use, or abuse, of coastal and marine ecosystems be taken into account in fisheries management. In many countries the key interacting sectors include agriculture, tourism, mining (including oil and gas), and shipping. Planning that engages the fisheries sector at the science-policy interface along with other sectors is considered best practice in stewardship (McConney et al. 2016; Compton et al. 2017). This is a huge area of concern for stewardship as it speaks to fundamental ideologies and beliefs about the pathways towards sustainable practices and development. These include the extent to which privatization (grabbing) of ocean and coastal space, rather than more communal and public approaches, is the most efficient and effective means of both generating and distributing wealth derived from marine resources (Bavinck et al. 2017).

10.6.1 Integrated Coastal Area Management and Marine Spatial Planning

Many small-scale fisheries, especially those for subsistence and those pursued by the poor, young, and women, occur along and near coasts, including: gleaning on beaches, mud flats, and mangroves; and in habitats such as lagoons, estuaries, deltas, flood plains, and similar areas. As a means of supporting fisheries stewardship and sustainable practices with a wider scope, the Code of Conduct for Responsible Fisheries (FAO 1995) calls for the integration of fisheries into coastal area management (Fig. 10.3).

In some locations, small-scale fisheries are marginalized in order to privilege development in other economic sectors, such as tourism or shipping, but also to allocate space for high-end residential expansion. Often these developmental activities contribute more to economic growth measured in conventional national accounts. They may also primarily benefit powerful elites and trample the access and use rights of fishers and other coastal users (Bennett and Dearden 2014).

Recent socioeconomic monitoring for coastal management (SocMon) studies in the Caribbean have indicated a growing sense of stewardship among communities adjacent to marine managed areas. Results suggest fairly high levels of understanding what these areas are for, and, therefore, appreciation for these management tools (Lovell and Spencer 2017). They note support for management measures to protect important reef species (e.g. parrotfish) and coastal ecosystems (e.g. coral reefs);

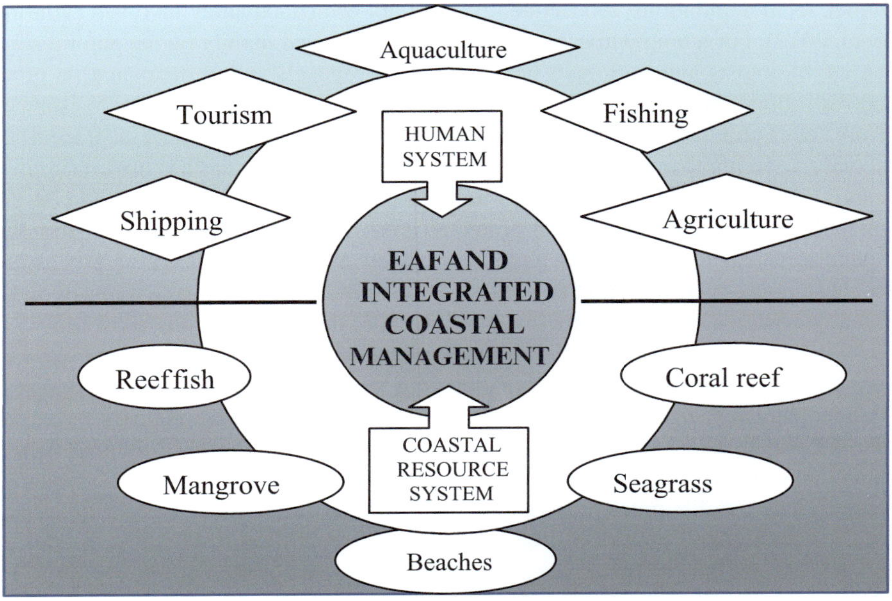

Fig. 10.3 The integration of fisheries into coastal area management helps to support stewardship

increasing beliefs that communities and NGOs should share local area management responsibility; learning-by-doing awareness-raising activities such as beach and coastal cleanups; and interest in engaging in ecological and socioeconomic monitoring. All of these are practical expressions of stewardship and foundations for sustainable practices.

Often, however, it is not easy to make socioeconomic or environmental trade-offs in order to decide what use is made of coastal and marine space. Integrated coastal area (or zone) management employs marine spatial planning (MSP) as the main tool for addressing the multiple types and layers of decision making (Pomeroy et al. 2014). Quantitative, computerized decision support systems, including geographic information systems, assist in the process. This integrated and participatory marine planning system is transdisciplinary (Ehler and Douvere 2009). Within this aspect of stewardship, special mention must be made of marine protected areas (MPAs) as being particularly promising for stewardship and sustainability, while also being potentially problematic for fisheries, even when overall objectives for sustainability are shared (Rice et al. 2012).

10.6.2 Marine Protected Areas and Their Networks

International agendas demand the expansion of MPAs across the globe. The Convention on Biological Diversity requires formal regulated protection of 10% of coastal and marine areas. This has generated environmental protection processes in which local populations, too frequently, have not been adequately involved (Jentoft et al. 2012). For a long while, research on MPAs focused mainly on the natural sciences; biologists and ecologists have led MPA proposals and implementation processes around the world. However, these initiatives have often met with resistance from local populations and created conflicts that limited their success. Recently, social research into environmental protection and protected areas has grown significantly. Frequently, this research has focused on the impacts and unwanted consequences of the implementation of protected areas on local populations. Sometimes, social scientists are involved to a greater or lesser extent in the inception processes of these areas, giving support to public administrations, local actors, NGOs or civil society organizations, and sometimes to several of these actors at the same time.

It is clear that the period prior to the implementation of a protected area plays a critical role in its success, but this step-zero period involves complex processes (Chuenpagdee et al. 2013). This constitutes a field where transdisciplinary research should play a strong role, as MPA design and implementation need a wider focus than that provided by social or natural scientists working independently. This means coping with the dilemmas present in complex scenarios in which, frequently, none of the stakeholder groups or institutions involved have full control of the process nor all the relevant information, and conflict may be present between different groups or institutions.

The success of institutional arrangements for MPAs depends on the ecological design of the protected area, on how it fits stakeholders' needs and goals, how these stakeholders participate, and how the inception process, as a whole, is built. The ethical and practical implications of scientists' work in these scenarios reinforce the need to provide the best available information about what an MPA can be for different stakeholder groups and institutions, and how to create the opportunities for the ecosystems to recover or increase yields in a sustainable way. Thus, the opportunity and need for transdisciplinarity in this arena is clear, especially for locally and community managed marine areas, which are types of MPAs. In this context, communities' understanding of the relationship between stewardship and sustainability needs to be appreciated in the engagement (see Box 10.6).

Recent interest in the human dimensions of marine protected areas offers opportunities for broader approaches to MPA governance, through transdisciplinary research and outreach. For a focus on stewardship, these should be put on the same level as ecological dimensions in order to highlight the conditions likely to contribute to wellbeing in fishing households and society. An MPA focus on fisheries management objectives, giving priority to poverty alleviation and food security, instead of more externally driven aims, could assist stewardship in some communities (Rosendo et al. 2011). Furthermore, MPAs can be designed to fit existing societal and natural realms being focused not only on environmental sustainability, but also on the creation of synergies that improve human wellbeing in the populations related

Box 10.6 Eastport Dialogue on Stewardship

Eastport is a small community of less than 2000 people in the eastern part of Newfoundland, Canada, about 150 km west of St. John's. It is the site of a small MPA (about 2.1 km^2), established in 2005, with an aim to conserve and protect lobster populations and endangered species, like wolffish (*Anarchichas lupus*). The idea to protect lobsters was initiated by the community and supported by the government. As such, stewardship seems to have been a strong trait of the people in the area. But what does stewardship really mean and how does it relates to what is going on in the area?

The 'Eastport Dialogue on Stewardship' was conducted in 2010 as an academic exercise to address these questions. As part of this, community participants discussed the two concepts: sustainability and stewardship. They all agreed that there were differences between the two terms but identified stewardship as the broader of the two. The group highlighted the importance of sustainability from an economic standpoint. Although sustainability was considered important, the group believed sustainability required first having a high level of stewardship. This sense of stewardship would then lead to activities for sustainability.

Source: Ivany (2011), contributed by R. Chuenpagdee

to the area. For this purpose, a deep knowledge of the existing societal challenges is necessary. This underscores the importance of transdisciplinarity and local stakeholder involvement in MPA processes from inception to maturity (Gerhardinger et al. 2018).

10.7 Conclusion

Stewardship and sustainable practices in small-scale fisheries feature prominently in the SSF Guidelines. The two, together, can guide the human propensity to intervene in social-ecological systems. They can shape positive outcomes and desirable futures for fisheries that address ecosystem health, poverty alleviation, food security, nutrition, livelihoods, and other facets of human wellbeing. Networking is crucial to increase communication flow and build capacity on different scales and levels. Initiatives are linked within and between the harvest and postharvest sectors. More than focusing only on immediate problem solving, a culture and ethic of stewardship enables us to proactively create opportunities through which to advance adaptively in times of crisis in a sustainable and resilient manner. It is this focus on our responsibilities for future generations that encourages sustainable practices and empowers leaders of collective action. Transdisciplinarity is key to all of the above. No single academic discipline, or combination of disciplines is sufficient. Technical skills along with other capabilities are required. In this chapter we have seen how transdisciplinarity underpins the steps of information gathering and shared learning that lead to stewardship with sustainable practices along an entire fisheries value chain and in wider society.

References

Angelstam P, Andersson K, Annerstedt M et al (2013) Solving problems in social–ecological systems: definition, practice and barriers of transdisciplinary research. Ambio 42:254–265

Bavinck M, Chuenpagdee R, Diallo M et al (2005) Interactive fisheries governance. Eburon, Delft

Bavinck M, Berkes F, Charles A et al (2017) The impact of coastal grabbing on community conservation – a global reconnaissance. MAST 16(1):8. https://doi.org/10.1186/s40152-017-0062-8

Bellmann C, Tipping A, Sumaila UR (2016) Global trade in fish and fishery products: an overview. Mar Policy 69:181–188

Béné C, Macfadyen G, Allison EH (2007) Increasing the contribution of small-scale fisheries to poverty alleviation and food security, FAO fisheries technical paper 481. FAO, Rome

Bennett NJ, Dearden P (2014) Why local people do not support conservation: community perceptions of marine protected area livelihood impacts, governance and management in Thailand. Mar Policy 44:107–116

Bennett NJ, Whitty TS, Finkbeiner E et al (2018) Environmental stewardship: a conceptual review and analytical framework. Environ Manag 61:597–614

Blackman K, Almerigi S (2017) Leading fisherfolk. Cave Hill Campus, Centre for Resource Management and Environmental Studies, The University of the West Indies, Barbados

Brinson A, Lee MY, Rountree B (2011) Direct marketing strategies: the rise of community supported fishery programs. Mar Policy 35:542–548
Bunce L, Pomeroy R (2003) Socioeconomic monitoring guidelines for coastal managers in the Caribbean (SocMon Caribbean). Global Coral Reef Monitoring Network
Cash DW, Adger W, Berkes F et al (2006) Scale and cross-scale dynamics: governance and information in a multilevel world. Ecol Soc 11:8. http://www.ecologyandsociety.org/vol11/iss2/art8/. Accessed 15 Apr
Chapin FS III, Kofinas GP, Folke C (eds) (2009) Principles of ecosystem stewardship: resilience-based natural resource management in a changing world. Springer-Verlag, New York
Chuenpagdee R, Pascual-Fernández JJ, Szeliánszky E et al (2013) Marine protected areas: re-thinking their inception. Mar Policy 39:234–240
Compton S, McConney P, Monnereau I et al (2017) Good practice guidelines for successful National Intersectoral Coordination Mechanisms (NICs). Report for the UNDP/GEF CLME+ Project (2015–2020). Centre for Resource Management and Environmental Studies, The University of the West Indies, Cave Hill Campus, Barbados. CERMES Technical Report 88:14
de la Torre-Castro M, Fröcklin S, Börjesson S et al (2017) Gender analysis for better coastal management–increasing our understanding of social-ecological seascapes. Mar Policy 83:62–74
de Mattos SMG, Wojciechowski MJ, Macnaughton AE et al (2017) Implementing the small-scale fisheries guidelines: lessons from Brazilian clam fisheries. In: Jentoft S, Chuenpagdee R, Barragán-Paladines et al (eds) The small-scale fisheries guidelines: global implementation. Springer, New York, pp 473–494
Eayrs S, Bert TM (2007) A guide to bycatch reduction in tropical shrimp-trawl fisheries. FAO, Rome
Ehler C, Douvere F (2009) Marine spatial planning: a step-by-step approach toward ecosystem based management. UNESCO/IOC, Paris
Fanning L, Mahon R (2017) Implementing the Ocean SDG in the Wider Caribbean: state of play and possible ways forward. IASS, IDDRI, TMG, Potsdam
FAO (United Nations Food and Agriculture Organization) (1995) Code of conduct for responsible fisheries. FAO, Rome
FAO (2009) Agreement on Port State Measures to prevent, deter and eliminate illegal, unreported and unregulated fishing. FAO, Rome
FAO (2011) International guidelines on bycatch management and reduction of discards. FAO, Rome
FAO (2015) Voluntary guidelines for securing sustainable small-scale fisheries in the context of food security and poverty eradication. FAO, Rome
FAO (2017) FAO working for SDG 14: healthy oceans for food security, nutrition and resilient communities. FAO, Rome
FAO Fisheries and Aquaculture Department (2009) Guidelines to reduce sea turtle mortality in fishing operations. FAO, Rome
Foley P, McCay B (2014) Certifying the commons: eco-certification, privatization, and collective action. Ecol Soc 19(2):28. https://doi.org/10.5751/ES-06459-190228
Folke C, Hahn T, Olsson P, Norberg J (2005) Adaptive governance of social- ecological systems. Annu Rev Environ Resour 30:441–473
Galaz V, Crona B, Österblom H et al (2012) Polycentric systems and interacting planetary boundaries: emerging governance of climate change–ocean acidification–marine biodiversity. Ecol Econ 81:21–32
Gerhardinger LC, Gorris P, Gonçalves LR et al (2018) Healing Brazil's Blue Amazon: the role of knowledge networks in nurturing cross-scale transformations at the frontlines of ocean sustainability. Front Mar Sci 4(395). https://doi.org/10.3389/fmars.2017.00395
Gibson C, McKean M, Ostrom E (2000) Forests, people and governance: some initial theoretical lessons. In: Gibson C, McKean M, Ostrom E (eds) People and forests: communities, institutions, and governance. MIT Press, Cambridge, MA, pp 227–242
Guanais JHG, Medeiros RP, McConney PA (2015) Designing a framework for addressing bycatch problems in Brazilian small-scale trawl fisheries. Mar Policy 51:111–118

Hosch G, Blaha F (2017) Seafood traceability for fisheries compliance: country- level support for catch documentation schemes, FAO fisheries and aquaculture technical paper 619. FAO, Rome

Ivany I (2011) Marine protected areas sustainability: issues of complexity and stewardship. Dissertation, Memorial University of Newfoundland

Jacquet JL, Pauly D (2007) The rise of seafood awareness campaigns in an era of collapsing fisheries. Mar Policy 31:308–313

Jacquet JL, Pauly D (2008) Trade secrets: renaming and mislabeling of seafood. Mar Policy 38:309–318

Jentoft S, Chuenpagdee R (eds) (2015) Interactive governance for small-scale fisheries, Global reflections. Springer, Dordrecht

Jentoft S, Pascual-Fernandez JJ, De la Cruz Modino R et al (2012) What stakeholders think about marine protected areas: case studies from Spain. Hum Ecol 40:185–197

Jentoft S, Chuenpagdee R, Barragán-Paladines MJ et al (eds) (2017) The small-scale fisheries guidelines: global implementation, vol 14. Springer, Heidelberg

Johannes RE (1978) Traditional marine conservation methods in Oceania and their demise. Annu Rev Ecol Syst 9:349–364

Kalikoski D, Franz N (eds) (2014) Strengthening organizations and collective action in fisheries – a way forward in implementing the international guidelines for securing sustainable small-scale fisheries, FAO workshop, 2013. FAO fisheries and aquaculture proceedings 32. FAO, Rome

Kent K, Himes-Cornell A (2016) Making landfall: linkages between fishing communities and support services. Coast Manag 44:279–294

Kittinger JN, Teh LC, Allison EH et al (2017) Committing to socially responsible seafood. Science 356(6341):912–913

Lovell T, Spencer R (2017) Socio-economic monitoring at the Northeast Marine Managed Area (NEMMA), Antigua. Climate Resilient Eastern Caribbean Marine Managed Areas Network (ECMMAN): Eastern Caribbean Integrated Coral Reef Monitoring Project Report No. 5

Martin SM, Cambridge TA, Grieve C et al (2012) An evaluation of environmental changes within fisheries involved in the Marine Stewardship Council certification scheme. Rev Fish Sci 20:61–69

McConney P, Medeiros R, Pena M (eds) (2014a) Enhancing stewardship in small-scale fisheries: practices and perspectives. Too Big To Ignore (TBTI) and Centre for Resource Management and Environmental Studies, CERMES Technical Report 73. The University of the West Indies, Cave Hill Campus, Barbados

McConney P, Pomeroy R, Khan Z (2014b) ENGOs and SIDS: environmental interventions in small island developing states. In: Garcia SM, Rice J, Charles A (eds) Governance of marine fisheries and biodiversity conservation: interaction and co-evolution. Wiley-Blackwell, Chichester, pp 360–373

McConney P, Fanning L, Mahon R et al (2016) A first look at the science-policy interface for ocean governance in the wider Caribbean region. Front Mar Sci 2:119

Medeiros RP, Serafini TZ, McConney P (2014) Enhancing ecosystem stewardship in small-scale fisheries: prospects for Latin America and the Caribbean. Desenvolvimento e Meio Ambiente 32:181–191

Millenium Ecosystem Assessment (2005) Millennium ecosystem assessment. Ecosystems and human wellbeing: a framework for assessment. Island Press, Washington, DC

MSC (2014) MSC Fisheries Standard and Guidance v2.0. Marine Stewardship Council, London

Neis B, Gerrard S, Power N (2013) Women and children first: the gendered and generational social-ecology of smaller-scale fisheries in Newfoundland and Labrador and northern Norway. Ecol Soc 18(4). https://doi.org/10.5751/ES-06010-180464

Olomola AS (1993) The traditional approach towards sustainable management of common property fishery resources in Nigeria. MAST 6:92–109

Parrill E, Vodden K (2014) "We Always Did Fish the Eels"- Qalipu Mi'kmaq Ecological Impacts in the American Eel Fisheries of Western Newfoundland. In: McConney P, Medeiros R, Pena M (eds) (2014a). Enhancing stewardship in small-scale fisheries: practices and perspectives. CERMES Technical Report vol 73, Too Big To Ignore (TBTI) and Centre for Resource Management and Environmental Studies, The University of the West Indies, Cave Hill Campus, Barbados. pp 8–15

Paterson B, Isaacs M, Hara M et al (2010) Transdisciplinary co-operation for an ecosystem approach to fisheries: a case study from the South African sardine fishery. Mar Policy 34:782–794

Pomeroy R, Watson L, Parks J et al (2005) How is your MPA doing? A methodology for evaluating the management effectiveness of marine protected areas. Ocean Coast Manag 48:485–502

Pomeroy R, Baldwin K, McConney P (2014) Marine spatial planning in Asia and the Caribbean: application and implications for fisheries and marine resource management. Desenvolvimento e Meio Ambiente 32:151–164

Rice J, Moksness E, Attwood C et al (2012) The role of MPAs in reconciling fisheries management with conservation of biological diversity. Ocean Coast Manag 69:217–230

Rosendo S, Brown K, Joubert A et al (2011) A clash of values and approaches: a case study of marine protected area planning in Mozambique. Ocean Coast Manag 54:55–65

Sanders JS, Gréboval D, Hjort A (2011) Marine protected areas: country case studies on policy, governance and institutional issues, FAO fisheries and aquaculture technical paper 556/1. FAO, Rome

Secretariat of the Convention on Biological Diversity and the Scientific and Technical Advisory Panel – GEF (2012) Marine spatial planning in the context of the convention on biological diversity: a study carried out in response to CBD COP 10 decision X/29, Montreal, Technical Series No. 68

Singleton RL, Allison EH, Le Billon P et al (2017) Conservation and the right to fish: international conservation NGOs and the implementation of the voluntary guidelines for securing sustainable small-scale fisheries. Mar Policy 84:22–32

Soliman A (2014) Duty of stewardship and fisheries governance: a proposed framework. MAST 13:11

Sterling EJ, Filardi C, Toomey A et al (2017) Biocultural approaches to well-being and sustainability indicators across scales. Nat Ecol Evol 1:1798–1806

Stratoudakis Y, McConney P, Duncan J et al (2016) Fisheries certification in the developing world: locks and keys or square pegs in round holes? Fish Res 182:39–49

Thompson MH (2008) Fostering sustainable behaviours in community-based co-managed fisheries. Mar Policy 32:413–420

UN (United Nations) (2016) First World Ocean Assessment. United Nations, New York. www.un.org/depts/los/global_reporting/WOA_RegProcess.htm. Accessed 12 Jan 2018

Worrell R, Appleby MC (2000) Stewardship of natural resources: definition, ethical and practical aspects. J Agric Environ Ethics 12:263–277

Chapter 11
Interplay Between Local and Global: Change Processes and Small-Scale Fisheries

Prateep K. Nayak and Fikret Berkes

Abstract Important but neglected, small-scale fisheries remain vulnerable to a range of direct challenges, despite long-standing recognition of their multiple contributions to the economy and society. Global drivers contribute to vulnerabilities at local and regional levels, creating adverse changes, but these changes in turn may act as drivers that impact sustainability at higher levels. Thus, global drivers, and local and regional drivers can potentially impact each other in a two-way process. In this chapter, we discuss the interplay between local and global to explore the realities of small-scale fisheries. We use two empirical cases from the Bay of Bengal, east coast of India, to focus on (1) vulnerabilities experienced by small-scale fisheries, and (2) their existing strengths. We discuss possible strategies that can be used to build on the strengths of small-scale fisheries to counter the various vulnerabilities they face. Vulnerability is seen as a multidimensional, complex, highly dynamic, and relative concept, the study of which is highly inter- and trans-disciplinary. To this effect, we take a three-dimensional view of vulnerability that includes wellbeing, capitals, and resilience as measures to understand it, and provide a preliminary framework to help articulate viability in the context of small-scale fisheries. Given the complexity associated with global change drivers, small-scale fisheries will continue to remain vulnerable; however, we emphasise that they also have certain strengths. Identifying these strengths and building on them for long-term viability is an option that has not been fully explored.

Keywords Small-scale fisheries · Global change response · Vulnerability · Viability · Wellbeing · Resilience

P. K. Nayak (✉)
School of Environment, Enterprise and Development, University of Waterloo, Waterloo, ON, Canada
e-mail: pnayak@uwaterloo.ca

F. Berkes
University of Manitoba, Winnipeg, MB, Canada
e-mail: Fikret.Berkes@umanitoba.ca

R. Chuenpagdee, S. Jentoft (eds.), *Transdisciplinarity for Small-Scale Fisheries Governance*, MARE Publication Series 21,
https://doi.org/10.1007/978-3-319-94938-3_11

11.1 Introduction

> Small-scale fisheries contribute about half of global fish catches. When considering catches destined for direct human consumption, the share contributed by the small-scale fisheries increases to two-thirds (FAO 2015, ix).

Important but neglected, small-scale fisheries remain vulnerable to a range of direct challenges: conflicts with large-scale fisheries and recreational fisheries, and competition for resources and space with aquaculture, oil and gas, mining, ports, and tourism (Chuenpagdee 2011). As well, they are vulnerable to global change drivers, where global change refers to planetary-scale changes in the earth system with worldwide impacts (Steffen et al. 2015). Many examples of global change are biophysical in nature, such as climate change and biodiversity loss. Further complicating the issue, economic and political factors may also drive global change. For example, increase in international prices for shrimp/prawn has driven the global expansion of shrimp culture (Kurien 1992). Such global drivers may impact the sustainability of small-scale fisheries at local and regional (sub-national) levels, and these changes in turn may impact sustainability at higher levels. Thus, global drivers, and local and regional drivers can potentially impact each other in a two-way process (Nayak and Berkes 2014).

Here we discuss the interplay between local and global to explore the realities of small-scale fisheries. The broader picture is that small-scale fisheries face multiple direct challenges, as well as challenges from global environmental change, as in climate change (Armitage et al. 2017). They remain vulnerable to these challenges, despite long-standing recognition of multiple contributions of these fisheries to economy and society, for example, to food security (Berkes 2015; Jentoft and Chuenpagdee 2015).

> Small-scale and artisanal fisheries, encompassing all activities along the value chain – pre-harvest, harvest and post-harvest – undertaken by men and women, play an important role in food security and nutrition, poverty eradication, equitable development and sustainable resource utilization (FAO 2015, ix).

The vulnerability of small-scale fisheries to drivers of change and how they respond to various shocks and stresses have been major themes in Too Big To Ignore (TBTI): Global Partnership for Small-Scale Fisheries Research (Jentoft and Chuenpagdee 2015; Jentoft et al. 2017; Johnson et al. 2018). Box 11.1 summarizes the I-ADApT approach to understand how small-scale fisheries respond to change and develop appropriate policy responses. Such responses would be based on learning from a diversity of small-scale fisheries case studies to devise ways to reduce vulnerabilities and increase viability.

Following Millennium Ecosystem Assessment, a driver is any natural or human-induced factor that directly or indirectly causes a change in an ecosystem (Alcamo et al. 2003). We use sustainability broadly as patterns of development that can promote human wellbeing, while conserving the environment (Kates et al. 2001). Our unit of analysis is the social-ecological system: integrated complex adaptive systems that include social (human) and ecological (biophysical) subsystems in a two-way feedback relationship (Berkes 2011). Understanding multi-level interactions in

Box 11.1 Reducing Vulnerability and Enhancing Viability through I-ADApT

Ratana Chuenpagdee and Alida Bundy, IMBeR Human Dimensions Working Group

Small-scale fisheries coexist with many other sectors, including large-scale, industrialized fisheries, and are often marginalized. Maintaining viable and sustainable livelihoods has become a major challenge to small-scale fisheries, especially with ongoing global change. One way to deal with such change is to understand how small-scale fishers, fish workers, and their organizations respond to various pressures and stresses affecting their communities, and learn how effective these responses are. Likewise, an understanding of institutional responses (e.g. rules and regulations, policies, markets, etc.) is also required. The analysis of the responses needs to be comprehensive and specific to the small-scale fisheries undergoing change.

A framework that can be used for such a systematic assessment is I-ADApT (**A**ssessment based on **D**escription and responses and **Ap**praisal for a **T**ypology). I is originally for IMBeR – Integrated Marine Biosphere Research (www.imber.org), which developed the framework. In the application to small-scale fisheries, the "I" can also refer to the locally based approach to the assessment, the community-based monitoring of change, and the bottom-up process of learning how best to respond and prepare for change. In this context, the I-ADApT framework guides small-scale fishing communities to examine the following questions:

- What are the main issues affecting small-scale fishing communities and what are their impacts on fishing communities and the broader social, ecological, and governance systems in the area?
- How do small-scale fishing communities cope with these issues and respond to the changes around them?
- What can be learned about the effectiveness of these responses and other coping and adaptation strategies?
- What types of policies and interventions are appropriate to help small-scale fishing communities become less vulnerable and more viable?

I-ADApT is being used to develop a typology of responses based on the appraisal of how effective these responses are in dealing with change and mitigating the impacts, in order to aid decision making. Documenting what factors affect the outcomes of these responses, and what risks and uncertainties are involved in their implementation is part of the typology. Such learning can provide guidance to small-scale fishing communities, managers, and policy-makers to develop effective and timely responses. Development of the typology requires learning from as many case studies as possible. Thus, small-scale fishers and other users of the I-ADApT framework are encouraged to complete the template (available for download at http://toobigtoignore.net/research-cluster/global-change-responses/) and submit them for inclusion in the typology development at toobigtoignore@mun.ca.

Sources: Bundy et al. (2016) and Guillotreau et al. (2018).

social-ecological systems requires attention to levels and scales. Following Gibson et al. (2000) and Cash et al. (2006), we use the term 'scale' to refer to space, time, and jurisdictional scales, and 'level' to refer to a specific point along a particular scale.

The interplay between local and global is inherently political, and scale dominance is at the heart of the discussions of multi-level governance. According to Collinge (1999, 568) "scale dominance is an important attribute of the scale division of labour within capitals, and concerns the power which organisations at certain spatial scales are able to exercise over organizations at other, higher or lower, scales." Long neglected in environmental governance (Young et al. 2008), scaling is a critical consideration for framing problems because it shapes the way in which they are analysed (Gupta 2008). Thus, the choice of scale is not politically neutral; it often determines whose perspective will be used in problem formulation and analysis. In the case of fisheries governance, the relatively recent emphasis on small-scale fisheries (as opposed to large-scale fisheries, usually privileged by governments) provides a window of opportunity to rebalance fisheries policies at the national and other levels.

Relevant to the interplay between local and global, complexity theorists argue that there is no one 'correct' level in a complex adaptive system in terms of analysis or intervention, and that all levels are important because they are interacting (Norberg and Cumming 2008). Hence, global drivers and national, regional, and local changes potentially impact each other. There is a literature on multi-level processes (e.g., Cash et al. 2006; Börzel and Heard-Lauréote 2009). But this chapter focuses on local-global interactions and the argument that there is a two-way process in the interplay between local and global, with global drivers bringing about local and regional change, and local and regional changes in turn, affecting global processes. Following a section on vulnerabilities of small-scale fisheries highlighting a case involving conflict with aquaculture, we discuss viability and strengths of small-scale fisheries using a conservation-development case. Both cases are from the eastern coast of India, Bay of Bengal region. We then turn to global change responses, specifically the Voluntary Guidelines for Securing Sustainable Small-Scale Fisheries, henceforth SSF Guidelines (FAO 2015) and the Sustainable Development Goals, henceforth SDGs (UNDP 2017). We conclude by highlighting the complex interplay between the local and global, as a multidimensional and iterative process.

11.2 Vulnerabilities of Small-Scale Fisheries

The Oxford Dictionary defines the adjective 'vulnerable' as "exposed to the possibility of being attacked or harmed, either physically or emotionally." Wolf et al. (2013) define vulnerability as the property of being vulnerable that allows for comparison (e.g. small-scale fisheries are more vulnerable than large-scale fisheries). Vulnerability can be ascribed to somebody or something (e.g. the small-scale fishers/fisheries) as an entity. Weeratunge et al. (2014) note that scholars of sustainable livelihoods analysis consider vulnerability as a lens, emphasising threats that endanger livelihoods. These definitions tend to focus largely on the social dimensions of vulnerability and treat people as the object of vulnerability. Other definitions direct our attention to a social-ecological system view of vulnerability. According to Adger (2006), vulnerability is the state of susceptibility to harm from exposure to stresses associated with environmental and social change and from the absence of capacity to adapt.

Wolf et al. (2013) describe vulnerability as 'place-based' and 'context-specific', with additional characteristics of 'top-down' and 'bottom-up' (Dessai and Hulme 2004), 'end-point' and 'starting-point' (Kelly and Adger 2000), 'biophysical' and 'social' (Brooks 2003), and 'outcome' and 'context' specific (O'Brien et al. 2004). Multiple faces of vulnerability, as depicted by these definitions, suggest that it reflects the characteristics of a wicked problem. According to Jentoft and Chuenpagdee (2009), "problems are wicked (as opposed to "tame") when they are difficult to define and delineate from other and bigger problems and when they are not solved once and for all but tend to reappear – with no technical solution, it is not clear when they are solved, and they have no right or wrong solution that can be determined scientifically." There are two ways to look at vulnerability: as a property of a system; or, as an indication of potential problems. In the latter case, vulnerability can be analyzed as a wicked problem. We are discussing vulnerability as the vulnerability to the long-term viability of SSF. Such a view makes it perfectly valid to argue that vulnerability is always wicked, as it has no end point and no stopping rule. As such, goals to 'eliminate vulnerability' are misleading or make little sense. Small-scale fisheries continue to be vulnerable because of constantly changing conditions. Even when fishers and managers think they have solved problems of vulnerability, it resurfaces in other strange ways, for example, when fishers think they have found good markets for their capture fisheries, big aquaculture moves in with its many ugly faces to reinforce new vulnerabilities.

We view vulnerability in the context of small-scale fisheries as multidimensional, complex, highly dynamic, and relative. As individual disciplines provide relatively narrow 'windows' with which to view complex issues (Jentoft et al. 2010), the study of vulnerability needs to be highly inter- and trans-disciplinary. Several complementary approaches offer a more nuanced understanding of vulnerability in the context of small-scale fisheries: *First, vulnerability can be seen as the absence of wellbeing*, which is defined as "a state of being with others, where human needs are met, where one can act meaningfully to pursue one's goals and where one enjoys a satisfactory quality of life" (McGregor 2008, 1). Wellbeing is typically characterised by its three aspects: material, relational, and subjective (Gough and McGregor 2007), which can influence the level of vulnerability within a specific context.

Second, capitals or resources are a basis of vulnerability that enable individuals to navigate their position when they are vulnerable. Here, vulnerability can be seen as resulting from a lack of access to capital assets – human, physical, natural, social, and financial (Bebbington 1997, 1999). A relatively small number of studies have been published on different types of capitals, especially social capital, in fishing communities (Weeratunge et al. 2014).

Third, building (or losing) resilience is a way of dealing with vulnerability. It has been argued that vulnerability comes from a loss of resilience, defined as the capacity of a system to absorb disturbance and reorganize, while undergoing change so as to still retain essentially the same function, structure, identity, and feedbacks (Walker et al. 2004). Reduced resilience increases the vulnerability of a system to smaller disturbances that it could previously cope with (Walker et al. 2004). Given that uncertainty cannot be eliminated or discounted, leading to the necessity of living with change and uncertainty, Berkes (2007) emphasizes the strategy of reducing vulnerability by building resilience. Vulnerability approaches in fisheries have been enhanced by studies exploring its complex interplay with resilience (Béné et al. 2011; Armitage et al. 2012; Coulthard 2012). Armitage et al. (2012) highlight the complementarities between wellbeing and resilience, arguing that development of hybrid approaches and innovative combinations of social and ecological theory are now necessary to understand complexity and multilevel changes, such as those taking place in small-scale fisheries.

The role of multilevel drivers is also important in influencing and shaping vulnerability in all three dimensions of wellbeing, capitals, and resilience. Vulnerability often sets in when resilience is weakened by external or non-place-based drivers acting on the capacity of local communities to cope with, or adapt to, problems. Such external forces or drivers may include the impacts of structural adjustment policies, conflicts, commodity price fluctuations, and market, tariff or trade policies, or impacts of global environmental change on local social-ecological systems. The small-scale fishery sector is often caught between the highly dynamic interplay of drivers at various levels, adding to vulnerabilities.

Using the three dimensions of vulnerability (wellbeing, capitals, and resilience) as a measure, we briefly discuss a range of vulnerabilities observed in the case of the Chilika Lagoon small-scale fishery. Connected to the Bay of Bengal, Chilika Lagoon is in Odisha State of India, and a Ramsar Site of international conservation importance. The regional biodiversity is an integral part of sustaining the culture and livelihoods of the roughly 400,000 fishers and their families, who live in more than 150 villages. People in these villages have been engaging in customary fishing occupations for generations by drawing from, and hence also building, resources/capitals for their overall wellbeing and resilience.

However, Chilika Lagoon has been caught up in global change processes involving multiple drivers in the last few decades. The growth of shrimp aquaculture in the 1980s has led to questions about livelihoods, access, and usage rights, and changed the rules of the game in the lagoon economy and society. An increase in the international market price for tiger shrimp (*Penaeus monodon*) and a corresponding change in state policies on promoting aquaculture provided further impetus. Aquaculture in Chilika Lagoon has also received a boost from the 2001 opening of a new 'sea mouth' to the Bay of Bengal, which has had a direct impact on biophysical processes and livelihood systems (Nayak 2017). Consequently, there has been substantial threat to the capital/resource base of the Chilika fishers, contributing to a sharp decline in wellbeing and resilience, and an extraordinary increase in the level of vulnerability. Table 11.1 outlines the dimensions of vulnerability that Chilika small-scale fishers are currently being exposed to by linking those vulnerabilities to global change drivers, access to capitals, community wellbeing, and resilience.

Table 11.1 Key dimensions of vulnerability of small-scale fisheries to global change in Chilika Lagoon

Absence of wellbeing	Lack of access to capitals	Loss of resilience	Areas of vulnerability in Chilika small-scale fisheries sector	Resulting vulnerabilities in Chilika small-scale fisheries sector
Material	Natural Financial Physical	Lack capacity to absorb disturbance and reorganize while undergoing change	Ecological Problems Economic crisis Physical resources	• Loss of primary income and increased indebtedness • Lack of asset holding • Decline in quality and quantity of food - food insecurity • Pollution and adverse ecological changes • Shrinkage in lagoon fishing area and fish diversity • Protracted court cases, extraordinary financial implications • Fishing is capital intensive therefore unaffordable • Migration – income is not financially rewarding
Relational	Human Social	Lack capability for self-organization, learning and adaptation	Social crisis Economic dependence Political issues	• Breakdown of joint family/family support system • Increased dependence on external market • Increase in inter-village conflicts • Competitive fishing practices – unsustainable fishery • Encroachment – lack of access to fish stock and fishing grounds • Migration – family members forced to live separately • Migration – long absence weakens fishing rights • Loss of political voice
Subjective	Human Social		Individual and community level	• Loss of customary skill sets and knowledge • High dropout from school and low enrolment • Fishers turned wage labourers from entrepreneurs • Non-fishing activities disconnect fishers from 'Mother Chilika' • Fishers find it difficult to return to fishing after migration • Adverse mental and physical health conditions

Source: Bebbington (1997), Gough and McGregor (2007), Walker et al. (2004), Weeratunge et al. (2014), Nayak and Berkes (2014) and Nayak (2017)

The table provides details on the vulnerabilities experienced in the small-scale fishery sector of Chilika Lagoon in the material, relational, and subjective aspects of fishers' lives. At the *material level* (which includes natural, financial, and physical capitals), aquaculture has led to serious ecological problems (e.g. water pollution, salinity changes, shrinkage in lagoon area, changes in fish composition), economic crisis (e.g. loss of fishery-based income, increase in levels of debt, food insecurity, financial implications of conflicts, and court cases), and deficiencies in physical resources (e.g. declining fish stock, inability to retain assets, breakdown of fishery infrastructure, loss of access to fishing grounds, occupational displacement leading to migration).

At the *relational level* (which includes human and social capitals), aquaculture has led to social crisis (e.g. weakening of the joint family system, which worked as a social insurance at times of crisis, isolation of individual family members due to migration by some, increased conflict, and violence), economic dependence (e.g. high reliance on external market for food, highly inequitable income levels), and issues of political power and rights (e.g. collapse of fish cooperatives as the institutional foundation of small-scale fisheries, loss of access due to encroachment of fishing areas, competition over shrinking resource base, loss of fishing rights due to migration-related absences, further loss of political voice and power).

At the *subjective level* (which includes social and human capitals), aquaculture related impacts have caused vulnerabilities at individual and community levels, for example, a gradual loss of fishing skills and knowledge, a lack of education through dropping out, a sense of disconnection from Chilika as 'mother', a personal sense of fear or harm from powerful non-fishers, a lack of capacity to go back to fishing, and deteriorating mental and physical health.

The multiple levels and types of vulnerabilities in Chilika have impacted resilience in the small-scale fishery system of the lagoon. Its ability to deal with pressures of multiple drivers and strength to respond to change has been adversely affected. Changes experienced have eroded solidarity among fishers and the ability for collective action, hence the capacity for self-organisation and adaptation. Overall, the Chilika small-scale fishery system continues to remain vulnerable to ongoing processes of global change (Nayak and Berkes 2010, 2014).

11.3 Building on Strengths to Increase Viability

Global change drivers affect all the various productive systems in the world and create problems that have no easy solution (Chuenpagdee 2011). Small-scale fisheries will remain vulnerable to multiple challenges, as in the Chilika case; however, they also have certain strengths that help make them viable. Many small-scale fisheries are by necessity flexible, adaptable, and able to respond to change, if given a chance – otherwise they would have disappeared long ago (Berkes 2015). Identifying these strengths and building on them for their long-term viability may be accomplished through capacity development, as opposed to the use of the 'deficit' model of development which assumes that there are deficiencies in the ability of the group in question (Bockstael 2017).

What strategies can be used to build on the strengths of small-scale fisheries to counter the various challenges they face? Just to the southwest of Chilika Lagoon, on the Bay of Bengal coast of Odisha State, India, there is the informative case of Samudram Women's Federation. The 2010 UNDP Equator Prize winner for conservation development, Samudram is a federation of 160 women's self-help groups across 50 villages, and represents over 6000 women fish workers and their families. It operates as a social enterprise, empowering its economically marginalized members while monitoring and conserving olive ridley turtle nesting sites and restoring their coastal habitat (UNDP 2017). The direct economic benefit from international financial support for nest monitoring and turtle habitat conservation is relatively minor, and falls far short of explaining why Samudram engages in conservation. Rather, several complementary and interacting factors – economic environmental, political, social, and cultural – seem to be at work (Zachariah-Chaligne 2016):

- Small income for nest monitoring during the turtle nesting season is important because it provides much needed cash in the fishing restriction season.
- Perhaps more importantly economically, Samudram's social enterprise has developed value-added marine products and facilitated direct access to wholesale fish markets, cutting out money-lenders and middlemen.
- Fishers use their local and traditional knowledge to appreciate the importance of biodiversity, the sea being their 'food basket'. They see marine turtles as an indicator of the 'health' of the sea, which they refer to as 'mother ocean'.
- They are able to use their connections through turtle conservation to have a say in management policy, in particular to limit large-scale fisheries from coming into their waters and hurting their livelihoods.

- Samudram has created a platform for empowering women fish workers so they are aware of their rights and take pride in their work.
- Samudram has invested some of its income into health and education, helping to start schools and adult education centres, thus contributing to an improved quality of life.
- India has folklore regarding sea turtles. The majority of the households follow Hindu religion, according to which, turtles are considered as one of the incarnations (*Avatars*) of Lord Vishnu.

The story of Samudram is consistent with article 8.1 of SSF Guidelines (FAO 2015), which states "gender mainstreaming should be an integral part of all small-scale fisheries development strategies." Samudram's conservation-development work can be used to pinpoint some of the strengths of small-scale fisheries, and the possibility to develop capacity by building on existing strengths (Table 11.2). Being materially poor does not necessarily mean an absence of wellbeing. At the *material level*, increase in financial capital is a major success story, fueled by product diversification and value-added marine products, such as pickled prawn, which help build capital. Active since 1995, Samudram has built, step by step, a capital base. It provides microfinance to fishers, and financial literacy programs through which households are encouraged to access mainstream credit through banks. Direct access to markets and capital to buy the needed equipment have resulted in greater resilience in the fishery to respond to change such as market demand. All of this has resulted in income diversification for the whole area, with fisher households

Table 11.2 Key dimensions and areas of strength of small-scale fisheries in response to global change in Samudram case

Increase in wellbeing	Better access to capitals	Building resilience	Areas of strength in Samudram small-scale fisheries sector	Resulting strengths in Samudram small-scale fisheries sector
Material	Natural Financial Physical	Capacity to absorb disturbance and reorganize while undergoing change Capability for self-organization, learning and adaptation	Coastal habitat conservation Economic development Mobilizing physical resources	• Gain in primary income and decreased reliance on informal credit through microfinance • Direct access to market and financial capital • Product and income diversification through value addition • Creation of household and community assets • Increased food security • Community-led conservation of ecosystem (turtle nest monitoring and habitat conservation) • Maintenance of resource base
Relational	Human Social		Social cohesion Greater economic independence Increasing political voice	• Strengthening of community support systems • Increase in social and human capital, and social resilience • Social and political empowerment, especially of women • Coordinated response to markets, leading to economic independence • Increase in inter-village collaboration and cooperation through network building • Women's education leading to self -esteem and political voice
Subjective	Human Social		Individual and community level	• Improvements in education, health, and economic wellbeing • Being a member of Samudram is reason for pride • Local and traditional knowledge at the forefront • Community members (mainly women) becoming entrepreneurs and innovators • Stronger community connection with 'Mother ocean' • Protection of sea turtles, as incarnation of Hindu God Vishnu

Source: Bebbington (1997), Gough and McGregor (2007), Walker et al. (2004), Weeratunge et al. (2014), Zachariah-Chaligne (2016) and United Nations Development Programme (2012)

retaining a greater part of their earnings. The fishing restriction season (due to turtle conservation) is compensated by cash income for nest monitoring and habitat conservation. There is no (short term) reduction of the resource base, but some reduction in resource access. Competition with large-scale fisheries that come into their inshore area is a continuing problem.

At the *relational level*, Samudram, as a robust institution with a 20-plus year record, seems to have been successful in strengthening community support systems and increasing human and social capital, and enhancing social resilience to deal with family and community problems. The emphasis has been on women's empowerment towards social and economic development, but benefits accrue to all fishers because women have a key role in keeping the community together and functioning (Jentoft 2000). Self-organization and education help build self-esteem, bringing formerly marginalized women into mainstream society. Exposure to mainstream media and institutions has boosted women's confidence to question laws and practices detrimental to women. Such political gains by women parallel social change in communities, facilitated by better education and health care.

At the *subjective level*, gains through improvements in education, health, and economic wellbeing has meant that being a Samudram member and a vehicle of social change is a reason to be proud. Small-scale fishers are often intimately familiar with their resources, and hold local and traditional knowledge about the aquatic environment and the fish they harvest, making them natural partners in ecosystem management and conservation, as in the Samudram case. Members take pride in the fact that thousands of turtles choose to come back to their beaches to breed and nest every year. India being a country of diverse religious and cultural icons, there are temples dedicated to sea turtles in some parts of Odisha, where paintings and idols depict Lord Vishnu as a turtle. Since Lord Vishnu is believed to be the preserver of life, many members of the community, proud of their cultural heritage, consider it their responsibility to protect turtles.

In summary, the Samudram small-scale fishery case, in contrast to the Chilika case, has been able to build on strengths through capacity development (Bockstael 2017). Improvements in wellbeing follow increased access to capital assets – human, physical, natural, social, and financial – and increased resilience to deal with social and ecological problems thus reducing vulnerability. Samudram members and the wider community benefit from increased income as a result of value-added products, improved marketing, access to microfinance, and better fish yields. They also benefit from women's empowerment, capacity development training, and better education and health services (Zachariah-Chaligne 2016; UNDP 2017).

Although the villages of the area are still relatively poor by international standards, Samudram's accomplishments are remarkable. This is in part due to their favourable location: presence of olive ridley turtle nesting beaches and the absence of sites suitable for aquaculture. In contrast to Chilika, Samudram is not threatened by aquaculture, as the open waters of the Bay of Bengal coast are not suitable for shrimp farming. What the Samudram Women's Federation case shows is that it is possible to build on the strengths of small-scale fisheries through locally directed capacity development, rather than the deficit model of development. This has been accomplished through good leadership, collective action, and self-organization, all of them bottom-up (rather than top-down) development factors.

11.4 Global Change Responses

Here we turn to global change responses and discuss two recent global instruments, SSF Guidelines (FAO 2015) and Sustainable Development Goals or SDGs (UNDP 2017), which are relevant for global change and governance of small-scale fisheries, to further our understanding about the interplay between local and global, often involving the intermediate levels as well. The SSF Guidelines were designed to complement the 1995 FAO Code of Conduct for Responsible Fisheries (which were not specifically about small-scale fisheries). SDGs came into effect in 2016 to replace the earlier Millennium Development Goals (MDGs) and to provide overarching goals, targets, and indicators for sustainable development. The SSF Guidelines are the first internationally agreed upon instrument dedicated to the small-scale fisheries sector. They are the result of a bottom-up participatory development process (2010–2013) based on a global effort that involved more than 4000 representatives of governments, small-scale fishers, fish workers and their organizations, researchers, development partners, and other relevant stakeholders, from more than 120 countries. The main objectives (Table 11.3) of the SSF Guidelines recognise prolonged marginalisation and high levels of vulnerability of small-scale fisheries and provide a basis to build capacity and respond to global change processes.

In order to achieve the objectives of responsible fisheries and sustainable development through securing small-scale fisheries against existing and potential vulnerabilities, the SSF Guidelines identify a number of key areas where action is needed (Table 11.4). The SSF Guidelines aim at providing guidance for the governance and development of small-scale fisheries in the context of global and national efforts to eradicate hunger and poverty. They are 'revolutionary' in the sense that they recognize the crucial importance of small-scale fisheries, globally, in providing two-thirds of the food destined for direct human consumption (FAO 2015, ix), and the important role they play in poverty eradication, equitable development, and sustainable resource utilization (FAO 2015, ix). To fulfill this role, small-scale fisheries need policy support at the national level, support they never had previously. Guideline 5.7 recommends that "**...**States should where appropriate grant preferential access of small-scale fisheries…" (FAO 2015, 6), and call for capacity development (FAO 2015, 17). If implemented at the national level, such recommendations would turn fisheries policies on their head. As such, the Guidelines address the issue of vulnerability (e.g. the Chilika case) and provide space to further develop existing strengths (e.g. the Samudaram case).

Table 11.3 Key objectives of SSF Guidelines recognising vulnerabilities and enabling conditions

(a) Enhance the contribution of small-scale fisheries to global food security and nutrition, and to support the progressive realization of the right to adequate food.
(b) Contribute to the equitable development of small-scale fishing communities and poverty eradication, and to improve the socioeconomic situation of fishers and fish workers within the context of sustainable fisheries management.
(c) Achieve sustainable utilization, prudent and responsible management, and conservation of fisheries resources consistent with the code of conduct for responsible Fisheries (the code) and related instruments.
(d) Promote the contribution of small-scale fisheries to an economically, socially, and environmentally sustainable future for the planet and its people.
(e) Provide guidance that could be considered by states and stakeholders for the development and implementation of ecosystem friendly and participatory policies, strategies, and legal frameworks for the enhancement of responsible and sustainable small-scale fisheries.
(f) Enhance public awareness and promote knowledge on the culture, role, contribution, and potential of small-scale fisheries, considering ancestral and traditional knowledge, and their related constraints and opportunities.

Table 11.4 A number of key areas identified by the SSF Guidelines where action is needed to achieve responsible fisheries and sustainable development

Key areas in SSF Guidelines	Action points
Responsible governance of tenure	Recognises secure tenure rights of small-scale fishing communities to the resources that form the basis for their social, economic, and cultural wellbeing
Sustainable resource management	Adopting measures for the long-term conservation and sustainable use of fisheries resources and for securing the ecological foundation for food production
Social development, employment, and decent work	Ensuring that small-scale fishing communities are empowered and can enjoy their human rights
Recognition of value chains, postharvest, and trade	Recognize the importance of processing, market, and trade related to small-scale fisheries, and pay due consideration to the impact of international trade and vertical integration on local small-scale fishers, fish workers, and their communities
Gender equality	Gender mainstreaming as integral to small-scale fisheries development strategies to challenge practices that discriminate against women
Attention to disaster risks and climate change	Highlight that combating climate change requires urgent and ambitious action, in accordance with the objectives, principles, and provisions of the United Nations

SDGs offer a broad framework and goals for human development. The framing of these goals followed a collaborative process to incorporate inputs from a broad range of stakeholders. The United Nations conducted the largest consultation program in its history to gauge opinions on what the SDGs should include, guided by an international working group with representatives from 70 countries. A series of 'global conversations' through 11 thematic and 83 national consultations, and door-to-door surveys were conducted to include many voices. Finally, a total of 17 Goals, 169 Targets, and 241 Indicators were formulated in 2016, comprehensively covering global social and environmental systems (UNDP 2017).

SDG Goal 14 is abbreviated as "Life below Water" and more fully described as "Conserve and sustainably use the oceans, seas and marine resources." SDG Target 14 combines ocean wellbeing and human wellbeing generally, and is about "... increase[ing] the economic benefits to Small Island Developing States and Least Developed Countries from the sustainable use of marine resources." Finally, SDG Target 14.b addresses small-scale fisheries: The Target is to "provide access for small-scale artisanal fishers to marine resources and markets." It is accompanied by a measurable indicator 14.b.1 to monitor "progress by countries in the degree of application of a legal/regulatory/policy/institutional framework which recognizes and protects access rights for small-scale fisheries."

This is the extent of coverage of small-scale fisheries in SDGs. Charles (2017, 48), who has analyzed both small-scale fisheries Guidelines and SDGs, comments that "implementation of the SDGs can be guided by the kind of innovative, holistic thinking found in the small-scale fisheries Guidelines." Further, "...implementation of SDG 14 should be closely informed by the content of the small-scale fisheries Guidelines – not only for small-scale fisheries, but indeed in terms of the future of fisheries and of oceans more broadly" (Charles 2017, 46).

11.5 Discussion and Conclusions

Small-scale fisheries are vulnerable to drivers of change, a major theme in the TBTI project. A common assumption is that the interplay between local and global always works to the disadvantage of the local. This is, in part, because organizations at certain levels are said to be able to exercise power over organizations at other levels – the phenomenon of scale dominance (Collinge 1999). However, we have argued elsewhere that not only can global drivers impact the sustainability of local social-ecological systems, but sustainability at higher levels can also be impacted by changes at local and regional social-ecological systems (Nayak and Berkes 2014).

For example, global drivers, such as shrimp aquaculture expansion (in turn driven by high profits in international markets), have a large and demonstrable negative impact on the sustainability of small-scale fisheries in Chilika. However, there is also a reverse process of influence, operating from the bottom up, with potential for triggering changes at higher levels. In the case of Chilika, the disintegration of small-scale fisheries contributes to national and global poverty and food insecurity, making it difficult for India to meet SDGs and other development goals. Thus, the effect of drivers could be seen as a two-way process in which global drivers and local changes can potentially impact each other, and sustainability at higher levels can be impacted by bottom-up processes (Nayak and Berkes 2014).

The evidence reviewed in this chapter and other TBTI publications (e.g., Jentoft et al. 2017; Johnson et al. 2018) suggest that the interaction between global and local is more than a straight line two-way process. Rather, the interaction may be multidirectional and iterative in nature, involving complex multi-level interplay (Börzel and Heard-Lauréote 2009), and continuous feedback between local and global. For example, global change responses and new policy initiatives, such as SSF Guidelines and SDGs, may create a reverse process of positive impacts at the local level, countering negative impacts on small-scale fisheries. Case studies have a role in understanding such impacts of change in small-scale fisheries. For example, the I-ADApT approach can help devise ways to reduce vulnerability and increase viability in the face of change. Box 11.1 and Tables 11.1 and 11.2 illustrate that transdisciplinary approaches are required and that disciplinary approaches are simply inadequate to deal with these issues.

Further, positive impacts are not restricted to SSF Guidelines and SDGs. In the Samudram case, we see that international conservation initiatives can trigger locally driven social and economic development if the conditions are right, providing additional reasons for using transdisciplinary approaches. Iterative feedback relations can lead to creative interplay between local and global, and offer an interesting opportunity to analyse global change responses and sustainability in the context of small-scale fishery systems.

Figure 11.1 reflects on the internal (i.e. multi-level) complexity involved in the interactions between local and global. One set of arrows indicates that drivers at higher levels (regional, national, and global) can create vulnerabilities at the local level (e.g. the Chilika case). But, another set of arrows shows that local small-scale fishery systems may have the ability to harness their strengths to respond to global change (e.g. the Samudram case). Overcoming vulnerabilities and bringing into play existing and newly acquired strengths can lead to the viability of small-scale fishery systems. We hypothesize that the three key dimensions of vulnerability are: increase/decrease in wellbeing, better/poorer access to capitals, and building/losing resilience (Tables 11.1 and 11.2). We use the term viability not just in an economic sense but also to include social, political, and ecological aspects of small-scale fisheries systems. Global change responses reviewed here are positive in that they provide opportunity for small-scale fisheries systems to assess their vulnerabilities and realise their strengths to develop adaptive capacity and resilience. The capacity to adapt is what makes small-scale fisheries viable.

Viewed this way, the prognosis for small-scale fisheries is not necessarily bleak. Clearly, the vulnerabilities are there and are dynamic, but many small-scale fishery systems have survived in the long-term, meaning that they have been viable. Some of the literature has commented on this (e.g., Jentoft 2000; Béné et al. 2011; Berkes 2015), and TBTI research has been adding details regarding the mechanisms involved, for example, as related to wellbeing (Johnson et al. 2018). Isaacs (2016) observed that the protein value of small pelagics, such as sardines, was being largely wasted as livestock feed. Using small pelagics directly as human food toward food security may require, among others, changes in national fishery policies to invest in post-harvest processing. In another TBTI study, DesRivieres et al. (2016) found that

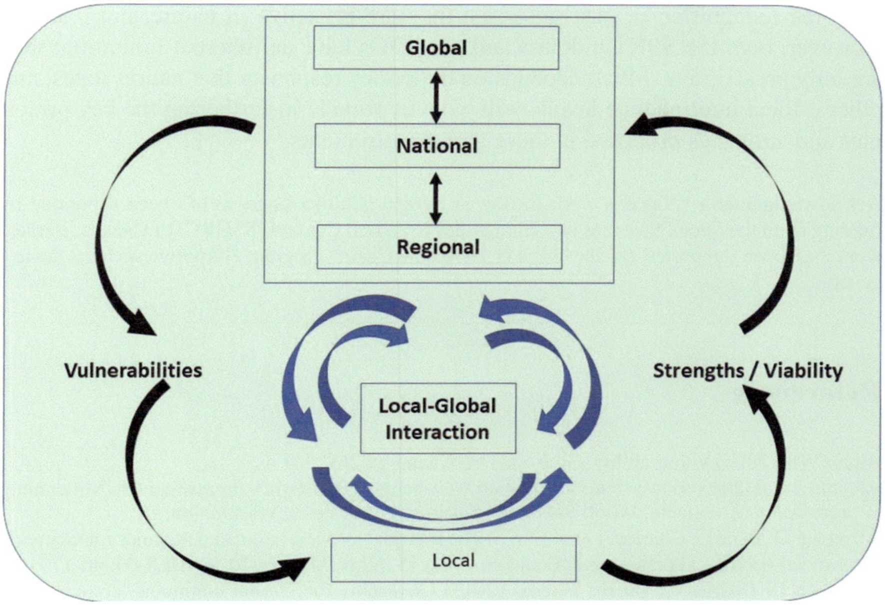

Fig. 11.1 Interaction between local and global as a multidirectional and iterative process characterised by complex interplay and continuous feedback

alternative food networks were being developed as a way to better include fish and seafood in the local food system. Both examples illustrate the kind of creative interplay of local-regional-national-global levels that Fig. 11.1 attempts to capture.

TBTI studies and some earlier work have documented what small-scale fishers themselves consider important for viability. They want the ability to collectively manage their own affairs; have security of access to their resources; improve market access, credit, and capacity development; and better health, education, and social services. Thus, wellbeing is a priority area (Johnson et al. 2018), capital provide the means for development, and resilience is all about maintaining or building options and flexibility – our three key dimensions of vulnerability. As illustrated by the Samudram example, viability requires building stronger fishing communities, with higher social capital, networks, trust relationships, and increased adaptive capacity (Jentoft 2000; Berkes 2015). Breaking the vicious cycle of vulnerability-poverty-vulnerability and improving the ability to respond to change depends on building such strong fishing communities.

Some higher-level policy help for small-scale fisheries is most welcome in this context. Together, the SSF Guidelines and the SDGs signify a reverse process (top-down) of global change responses to help deal with the outcomes of the earlier impacts of higher level drivers on small-scale fishery systems. Unlike other global drivers that have created adverse local impacts, these instruments are designed to be constructive, positive, enabling, and inclusive. They are a set of responses emanating from the recognition of past mistakes; they are proactive in nature, not reactive. However, both the SSF Guideless and the SDGs have an inherent limitation: they are only prescriptive. Much depends on the policy responses that nation states, and other critical intermediate levels, will have to provide in furthering the key principles and strategies proposed in these global instruments.

Acknowledgements Nayak's work in Bay of Bengal (Chilika Lagoon) has been supported by funding from the Social Sciences and Humanities Research Council (SSHRC) of Canada. Berkes' work has been supported by the Canada Research Chairs program (http://www.chairschaires.gc.ca).

References

Adger WN (2006) Vulnerability. Glob Environ Chang 16:268–281

Alcamo J et al (2003) Ecosystems and human Well-being: a framework for assessment, Millennium ecosystem assessment. World Resources Institute/Island Press, Washington, DC

Armitage D, Béné C, Charles A et al (2012) The interplay of well-being and resilience in applying a social-ecological perspective. Ecol Soc 17(4):15. https://doi.org/10.5751/ES-04940-170415

Armitage D, Charles A, Berkes F (eds) (2017) Governing the coastal commons: communities, resilience and transformation. Earthscan/Routledge, London/New York

Bebbington A (1997) New states, new NGOs? Crisis and transition among Andean rural development NGOs. World Dev 25:1755–1765

Bebbington A (1999) Capitals and capabilities: a framework for analysing peasant viability, rural livelihoods and poverty. World Dev 27:2021–2044

Béné C, Evans L, Mills D et al (2011) Testing resilience thinking in a poverty context: experience from the Niger River basin. Glob Environ Chang 21:1173–1184. https://doi.org/10.1016/j.gloenvcha.2011.07.002

Berkes F (2007) Understanding uncertainty and reducing vulnerability: lessons from resilience thinking. Nat Hazards 41:283–295

Berkes F (2011) Restoring unity: the concept of social-ecological systems. In: Ommer RE, Perry RI, Cochrane K, Cury P (eds) World fisheries: a social-ecological analysis. Wiley-Blackwell, Oxford, pp 9–28

Berkes F (2015) Coasts for people. Interdisciplinary approaches to costal and marine resource management. Routledge, New York/London

Bockstael E (2017) Critical capacity development: an action research approach in coastal Brazil. World Dev 94:336–345

Börzel TA, Heard-Lauréote K (2009) Networks in EU multi-level governance: concepts and contributions. J Publ Policy 29(2):135–151

Brooks N (2003) Vulnerability, risk and adaptation: a conceptual framework. Working Paper 38. Tyndall Center for Climate Change Research, Norwich

Bundy A, Chuenpagdee R, Cooley SR et al (2016) A decision support tool for response to global change in marine systems: the IMBER-ADApT framework. Fish Fish 17:1183–1193

Cash DW, Adger W, Berkes F et al (2006) Scale and cross-scale dynamics: governance and information in a multilevel world. Ecol Soc 11(2):8. http://www.ecologyandsociety.org/vol11/iss2/art8/

Charles A (2017) The big picture. Samudra Reports 76:45–48

Chuenpagdee R (2011) Too big to ignore: global research network for the future of small-scale fisheries. In: Chuenpagdee R (ed) World small-scale fisheries. Eburon, Delft, pp 383–394

Collinge C (1999) Self-organization of society by scale: a spatial reworking of regulation theory. Environ Plann D 17:557–574

Coulthard S (2012) Can we be both resilient and well, and what choices do people have? Incorporating agency into the resilience debate from a fisheries perspective. Ecol Soc 17(1):4. https://doi.org/10.5751/ES-04483-170104

DesRivières CP, Chuenpagdee R, Mather C (2016) Reconnecting people, place and nature: examining alternative food networks in Newfoundland's fisheries. Agric Food Secur 6:33

Dessai S, Hulme M (2004) Does climate adaptation policy need probabilities? Clim Pol 4:107–128

FAO (Food and Agriculture Organization) (2015) Voluntary guidelines for securing sustainable small-scale fisheries. United Nations, Rome

Gibson C, Ostrom E, Ahn TK (2000) The concept of scale and the human dimensions of global change: a survey. Ecol Econ 32:217–239

Gough I, McGregor JA (eds) (2007) Wellbeing in developing countries: from theory to research. Cambridge University Press, Cambridge

Guillotreau P, Bundy A, Perry RI (2018) Global change in marine systems: integrating natural, social and governing responses. Routledge, London

Gupta J (2008) Global change: analyzing scale and scaling in environmental governance. In: Young OR, King LA, Schroeder H (eds) Institutions and environmental change. MIT Press, Cambridge, MA, pp 225–258

Isaacs M (2016) The humble sardine (small pelagics): fish as food or fodder. Agric Food Secur 5:27

Jentoft S (2000) The community: a missing link of fisheries management. Mar Policy 24:53–59

Jentoft S, Chuenpagdee RC (2009) Fisheries and coastal governance as a wicked problem. Mar Policy 33:553–560. https://doi.org/10.1016/j.marpol.2008.12.002

Jentoft S, Chuenpagdee R (eds) (2015) Interactive governance for small-scale fisheries: global reflections, MARE Publication Series. Springer International Publishing, Cham

Jentoft S, Chuenpagdee R, Bundy A et al (2010) Pyramids and roses: alternative images for the governance of fisheries systems. Mar Policy 34:1315–1321

Jentoft S, Chuenpagdee R, Barragan-Paladines MJ et al (eds) (2017) The small-scale fisheries guidelines. Global implementation, MARE Publication Series. Springer International Publishing, Cham

Johnson DS, Acott TC, Stacey N et al (eds) (2018) Social wellbeing and the values of small-scale fisheries, MARE Publication Series. Springer International Publishing, Cham

Kates R, Clark WC, Corell R et al (2001) Sustainability science. Science 292:641–642

Kelly PM, Adger WN (2000) Theory and practice in assessing vulnerability to climate change and facilitating adaptation. Clim Chang 47:325–352

Kurien J (1992) Ruining the commons and responses of the commoners: coastal overfishing and fishworkers' actions in South India. In: Ghai D, Vivian J (eds) Grassroots environmental action. Routledge, London, pp 221–258

McGregor A (2008) Well-being, poverty and conflict. Briefing Paper 1/08. ESRC Research Group on Wellbeing in Developing Countries, University of Bath, Bath, UK

Nayak PK (2017) Fisher communities in transition: understanding change from a livelihood perspective in Chilika Lagoon, India. MAST 16:13

Nayak P, Berkes F (2010) Whose marginalisation? Politics around environmental injustices in India's Chilika Lagoon. Local Environ 15(6):553–567

Nayak P, Berkes F (2014) Linking global drivers with local and regional change: a social-ecological system approach in Chilika Lagoon, Bay of Bengal. Reg Environ Chang 14:2067–2078

Norberg J, Cumming GS (eds) (2008) Complexity theory for a sustainable future. Columbia University Press, New York

O'Brien KL, Eriksen S, Schjolden A et al (2004) What's in a word? Interpretations of vulnerability in climate change research. Center for International Climate and Environmental Research Series. CICERO working paper 04. Oslo, Norway

Steffen W, Broadgate W, Deutsch L et al (2015) The trajectory of the anthropocene: the great acceleration. Anthr Rev 2:81–98

UNDP (United Nations Development Programme) (2017) http://www.undp.org/content/undp/en/home/sustainable-development-goals.html. Accessed 15 Apr 2018

United Nations Development Programme (UNDP) (2012) Samudram Women's Federation, India. Equator Initiative Case Study Series. New York. http://www.equatorinitiative.org/wp-content/uploads/2017/05/case_1348165017.pdf. Accessed 15 Apr 2018

Walker B, Holling CS, Carpenter SR et al (2004) Resilience, adaptability and transformability in social-ecological systems. Ecol Soc 9(2):5 http://www.ecologyandsociety.org/vol9/iss2/art5/print.pdf

Weeratunge N, Béné C, Siriwardane R et al (2014) Small-scale fisheries through the wellbeing lens. Fish Fish 15(2):255–279

Wolf S, Hinkel J, Hallier M et al (2013) Clarifying vulnerability definitions and assessments using formalisation. Int J Clim Change Strateg Manag 5(1):54–70. https://doi.org/10.1108/17568691311299363

Young OR, King LA, Schroeder H (eds) (2008) Institutions and environmental change. MIT Press, Cambridge, MA

Zachariah-Chaligne A (2016) Orissa, India: the role of Samudram Women's Federation in reducing poverty and protecting resources. Community Conservation Research Network. http://www.communityconservation.net/odisha-india/. Accessed 15 Apr 2018

Chapter 12
Enhancing the Stewardship in Trat Bay, Eastern Thailand: A Transdisciplinary Exercise

Suvaluck Satumanatpan, Ratana Chuenpagdee, Wichin Suebpala, Thamasak Yeemin, and Kungwan Juntarashote

Abstract Small-scale fisheries often exist among other activities, such as large-scale fisheries, urban development, tourism industry, and transportation that may be competing for resources and/or for space. To a varying degree, and depending on the practices, all of these activities affect the health of ecosystems and contribute to degradation of the environment. This is particularly problematic in ecologically important and biologically sensitive areas. Governance of the multiple-use system is challenging because of the complexity of the problem and the general lack of data to determine the levels of impacts. Further, sectors may disagree about what causes degradation and will likely support interventions that have the least consequences on their activities. Sectors that are not well organized and have little political influence or economic power are prone to being disadvantaged by decisions that inhibit their opportunity for viable livelihoods. Small-scale fisheries frequently fall into this category, despite the fact that impacts from their practices may be less than other sectors. Importantly, insufficient consideration is given to the social and

S. Satumanatpan (✉)
Faculty of Environment and Resources Studies, Mahidol University,
Nakorn Pathom, Thailand
e-mail: suvaluck.nat@mahidol.ac.th

R. Chuenpagdee
Department of Geography, Memorial University of Newfoundland, St. John's, NL, Canada
e-mail: ratanac@mun.ca

W. Suebpala
Marine Biodiversity Research Group, Faculty of Sciences, Ramkhamhaeng University,
Bangkok, Thailand

T. Yeemin
Marine Science Association of Thailand, Faculty of Sciences, Ramkhamhaeng University,
Bangkok, Thailand

K. Juntarashote
Department of Fisheries Management, Kasetsart University, Bangkok, Thailand
e-mail: kungwan.j@ku.ac.th

R. Chuenpagdee, S. Jentoft (eds.), *Transdisciplinarity for Small-Scale Fisheries Governance*, MARE Publication Series 21,
https://doi.org/10.1007/978-3-319-94938-3_12

cultural connectivity that small-scale fisheries have to their surroundings and the importance of such connectivity on conservation and stewardship. A holistic perspective and innovative thinking are required to level the playing field for all sectors and stakeholders and to encourage collaboration and synergies. The chapter presents a case study of Trat Bay in the east coast of Thailand as an exercise in transdisciplinarity to enhance interactions and discuss stewardship options in the area.

Keywords Stewardship · Small-scale fisheries · Interactive governance · Transdisciplinary · Royal Ordinance on fisheries · Coastal pollution · Thailand

12.1 Stewardship and Small-Scale Fisheries

Stewardship is one of those concepts that are often used, but are not always well defined. Some may relate the term to the biblical usage, for example, with reference to a three-tiered relationship between God as the owner, human as the steward, and nature as property to be guarded (Roach 2000). In common expression, stewardship is a set of values that foster human duty, responsibility, and care towards nature as a result of interaction and relationship with it (see Rogers 2015). While part of personal virtue, stewardship is frequently expressed at a community level where relationships between people and the environment can be easily observed and promoted. As suggested by Roach et al. (2006), stewardship is sometimes motivated by common concerns in the community, made possible by the sense of place, volunteerism, and local knowledge. Thus, when Ivany (2011) examined what stewardship meant to a small community in eastern Newfoundland, terms like community spirit, respect for elders, leadership, and future generations were added.

The connection between stewardship and sustainability may seem obvious, but is often not explicitly made clear in the discussion about sustainable development. As seen in McConney et al. (Chap. 10, this volume), many references are made to the stewardship concept in the deliberation about the United Nations (UN) Sustainable Development Goals (SDGs), even though the term itself is not mentioned in the document 'Transforming Our World' (United Nations 2015). The call for stewardship is also implicit in the statement about Goal 14, which aims to conserve and sustainably use the oceans, seas, and marine resources. The importance of stewardship was also reflected during the 1st UN Ocean Conference held in June 2016 in New York, where countries and civil societies pledged support and made commitments about what they would do to achieve SDG 14, and other linked goals.

The reference to stewardship in the context of fisheries is also not explicit in these initiatives. Small-scale fisheries, in particularly, are only directly mentioned in the context of Target 14.b related to access to resources and markets. However, other targets, like prevention of pollution (14.1), avoiding ecosystem impacts (14.2), and the emphasis on increasing benefits from fisheries in Small Island Developing States and least developed countries in Target 14.7, among others, are also relevant to small-scale fisheries. In a similar vein, the term stewardship is not used in the Voluntary Guidelines for Securing Sustainable Small-Scale Fisheries in the Context of Food Security and Poverty Eradication (SSF Guidelines; FAO 2015), although sustainability is one of the main principles and conservation is one of the main management objectives. This neither implies the absence nor the lack of importance of stewardship in small-scale fisheries sustainability. In all likelihood, stewardship is already assumed as one of the values or properties that should be fostered in order to achieve sustainability and conservation, as suggest by Chapin et al. (2009). The implicitness of the stewardship concept in policy dialogue is perhaps not an issue, except in the discussion about governing interventions, which are mostly dominated by conservation ideology. A question needs to be raised about whether it is conservation or stewardship that is the best route to sustainability.

Generally speaking, efforts to achieve sustainable resource use and conservation are mostly concentrated on applying rules, laws, and regulatory measures, in order to deter or encourage certain behaviors, like elimination of destructive, illegal fishing practices, or the establishment of marine protected areas. Studies show, however, that these tools are not always effective in the long run (see, for instance, Agardy et al. 2003; Bennett and Dearden 2014; Satumanatpan et al. 2014; Satumanatpan and Chuenpagdee 2015), especially when they are imposed from the outside and have little or no engagement with communities. Further, sustaining these efforts require change at the 'meta-order,' where values, images, and principles matter (Jentoft et al. 2012; Song et al. 2013). As a concept, stewardship aligns well with these meta-level elements because of its inherent quality and the intangibility of its value. They are, in effect, difficult to understand and articulate, which explains the omission from documents like SDGs and SSF Guidelines. As such, the importance of stewardship in small-scale fisheries is a good topic to examine through the transdisciplinary perspective. This chapter illustrates this, using the case study of Trat Bay in the east coast of Thailand, where a transdisciplinary workshop was organized during July 31 to August 3, 2017. While it was not set up as a study on stewardship, the connection between stewardship and sustainability of small-scale fisheries in the area became clear through the way the participants identified problems and articulated solutions.

In the following, we provide general background about Trat Bay and details about the workshop. Next, we present the results of the governability analysis of Trat Bay, starting with the characteristics of the natural, social, and governing systems, along with the wicked problems associated with them. The chapter concludes with the participants' recommendations about what needs to be done to improve governability of Trat Bay.

12.2 Trat Bay Situation Overview and the Transdisciplinary Workshop

Trat Bay is located in Trat Province, a coastal province in the most eastern part of Thailand (Fig. 12.1). Situated in the upper Gulf of Thailand, bordering with Cambodia, Trat Province is one of the main commercial hubs in the region and is a popular tourism destination among Thais and foreigners, with its rich history, cultural heritage, and natural ecosystems, augmented by several archipelagos, corals, and reef areas.

The natural setting of Trat Bay makes for an ideal fishing ground. Fisheries, therefore, play an important role to the economy of the province, contributing about 14% to the total gross domestic product and supporting over 4000 households in the area (Trat Provincial Office 2017). Like other coastal provinces in the country, Trat is going through a major governance reform as a result of the new Royal Ordinance on Fisheries, which came into effect in 2015. Under pressure from the European Union and its 'yellow card' barring Thai seafood exports to member countries, the Government of Thailand rushed into issuing the new act in order to address issues related to fisheries unsustainability, and safety and working conditions of fishery workers. The Royal Ordinance on Fisheries aligns well with the SSF Guidelines in its aim to prevent overfishing and overcapacity, deterring of illegal, unregulated, and

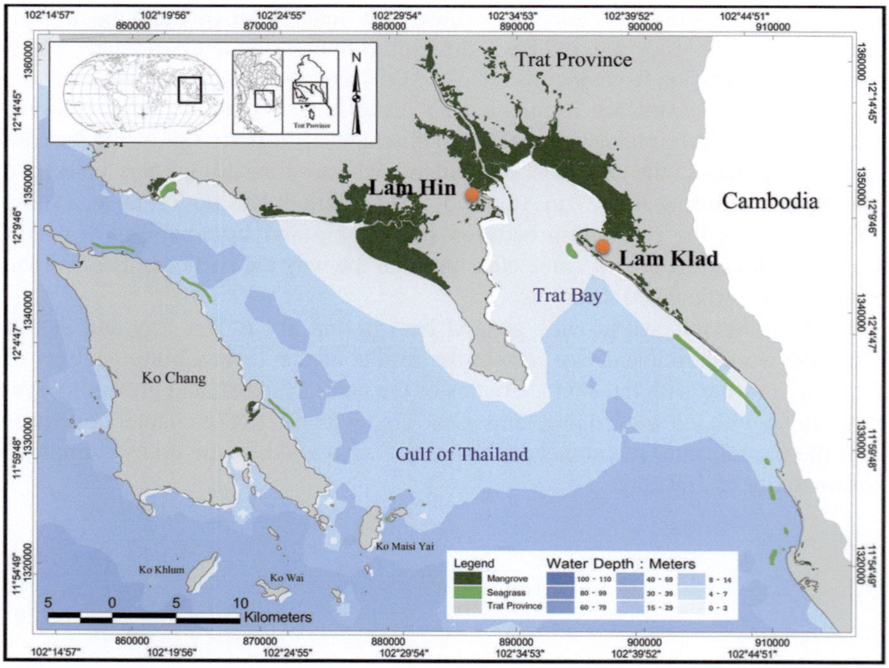

Fig. 12.1 Map of the Trat Bay, Trat Province

unreported (IUU) fishing, and improving welfare and working conditions of fishing labor (Chuenpagdee et al. 2017).

An important provision in the Royal Ordinance on Fisheries (2015) is the 'partial' decentralization of decision making to the Provincial Fisheries Committee. The term partial is used here to acknowledge that decision making is not taking place entirely at the provincial level, but there are mechanisms for the local level committee to provide inputs. In principle, the Royal Ordinance on Fisheries provides guiding principles and general rules to govern fisheries in the whole country. The Provincial Fisheries Committee can, however, propose modifications to these rules based on the consultation with fishers and to better align with the ecological and socioeconomic characteristics of the fisheries in each province. The Fishery Department reviews proposals and makes recommendations before the proposed rules could come into effect. Trat Province was the first in the country to take advantage of this provision, and through extensive consultation with local fishers, both small-scale and large-scale, as well as the communities-at-large, made several propositions to amend the regulations (see details in Chuenpagdee et al. 2017). Many of these recommendations reflect conservation and stewardship values that seem prominent in the province. For instance, they asked for an extension of protected areas in mangrove forests and clam beds, and prohibition of unsustainable fishing gears such as trawling, beach seining, gillnet, small-mesh crab trap, and light luring for anchovy, within three nautical miles from shore.

The community-based consultation process and negotiations with the government do not always go smoothly, however. The decisions made at the consultation reflect the influence of stakeholder groups present in the certain meetings. The large-scale industrialized fisheries, for instance, were absent from the earlier meetings, and thus did not have a major voice in the decision about the extension of restricted fishing zone (from 3 to 5 nautical miles). They came into the discussion later and were able to convince the Provincial Fisheries Committee to submit a new proposal to revert the restriction back from 5 to 3 nautical miles. The opportunity for ongoing discussion and negotiations about the fisheries reflects the interactive quality of the governance process, but at the same time, it creates frustration and uncertainty among fishers, also affecting their decisions about future investment. The Fisheries Department responsible for the implementation of the Royal Ordinance on Fisheries also felt the pressure to improve the situation, particularly the engagement process. For this reason, they welcomed the idea of conducting a 'transdisciplinary' training workshop that could help build capacity within the government departments, and at the same time foster relationship with fisheries stakeholders. The 'Transdisciplinary Policy and Practices for Natural Resource Sustainability' workshop was, therefore, organized by Too Big To Ignore (TBTI) Global Partnership for Small-Scale Fisheries Research, in collaboration with the Marine Science Association of Thailand in Banpu Resort, Trat Province from July 31 to August 3, 2017. The broader focus on natural resources, instead of just fisheries, was meant to encourage the participants to look holistically at the issues and the problems in the area. The transdisciplinary perspective (Lang et al. 2012; Said et al. Chap. 22, this volume) was used as a way to help guide discussions about the identification of problems and the articulation of possible solutions. The workshop also introduced interactive governance theory (Kooiman et al. 2005) and the governability assessment framework (Bavinck et al. 2013) to comprehensively examine all aspects of the fisheries systems that need to be considered to improve the overall quality of fisheries governance in the area.

The workshop was opened to anyone interested in the discussion about fisheries and natural resource governance, and to those with good knowledge about Trat Bay. However, the majority of the participants (16 out of 26) were from the Department of Fisheries, the Department of Marine and Coastal Resources, and the National Parks. This could be because of the 'training' nature of the workshop and the academic-oriented focus. Nevertheless, four participants were from fishing communities and local environmental organizations. The other participants included three academics and two persons working in the local area for a private company. While only seven of the participants were women, they represented all three groups: governments, community organizations, and academic. The majority of the participants either work or live in Trat Bay.

The workshop began with a discussion about the issues and challenges facing Trat Bay and whether they were considered to be wicked problems. Working in small groups, the participants were asked to describe the natural, social, and governing systems in the area, following the governability assessment template (see Chuenpagdee and Jentoft Chap. 23, this volume), and discuss aspects of these characteristics in terms of their contribution to making the system more or less governable. The workshop included a one day field visit to two main fishing communities in Trat Bay, Lam Klad and Lam Hin, which was an opportunity for the participants (split into two teams) to discuss issues with the local people, mostly small-scale fishers, and to ground truth their knowledge. The last day was for report back and a general discussion about the next steps to promote sustainability of fisheries and natural resources in Trat Bay. All authors of the chapter participated in the workshop, with the first two authors serving as facilitators, and the others as resource persons.

12.3 Governability Analysis of Trat Bay

As posited by interactive governance theory, the governability of a fishery system depends on many factors, including the characteristics of the natural and the social systems that are being governed, the features and the capacity of the governing system, as well as how these systems interact (Chuenpagdee and Jentoft 2009). Further, some problems facing governance are considered wicked for the reasons explained by Rittel and Webber (1973). For instance, they are difficult to define and determine the sources or the courses. The problems are unique even though they may seem familiar, and thus there are not always solutions, but resolutions. They are social in nature, not technical, which means that reaching agreement on how to solve them is also a problem. Because most fisheries problems have these characteristics, Jentoft and Chuenpagdee (2009) conclude that fisheries governance is a wicked problem, and posit that a holistic and systematic assessment of the fisheries systems, using the governability assessment framework, can help reveal what makes the system more or less governable.

The following sections present various aspects of the natural, social, and governing systems that affect the governability of Trat Bay. The presentation is based on the discussions at the workshop, coupled with field observation and informal interviews, and supplemented by literature. The discussion about the wicked problems related to the governance of Trat Bay is presented at the end to conclude the governability analysis.

12.3.1 A Rich and Productive Ecosystem Supporting Diverse Fisheries

Trat Bay is rich with mangrove, seagrass beds, marine mammals, some fish and shellfish species, and other marine organisms. Twenty-six species of mangroves grow in a close to 100 km^2 area, eight of which are edible plant species, reflecting the importance of mangroves for local food security (Mansilp 2013). Trat River and other canals supply lots of nutrients, leading to high diversity of phytoplankton, thus enhancing marine primary productivity in the bay. Economically important benthos species like crabs, shrimps, clams, and other bivalves are abundant. Undulated surf clam (*Paphia* spp.), hard clam (*Meretrix* spp.) and mangrove clam (*Geloina* spp.) are particularly popular (Munprasit et al. 2009; Chanrachkij 2013) and help draw tourists to the area.

With high productivity and complex marine and coastal resources, Trat Bay is one of the most important fishing grounds in Thailand, supporting more than 2515 fishing households in the province, the majority of which are small-scale (NSO 2013). In 2015, total landings were about 36,500 tonnes, or 4% of the total landings in the country (DoF 2017). The diversity in marine species that can be harvested also means a wide array of fishing methods are used, including artisanal and low impact gears like falling nets, traps, bamboo stake traps, hand lines, and the traditional Acetes scoop nets. The Acetes scoop net, or 'hand' push net, is noteworthy because of its resemblance to the regular push nets, banned under the new Royal Ordinance on Fisheries. According to the local people, Acetes scoop nets have long been used in the area to catch Acetes shrimp (*Acetes spp*. and *Mesopodopsis spp*.), used as raw materials for shrimp paste, for which the area is famous. They also argue that unlike other bottom touching gears, the Acetes scoop net is highly selective and has little or no bycatch or discards (Chuenpagdee et al. 2017). In addition to the traditional gears, other larger, commerical operations take place, with trawlers targetting demersal fish species and purse seines and falling net catching anchovy. Drift gillnets are also used to catch a wide range of pelagic fish, such as mackerels, mullets, and threadfins. Finally, shellfish collection occurs along the shores (often involving women), in mangrove forests, or in the open water area using motorized fishing boat with dredge.

Trat Bay also serves as an important feeding ground for several marine mammals such as Irrawaddy dolphins (*Orcaella brevirostris*), Indo-Pacific humpback dolphins (*Sousa chinensis*) and Finless porpoises (*Neophocaena phocaenoides*). During 5 years of boat surveys an average of 423 dolphins were estimated ranging within 12 km of the coast of Trat Bay to Klong Yai near the Cambodian border (Hines et al. 2015). Since fishing takes place in the migratory route and feeding area of those marine mammals, they risk being entangled by fishing gears, especially gillnets (Junchompoo et al. 2014; Whitty 2014; Hines et al. 2015). The issue of gear entanglement is a concern among conservation organizations and researchers, and various projects have been initiated to address the problem (Hines et al. 2015; DMCR 2016). Several members of local communities and environmental groups have been able to develop research skills, scientific knowledge, and stewardship ethics through their involvement in the projects. Examples of these include the conservation and stewardship practices demonstrated in the 'crab bank' project, which involves holding gravid crabs in a tank until they release the eggs before selling them. While there is not sufficient evidence to suggest that returning the eggs to the sea helps enhance the stock, the project helps bring awareness about the importance of conservation and the protection of stocks (Thiammueang et al. 2012). The Lam Klad community also engages in other restoration efforts, including making artificial seagrass (known locally as '*sang*'; Fig. 12.2) as spawning and feeding grounds for fish and aquatic animals. These community-based stewardship initiatives reflect people's understanding about the interdependency between habitats and marine species.

Fig. 12.2 Artificial seagrass

12.3.2 Dynamic Social Systems Affecting Sustainability

The rich and diverse marine and coastal ecosystems in Trat Bay support a range of activities and contribute to social and economic development in the area. Not all of them are sustainable, however, and heavy pressure from some of these activities has resulted in environmental degradation and resource decline; for instance, excessive use of destructive gears, like trawls and dredges. Deterioration of the fishing grounds has affected the abundance and natural reproduction of the undulated surf clam (Chanrachkij 2013). Many mangrove forests have been cleared for the timber industry, shrimp farming, and urban development. While some success has been observed in the restoration of mangroves, decline in fisheries resources is difficult to reverse.

Within the fisheries sector, the most complicated problem seems to be related to tension between fishers using different gears. The conversation with the communities during the field excursion revealed that fishers in Lam Klad village on the east side of the bay were highly unhappy with those in Lam Hin village on the other side who share the fishing grounds (Fig. 12.1). Lam Klad fishing people strictly adhere to the regulation that prohibits use of destructive fishing gears, like trawls and push nets, and have set up their own rules to enforce specific area closures. While acknowledging resource degradation as one of the problems requiring attention, some fishers in Lam Hin pointed out that the new rules, especially the banning of push nets and the restricted fishing zone, have major consequences on their livelihoods and that more discussion would be beneficial before such rules came into effect. As noted in Chuenpagdee et al. (2017), there is still no consensus about the impact of push nets, with some of the fishers arguing that their operation takes place in muddy areas with low habitat value.

The different perspectives on many of these issues, along with the fact that resource degradation continues, with linkages to other causes beyond fisheries, suggest that small-scale fishing people in Trat Bay are facing a series of wicked fisheries problems (see Sect. 3.4). According to the workshop participants, the situation is made worse by the incoherent current policies and regulations; more discussion is required to better align them with the principles and values expressed by the local communities. An example of how the Royal Ordinance on Fisheries adds complexity to the social relationships was revealed during the workshop when participants spoke about how some small-scale seabass farmers used to rely on bait fish from the local push net fisheries to feed seabass. Since the push net ban came into effect, they have to buy bait fish from other sources, often at higher prices. This has affected the viability of the seabass farmers, and some of them had to change to fishing for mud crab or collecting shellfish in mangrove forests. According to interviews with small-scale fishers during one of the field visits, the ban on push net has led to three distinct patterns for push net fishers. Some of them moved into trawl fisheries (beam trawls), some turned to small-scale fishing using crab or fish gillnets, while others had to temporarily stop fishing. At the community level, the ban also results in reduced income to small businesses, like ice makers, food vendors, and grocery stores. With continued decline in fisheries catches and income, the conclusion from the local communities is that the ‘economic slump’ has never been so dramatic in

Trat Bay. Such economic hardship affects the governability of the social system, and lowers the willingness of the community to contribute to stewardship ethics and practices (Pollnac et al. 2001; Morgan 2016; Pascoe et al. 2015; Satumanatpan and Pollnac 2017).

12.3.3 Complex Governing System Lowering Governability

As already mentioned, the most important change in the fisheries governing system is the policy and regulatory reform under the recent Royal Ordinance on Fisheries. But this new law is not the only instrument being implemented in the coastal area. The second legislation, the Promotion of Marine and Coastal Resources Management Act (2015), came into effect soon after. These two acts have the same main provision in setting up a National Committee to be responsible for approving policy and management plans for fishery and coastal resources, respectively. In both cases, a provincial level committee needs to be established, with similar mandates, including promoting community participation in fishery and coastal resources governance. Through these new laws, support to local community projects and activities related to conservation and restoration of marine and coastal resources, such as materials for the artificial reef, seagrass rehabilitation, and fishing gears, can be provided.

Under the Royal Ordinance on Fisheries, formal participation of local communities in fishery management is through the registered local fishing community organizations, whose representatives are appointed as members of the 'Provincial Fisheries Committee' (Fig. 12.3). In order to promote fairness and comprehensiveness, the registered local fishing community organizations must cover a range of sectors, such as coastal and offshore fisheries, inland fisheries, aquaculture, and processing. However, the condition that a minimum of 30 members are required to register an organization is an impediment to some fishing communities, limiting their opportunity to participate in the fisheries governance.

The Provincial Fishery Committee is supposed to work closely with the National Fisheries Policy Committee, by bringing up the issues from the local areas for consideration at the national level. The National Fisheries Policy Committee is equally complex in the structure and function, which adds to the level of complexity of the overall governing system. Coordination between these two committees is not as smooth as it could be, partly because it is a relatively new governing structure, with certain provisions requiring further deliberation before they can be fully operationalized. While the Royal Ordinance on Fisheries seems to be compatible with the Promotion of Marine and Coastal Resources Management Act, it is not possible to determine at this time whether there is a real synergy between the two.

At the community level, both positive and negative factors are at play. On the positive side, the community-based consultative process has been well practiced in Trat province and has resulted in several decisions that align best with the characteristics of the small-scale fisheries in the area. Several conservation efforts and stewardship initiatives have been implemented, including the preservation of tradi-

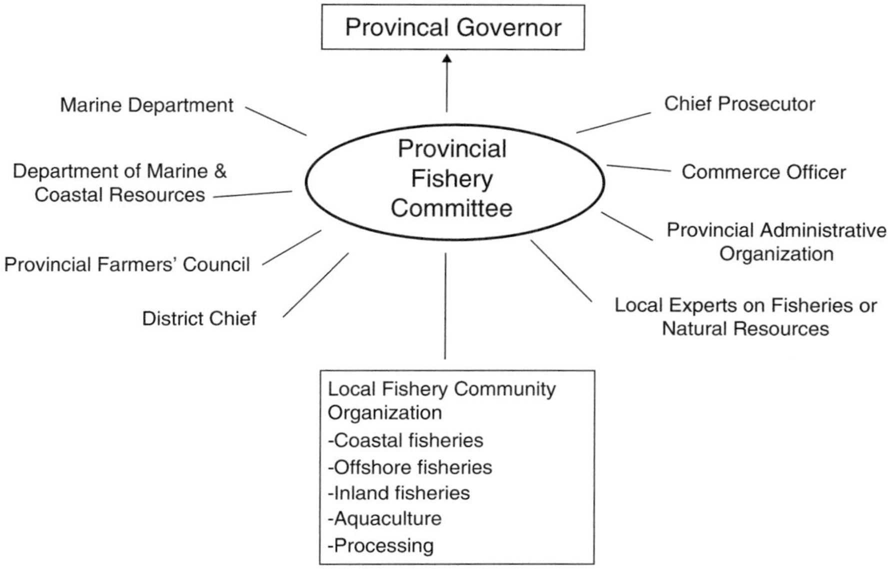

Fig. 12.3 Governing structure for fisheries governance in Trat Province

tional fishing practices, like the Acetes scoop net fishery. The presence and roles of many locally-based organizations, as well as the sustained interests from environmental organizations and research institutes, have made it possible for Trat Bay to maintain the viability of small-scale fisheries, contributing thus to enhancing the governability of fisheries and coastal resources in the area.

On the downside, a question has been raised about the effectiveness of the governing system, especially in its monitoring, control, and surveillance role. The local communities indicated during the field visit, for instance, that the level of compliance with rules and regulations is generally low and that few actions have been taken even after the fisheries enforcement officers are informed about the violation (e.g. illegal fishing from trawling, dredging, and push netting). The plan to install the vessel monitoring system, the new vessel registration system, and the improvement in the data collection system to capture all landings, are considered positive initiatives that should help promote fisheries sustainability in Trat Bay.

12.3.4 Wicked Problems in Trat Bay Within and Beyond Fisheries

Table 12.1 summarizes the key characteristics of the natural, social, and governing systems in Trat Bay, in terms of the diversity, complexity, dynamics, and scale. On their own, these characteristics create major challenges for governance, as

Table 12.1 Key characteristics of the Trat Bay governance system

Properties	Natural system	Social system	Governing system
Diversity	Trat Bay has diverse ecosystems, with mangrove, seagrass beds, mud flat, and sandy beaches, providing habitats and feeding grounds for a wide range of marine fish and marine mammals.	A number of resource users exist in Trat Bay, including small-scale fishers, large-scale fishers, aquaculture farmers, rice farmers, fruit farmers, and field workers. Many groups also participate in post-harvest and non-fisheries related activities, such as fish collectors, traders, retailers, storeowners, and tourism business.	Through the Royal Ordinance on fisheries, the Provincial Fishery Committee is directly responsible for fisheries governance in Trat Bay. The committee includes a wide range of stakeholders (Fig. 12.3).
Complexity	Diverse coastal habitats in the bay support complex marine life, which are related to each other mostly through a food-web system.	Resources users of Trat Bay are connected through multiple relationships. For instance, fishers are often engaged in farming, and tourism business depends on fisheries. Some fishers depend on others for bait fish. Many depend on fish collectors and buyers for trades.	The appointed members of the Provincial Fishery Committee have relationships with each other, in addition to the role they play in the committee.
Dynamic	Trat Bay natural system has gone through many changes, due partly to the heavy exploitation and the use of destructive fishing gears and over fishing. Mangrove forests were cleared for timber industry, shrimp farming, and urban development.	Changes in the fishing practices have been constant since the release of the Royal Ordinance on fisheries, but also before as a result of resource decline. In other sectors, the changes in land use (DMCR 2018) bring new people to the area.	The provision in the Royal Ordinance on fisheries about community consultation and proposals by the Provincial Fisheries Committee for the amendment of the law creates a highly dynamic system. The Promotion of Marine and Coastal Resources Management Act enacted soon after the Royal Ordinance on fisheries adds to the dynamics, through their interactions.
Scale	Because of its shallowness, Trat Bay covers a small area, less than 12 meters deep. The system is not bounded, however, with Trat River bringing high nutrients to Trat Bay.	Resource degradation and conflicts between resource users are not confined within Trat Bay, but involve fishers from other villages.	The boundary in the governing system is highly permeable. The Royal Ordinance on fisheries is the national level institution that applies to the provincial level, where other governing systems exist at the sub-district and village levels.

discussed. When considering them in the context of the wicked problems, the system becomes highly ungovernable, as explained below.

On the natural front, the geophysical characteristics of Trat Bay, with its shallow depth of less than 12 m and huge amount of freshwater discharge from Trat River (Hydrographic Department 2015), lead easily to a high level of siltation. When coupled with clearing of vegetation, including mangrove forests, and conversion of land to other uses like housing, rice fields, and shrimp ponds, other problems arise related to domestic waste, like trash and pollution. Urbanization and coastal development, in particular, have been without proper management or planning. The area is affected by untreated wastewater generated from the communities living in and around the bay. Runoffs and high sediment loads, particularly from gemstone mining, factories, and agriculture add to coastal erosion and siltation problems, as well as habitat degradation in river and coastal systems. Coastal erosion has been rather severe in some areas of Trat Bay, with more than 20 km of eroded shorelines (DMCR 2014). Shellfish dredging is another activity adding to the erosion problem, since it leads to the change of substrate and consequently a drastic decline of seagrass beds (DMCR 2015). Taken together, the overall environmental quality in Trat Bay has been declining, which has negative consequences on living aquatic organisms, including commercially important fish and shellfish species. The multiplicity, the connectivity and the broad scale of the problems, beyond fisheries, make them difficult to address and qualify them as wicked problems.

The wickedness of the problem intensifies when considering the weak public support of the governing institutions, which could be due to the lack of communication about their intention, lack of transparency in their conception, and lack of understanding about the local context. The Royal Ordinance on Fisheries was overwhelmingly considered to be the most problematic law, with several provisions that neither align with the reality on the ground nor with the values and interests of the local communities. Another new legislation that faces serious public scrutiny is the one issued by the Marine Department, requiring owners of any built structures in coastal waters to pay permit holder fees, calculated based on the size of the structure. This rule applies to small-scale fishers who have cage culture as part of their fishing livelihoods, as well as others whose houses are located along the coast. This regulation basically applies to any building and construction that encroaches over, into, and under any river, canal, marsh, reservoir, lake, or any sea within Thai waters, which must obtain permission from the Marine Department and pay fees. Some regulations are considered too stringent, while others are inconsistent with conservation and sustainability goals. Push net bans and the closure of the upper Gulf of Thailand from gillnets with smaller mesh size than 5 centimeters belong to the former. While these restrictions were considered important, the participants felt that they should not be applied indiscriminately and without sufficient scientific data to support them. The added complication in the case of the push net ban is related to the claim from some push netters who did not receive compensation for abolishing their use, as explained in Chuenpagdee et al. (2017). The lack of agreement and criticism about these banning regulations suggest that more groundwork and proper engagement with the local fishers are required. Examples of the regulations that seem to be too lenient are related to the use of light luring device in falling nets and lift nets, and the use of beam trawls, which target shrimps. Both of these fishing practices are considered by the workshop participants to be harmful and destructive. Thus, they were puzzled at the rationale for allowing these gears to continue to operate.

The opportunity for fishers and other local key actors to participate in the decision making through the Provincial Fisheries Committee is considered a positive aspect of the Royal Ordinance on Fisheries. The participants pointed out, however, that more could be achieved with the provincial committee. For instance, the committee should be more empowered to deal with the local issues than it is currently. Such empowerment follows the subsidiarity principle, which bestows decision-making authority to the lowest possible organization (Jentoft et al. 2011). With respect to community consultation, which is another key feature of the new law, the workshop participants viewed it as crucial as a way to provide inputs to the Provincial Fisheries Committee. The current process leaves ample room, however, for certain interest groups to influence the discussion at each consultation. The continued discussion and reconsideration about the spatial extent of the coastal area, and thus the definition of coastal fisheries, is an example of the dynamics between the stakeholders and the governing actors, which contributes to the wickedness of fisheries governance.

12.4 Ways Forward to Enhancing Stewardship

The governability assessment identifies factors inhibiting the ability of the governing system to perform its function. But, it also reveals opportunities that can be explored when considering solutions to addressing the problems. The discussion about the fisheries system and the wicked problems during the workshop highlighted stewardship as one of the possible avenues for promoting environmental sustainability and community wellbeing. As articulated by McConney et al. (Chap. 10, this volume), stewardship plays a central role in both the SSF Guidelines and the SDGs. Trat Bay communities are in a good position to contribute to achieving the goals and targets related to small-scale fisheries set out in these international instruments, including securing sustainable fishing livelihoods, providing food security, and alleviating poverty. In addition to tapping into their existing assets, especially the traditional stewardship practices, other capacities also need to be developed. Several suggestions were made and discussed during the transdisciplinary workshop, which can be summarized as follows.

12.4.1 Innovative Policies Aligning with the Local Context

The workshop participants were very sympathetic with push netters and other fishers using destructive gears, partly because the majority of them are from the area. With sustainable resources and viable community as common goals, rather than using controls and regulations, several interesting ideas have been proposed. These include using appropriate incentives to modify their behavior, collecting a 'user fee' from outside fishers and using the fund for local initiatives. As already mentioned, small-scale fishers in Trat Bay are not only involved in fisheries; many of them also farm seabass. Fisheries policies and regulations need to take into consideration the multiple sources of livelihoods (Torell et al. 2010, 2017) on which small-scale fishing people and the coastal communities depend.

12.4.2 Strengthening Local Organizations

Fisheries cooperative is suggested as a good option for promoting viable livelihoods of the communities (Amarasinghe and Bavinck 2011). The cooperative can help fishers fetch better prices for their catches, facilitate discussion about how to mitigate gear impacts, and promote environmental awareness, including among youth. Having high debt is a common feature in many fishing communities and the cooperative can provide low-interest loans to its members, so that they can be liberated from moneylenders. Finally, the Department of Fisheries is developing a registration system for all vessels, large and small, thus, the cooperative can also assist in data collection. This is not only to identify small-scale fisheries or to differentiate between commercial and artisanal as planned by the government alone. Working with its members, the cooperative can use this opportunity as a way to improve information and knowledge about this sector, which is required for better governance (see Agapito et al. Chap. 20, this volume). Other local level organizations, both formal and informal, are also seen as essential actors in the governance of Trat Bay fisheries. Better coordination of their efforts would be helpful, not only in elevating the issues affecting the local communities but also in attracting funding support from the governments.

12.4.3 Building Alliance with Other Sectors

It is clear from the governability analysis of Trat Bay that problems with fisheries are not confined within the fisheries sector. Likewise, solutions can be found elsewhere. Collaboration between communities, governments, local organizations, and private sector needs to be fostered. The private sector, in particular, can provide funding to support some of the initiatives, but some cautions are needed, for instance, to avoid private agenda setting. The tourism business is another important ally, which can provide additional income to local small-scale fishers. Trat Bay is increasingly popular among Thai tourists looking to experience nature and fishing livelihoods. For instance, small-scale fishers can take tourists on a 30 min boat ride from Lam Hin village to beautiful mangrove forests with fascinating root systems, which they can walk on and get a 'foot massage'. Tourists can also enjoy seafood at the local restaurants as well as buy local products made by the women of small-scale fisheries households, like shrimp paste, pickled clams, and dried squid. Some of the small-scale fishing families also offer their houses for tourists as part of the 'home-stay' program.

In addition to these suggestions, several participants made voluntary commitments to do various things. For instance, a fisheries officer holding a senior level position in the provincial fisheries department was very positive about the training and believed that similar conversations should take place in other coastal areas. One participant who is a founder of a local environmental organization wanted to initiate discussion to address trash problems and waste pollution in the bay.

Overall, the study shows not only the importance of stewardship for sustainable small-scale fisheries, but also that it is a responsibility of everyone. The transdisciplinary workshop helped them realized that the foundation for improving governance is about having good and comprehensive knowledge about the fisheries system, from ecological, social, and institutional perspectives. It is about recognizing that some features and aspects of these systems may contribute to or inhibit stewardship. The higher the diversity, complexity, and dynamics of the systems are, the more likely that problems will be wicked. The critical first step is, therefore, to bring the key stakeholders together to discuss the problems, in a well-facilitated manner, to share knowledge and experience and exchange ideas in respectful ways, and to work collaboratively in coming up with solutions that are acceptable, and thus feasible to implement. Governance reform is not a task that can be done once and for all. The experiences with the implementation of the new Royal Ordinance on Fisheries, as illustrated in the case of Trat Bay, are valuable lessons for the governing system to consider and improve the law and legislation such that they align well with the community values and local norm. It is also an opportunity to encourage capacity development programs that enhance stewardship ethics among resource users and coastal communities.

References

Agardy T, Bridgewater P, Crosby MP et al (2003) Dangerous targets? Unresolved issues and ideological clashes around marine protected areas. Aquat Conserv Mar Freshwat Ecosyst 13(4):353–367

Amarasinghe O, Bavinck M (2011) Building resilience: fisheries cooperative in Southern Sri Lanka. In: Jentoft, Eide (eds) Poverty mosaics: realities and prospects in small-scale fisheries. Springer, Dordrecht

Bavinck M, Chuenpagdee R, Jentoft S et al (eds) (2013) Governability of fisheries and aquaculture: theory and applications. MARE publication series, vol 7. Springer, Dordrecht

Bennett NJ, Dearden P (2014) Why local people do not support conservation: community perceptions of marine protected area livelihood impacts, governance and management in Thailand. Mar Policy 44:107–116

Chanrachkij I (2013) Monitoring the undulated surf clam resources of Thailand for sustainable fisheries management. Fish People 11(3):33–44

Chapin FS III, Kofinas GP, Folke C (eds) (2009) Principles of ecosystem stewardship: resilience-based natural resource management in a changing world. Springer-Verlag, New York
Chuenpagdee R, Jentoft S (2009) Governability assessment for fisheries and coastal systems: a reality check. Hum Ecol 37:109–120
Chuenpagdee R, Juntarashote K, Satumanatpan S et al (2017) Aligning with the small-scale fisheries guidelines: policy reform for fisheries sustainability in Thailand. In: Jentoft S, Chuenpagdee R, Barragan-Paladines MJ et al (eds) The small-scale fisheries guidelines: global implementation. Springer, Cham
DMCR (Department of Marine and Coastal Resource) (2014) Situation of coastal erosion: past to present. DMCR, Bangkok
DMCR (2015) Report on the surveys and status assessment of marine and coastal resources: coral reefs and seagrass beds in fiscal year 2015. Marine and Coastal Resources Research and Development Institute. DMCR, Bangkok
DMCR (2016) Project on the preparation marine and coastal protected areas of Aow Trat (Laem Klad to Klong Yai). DMCR, Bangkok
DMCR (2018) Report on socio-economic survey of Trat bay (Muang District). Marine and Coastal Resources Research and Development Center the Eastern Gulf of Thailand, Rayong Province
DoF (Department of Fisheries) (2017) Fisheries statistics of Thailand 2015. Department of Fisheries, Ministry of Agriculture and Cooperatives, Bangkok
FAO (Food and Agriculture Organization of the United Nations) (2015) Voluntary guidelines for securing sustainable small-scale fisheries in the context of food security and poverty eradication. Food and Agriculture Organization of the United Nations, Rome
Hines EM, Strindberg S, Junchumpoo C et al (2015) Line transect estimates of Irrawaddy dolphin abundance along the eastern Gulf Coast of Thailand. Front Mar Sci 2:63. https://doi.org/10.3389/fmars.2015.00063
Hydrographic Department (2015) Nautical chart of the Gulf of Thailand Section 5533 III. Trat Province. Map [ca 1:50,000], Royal Thai Navy. Bangkok, Thailand
Ivany I (2011) Marine protected areas sustainability: issues of complexity and stewardship. Masters' thesis, Memorial University of Newfoundland
Jentoft S, Chuenpagdee R (2009) Fisheries and coastal governance as a wicked problem. Mar Policy 33(4):553–560
Jentoft S, Eide A, Bavinck M et al (2011) A better future: prospects for small-scale fishing peoples. In: Jentoft, Eide (eds) Poverty mosaics: realities and prospects in small-scale fisheries. Springer, Dordrecht
Jentoft S, Pascual-Fernández JJ, de la Cruz-Modino R et al (2012) What stakeholders think about marine protected areas: case studies from Spain. Hum Ecol 40(2):185–197
Junchompoo C, Monanunsap S and Penpein C (2014) Population and conservation status of Irrawaddy dolphins (Orcaella brevirostris) in Trat Bay, Trat Province, Thailand. Proceedings of the Design Symposium on Conservation of Ecosystem (The 13th SEASTAR2000 Workshop), Vol. 2, pp 32–38
Kooiman J, Bavinck M, Jentoft S et al (eds) (2005) Fish for life: interactive governance for fisheries. Amsterdam University Press, Amsterdam
Lang DJ, Wiek A, Bergmann M et al (2012) Transdisciplinary research in sustainability science: practice, principles, and challenges. Sustain Sci 7(Suppl 1):25–43
Mansilp P (2013) The surveying on mangrove structure for prepare the community food bank in Trat Province. Technical Report. Office of Mangrove Conservation. Department of Marine and Coastal Resources, Bangkok
Morgan R (2016) Exploring how fishermen respond to the challenges facing the fishing industry: a case study of diversification in the English channel fishery. Reg Stud 50(10):1755–1768
Munprasit R, Phuttharaksa K, Charoensombat B et al (2009) Habitation factors of hard clam (Meretrix spp.) along the coast of Laem Klat Subdistrict of Trat Province. Technical Paper No. 6/2009. Department of Fisheries, Bangkok

NSO (2013) Preliminary report: 2013 agricultural census. Ministry of Information and Communication Technology, Bangkok

Pascoe S, Cannard T, Jebreen E et al (2015) Satisfaction with fishing and the desire to leave. Ambio 44:401–411

Pollnac RB, Pomeroy RS, Harkes IHT (2001) Fishery policy and job satisfaction in three southeast Asian fisheries. Ocean Coast Manag 44:531–544

Promotion of Marine and Coastal Resource Management Act (2015) Royal Gazette Volume 132, Part 21, dated 26 March B.E. 2558 (2015)

Rittel HWJ, Webber MM (1973) Dilemmas in a general theory of planning. Policy Sci 4:155–169

Roach CM (2000) Stewards of the sea: a model for justice? In: Coward H, Ommer R, Pitcher T (eds) Just fish: ethics and Canadian marine fisheries, vol 23. ISER Books, St. John's, pp 67–82

Roach CM, Hollis TI, McLaren BE et al (2006) Ducks, bogs, and guns: a case study of stewardship ethics in Newfoundland. Ethics Environ 11(1):43–70

Rogers VL (2015) Synergies for stewardship and governance of multiple-use coastal areas: a case study of Koh Chang, Thailand. Master's thesis, Memorial University of Newfoundland

Royal Ordinance on Fisheries (2015) Royal Gazette Volume 132, Part 108, dated 13 November B.E. 2558 (2015)

Satumanatpan S, Chuenpagdee R (2015) Assessing governability of environmental protected areas in Phetchaburi and Prachuap Kirikhan, Thailand. MAST 14:1–19. https://doi.org/10.1186/s40152-015-0035-8

Satumanatpan S, Pollnac R (2017) Factors influencing the well-being of small-scale fishers in the Gulf of Thailand. Ocean Coast Manag 142:37–48

Satumanatpan S, Senawongse P, Thansuporn W et al (2014) Enhancing management effectiveness of environmental protected areas, Thailand. Ocean Coast Manag 89:1–10

Song AM, Chuenpagdee R, Jentoft S (2013) Values, images, and principles: what they represent and how they may improve fisheries governance. Mar Policy 40:167–175

Thiammueang D, Chuenpagdee R, Juntarashote K (2012) The "crab bank" project: lessons from the voluntary fishery conservation initiative in Phetchaburi Province, Thailand. Kasetsart J Nat Sci 46(3):427–439

Torell E, Crawford B, Kotowicz D et al (2010) Moderating our expectations on livelihoods in ICM: experiences from Thailand, Nicaragua, and Tanzania. Coast Manag 38:216–237

Torell E, McNally C, Crawford B et al (2017) Coastal livelihood diversification as a pathway out of poverty and vulnerability: experiences from Tanzania. Coast Manag 45(3):1–20

Trat Provincial Office (2017) Trat development plan, 2018–2021. Division of Position and Data for Development of Trat Province

United Nations (2015) Transforming our world: the 2030 agenda for sustainable development. United Nations Publication A/RES/70/1

Whitty T (2014) Conservation-scapes: An interdisciplinary approach to assessing cetacean by Catch in Smallscale Fisheries. Ph.D. dissertation, Scripps Institution of Oceanography. p 242

Part V
Defending the Beach

Chapter 13
Strategies and Policies Supporting Small-Scale Fishers' Access and Conservation Rights in a Neoliberal World

Evelyn Pinkerton

Abstract Around the globe, small-scale fishers are systematically excluded from access to fish or the benefits of fishing, yet in many cases they persist. This chapter considers key strategies used by small-scale fishers' organizations, as well as some national or sub-national governments which support them, to maintain their access in the neoliberalizing world whose ideology runs contrary to their own. Equally important are their conservation rights and their rights to the benefits of fishing, such as fair prices for fish or fishing licenses. Strategies include local small-scale fishers forming organizations which buy up or are allocated pre-existing privatized fishing quotas and then leasing them out to members at affordable prices; governments leasing out quotas annually; governments prohibiting license transfer via the market or limiting licenses to local, owner-operated fishers; communities making government protect local habitat from large-scale overfishing or habitat-destroying developments; fishers' organizations bypassing fish processors to get fair prices; and greater government oversight of socially destructive neoliberal practices. These strategies and policies illustrate the need for governance informed by interdisciplinary thinking beyond the narrow perspective of neoliberal economics – thinking which considers equitable distribution, legitimacy, the importance of livelihoods, the health of local ecosystems, and many socio-economic, cultural, and ecological issues.

Keywords Neoliberalism · Individual Transferable Quotas · Access rights

E. Pinkerton (✉)
School of Resource & Environmental Management, Simon Fraser University, Burnaby, BC, Canada
e-mail: epinkert@sfu.ca

R. Chuenpagdee, S. Jentoft (eds.), *Transdisciplinarity for Small-Scale Fisheries Governance*, MARE Publication Series 21,
https://doi.org/10.1007/978-3-319-94938-3_13

13.1 Introduction: Why Small–Scale Fishers Tend to Lose Access to Fish and Fishing Benefits

Although much has been written about the contribution of small-scale fisheries to local and national economies and wellbeing, this understanding is not universal. Some economists at the World Bank and the Food and Agriculture Organization (FAO) saw small-scale fisheries as part of the problem of 'The Sunken Billions'. This was their term for the US$50 billion/year difference between potential and actual net economic benefits from marine fisheries which they believed could be gained by eliminating "excess competition over limited fish resources [which] results in declining productivity … economic inefficiency, and depressed fisher incomes" (World Bank and FAO 2009). For them, privatization and concentration of fishing effort into fewer, larger, and presumably more efficient vessels was the answer and Individual Transferable Quotas (ITQs) were often the mechanism to bring it about. ITQs turned access rights to fish into tradable commodities, based on the idea that the market was a better instrument than government for allocating access rights. Although this notion has been widely contested (e.g., Pinkerton and Davis 2015), neoliberal economic thinking became widespread in the 1980s and anthropologist Einarsson (2012) noted that, "some of the [Icelandic] economists who are responsible for the introduction of the ITQ system [in Iceland] did, before the meltdown, quite honestly express [the view] that the privatization of the commons inevitably causes smaller communities to lose out. They even questioned whether fisheries-dependent communities are actually part of the fishing industry proper. The exclusion of these communities was seen by them to be not just logical, but also justifiable, rational and necessary." Similarly, Fiona McCormack (2018, 67) showed that, as New Zealand adopted ITQs, the expropriation of small-scale fishers was considered necessary to increase efficiency and 1500–1800 fishers were excluded from the fishery. Likewise, British Columbia automatically excluded all small-scale fishers who had earned less than $2500 a year when implementing its limited entry policy, a precursor to ITQs (Pinkerton and Davis 2015). Not surprisingly, economist Stephen Marglin (1991) concluded that artisanal production is nearly always considered an obstacle to capitalist profit. Thus, small-scale fishers all over the world have often been deprived of access via policies which deliberately exclude them, rather than through their presumed 'inefficiency' or inability to survive (Sabau and van Zyll de Jong 2015). Table 13.1, based on a Special Issue of 19 papers on neoliberalism and small-scale fisheries globally (Pinkerton 2017), connects the neoliberal principles – privatization, deregulation, and devolution of risks and responsibilities to the private sector – to their expression in the implementation of ITQs, both in terms of the results predicted by neoliberal economists and the actual impacts of ITQs and related neoliberal policies on small-scale fishers, their communities, and the public welfare of the countries in which they are located.

The goal of this chapter it to illustrate a number of strategies used to combat the loss of access, benefits, and conservation rights suffered by small-scale fishers through neoliberal policies such as these, as well as policies adopted by national and sub-national governments which facilitate access for small-scale fishers. In doing

Table 13.1 How neoliberal economic principles were translated into ITQs, the results of implementing ITQs claimed by neoliberal economists, and actual results (Pinkerton 2017)

Neoliberal principle exemplified in ITQs	Claimed results of implementing principal through ITQs	Actual results of implementing ITQs and related neoliberal policies
Privatization of public goods	*public welfare increases through more profitable enterprises	*inequitable initial allocation hugely raises cost of entry for all of future entrants, who must buy what 1st entrants got gratis
	*capital accumulation	*ITQs become overcapitalized
	*creates incentive to conserve	*leasing out of ITQs at high prices creates barely profitable enterprises which lease in ITQs
		*poor fit for small multi-species fisheries who often don't qualify for initial allocation because of small catches in qualifying years
		*ITQ owners benefit at expense of crew & rental skippers who get much lower wages; longstanding share system ended
		*lowered capacity & flexibility to take precautionary approach
		*ITQs undermine moral values that lead fishers to act in public spirited ways
Deregulation= market rule	*ITQs will gravitate to most efficient	*ITQs gravitate to investors & processors with most capital, who concentrate quota, achieve market power; state regulations to protect SSFs' access are ineffective
		*state removes social, political, environmental barriers to capital accumulation, suppresses labour movements
		*state forces out small-scale fishers via regulations (deeming them inefficient) & via expensive monitoring requirements
		*free transferability of ITQs out of community, region, country
		*loss of sovereignty as fishing & fish processing jobs leave country
Devolution of responsibility & risk	*public benefits	*fishing dependent communities emptied out or become dependent on state funds as traditional livelihoods lost
		* rising costs to state of supporting unemployed former fishers
	*corporate social responsibility	*costs of monitoring fishery downloaded onto fishers
		*speculative finance in ITQs, licenses, fish plants not growth inducing; state subsidizes investment but corporations don't honour employment agreement

(continued)

Table 13.1 (continued)

Neoliberal principle exemplified in ITQs	Claimed results of implementing principal through ITQs	Actual results of implementing ITQs and related neoliberal policies
	*greater safety	*lowered safety at sea as indebted lessees take greater risks in bad weather
		*embrace of audit culture & corporate social responsibility = quantified assessment for improved efficiency & lowered costs, via accepted bureaucratic standards, missing community & small-scale fisheries' values
		*transformation of fishers' consciousness into solely self-interested *homo economicus* instead of other-directed community citizens

so, it is important to recognize that the thinking behind neoliberal economics is asocial or even anti-social and does not match what has been recorded on the ground by other social scientists, including other economists. Much broader transdisciplinary thinking is needed to understand how small-scale fisheries can be efficient, effective, sustainable, resilient and serve a more general social purpose of great value to public welfare and to government agencies, one which is not dedicated to capital accumulation and profit making. The perspective of several social sciences, as well as natural sciences, is illustrated in the strategies and policies discussed below, whose intention is to support small-scale fisheries because of their contributions to public welfare, community wellbeing, and ecological sustainability.

The eight overall general strategies and policies discussed below are not exclusive of one another but often include elements of the other strategies as well. Each section below focuses on the unique contribution of each policy or strategy, while Table 13.2 notes the overlap in strategies and policies, which are often based on a combination of geographic location, occupation, scale of operations, and political or ecological values such as the wellbeing of local communities and ecosystems.

13.2 Strategy 1. Local or National Institutions Hold and Lease Out Access Privileges According to Place-Based and Sustainability Criteria

This section considers institutions at vastly different scales – local, regional, state, and national – all of which hold and lease out access privileges to small-scale fishers in North America, Europe, and Africa. Their purpose is not only to create conditions for small-scale fishers' access but also to avoid excessive acquisition of access by those with the most capital. In these institutions, access privileges are conceptualized not as commodities which can be traded for private gain but rather as public goods which can be used but not owned by local small-scale fishers. These institutions are located in Massachusetts and Alaska (USA), Denmark, and Namibia.

Table 13.2 Strategies and policies used by NGOs, states, communities, individuals and Indigenous organizations to support small-scale fisheries access rights or privileges

Strategy or policy ➔ case studies	Hold licenses or IQs leased to local SSFs	Non-market license transfer	License transfer to local only	Owner-operator & fleet separation	SSFs resist habitat loss	SSFs outlaw destructive gear	Local conservation rights	Direct marketing	Resistance to neo-liberal state policies	Provide loans or support to local SSFs
CCFT	√	√	√	√	√	√	√			
Thorupstrand guild	√	√	√	√	√	√	√			
Namibia state	√	√								
Alaska CDQs	√	√	√	√			√		√	√
Maine lobster	√	√		√						
France	√	√	√	√						
Norway	√	√		√					√	
Atlantic Canada	√	√		√					√	
Buen Hombre			√		√	√	√			√
Lake Chiuta			√		√		√			
Lummi & allies					√		√		√	
Haida							√		√	
CSFs							√	√	√	√
Knutson				√				√	√	√

13.2.1 The Cape Cod Fisheries Trust in Massachusetts, USA

The Cape Cod Fisheries Trust (CCFT n.d.) was formed in 2012 when a group of fishers from this region came together with the Cape Cod Commercial Fishermen's Alliance, Inc. (CCFA n.d., incorporated in 1996) to figure out how to cope with some of the unforeseen consequences of scallop and groundfish management moving into ITQs (called 'catch shares'). That move meant that large-scale fishing businesses – those with more capital for buying ITQs – had a distinct advantage over the local small-scale fishers. With support from lenders and donors, the Trust buys ITQs, much the way land trusts buy land, and keeps it in the community by leasing it to local fishers. In 2015, the Trust team published a handbook intended for communities who want to build a 'permit bank' (so called because it allows access to local small-scale fishers who could not afford the rising price of fishing permits or ITQs) or who have launched a permit or license bank and are learning to manage it as a non-profit business (CCFT n.d.). The Trust is committed to sustaining the tradition of small-scale fishing, working with more than 130 independent fishing businesses and 300 local fishing families. In 2014, the CCFT leased out approximately 390,000 quota pounds, usually at c. 50% of market rates (e.g., what it would have cost them to lease ITQs on the open market), to local Cape Cod fishers and saved them more than $600,000. Due to affordable CCFT leasing, 93 captains and crew brought fresh seafood worth $2.75 million into the local economy.

The Trust plays other ecologically and socially positive roles as well. When the local small boat scallop fleet recognized that it needed to protect inshore waters from over depletion caused by big boats, the Trust began working to protect key habitats, to promote a robust program to better monitor the fishing fleet with both human observers and cameras, bringing in restrictions on bycatch dumping, and initiating the development of a fishery ecosystem plan in New England. Fishers who installed cameras on their boats and did 100% coverage of their cod fishery were able to lease ITQs for about 30% of market rates.

The Trust has also provided more than 100,000 portions of local seafood, mostly of 'under-appreciated' (non-market or under-utilized) fish such as skate, dogfish, and mackerel to The Family Pantry of Cape Cod to offer healthy nourishment for the needy. The Trust also provides loan programs, business planning and economic development support to fishing businesses and, as the first permit bank in the US, guides other communities to form their own permit banks. A permit bank in British Columbia, Canada (Edwards and Edwards 2017; Ecotrust n.d.) illustrates a similar combination of social and ecological values driving the bank's operation. As both the US and Canadian examples illustrate, however, this strategy is most appropriate for jurisdictions where there is sufficient wealth linked to progressive ideas that volunteer fund-raisers or supportive non-governmental organizations can recruit funding to purchase very expensive ITQs from the permit holders who now control (without legally owning) what was formerly a public resource.

13.2.2 The Thorupstrand Coastal Fishermen's Guild in Denmark

The implementation of ITQs in Denmark in 2007 rapidly reduced vessels and fishers by 50% in the first 4 years, a reduction experienced most intensely in the fishing-dependent coastal communities where small-scale vessels up to 15 m in length used gillnets and Danish seines, in contrast to the offshore large-scale otter trawl fleet which expanded and concentrated ownership of ITQs. Policies designed to protect the inshore (within 25 nautical miles from shore) vessels in these communities were ineffective and Host (2015) argues persuasively that it is virtually impossible for market-based governance to include social objectives. As discussed by Hojrup (2011), the small coastal community of Thorupstrand, observing the radical loss of fishing boats in a neighboring community, formed a cooperative of 20 families in 2007 who pooled their own ITQs, purchased more ITQs with bank loans, and leased these ITQs out on equal terms to themselves and crew (termed 'share fisher' because of the traditional system of paying crew a share of the vessel earnings instead of a wage, as is done under the new ITQ system). A major goal of the cooperative or guild was to enable crew and young people to enter or retain their place in the fishery. Their success in doing this is evidenced by the fact that only 33% of Thorupstrand fishers are over age 40 compared to the national average of 77% (Dinesen et al. 2018). Each new guild member paid an entry fee of 13,200 Euros or c. US$16,000 which it could reclaim if it left the guild. These fees were used to repay bank loans to purchase more ITQs in the region. ITQs are distributed equally but flexibly in several rounds annually as members ask for the amounts they want, and quota is then later redistributed to adjust to actual catches (Dinesen et al. 2018).

The small-scale Thorupstrand fishery lands its fish daily as the highest quality fish and is more efficient than the trawl fishery as measured by obtaining a higher value for the fish per unit of fuel consumed in the fishery. It also has a lower discard rate and lower seabed impact than the trawl fishery (Dinesen et al. 2018). The Thorupstrand Winch and Packing Association is another cooperative which manages and maintains facilities on the beach for offloading, sorting, gutting, packing, and icing the fish, and it obtains a 6% share of the overall value of all catch landed by the Thorupstand Guild.

Because it is one of the few remaining coastal fishing communities in Denmark, Thorupstrand enjoys high tourist interest and stands as a reminder that centuries of dependence of Danish coastal communities on fishing has been handed over to private interests. Some Danish scholars suggest that it would be in the national interest for government funds to be used to implement the Thorupstrand model more broadly, since an assortment of bankers, investors, lawyers, accountants, and fish processors have been able to bypass all regulations intended to protect small-scale fishers and concentrate access to fish into fewer and fewer large-scale vessels (Host 2015; Dinesen et al. 2018). This model exemplifies a successful strategy exercised at a very small, local scale, compared to the Cape Cod model at a larger regional scale.

13.2.3 *The Namibian Government Holds and Leases Out IQs on a Temporary Basis*

Unlike the situation in Denmark and Massachusetts, USA, the Namibia government did not adopt ITQs or tradable fishing permits, and therefore Namibian fishers did not have to adopt expensive strategies to counter national or regional policies. Instead, as Levy (2010) recounts, the Namibian government conceptualized access as a public good which is temporarily allocated to parties as individual quotas (IQs) which must meet criteria serving public interests.

At the time of Namibian independence from South Africa in 1990, Namibian waters had been drastically overfished by foreign fleets. So the Namibian government set Total Allowable Catches (TACs) for fish and the government leased out IQs to individual fishers or companies for set periods. This allowed government to reallocate quota – and charge lower lease fees – to parties it considered to operate more in the national interest. Thus, when local small-scale fishers became able to set up viable enterprises, they could get preferred access over foreign vessels. In the meantime, the state charged lower lease fees to vessels carrying many Namibian crew, as well as allocated access privileges to those parties with the most sustainable fishing practices. As the TAC was monitored and enforced, fish stocks became more abundant. Nearly 14,000 Namibians now work as fishers or in fish processing plants, while none did before independence. As of 2003, Namibians held 162 out of the 163 permits to access fisheries. Meanwhile, the moderate lease fees cover the costs of managing this public resource.

Considering the United Nations' Law of the Sea (UNCLOS) requirements, these actions meant that Namibia had to make at least some investment in setting TACs and patrolling its Extended Economic Zone for compliance. However, Article 62(3) of UNCLOS allows coastal states considerable discretion to "take into account all relevant factors, including, *inter alia*, the significance of the living resources of the area to the economy of the coastal State concerned and its other national interests," so the Namibian example should not be seen as prohibitively costly or difficult by other small coastal states.

13.2.4 *The Western Alaska Community Development Quota (CDQ) Program, USA*

The Community Development Quota (CDQ) program has elements of the Danish, Namibian and Cape Cod situations. It was initially a seemingly minor concession to onshore (vs offshore) allocations and to opponents of the controversial privatization of various species in Alaska. It provides an illustration of how significant such an allocation to fishing communities can become.

As discussed by the National Research Council study of the CDQ program (NRC 1999), Langdon (2015), and the US National Oceanic and Atmospheric

Administration (n.d.), when allocation of groundfish access to offshore and onshore sectors in Alaska was being debated by federal and state agencies in the 1980s and 1990s, the opportunity arose for the Bering Sea Fishermen Association to lobby for a modest allocation of pollock for the fishing-dependent communities in Western Alaska as part of the onshore allocation. Pollock, a groundfish living almost exclusively in the Bering Sea, is the most abundant and valuable fish in this region, contributing an average annual harvest of 1.2 million metric tons and the largest portion of the $1.96 billion groundfish (2008) value.

Through earlier limited entry programs and the high capital investment needed, the largely Indigenous rural communities on the Bering Sea had lost their traditional access to commercial fisheries. Because historically Alaska fishing allocations had been dominated by Seattle corporate interests which owned the offshore catcher-processors taking the majority of the pollock and contributing no jobs to Alaska, the Alaska governor supported the CDQ idea. As a speaker at the United Nations Earth Summit in Rio de Janeiro, his views were captured by a section of the Summit's final document which stated: "Coastal States should support the sustainability and special needs of small-scale artisanal fisheries, local communities and indigenous people, to meet human, nutritional and other development needs." (Treadwell 2012). Furthermore, this international declaration helped him persuade a reluctant US Secretary of Commerce to support the CDQ idea.

So when ITQs were introduced in the Bering Sea and Aleutian Islands pollock fishery in 1992, 7.5% of the pollock TAC (later increasing to c. 10% of pollock and other groundfish, halibut, sablefish, and crab when these fisheries were ITQed) was allocated to the Western Alaska Community Development Quota (CDQ) Program. This program includes 65 villages (total population c. 27,000) which formed themselves into six non-profit corporations (CDQ groups) that manage and administer the CDQ allocations, investments, and economic development projects. After the program was judged to be successful by a National Research Council study (1999), it was institutionalized, and federal oversight was largely delegated to the state's regional board.

Since pollock was an offshore species not traditionally fished by adjacent communities, the CDQ groups initially leased out their pollock allocation to catcher-processor partners and used these lease fees to build onshore processing plants for halibut, salmon, Pacific cod, and other species traditionally fished. Lease fees have been invested in smaller vessels, infrastructure, processing capacity, specialized gear, fishing licenses, and part or full ownership of existing fishery businesses including pollock, halibut, sablefish, crab, and other groundfish. Revenue from such investments has exceeded lease fee income since 2004. In 2013, the aggregated revenue from all CDQ groups was $248.7 million, of which only 23% was derived directly from CDQ lease fees. Other CDQ community projects include seafood branding and marketing, quality control training, construction and staffing of equipment maintenance and repair facilities, student scholarships, salmon assessment and enhancement projects, and other environmental programs. In 2011, the six CDQ groups held approximately $938 million in assets and invested more than $176 million in CDQ communities and in fisheries activities, including financial services that support small-scale operations targeting salmon, herring, halibut, or other nearshore species. CDQ revenue supports permit brokerages and revolving loan programs that are intended to retain limited entry salmon permits within CDQ communities, providing the financing necessary for resident fishers to purchase new boats and gear. In 2011, the economic activity generated by the CDQ Program contributed over $4.5 million in state and regional taxes and fees in addition to the aggregated community capital investments of approximately $30 million. The CDQ Program provides a means to support and sustain fisheries-based economies in western Alaska that are deeply rooted in subsistence fisheries, traditional artisanal fisheries and major commercial operations in the Bering Sea and Aleutian Islands. Possibly the most isolated and marginalized fishing communities in the US have become among the most viable. Communities which had formerly been forced into depending on government transfer payment because of limited access to commercial fisheries are now substantial contributors to government revenues.

13.3 Strategy 2. State Uses Non-market Mechanisms to Limit and Transfer Permits

The state of Maine, USA, has allocated a limited number of lobster permits to owner-operator fishers since 1989 but, like Namibia, does not give them the right to sell these permits. Unlike Namibia, however, the permits cannot be terminated by the state at will, but only when a fisher retires. But the state of Maine can decide not to reissue a permit after a fisher retires if it deems that fewer permits would be desirable. Normally an aspiring lobster fisher has to wait in line to be offered an available permit (Acheson 2003), but the state also has an apprenticeship and even a student program for young people who can own and fish a small number of traps while gradually building up their experience, official training hours, and trap ownership, while working for a full-time lobsterman, often a family member (Ounanian 2016). It is useful to note that the effort of both apprentice and full-time lobster fishers is limited by the number of traps they are allowed to fish and that this gear limitation is considered an effective way of limiting fishing effort, alongside limited entry.

The state of Louisiana, USA, issues permits but has no limited entry program for its small-scale shrimp fishery and, therefore, there is no sale of access privileges. A fisher wanting to fish shrimp simply applies for a permit (Durrenberger 1997).

13.4 Strategy 3. Local Bodies Limit Sale or Allocation of Access Privileges to Local Fishers

This type of strategy uses the geographic criteria of local residence and/or the criteria of local decision-makers as a way of regulating who can obtain access privileges. For example, the European Common Fisheries Policy allocates catch limits by species to each country, which in turn allocates catch limits to local Producer Organizations, which then allocate individual quotas (IQs) to their members. Transfer of IQs is usually regulated by the Producer Organization. Although IQs have become ITQs in some European countries, the method is often a public auction, and purchase is often restricted to other members of the local Producer Organization. France, however, does not allow transferability of IQs via the market (Frangoudes and Bellanger 2017). An IQ no longer being fished goes back to the Producer Organization, which re-allocates it to another local fisher.

13.5 Strategy 4. State Prohibits Access Privileges Going to Non-fishers

In many places the geographic criteria for prohibiting access mentioned above are supplemented by or replaced by occupational criteria. Norway was a leader in this strategy, as famously illustrated in Ottar Brox's account of the Norwegian 'Battle of Trollfjord' of 1890, in which tens of thousands of non-motorized small-scale fishers in Norway forced large fishing steamships out of their traditional cod fishery and retained their traditional access privileges. This was the beginning of Norway's fleet separation policy in which processing companies are not allowed to own fishing permits, also called licenses (Brox 1993; Bavinck et al. 2015). Norway had a similar policy with farmland, which could not be sold to non-farmers.

Canada's eastern provinces developed a policy similar to Norway's. The Atlantic Fleet Separation Policy of 1979 prohibited fish processing companies from holding licenses in the less than 65 feet inshore fishery, although in some areas it has been difficult to implement (Barnett et al. 2017). However, this policy survived a recent court challenge (Elson v Canada 2017). The license is on the fisher, not the vessel, which can only be owner-operated (and is also non-transferable via the market, as in Strategy 2 and 3).

Alaska's owner-on-board salmon policy, requiring that the license owner be present on the fishing vessel but need not operate the vessel, is only partially effective. It has not prevented the "emergence of a class of wholly absentee quota shareholders, who hold only nominal interest in the vessel upon which their IFQ is fished, do not share in the risk of fishing, and continue to profit from the fishery while residing far away from the actual fishing grounds." This occurred because the prohibitions targeted corporate ownership but not initial ITQ recipients. It is now the latter who are becoming the new landlords (Szymkowiak and Himes-Cornell 2015).

13.6 Strategy 5. Successful Resistance by Artisanal Fisheries to Overfishing by Larger Gear and Habitat Destruction by Development Projects

13.6.1 Dominican Republic Community Protects Local Waters from Outsiders' Destructive Gear

In the Dominican Republic, the community of Buen Hombre had successfully educated its fishers since the early 1900s to practice a conservation ethic, observing fishing rules which resulted in the conservation of the reef and mangrove ecosystems fronting their village of about 800. The 45 adult fishers fished in egalitarian teams which shared both the costs and benefits of fishing, and had organized themselves into a voluntary association which worked to increase fish prices, reduce spoilage, and educate new members.

Their successful co-existence with the reef and mangrove ecosystem was threatened, however, by outsiders who began to fish their reef with illegal small-mesh nets which depleted local populations by taking all sizes and species of fish instead of avoiding smaller fish as the locals did. In addition, these outsiders harvested coral for sale to tourists, chased juvenile fish out of their mangrove nursery into their nets, and used compressors to dive for long periods, thus also targeting the largest and most productive fish.

The community was unsuccessful in persuading the outsiders to leave or in convincing the government that it was outsiders who were overfishing until a research team of American anthropologists was able to provide underwater photographs of the gear being used by outsiders and Landsat Terrain Mapping satellite imagery which visually demonstrated changing patterns of fishing behaviour and distribution of site use. The community could be said to have gained a formal 'conservation right' when the Dominican government issued badges to two Buen Hombre fishers chosen by other fishers' association to patrol their local waters and arrest offenders (Stoffle et al. 1994).

The role of outside university researchers in documenting and legitimizing the conservation efforts of small adjacent fishing communities can be key to their success, as was also demonstrated in the effort of the Eastport Peninsula Lobster Protection Committee in Newfoundland, Canada. Combining the interdisciplinary knowledge and support of both university researchers and government scientists with their own knowledge, local fishers successfully developed and promoted a unique approach to lobster conservation based on exclusive harvesting rights and a diverse array of conservation initiatives, including closed areas (Davis et al. 2006).

13.6.2 Malawi Fishers' Committees Protect Local Waters from Outsiders' Destructive Gear

Steve Donda (2017) documents a similar occurrence on Lake Chiuta in Malawi, where the 200 km^2 lake was sustainably fished by about 150 local artisanal fishers, mainly from nine villages, using low-impact technology such as traps and non-motorized dugout canoes. Submerged aquatic vegetation makes the use of motorized vehicles impractical and even impossible in parts of the lake, which are penetrable only by canoes.

Starting in the mid-1980, fishers from outside the lake communities began to fish Lake Chiuta using non-selective, higher impact gear such as small-mesh nets which took all sizes and species of fish and damaged fish habitat by dragging on the bottom. Gradually this increased to 300 outsiders whose gear often damaged local fishing gear, who lowered fish prices by selling their catch more cheaply, and who made lake water undrinkable by increasing its turbidity. Local fishers' committees organized themselves and pressured their senior chief to take action; they all met with government officials and eventually got a formerly unregulated lake to be able to legally evict outsiders. The committees took on the formal roles of regulating acceptable gear and fishing practices, involving voluntary patrolling of the lake, and became the conduit of communication with government. Catches stabilized at a higher level. The resulting system involved high levels of civic pride and honour in voluntary monitoring and enforcement of rules.

In both Malawi and the Dominican Republic, national governments were persuaded by local committees to lend moral and legal support, but not resources, to exclude the use of non-sustainable and ecologically-destructive gear and practices. In these cases, governments recognized and valued the fact that small-scale fishers in local communities were playing key roles in ecological protection and sustainable fishing practices.

13.6.3 Lummi Tribe and Coalition in Washington State, USA, Defeats Attempt to Build Habitat–Destroying Coal Port Terminal

It is significant that many of the examples above involved local small-scale fishers protecting fish *habitat* and local ecosystems from destructive gear used by outsiders, although the discussion so far has focused more on the direct threats of such gear to fish stocks. In a case explored by Allen et al. (2017), it was not fishing gear, but rather a proposed coal port at Cherry Point, Washington, which local Lummi tribal small-scale fishers knew would destroy the remaining habitat of the herring and crab which supported their small-scale fisheries, in addition to blocking their access to these fishing grounds. Other species such as marine mammals would also be damaged by this largest coal port terminal in North America and the planned transport of 17.6 million tons of coal. In order to stop the approval of this proposal, the Lummi Tribe, together with a number of environmental and faith-based advocacy groups coordinated opposition to the project. The coalition organized a letter-writing campaign to the US Army Corps of Engineers, urging it to deny the permit. Some members of the coalition had a statewide (Washington Environmental Council) or even nationwide (Sierra Club) sphere of influence and action. Quite independently and spontaneously other local groups arose, motivated by the variety of threats that the project posed, ranging from train noise and air pollution to harm to fisheries to the global warming that would result from burning the coal in Asian power plants. The Lummi later stated that, although they held treaty-protected access rights to fish and to fish habitat protection affirmed through the *US v Washington* 'Boldt decisions' of 1974 and 1980, they believed that it was the collective opposition of these dozen or so diverse groups, in addition to their own legal rights, which won the day. Overall these groups represented environmental, faith-based, political party, think tank, medical, community rights, and general political activist groups, some of which did not coordinate their actions with the Lummi-centered coalition. They organized in different ways, ranging from devising and distributing bumper stickers and yard signs, to organizing letter-writing campaigns for letters to the editor or to the decision-making agency, to a community rights initiative campaign that would have banned the transportation of coal through Bellingham. Even though they were largely uncoordinated in their methods and actions, they truly were 'stronger together' (Allen et al. 2017). The evidence of massive political support no doubt played a part in the Army Corps of Engineers' eventual decision not to grant a permit.

This kind of social movement-like action by groups that to some extent acted in coalition, but often arose and acted independently and spontaneously, is emblematic of the Indigenous/non-indigenous opposition to oil and gas pipelines in British Columbia (seen as destructive to fish habitat, in addition to many other values) which has frequently resulted in the withdrawal of applications or denial of permits. This case illustrates the fact that, in situations where small-scale fishers are not in isolated communities, but are embedded in a more urbanized society, public education and the formation of coalitions with allies may be critical to protecting fish habitat. It also illustrates the fact that groups with diverse motivations may happen to act together around a cause that engages all of their motivations.

13.7 Strategy 6. Local Governing Body Exercises Conservation Rights in British Columbia, Canada

Jones et al. (2017) recount how the Haida Nation in British Columbia, Canada, was able to use a court injunction in 2015 to stop the Canadian Department of Fisheries and Oceans (DFO) from opening a herring fishery in a protected area the Haida co-managed with Parks Canada and DFO. The grounds on which the Haida kept the herring fishery closed were that (a) persistent low stock abundance would not support a fishery; (b) the Haida held presumed aboriginal title and thus the right to be consulted and accommodated about fishing; and (c) the co-management agreement governing the National Marine Conservation Area Reserve where the herring fishery would have happened should require Haida consent to a fishery there. The judge ruled in favour of the Haida, largely based on the existing co-management agreement which, the judge argued, would be damaged by such a unilateral government action. It is useful to conceptualize the Haida action as the exercise of local conservation rights. The term 'local' is significant here, since Haida negotiation attempts in coast-wide forums incurred powerful opposition from outside fishing interests which desired access to the Haida Gwaii herring. However, the co-managed National Marine Conservation Area Reserve in southern Haida Gwaii provided a favourable context in which local conservation rights could be exercised because of the conservation commitments in the agreement governing this area. Like the Dominican Republic, Malawi, and the Lummi Tribe in the USA, which also exemplified the exercise of conservation rights, the Haida example reminds us that protection of local stocks and their habitat is a powerful way to exercise conservation rights. In the case of the Haida and the Lummi, opposed by powerful interests, the support of other parties made a difference, even though Haida and Lummi rights were legally affirmed. In the Lummi case, the support came from non-Lummi local community members, while in the Haida case, support came from the existence of a co-management board with a history of successful collaboration. Thus each of these cases illustrates different ways and contexts in which local conservation rights are successfully exercised.

13.8 Strategy 7. Alternative Marketing Strategies by Small–Scale Fishers Bypass Corporate Fish Processors and Gain Market Power

Many of the strategies discussed above concern how small-scale fishers (a) are able to acquire and maintain fishing access privileges or rights which are attached to local areas and restricted to local residents or (b) are able to keep local fish stocks from being overfished and fish habitat from being damaged by outsiders, in situations where the restriction of access to local residents is also being asserted. Some of these strategies also involved a degree of market protection in the provision of funds or loans or lower lease rates for local fishers to acquire access privileges. An equally or even more important aspect of control over fisheries access is some influence over prices at which raw fish are sold. If fishers cannot make enough money selling their fish, they may not be able to continue enjoying their access. Pinkerton and Weinstein (1995) conceptualized this as the right to optimum value from fishing activity.

Some small-scale fishers who do not enjoy any of the privileges or rights discussed above have nonetheless been able to exercise some market power through bypassing the corporate buyers who in many cases are able to control fish prices, either through oligopsonistic practices (collusion among a few buyers to hold the price down), or through control of infrastructure such as moorage and ice, a 'bonus' to the skipper which is not shared with the crew, or through pre-season loans tying fishermen to one buyer, who can then avoid price competition. Two strategies, which allow small-scale fishers some market power in the face of these challenges, are discussed below.

13.8.1 Direct Marketing of Higher Quality Fish for a Better Price, Alaska and Washington State, USA

Knutson (2017), an anthropologist and Alaska fisher who markets his Alaska-caught fish in Washington State, USA, discusses an individual family's and later a larger small-scale fishers' organization's effort to market fish directly, using public cold storage, wharfs, and niche markets where available, including farmers' markets, to bypass the corporate power and much lower price (1/2–1/3 of the farmers' market price) of the buyer-processors. Farmers markets were important alternative outlets because of the political power of processors to dominate markets, intimidate most retailers, manipulate prices during supply crises to drive fishing cooperatives out of business, and blacklist fishers who delivered fish to independent buyers. It took the work of a family member to receive and deliver the fish sent by the fisher via air or ferry, to local markets.

It often took concerted public pressures to keep public space such as cold storage, and moorage available, allowing niche marketing. It also took employment of more people (two deckhands) per unit of fish sold to clean the fish in order to get a better price for a higher quality product. Knutson emphasizes the amount of damage sustained by fish (in this case salmon) when caught and processed via the mass production methods of the large-scale seine fleet and the corporate processors, in contrast to the careful individual handling of the family gillnetter. He notes the independent producers' ability to also freeze and ship the lower-priced pink and chum salmon into independent markets at reasonable prices so that the salmon they produce is not just for elites.

13.8.2 Community Supported Fisheries (CSFs): Social Enterprises That Use Market Power to Support a Broader Range of Benefits

Witter and Stoll (2017) explain the arrangements through which conscientious consumers concerned about labour injustices, overfishing, and mislabelling bypass up to 10 steps in the usual value chain by working through a single marketer instead of a corporate processor. In one common practice, consumers pay any chosen sum in advance to an alternative marketer, are informed when fish is delivered, and choose what and when to buy by reserving online and picking up the fish at a specified time and place. Some alternative marketers deliver on request to local health food stores in different parts of a city; both alternative marketer and health food stores provide this service without charge. Consumers learn the identity of the fisher through the labelling of the fish and pay a bit more for a higher quality product. Payment in advance gives the fisher some cash flow security during the fishing season. In this way fishers enjoy an up to 30% better price for their fish than they could achieve through regular marketing channels, and receive it in advance.

However, community supported fisheries are called *social* enterprises because they also support non-market values and could be considered part of the social economy (Restakis 2006). One is simply the value of buying locally which avoids unnecessary production of greenhouses gases through extra transportation costs. Another value is supporting local businesses or even regional businesses, such as students at the University of New Hampshire, USA, and New Hampshire hospitals insisting that their cafeterias' fish come from New Hampshire fishers (Tolley and Hall-Arber 2015). Other non-market values include transparency of costs and markets, traceability of product, education and advocacy about better fisheries management, development of markets for underutilized species, independence, and the creation of support networks which sometimes lead to cooperatives.

13.9 Strategy 8. State Regulation or Re-regulation Dampens Neoliberal Control Mechanisms

In Iceland, unlike most other countries engaging in such practices, there have been 26 convictions of bankers and financiers since 2010 who are being held accountable for the collapse of the Icelandic economy, originally precipitated by speculative finance in ITQs (Einarsson 2012). The popularity of Iceland's Pirate Party in 2016 revealed public anger (Robinson and Valdimarsson 2016). The Panama Papers revelations in April 2016 that Iceland's prime minister held tax-avoiding offshore financial accounts led to his resignation in the face of public condemnation. These could be considered examples of 'push back' against policies which have privatized access rights and drastically reduced the access of small-scale fishers, once a country recognizes the effects of such policies. In such circumstances, states can change course by reducing the offending activities by 2–3% a year, such as Copenhagen did with urban automobile parking, which eliminated traffic congestion and created public access to space in the urban core since the 1970s (Grescoe 2012) followed by 9 other European cities (Stecker 2011). A nascent social movement in New Zealand and one in Europe led by Greenpeace (Frangoudes and Bellanger 2017), which are attempting to claim access for small-scale fishers in the context of established ITQs, are apparent moves in that direction.

In Norway, as Jentoft and Johnsen (2015) explain, an owner-operator policy and, since 1989, a Namibian-style allocation of Individual Vessel Quotas by the state, has so far inhibited market-driven ITQs. Norwegian small-scale fishers have so far successfully opposed even the transfer of IQs between small-scale fishers or within local areas, such as is permitted in France. The authors argue that the reluctance of the Norwegian state to violate the long-term social contract with their small-scale fishers may succeed in preventing the enormous loss of coastal community and small-scale fishers' wellbeing that would result from a move to transferable ITQs, in this case called Individual Vessel Quotas. However, they worry that the system may create its own ideology and limit thinking which had previously focused on equitable distribution, retention of youth who are now leaving, and the management power retained by the state needed to deal with such wicked problems (Johnsen and Jentoft 2018).

13.10 Conclusions

This discussion has illustrated multiple problems which occur when market governance of fisheries is attempted through Individual Transferable Quotas and related neoliberal policies. The main focus of the discussion, however, was on the strategies that organizations or communities of small-scale fishers, as well as some governments, used in their attempts to redress or avoid those problems and to create viable alternatives. The broad range of concerns relevant to the governance of fisheries such as equitable distribution of access and benefits, the legitimacy of policies and regulations, viable livelihoods, viable coastal communities, retention and use of cultural heritage and local knowledge, and ecosystem-based management are simply ignored in the narrow neoliberal framing of problems. Even neoliberal concerns such as efficiency can be poorly applied to local situations without paying attention to the local knowledge of people in fishing communities and situations where small-scale fisheries may be more efficient than large-scale. Overall, the cases discussed above show that government attention and regulation is required to achieve the broader goals of fisheries management.

Five of the overall strategies involved government policies or regulations which directly enabled small-scale fisheries to survive. Other strategies involved the ability of local or regional organizations of small-scale fishers to address difficult challenges to their survival, and all of these would have benefitted from government support. Table 13.2 lays out ten examples of policies or strategies used by governments and by fishers' organizations to support the access of small-scale fishermen to fish and to receive the benefits of fishing. The table reveals the large degree of overlap in the use of certain strategies: keeping licenses in local areas, keeping licenses affordable, preventing transfer of licenses via the market, allowing only owner-operators to own licenses, asserting local conservation rights to prevent habitat destruction or use of destructive gear, direct marketing to obtain optimum value from fishing, and providing loans or support to local small-scale fisheries.

These strategies concerned both access rights asserted by fishermen's organizations and the privilege to fish accorded by governments which seek to make access contingent on fishing practices leading to sustainable social and ecological outcomes. Many countries consider fish a public good to which access is accorded in relation to the benefits accruing to adjacent fishing communities or the nation, as was the case in Namibia. In cases such as Buen Hombre in the Dominican Republic, Lake Chiuta in Malawi, and the Haida in British Columbia, government or the courts played a useful role in recognizing the value of local conservation rights and granted or delegated formal protection against outside fishers under their authority. In a neoliberalizing world, the rights to protect fish habitat and stocks from destructive developments and to prevent complete domination by corporate parties in controlling raw fish markets could be considered as important as, or even more important than, access rights. It is also useful to note that interdisciplinary knowledge is often built through coalitions between multiple actors who shared intersecting interests in conservation and access rights, as occurred in the Dominican Republic, in Newfoundland, and with the Lummi and their allies in Washington State.

Overall, these examples illustrate some of the unrecognized ways in which small-scale fisheries contribute to social and ecological welfare, inviting a more interdisciplinary consideration of these issues. Small-scale fishers who have gained access rights or privileges through their own struggles have played both ecologically and socially positive roles in local marine or lake ecosystems, and in contributing to the wellbeing of their communities and even their nations. This should not be surprising, considering that small-scale fishers are not in the business of private capital accumulation, but are concerned rather with food security, livelihoods, and community wellbeing. When they have these, their communities are net contributors to the larger public and ecological welfare, and will seldom if ever be a drain on public resources, as is so dramatically illustrated by the Alaska CDQ program. They are instead a substantial boon to public welfare which needs to be recognized and protected.

Acknowledgements The author is grateful to the many community members who reviewed portions of this manuscript, to Steve Langdon who reviewed the Alaska portion, and to Svein Jentoft and two anonymous reviewers for excellent suggestions, which have improved it. The author is also grateful to the Social Sciences and Humanities Research Council of Canada for supporting this research, especially through the Partnership Grant 'Too Big To Ignore' awarded to Dr. Ratana Chuenpagdee.

References

Acheson J (2003) Capturing the commons: devising institutions to manage the Maine lobster industry. University Press of New England, Hanover

Allen M, Bird S, Breslow S, Dolsak N (2017) Stronger together: strategies to protect local sovereignty, ecosystems, and place-based communities from the global fossil fuel trade. Mar Policy 80:168–176

Barnett AJ, Messenger RA, Wiber MG (2017) Enacting and contesting neoliberalism in fisheries: the tragedy of commodifying lobster access rights in Southwest Nova Scotia. Mar Policy 80:60–68

Bavinck M, Jentoft S, Pascual-Fernandez J et al (2015) Interactive coastal governance: the role of pre-modern fisher organizations in improving governability. Ocean Coast Manag 117:52–60

Brox O (1993) Let us now praise dragging feet! In: Åkerman N (ed) The necessity of friction. Springer, New York, pp 123–134

CCFA (Cape Cod Fishermen's Alliance, Inc.) (n.d.) The giving common. http://givingcommon.guidestar.org/FullPDF.aspx?OrgId=1076642. Accessed 15 Apr 2018

CCFT (Cape Cod Fisheries Trust) (n.d.) http://www.capecodfishermen.org/fisheries-trust

Davis R, Wahlen J, Neis B (2006) From orders to borders: toward a sustainable co-managed lobster fishery in Bonavista Bay, Newfoundland. Hum Ecol 34:851–867

Dinesen GE, Rathje IW, Hojrup M et al (2018) Individual Transferable Quotas, does one size fit all? Sustainability analysis of an alternative model for quota allocation in a small-scale coastal fishery. Mar Policy 88:23–31

Donda S (2017) Who benefits from fisheries co-management? a case study of Lake Chiuta, Malawi. Mar Policy 80:147–153

Durrenberger EP (1997) Fisheries management models: assumptions and realities or, why shrimpers in Mississippi are not firms. Hum Organ 56:158–166

Ecotrust (n.d.) Pacific Coast Fishermen's Conservation Company. http://ecotrust.ca/project/pacific-coast-fishermens-conservation-company/. Accessed 15 Apr 2018

Edwards DN, Edwards DG (2017) License banks as a tool to mitigate corporate control of fisheries: a British Columbia groundfish example. Mar Policy 80:141–146

Einarsson N (2012) From fishing rights to financial derivatives. Individual Transferable Quotas and the Icelandic economic collapse of 2008. In: Hojrup T, Schriewer K (eds) European fisheries at a tipping point. Murcia, Editum, pp 204–255

Elson v. Canada (Attorney General), 2017 FC 459 (CanLII)

Frangoudes K, Bellanger M (2017) Fishers' opinions on marketization of property rights and the quota system in France. Mar Policy 80:107–111

Grescoe T (2012) Straphanger: saving our cities and ourselves from the automobile. Henry Holt and Co., New York

Hojrup T (2011) The Common Fisheries Policy has to recognize the need for common goods for coastal communities: common community quotas for sustainable life modes in coastal fisheries: the alternative to privatization of fishing rights in the home waters of Europe. Narayana Press, Gylling

Host J (2015) Governing through markets: societal objectives, private property rights, and small-scale fisheries in Denmark. In: Jentoft S, Chuenpagdee R (eds) Interactive governance for small-scale fisheries: global reflections. Springer International Publishing, Dordrecht, pp 319–336

Jentoft S, Johnsen JP (2015) The dynamic of small-scale fisheries in Norway: from adaptamentality to governability. In: Jentoft S, Chuenpagdee R (eds) Interactive governance for small-scale fisheries: global reflections. Springer International Publishing, Dordrecht, pp 705–723

Johnsen JP, Jentoft S (2018) Transferable quotas in Norwegian fisheries. In: Winder G (ed) Fisheries, quota management, and quota transfer. Rationalization through bio-economics. Springer International Publishing, Dordrecht, pp 121–139

Jones R, Rigg C, Pinkerton E (2017) Strategies for assertion of conservation and local management rights: a Haida Gwaii herring story. Mar Policy 80:154–167

Knutson P (2017) Escaping the corporate net: pragmatics of small boat direct marketing in the US salmon fishing industry of the northeastern Pacific. Mar Policy 80:123–129

Langdon S (2015) A tale of two oceans: Alaska native coastal communities and fisheries policies in the Gulf of Alaska and the Bering Sea. Paper presented at the Canadian Association of Geographers' Conference. Vancouver, BC, 3 June 2015

Levy S (2010) Catch shares management. BioScience 60(10):780–785

Marglin S (1991) Understanding capitalism: control vs. efficiency. In: Gustaffsson B (ed) Power and economic institutions: reinterpretations in economic history. Edward Elgar, Aldershot
McCormack F (2018) Private oceans: the enclosure and marketisation of the seas. Pluto Press, London
National Oceanic and Atmospheric Administration (n.d.) Community Development Quota Program. https://alaskafisheries.noaa.gov/fisheries/cdq. Accessed 19 Sept 2017
NRC (National Research Council) (1999) The community development quota program in Alaska. National Academy Press, Washington, DC
Ounanian K (2016) In place of fishing: coastal communities in transition. Dissertation, University of Rhode Island and Aalborg University, https://doi.org/10.5278/vbn.phd.engsci.00101
Pinkerton E (2017) Hegemony and resistance: disturbing patterns and hopeful signs in the impact of neoliberal policies on small-scale fisheries around the world. Mar Policy 80:1–9
Pinkerton E, Davis R (2015) Neoliberalism and the politics of enclosure in North American small-scale fisheries. Mar Policy 61:303–312
Pinkerton E, Weinstein M (1995) Fisheries that work: sustainability through community-based management. A report to the David Suzuki Foundation, Vancouver, BC. http://davidsuzuki.org/publications/reports/1995/fisheries-that-work/. Accessed 15 Apr 2018
Restakis J (2006) Defining the social economy – the BC context. British Columbia Cooperative Association, Vancouver
Robinson E, Valdimarsson O (2016) This is where bad bankers go to prison. Bloomberg. http://www.bloomberg.com/news/features/2016-03-31/welcome-to-iceland-where-bad-bankers-go-to-prison. Accessed 15 Apr 2018
Sabau G, van Zyll de Jong M (2015) From unjust uneconomic growth to sustainable fisheries in Newfoundland. The true cost of closing the inshore fishery for groundfish. Mar Policy 61:376–389
Stecker T (2011) Reducing parking spaces helps cities cut auto emissions. Scientific American. https://www.scientificamerican.com/article/reducing-parking-cut-auto-emission/. Accessed 15 Apr 2018
Stoffle B, Halmo D, Stoffle R et al (1994) Folk management and conservation ethics among small-scale fishers of Buen Hombre, Dominican Republic. In: Dyer CL, McGoodwin JR (eds) Folk management in the World's fisheries: lessons for modern fisheries management. University Press of Colorado, Niwot, pp 115–138
Szymkowiak M, Himes-Cornell A (2015) Towards individual-owned and owner-operated fleets in the Alaska Halibut and Sablefish IFQ program. MAST 14:19. https://doi.org/10.1186/s40152-015-0037-6
Tolley B, Hall-Arber M (2015) Tipping the scale away from privatization and towards community-based fisheries: policy and market alternatives in New England. Mar Policy 61:401–409
Treadwell M (2012) Comments at the 20th anniversary celebration of the CDQ program. Juneau, Alaska
Witter A, Stoll J (2017) Participation and resistance: alternative seafood marketing in a neoliberal era. Mar Policy 80:130–140
World Bank, FAO Rome (2009). The sunken billions: the economic justification for fisheries reform. Washington, DC. http://siteresources.worldbank.org/EXTARD/Resources/336681-1224775570533/SunkenBillionsFinal.pdf. Accessed 15 Apr 2018

Chapter 14
The Small-Scale Fisheries of Indigenous Peoples: A Struggle for Secure Tenure Rights

Svein Jentoft, Natasha Stacey, Jackie Sunde, and Miguel González

Abstract The UN estimates that there are about 370 million indigenous people worldwide. Indigenous people often find their natural resources, cultures, and communities under pressure. In many instances, they are victims of systemic discrimination and human rights abuse. Indigenous people who draw their livelihood from small-scale fisheries are no exception to this rule. The recognition of their terrestrial and marine tenure rights is often lacking, which has repercussions for their short and long-term wellbeing. In this chapter, we explore the political and legal foundation of indigenous small-scale fisheries, drawing from international and domestic law, and learning from situations in four countries: Norway, Australia, South Africa, and Nicaragua. What institutional reforms would facilitate the self-determination and sustainable economic development of indigenous small-scale fisheries, given that they are not only a marginalized group within their countries, but also within their industry? What prospects exist for the international legislation having real influence on the livelihoods of small-scale fishers and fish workers in indigenous communities? What role can customary law play in this respect?

Keywords SSF Guidelines · Indigenous peoples · Tenure rights · Customary law · ILO 169

S. Jentoft (✉)
Norwegian College of Fishery Science, UiT- The Arctic University of Norway, Tromsø, Norway
e-mail: svein.jentoft@uit.no

N. Stacey
Research Institute for the Environment and Livelihoods, Charles Darwin University, Darwin, Australia
e-mail: natasha.stacey@cdu.edu.au

J. Sunde
ABALOBI, Cape Town, South Africa
e-mail: jsunde@telkomsa.net; http://www.abalobi.info

M. González
Department of Social Science, York University, Toronto, ON, Canada
e-mail: migon@yorku.ca

R. Chuenpagdee, S. Jentoft (eds.), *Transdisciplinarity for Small-Scale Fisheries Governance*, MARE Publication Series 21,
https://doi.org/10.1007/978-3-319-94938-3_14

14.1 Introduction

The UN estimates that there are about 370 million indigenous people worldwide, which constitutes less than 5% of the total population globally. According to the same source, they speak an overwhelming majority of the world's 7000 languages, and represent 5000 different cultures. Indigenous people account for 15% of the world's poorest population (UN 2017).

Indigenous people often find their natural resources and communities are under pressure, which leaves them socially and politically marginalized, and with the risk of becoming culturally extinct (Jentoft et al. 2003; Anaya 2004; Tobin 2014; Stacey et al. 2017). Their institutions are often not in a position to protect their interests and to voice their concerns and interests effectively. The formal recognition of their terrestrial and marine tenure rights are often missing, and they are frequently victims of systemic discrimination and human rights abuse.

Those among the indigenous people who draw their livelihoods from small-scale fisheries are no exception to this rule. They are often a minority within the sector, and are, therefore, often subject to majority power. In recognition of this, the Voluntary Guidelines for Responsible Governance of Tenure of Land, Fisheries and Forests in the Context of Food Security (Tenure Guidelines) (FAO 2012), and the Voluntary Guidelines for Securing Small-Scale Fisheries in the Context of Food Security and Poverty Eradication (SSF Guidelines) call them out as a group requiring special attention and responses due to their particular circumstances (FAO 2015).

This chapter describes the diverse ethno-political realities that exist where both hard and soft-law instruments are supposed to take effect.[1] Included are examples from Norway, Australia, South Africa, and Nicaragua, where the survival of indigenous small-scale fisheries communities is far from secure. As we shall see, regardless of contexts, indigenous small-scale fisheries peoples are under pressure, and must fend for their livelihoods and communities in a political landscape tilted against them. Implementation of soft legal mechanisms, such as the SSF and Tenure Guidelines, takes place in 'zero-sum' societies (Thurow 1981), where power relations in most instances are not conducive to indigenous peoples. Governments are not necessarily as supportive as one would expect, given their endorsement of these legal instruments. What challenges and opportunities exist for indigenous peoples to secure their territorial and resource rights to territories within a global and domestic legal framework? What prospects exist for the international legislation having real influence on the livelihoods of small-scale fishers and fish workers in indigenous communities, which are commonly governed within a framework of customary law?

[1] "Soft law refers to rules that are neither strictly binding in nature nor completely lacking legal significance. In the context of international law, soft law refers to guidelines, policy declarations, or codes of conduct which set standards of conduct. However, they are not directly enforceable. Hard law refers to binding laws. In the context of international law, hard law includes treaties or international agreements, as well as customary laws. These documents create enforceable obligations and rights for countries (states) and other international entities." https://definitions.uslegal.com/s/soft-law/

14.2 The Legal Status of Indigenous Fishing Peoples

The International Labor Organization (ILO) 169, (1989) defines indigenous peoples as: "(a) tribal peoples in independent countries whose social, cultural, and economic conditions distinguish them from other sections of the national community, and whose status is regulated wholly or partially by their own customs or traditions or by special laws or regulations; (b) peoples in independent countries who are regarded as indigenous on account of their descent from the populations which inhabited the country, or a geographical region to which the country belongs, at the time of conquest or colonization or the establishment of present state boundaries and who, irrespective of their legal status, retain some or all of their own social, economic, cultural and political institutions (ILO 1989, C169, Article 1).

Twenty-two countries ratified this convention, among them Norway and Nicaragua, whereas South Africa and Australia did not. The more than 40 articles that make up the convention talk about the responsibilities of governments to protect the social, economic, and cultural rights, including customs, traditions and institutions, with respect to their social and cultural identity. Notably, this shall be done in cooperation with the indigenous peoples themselves. Without going into further detail, Part II of the ILO 169 addresses the safeguarding of the rights of ownership and possession of land, territories, and natural resources, and emphasizes that indigenous peoples have a right to stay in the areas that they have traditionally settled and used.

The ILO 169 sets the stage for subsequent international and national legislation regarding indigenous peoples, such as the 2007 UN Declaration on the Rights of Indigenous Peoples (UNDRIP), whose principles are also included in the Tenure Guidelines and the SSF Guidelines. The Tenure Guidelines has a whole section devoted to indigenous peoples. Paragraph 9.4 says that states "should provide appropriate recognition and protection of the legitimate tenure rights of indigenous peoples and other communities with customary tenure systems."[2]

The SSF Guidelines, which draws heavily on the Tenure Guidelines, have among their guiding principles 'respect of cultures', including existing forms of organizations, traditional knowledge, and practices of small-scale indigenous communities (See Box 14.1). Paragraph 3.6 in the SSF Guidelines makes states responsible for "ensuring active, free, effective, meaningful, and informed participation of indigenous small-scale fishing communities in the whole decision-making process, taking existing power imbalances between different parties into consideration." Paragraph 5.4 says: "Local norms and practices, as well as customary or otherwise preferential access to fishery resources and land by small-scale fishing communities including indigenous peoples and ethnic minorities, should be recognized, respected and protected in ways that are consistent with international human rights law." Moreover,

[2] Other international legislations have provisions regarding indigenous peoples, such as the International Covenant on Civil and Political Rights, Article 27 (1966), and the Convention on Biodiversity Conservation, Article 8 (1992). The Code of Conduct for Responsible Fisheries mentions indigenous peoples in paragraph 7.6.6. (1995).

Box 14.1 Indigenous Peoples in the SSF Guidelines

- Guiding Principles emphasize "respect of cultures", which involves "... recognizing existing forms or organization, traditional and local knowledge and practices of small-scale fishing communities, including indigenous peoples."
- Paragraph 3.6 talks about consultation and participation and the need to take into account the UN Declaration on the Rights of Indigenous Peoples "in the whole decision-making process related to fisheries resources and areas where small-scale fisheries operate, and taking existing power imbalances between different practices into consideration."
- Paragraph 5.4 says states should protect all forms of legitimate tenure rights, "including those of indigenous peoples. States should recognize the role of small-scale fisheries communities and indigenous peoples to restore, conserve, protect, and co-manage local aquatic and coastal ecosystems."
- Paragraph 6.2 says the need for preferential treatment of indigenous peoples is emphasized so as to "ensure equitable benefits."
- Paragraph 11.6 says: "All parties should ensure that the knowledge, culture, traditions, and practices of small-scale fishing communities, including indigenous peoples, are recognized and, as appropriate, supported and that they inform responsible local governance and sustainable development processes."

"states should recognize the role of small-scale fishing communities and indigenous peoples to restore, conserve, protect and co-manage local aquatic and coastal ecosystems" (Para. 5.5).

The questions addressed in the following sections are whether these governance principles reach indigenous fisheries communities to the extent one could expect, and whether they make a difference in the daily lives of people as they go fishing. This, we argue, largely depends on the local realities at the receiving end of these principles. Indigenous communities are not a legal tabula rasa. Instead, they are legal systems in their own right, often governed by indigenous institutions according to customary norms. The processes of implementing indigenous rights are therefore not straightforward, but may get entangled in legal pluralism and, thus, conflict (von Benda-Beckmann et al. 2009; Bavinck and Gupta 2014).

14.3 Indigenous Peoples – Recognition of Fishing Tenure

For indigenous peoples, the issue now is about unfulfilled expectations as to what this legislation would deliver, and about broken political promises. Whether recent 'soft law' instruments, such as these two voluntary guidelines and the UNDIP, can help to rectify what decades long 'hard law' conventions, such as ILO169, have failed to deliver, is an open question. More attention should, therefore, now be paid to the processes and mechanisms that facilitate (or prevent) effective implementation. Why have these otherwise impressive legal achievements not reached the indigenous peoples, whose small-scale fisheries are in a precarious state? Why are small-scale fisheries people in many parts of the world still struggling, not only to make ends meet, but also to confront the real prospect of social and cultural extinction? It is to these questions that more attention should be directed. A linear process, in which international codes trickle down smoothly to national and local settings, is unrealistic. More likely, it would take bottom-up initiatives from indigenous communities and organizations, who, in the process, may meet resistance from other fisheries stakeholders, including government (Jentoft 2014). The four countries provide contextual insights into the way international and domestic legislation works itself into the world of indigenous small-scale fishing people.

14.3.1 Norway[3]

For the indigenous Sami settled along the coast and fjords of northern Norway, small-scale fishing has for centuries been a way of life and a source of income and subsistence (Paine 1965; Pedersen 2012). Although the number of Sami fishers has declined in recent decades, small-scale fisheries are still important (Broderstad and Eythórsson 2014). In contrast to many other indigenous peoples around the world, the Sami in Norway benefit from the rights and services provided by an advanced welfare state, be it in education or healthcare, and their social and economic wellbeing cannot be classified as poor. Still, their tenure rights to natural resources, including fisheries, are an ongoing struggle to maintain their small-scale fishing, on which the sustainability of their coastal communities, and hence their distinct Sami culture, depends. In a report solicited by the government due to claims that Sami fisheries were negatively affected by a new quota system, Smith (1990) suggested that Sami small-scale fishing could soon become extinct.

By ratifying the ILO 169, the UN Covenant on Civil and Political Rights, the Convention of Biodiversity Conservation, and endorsing the UNDRIP, Norway committed to respecting the rights of the Sami as an indigenous people and to not undermine the material basis for Sami culture. In Norway, there are around 60,000 Sami, but Sami people are also in northern Sweden, Finland, and Russia. Sami rights and state responsibilities are also confirmed in domestic fisheries law, but most prominently in Paragraph 108, amended to Norway's Constitution in 1988, which says: "It is the responsibility of the authorities of the State to create conditions enabling the Sami people to preserve and develop their language, culture and way of life." (One may think that this was about time, given that the Constitution itself was enacted in 1814).

---

[3] This section draws heavily on Jentoft and Ulfsdatter Søreng (2017).

As a follow up to the constitutional amendment, the Sami Parliament was established in 1989. The government also initiated investigation into Sami land rights (Minde 2005), which eventually resulted in the Finnmark Act. Here, Sami rights to ancestral land and terrestrial resources in that county (which is Norway's biggest) are confirmed, with the outcome that jurisdiction over former state land was transferred to a new regional institution, the "Finnmark Estate", to which the Sami Parliament and the County Assembly appointed the Board of Directors.[4]

Established in 2006, The Coastal Fisheries Committee (CFC) was mandated to inquire whether similar historical rights pertained to ocean space and marine resources. In a series of local hearings, the CFC documented historical usage and people's perceptions of traditional fishing rights. Based on the information received, and with reference to Norwegian and international law, the CFC concluded that fishers living in Finnmark indeed have a historical right to fish, and that the Norwegian government should legally recognize and formally implement this right. The CFC's report included a complete fisheries law proposal for Finnmark (NOU 2008, 5), including what a regional co-management system should look like. However, the Ministry of Fisheries and Coastal Affairs appeared to be negative toward this proposal, and argued that the existing Norwegian fisheries management system already sufficiently addressed indigenous rights, and that Sami fisheries should be managed within the existing legal system. Therefore, the Ministry decided not to support the idea of historical Sami fishing rights or that they merited a separate law. Nor did it agree with the creation of a co-management institution for Sami fisheries. As could be expected, the government's reaction met reactions. The Sami Parliament and the chair of CFC conveyed great disappointment, and the media was flooded with letters from angry Sami small-scale fishers. Even The United Nations Special Rapporteur on the rights of indigenous peoples, James Anaya, warned that a proposal to repeal key laws and policies related to Sami people in Norway could "constitute an enormous setback for the recognition and protection of human rights in the country."

However, the Norwegian Fishers' Association (NFA), which had been highly critical of the CFC report, was supportive of the Ministry's conclusion. When the Sami Parliament, after passionate debate, voted on the Ministry's alternative proposal, which included the establishment of an advisory board, The Fjord Fisheries Board (FFB), plus a number of other measures pertaining to fishing and fish processing, a small majority of the Sami Parliament delegates supported it. Those who voted for the Ministry's proposal argued that it was the best they could hope for, at least for now. Thus, the government's proposal prevailed, and continued to do so when the Norwegian parliament passed a reformed version of the Marine

[4] § 5 in the Finnmark Act states: "Through prolonged use of land and waters areas, the Sami have collectively and individually acquired rights to land in Finnmark. This Act does not interfere with collective and individual rights acquired by Sami and other people through prescription or immemorial usage." The confirmation of rights happens through the Finnmark Commission "which, on the basis of current law, shall investigate rights of use and ownership to the land." If there is a dispute on the conclusions of the Commission, this can be brought to the uncultivated Land Tribunal, and in the last instance be appealed to the Supreme Court of Norway.

Resource Act in 2013, which confirmed the government's position on the Sami rights issue. In other words, no new law was enacted, as the CFC had proposed, but rather, a modest amendment was included in the existing legislation. Paragraph 7 in Law on the Management of Living Marine Resources added a section, g), stating that the ministry shall assess what kinds of management measures are necessary to secure a sustainable management of living marine resources "and that these help secur[e] the material basis for Sami culture." Likewise, paragraph 11, which talks about quotas and other kinds of regulations, emphasizes that the Sami use of marine resources and what they mean for Sami local communities shall be given considerable weight (our translation)."

How effective these measures are and how sustainable the government's position is on the matter of Sami tenure, remains to be seen, as there might be future lawsuits. Still, the amendment and the Fjord Fisheries Board (FFB) are concrete steps in meeting Sami goals, and also move in the directions advocated by the SSF Guidelines. Nevertheless, the FFB is heavily criticized, with some of the harshest criticism coming from the Norwegian Fisheries Association (NFA), where large-scale fisher interests dominate. This is despite the fact that FFB's mandate is limited to assessing the situation for Sami small-scale fisheries. Notably, the mandate does not prohibit the FFB from taking initiatives on matters that are within its mandate. Veto power nonetheless rests with the government. The Sami Parliament has repeatedly pointed out that management systems that do not entail autonomous decision-making powers for indigenous people are in conflict with international law. So far, however, it has decided to live with the new arrangement.

14.3.2 Australia

Identified as Aboriginal or Torres Strait Island (ATSI) people, indigenous people in Australia comprise approximately 3% (or about 650,000) of the total population. The highest proportion per capita live in the Northern Territory (NT), comprising around 25.5% (approximately 58,248) of a total population of approximately 228,833 people (ABS 2016). The majority of indigenous people live in remote or very remote locations, most of which are part of the indigenous estate, Aboriginal owned and managed land and sea, held under some form of indigenous tenure. This is estimated to range from 23% (Altman 2012) to 52% (Renwick et al. 2017) of the Australian continent. Indigenous people view the seas, islands, and coastal environments as part of their 'sea country' or 'saltwater country' and often refer to themselves as 'saltwater people' (Smyth 2007).

There is no overarching recognized legal framework for indigenous customary fishing rights in Australia. Instead, rights have been intricately tied to developments in land and native rights with different local arrangements across states and territories (Schnierer and Egan 2016).[5] This has meant that in most cases indigenous customary fishing is exempt from fisheries management frameworks and laws. The effect is that indigenous fishing has not been recognized and subsequently resulted in low numbers of indigenous people in fisheries and associated businesses (Fleming 2015; Productivity Commission 2016). In most cases,[6] the definition of indigenous customary fishing does not include fishing for commercial purposes, even though it is recognized traditionally that Aboriginal people have fished commercially (i.e., sale, exchange or barter of fish) according to traditional laws and customs. As such, land or native title-holders who fish commercially have been subject to the same commercial fishing laws and regulations as the rest of the population.

The establishment of indigenous commercial fishing enterprises is the NT is closely linked to recent legal recognition of rights in the coastal zone. In 2008, the High Court of Australia declared that a group of traditional owners[7] in the Blue Mud Bay region of north east Arnhem Land in the NT have exclusive access rights to intertidal waters (to the water and land between the high and low tide – approximately 1 nautical mile) overlying inalienable Aboriginal freehold land granted under Commonwealth legislation, the Aboriginal Land Rights (Northern Territory) (ALR) Act (1976) (Northern Land Council 2017; Sanders and Xuereb 2016). The ALR Act was the first Act by Australia to legally recognize a system of Aboriginal land ownership and established the concept of inalienable freehold title. This finding subsequently applied to almost the entire coastline of the NT.

The recognition of Aboriginal exclusive access rights for indigenous people in the intertidal zone, allowed for the NT government to amend the NT Fisheries Act (1988) in 2015 to allow Aboriginal people to fish commercially (on a very small-scale) under an Aboriginal Coastal Fishing license (ACL). The application must specify the area to be fished in, applicants must have an interview with the Government Fisheries Indigenous Development Unit, and workers are encouraged to complete a Vocational Certificate II in Fishing Operations, and develop a business plan. The license holder must also submit a monthly logbook to NT Fisheries Department which details the species caught, total number, weight, and kilograms sold.

One of the first fishing operations established under the new regulations was by a group of three senior Traditional owners (TOs) from coastal outstations near the township of Maningrida, one of the largest Aboriginal communities in Arnhem Land, located approximately 500 km from Darwin. The Traditional Owners were motivated to provide a regular supply of fish for the community and employment opportunities for family members. The fishing business is supported by a local Aboriginal Corporation, which provides administration, transportation, and infra-

[5] One exception is the Torres Strait Islands where legal rights for customary and commercial fishing are recognized under Native Title legislation (Lalancette 2017).

[6] Torres Strait is the exception (Lalancette 2017).

[7] Traditional Owners are defined as "local descent group of Aboriginals who have common spiritual affiliations to a site on the land...and are entitled by Aboriginal tradition to forage as of right over that land" (Aboriginal Land Rights Northern Territory Act 1976, Australia: Part 1, Section 3).

structure assistance. The business is helped by family members who work catching and selling the fish via participation in the Government Community Development (Employment) Program (CDP). The NT government also provides training and mentoring support to the fishing business. Since then, approximately twenty TOs have applied for an ACL in the NT. While in its infancy, the recognition of exclusive Aboriginal rights to the intertidal zone and the implementation of the coastal licensing system have potential to deliver significant economic and social benefits to coastal communities through employment, and affordable fresh fish into local diets.

The establishment and successful operation of these small enterprises provide a pathway to further livelihood development opportunities in the fishing and seafood industry, such as through other high value capture fisheries (e.g. beche-de-mer, barramundi, and mud crab). The majority of these licenses are held by external non-indigenous people, and some operate in the intertidal zone now under Aboriginal ownership.

The recognition of Aboriginal legal rights in coastal waters has led to significant changes in fisheries management in the NT. Future studies will determine how effective the ACL system is for improved Aboriginal wellbeing, what adjustments will be needed to sustain commercial fishing-based livelihoods in remote coastal settlements in the future and Aboriginal aspirations towards self-determination.

14.3.3 Nicaragua

Historically, small-scale fisheries on the Caribbean coast have been fundamental to the region's development, due to their contribution to employment, food security of coastal and inland communities, and for their contribution to the country export's market (PNUD 2005, 166). In 2015, indigenous small-scale fisheries contributed 71% of all landings registered in the Caribbean coast. Nationally, small-scale fisheries represent 32% of the total volume of seafood production (INPESCA 2016, 9). There is also substantive contribution of indigenous fisheries to high value species for the export markets. Indigenous artisanal lobster fisheries, for example, which are conducted mostly through diving, contributed 63% of the 5.5 million pound catch registered in 2015 (INPESCA 2016, 21). Despite this, the participation of indigenous and Afro-descendant peoples in critical decision making over regional fishery policies has been, for the most part, marginal.

In the Latin American context, Nicaragua has one of the broadest legal frameworks regarding the protection of the rights of indigenous and Afro-descendant peoples. The former group consists of the Miskito, the Ulwa, and the Rama people, whereas the latter consists of Garifuna and Creole people. The country's political constitution, passed in 1987, recognized for the first time in the country's history the multiethnic nature and the existence of indigenous peoples and ethnic communities. Article 180 guarantees "these communities the benefits of their natural resources, the effectiveness of their forms of communal property, and the free election of their authorities" (Nicaragua National Assembly 1987).

Article 181 inaugurates an unprecedented political administrative autonomy regime that grants limited rights of self-government to the residents of the Atlantic (Caribbean) coast of the country. According to this article: "the state shall organize by means of a law the regime of autonomy for indigenous peoples and ethnic communities of the Atlantic coast." It also asserts, "the concession and contracts of rational exploitation of the natural resources granted by the state in the autonomous regions of the Atlantic coast must have the approval of the corresponding Regional Autonomous Council" (Nicaragua National Assembly 1987). In turn, the Statute of Autonomy (Law 28), which was also passed in 1987, granted the inhabitants of the coast the rights to participate in the promotion, exploitation and sustainable use of their natural resources, and the protection of individual and collective property rights, including their preservation and inheritance (articles 10 and 11, Autonomy Statute, 1987). Indigenous and Afro-descendant peoples of the coast comprise around 3% of the national population and reside in two autonomous regions inaugurated by Law 28 in 1990.

Although the above mentioned rights were included in the Constitution, along with the corresponding operational legislation, for more than three decades their realization has been a complex and challenging process (Frühling et al. 2007). In part, this is due to the fact that the national administrations that governed Nicaragua since 1990 were not committed to strengthening the institutions of regional self-government. During this same period, the country ended a civil war that had lasted almost 10 years (it began in the early 1980s) and consequently the efforts of the governments focused mainly on the reconstruction of economic and social infrastructure, and the reactivation of the country's productive capacity. In relation to fisheries, the country quickly adopted a model of expansion of the fishing effort through the increase of the industrial fleet. The country also privatized the state owned fishing infrastructure that had once provided support to small-scale fisheries through a secure market for landings, fuel, and equipment (Stahler-Sholk 1997). The new model has been extractive in nature, with little value added to seafood production for exports. A case in point is the dangerous techniques of lobster catch through scuba diving, which has caused irreparable neurological damage to hundreds of indigenous fishing workers due to decompression accidents (González 2018). Lobster catching through diving has been officially banned; however, the legislation has not yet been enforced (Nicaragua National Assembly 2013).

In addition to the rights guaranteed in the Constitution and the autonomy regime, the last decade has also witnessed the country's ratification of other rights of equal or greater importance for indigenous peoples, both at the international and domestic level. Internationally, the country is now a signatory to ILO Convention 169 (in 2010) and of the 2007 United Nations Declaration on the Rights of Indigenous Peoples (in 2008). Domestically, through Law 445, an important legislation passed in 2003 to demarcate and title indigenous and Afro-descendant communal lands; Nicaragua has managed to title 23 territories that together make up 54.7% of the autonomous regions of the North and South Caribbean (URACCAN 2017). Out of the 23 titled territories, seven of them concentrate 95% of all coastal and maritime areas under collective ownership, covering around 155.8 thousand hectares.

The demarcation and titling of indigenous territories has provided legal security over the communal property of indigenous and Afro-descendant peoples. The rights to small-scale fishing, to the customary use of traditional fishing grounds and the notions of maritime tenure rights, have all been included in the wording of the official titles (Procuraduría General de la República 2018). However, the entitlements by themselves have not been able to stop the massive incursion of colonist farmers who immigrate to the coast from other regions of the country, in search of better life opportunities (Larson et al. 2016). This nonindigenous peasant population participates in illegal occupations of indigenous lands and consequently encroach upon fishing areas of exclusive use of indigenous and Afro-descendant communities. During the last 5 years these occupations have increased, often violently, which has caused deaths, and forced displacements and destruction in practically all the titled indigenous Afro-descendant territories. For these episodes, which violate the rights of indigenous peoples to the use of their natural resources and their physical integrity, several national human rights organizations have requested the Inter-American Commission on Human Rights to intercede with the Nicaraguan state, in order to prevent further harm to these communities.[8] The illegal colonist incursion and the inaction of state authorities to stop and mediate these conflicts threaten to further escalate tensions over the use of natural resources. Thus, it weakens the potential positive effect that the legislation confers to indigenous and Afro-descendant peoples. In such a scenario, the effect that international legal instruments can have on the rights of indigenous peoples becomes even more important.

14.3.4 South Africa[9]

The indigenous peoples of South Africa have experienced a lengthy history of systemic discrimination and violations of their human rights, from the early imposition of Dutch and, subsequently, British colonial rule, to the relatively more recent racially-based system of segregation imposed by the apartheid regime. Many indigenous small-scale fishing and coastal communities, from diverse ethnic and tribal groups, have been through several waves of dispossession from their ancestral lands, coastal forests, and traditional waters upon which they depended for their livelihoods. The Constitution, introduced with the transition to democracy in 1994, aimed to address this discriminatory legacy by recognizing the customary law and systems of tenure that governed these communities. Section 39 of the Constitution recognizes the rights and freedoms conferred on persons under customary law, as far as they are consistent with the Bill of Rights (Constitution of SA 1996, Section

[8] To this date, the IACHR has issued Nicaragua three rulings against the Nicaraguan state on issues related to the collective rights of indigenous peoples and due judicial process. The first one, Awastingni Vs. Nicaragua in 2001; Yatama Vs. Nicaragua in 2005, and more recently Acosta Vs. Nicaragua in 2017.

[9] This section draws on Sunde (2017).

39). It makes provision for the enactment of legislation that would address security of tenure and restitution of land lost due to racially discriminatory laws. The post-apartheid state has, however, avoided clarifying the term 'indigenous' and has instead preferred to refer to all black African communities as indigenous on the basis of their precolonial presence and their subordination to the colonial and apartheid regimes.

Despite the introduction of a suite of legislative reforms aimed at transforming the racially-based, discriminatory nature of access to and control of marine resources, artisanal and small-scale fishers, most of whom were indigenous peoples, were excluded from the new reforms. The new legislation introduced in 1998 favored the allocation of fishing rights to commercial fisheries, making no provision for the allocation of fishing rights to small-scale fishers. The Marine Living Resources Act (MLRA) of 1998 was silent on the preexisting customary tenure rights of small-scale fishers and on the restitution of any marine tenure rights lost during colonialism or apartheid.

In 2005, a group of small-scale fishers embarked on legal action, arguing that the failure of the new legislation to recognize them was unconstitutional. In 2007, the Equality Court ordered the then minister responsible for fisheries to develop a new policy that would accommodate the socioeconomic rights of traditional fishers. During the process of consultation and negotiation for the new policy, those indigenous small-scale fishing communities who continued to live within customary systems of law and governance, together with their supporters, lobbied very strategically and deliberately for the recognition of their customary fishing rights in terms of the Constitution (Sunde 2014). As fisheries law and policy making in a context of legal pluralism was and continues to be a very new area for the fisheries administration, references to customary law unsettled the fisheries administration, which feared wide sweeping claims to the coastal commons. The fisheries administration ultimately and reluctantly agreed to use the language of the Constitution in the policy, but appeared to doubt the success of any claim on customary grounds. In 2014, the MLRA was amended to accommodate the requirement of the Equality Court that there be legal recognition of small-scale fishers and that the amendment create legal space for customary rights. Notwithstanding this very clear constitutional and legislative framework, both the Department of Agriculture, Forestry and Fisheries (DAFF), and the Department of Environment continued to ignore the claims of small-scale fishing communities who used customary claims in their requests that their right to fish be recognized. In 2010, three fishers from the indigenous communities of Dwesa-Cwebe in the Eastern Cape were arrested for fishing in a Marine Protected Area and were charged for contravening the Marine Living Resources Act, despite the fact that they were arrested on their own land, which had been returned to their them following their successful land claim. These communities regard their ancestral lands, the forests, and the coastline and associated natural resources as their common property, upon which they depend for their livelihoods. They have a local system of customary law, which gives rise to a system of rights and governance processes that regulate access to and control of their natural resources. This case was heard in 2015 in the Mthatha High Court, where the fishers' defense argued that the law governing marine resources did not extinguish their preexisting customary rights and if it did, this was unconstitutional. They drew on international and foreign law from Canada and Australia, as well as citing the Voluntary Guidelines on Tenure and the SSF Guidelines to emphasize the need to recognize customary fishing rights. They also acknowledged that the recognition of customary rights did not mean that these rights could not be regulated. In 2016, the judgement in this case was handed down confirming that where a community has a customary fishing right based on their customary system of law it is understood that this right is not extinguished by any permitting system.[10]

Despite the fact that several hundred customary communities of small-scale fishers living along the eastern seaboard still have neither recognition of their customary fishing rights nor a statutory small-scale fishing permit, very few of these communities have used this judgement to advocate for the recognition of their own customary systems. This is due largely to the lack of information and awareness amongst both the fishers and their NGO partners about customary rights in a post-colonial context, where customary law is overshadowed by statutory law. However, this situation is slowly changing, as the social movement of small-scale fishers is strengthened and a number of indigenous communities have begun citing this judgement in their advocacy actions aimed at achieving recognition of their customary rights.

14.4 Discussion

With the ILO 169, the UN Covenant on Civil and Political Rights, the UN Declaration on the Rights of Indigenous Peoples (UNDRIP), and most recently the Tenure and SSF Guidelines, the legal status of indigenous peoples in small-scale fisheries has much improved. Domestic legislative reforms at constitutional and sectoral levels have proven to be useful tools to empower indigenous peoples in their long struggle for recognition and justice. Yet, their poverty situation largely prevails, discrimination has not ended, and the pressure on their communities and culture is still on – it may in fact have exacerbated. Thus, at the local level, in the daily lives of indigenous peoples, there are still barriers to achieving equity and meaningful arrangements consistent with the soft and hard law principles. Although the poverty situation differs in the four reported cases in this chapter, they largely share the same struggle to make their rights enshrined in international law operative at the local level.

Indigenous small-scale fishing people continue to fend for their rights in courts, as well as in the governance processes, where rules are determined and implemented. We see that, for instance, in all four cases in this chapter. Thus, the execution of the 'meta-governing' principles (Kooiman 2003), in the international

[10] Gongqose and Others Vs. Minister of Agriculture, Forestry and Fisheries and Others (2016) ZAECMHC 1.

legislation as summarized in Section 2, and in the national legislation presented in the in-country presentations, is far from straightforward. On their way from the international fora where these principles are established, to the local communities where the principles are meant to take effect, they must pass a number of hurdles at 'second' and 'first' orders of governing (Jentoft 2014). The second order would, in this context, involve institutions and government authorities who would have to operationalize these principles into legal and organizational mechanisms. The first order involves the routine governing actions at the ground level, where small-scale fishers are in contact, and often in competition, with other stakeholders for space and resources.

To pass the first hurdle, the consensus reached by FAO's Committee of Fisheries in 2014 about overarching meta-order principles in the SSF Guidelines, must, therefore, be institutionally integrated and contextualized at domestic levels. Institutional reform may be needed, such as new organizations and legislation, as seen in the countries presented in this chapter. The SSF Guidelines recognize the relevance of living customary law, but for them to be part of state law, as we see in the case of South Africa, they must first undergo a process of 'authentication' at a second order (Ørebech 2016, 146). At this point, the new institutional arrangements, rules, and regulations in support of indigenous fishers must be enforced by governing agencies, who would be obliged to change their routines and practices accordingly in their daily management work (first order hurdle). From the moment a new legislation is introduced at second order, until it is actually implemented in terms of concrete management actions, can be a long process, as we have seen in the case of Australia, where it took almost 10 years from legal recognition until new legislative licensing arrangements were in place.

At each order, the set of stakeholders differs. At a second order, the implementation of the SSF Guidelines requires the involvement of government officials and managers, who often "reject the law with profound intensity and vehemence" (Ørebech 2016, 140). Again, we see examples of this in this chapter: in Norway where the Sami Parliament representatives met 'a wall' in the Ministry (Jentoft and Ulvsdatter Søreng 2017), or in South Africa, where officials were extremely reluctant to acknowledge the existence of customary fishing rights and that any such claims would be legitimate. At the first order, management decisions that would favor indigenous fishers particularly, for instance with regard to quota allocation or spatial demarcation, would affect nonindigenous large and small-scale fishers or non-fishery actors. These actors may feel their interests threatened, and may decide to disregard such rules and regulations. In Norway, the CFC insisted that any positive discrimination in favor of the Sami should comprise local communities rather than individuals, given that most communities have a mixed ethnic population. Still, this was not sufficient to temper the most vociferous resistance within the overall Norwegian fisheries sector.

In Nicaragua, for instance, indigenous peoples' tenure rights are in conflict with the interest of peasant colonizers, who are as poor and marginalized as themselves. A lasting solution requires the state to lead the final stage of the demarcation and titling process, which would consist of clearing the titles from third parties, who might claim rights over indigenous collective property. However, a weak governance system makes it challenging to realize the tenure rights and management control of indigenous and Afro-descendant small-scale fishing peoples. Although the formal rights contained in the laws of the country are progressive and advanced, one issue is the overlapping legal mandates that exist between the different authorities, communal-territorial, municipal, regional, and national, which frequently create contradictions hindering the exercise of effective self-governance (Finley-Brooks and Offen 2009). This is particularly noticeable in terms of fisheries and natural resource management, with the negative effect of local authorities not being able to halt unsustainable forms of exploitation and the illegal use and occupation of indigenous communal property.

With the tenure rights of indigenous small-scale fishers, the implementation challenge is extra complex, as it also includes ethnic relations and boundaries, and a long history of colonization, discrimination and abuse. Important as it is, 'hard' (ILO 169) and 'soft' law instruments (UN Declaration of the Rights of Indigenous Peoples, Tenure and SSF Guidelines), help to clarify both the rights and responsibilities of involved parties. (In the case of Australia there was no real influence of these hard and soft law legal instruments – it was based on Australian common law challenge). Yet, such instruments do not automatically reach fisheries governance systems and practices. Established power relations are challenged, but not easily restructured. A history of discrimination and racial prejudice takes more than legal reform to transform. Positive discrimination favoring indigenous peoples usually triggers resistance by those who fear that they may be losing out. The ultimate test is how these legal instruments take effect at the local level, if they will deliver noticeable results for indigenous small-scale fisheries people.

In the four examples in this chapter, reforming governance institutions and actions to make them more sensitive to the needs of indigenous small-scale fisheries has proven to be a slow and cumbersome process. Such transformations challenge the inherent values and established principles of governance systems. They also affect the minds of those who inhabit these institutions, and who over the years have come to take them for granted. We see this perhaps most prominently in the case of South Africa, which in 1994 abolished apartheid, but where the discrimination still lives on in the context of fisheries. For indigenous small-scale fisheries peoples, social justice has much to do with distribution of access to marine resources. But, it is also about other issues, such as the recognition that indigenous fishing can be both customary and commercial. In the word of one commentator: "Until such sectors are acknowledged there is no way to neither allocate quota and funding, nor can roles in fisheries management be clearly defined" (Durette 2007, 23).

Justice is also about restoration that is making up for past injustice, like with the post-apartheid fisheries legislation, which recognizes customary law and community-based systems of tenure. The demarcation and titling of the land of indigenous and Afro-descendant communities in Nicaragua are meant to settle a historical debt. The titling process has allowed indigenous communities to collectively identify and map the areas where they have exercised their customary tenure fishing rights, both in marine coastal areas as well as in inland waters. This represents a fundamental achievement in the long struggle for justice of indigenous and Afro-descendant peoples. Norway was first to ratify ILO 169, but Sami small-scale fisheries are still struggling to secure their tenure rights, despite constitutional amendments. As in Australia, indigenous small-scale fishers have benefited from recent governance reforms, but there is still a way to go before justice is fully restored. Australian Aboriginals in the Northern Territory now have exclusive rights to the intertidal zone, but it remains to be seen how this plays out within the overall fisheries management system and how it will affect existing commercial fishing and licenses. A new permit system will be established to support these new rights in the intertidal zone. For indigenous people, frustration still remains as to which law to follow – customary law or 'white law'.

What these cases demonstrate is that the restoration of indigenous peoples' tenure rights is a work in progress. Conflict resolution pertaining to indigenous tenure rights claims characteristically suffers from 'translation' problems (Ween 2006). Norway is not alone among the four countries represented in this chapter in having a legal profession that does not speak the indigenous language. Lawyers are not trained in customary law, and they often lack the cultural competence needed to identify and conceptualize indigenous peoples' own justice principles. Consequently, the indigenous rights perceptions, and the cultural conceptions that underpin them, are underdeveloped in courts, within the political establishment, and in fisheries management agencies.

Recognizing indigenous fishing communities' customary rights is a necessary step in *de jure* and *de facto* securing their tenure. However, customary indigenous systems of fisheries law have, in many instances, been eroded by the imposition of statutory law and regulations. Restoring such legal systems as part of a policy to secure the governance of indigenous small-scale fisheries tenure rights is therefore a major governability challenge. Customary norms and practices may not always meet the constitutional standards. Whether fisheries governance systems can live with the exceptionalism of customary law is a question without an easy answer (Wicomb 2017). In the Norwegian case, the answer was negative when the government rejected the proposal of a separate law for Sami marine tenure rights, although a compromise was struck with an amendment to established fisheries law.

However, evidence from the Pacific suggests that even without formal legal recognition, customary systems can achieve successful resource management on their own (Vierros et al. 2010). Nevertheless, to prevent further erosion, formal recognition strengthens customary institutions and the role that they can play. In New Zealand, it has been observed that by empowering traditional organizations to participate fully in management and by developing a shared language about and understanding of customary rights, significant gains can be made (Cassidy 2000). Nicaragua provides a contrasting narrative. Here, state law is very supportive of tenure rights of indigenous and ethnic minorities, but local communities remain vulnerable, as the state itself is reluctant to do what it takes to secure them (Gonzalez and Jentoft 2011). When indigenous tenure rights stand in the way of powerful economic interests, as with a new inter-oceanic canal project that was initiated recently, they prove to be less than secure (Gonzalez 2017).

14.5 Conclusion

It would appear from these four countries and from a review of the international literature on approaches to recognition of customary marine tenure of indigenous peoples that there is merit in using legislation to reaffirm constitutionally recognized customary tenure rights. Lack of effective implementation should not be an argument against such recognition. In case of violation of established law, it provides local people with the tool to legally defend their interests, as happened in South Africa in 2005, when small-scale fishers took the government to court. In Norway, the leader of the CFC has hinted that the Sami Parliament should do the same thing. It would also be necessary to register notification of the different responsibilities (statutory and customary) in marine resources management and development and to regulate certain aspects of customary marine resource use and governance rights. Such notification arranges for establishment and identifications of institutions responsible for providing support and advice to customary fisheries and to make provision for the budgetary allocations for such support.

Statutory regulations should not, however, extend to determining the content of customary systems or rules. Rather, it should confirm the devolution of powers and establish and promote the mechanisms that will enhance and provide for dialogue and cooperative governance within a human rights framework. Several documented cases, like those in this chapter, in which indigenous fishing rights have been recognized, provide guidance on the content and process of implementation and what it means to give substance to legal recognition. Most importantly, the experience elsewhere highlights that this may take many years, even decades. Also in Nicaragua, the process of having demarcated customary areas through participatory processes, including indigenous knowledge, and then having records of ownership officially recorded, have been time consuming, as it has proven to be an uphill battle involving conflicts over which territories belong to which communities. But, this should not refrain governments and indigenous communities and organizations from embarking on such a process. It is a process in which the state often remains passive unless there is pressure from the indigenous communities themselves and from civil society organizations.

The synergy between a participatory, community-based resource management approach and an approach to customary governance is clearly evident. In both instances, implementation may well start with the community, working from the bottom up, enabling the community to share its indigenous knowledge of the resources, habitat and ecosystem, and the coproduction of new knowledge with science and for an iterative process of planning to commence from this point. However, statutory recognition of customary marine resource use and governance rights, as legitimate rights arising in law, with associated procedural rights such as the right to free, prior, informed consent, would be instrumental. Without it, there is a danger that in balancing protection of the environment against the indigenous communities' right to fish and harvest resources, the state may fail to adhere to the standards of consultation and participation in decision making arising from their status. This may well result in discrimination and violation of the human rights principles contained, for instance, in the SSF Guidelines, underpinned by those embedded in international and national legislation. The global implementation of the indigenous rights provides an important opportunity to raise awareness to all right holders and stakeholders of the tenure rights of indigenous small-scale fishing communities and to share lessons and insights on best practice.

Acknowledgements Svein Jentoft work on this chapter took place while on sabbatical at FAO from January until June 2018. He wishes to acknowledge the support of FAO and Daniela Kalikoski. Natasha Stacey acknowledges the Aboriginal coastal license holders and Bawinanga Aboriginal Corporation for support in documenting the development of their fishing business.

References

ABS (Australian Bureau of Statistics) (2016) ABS Census 2016. http://www.censusdata.abs.gov.au/census_services/getproduct/census/2016/quickstat/7?opendocument. Accessed 28 Mar 2018

Altman JC (2012) People on country as alternative development. In: Altman JC, Kerins SP (eds) People on country, vital landscapes, indigenous futures. The Federation Press, Sydney, pp 1–25

Anaya J (2004) Indigenous peoples in international law. Oxford University Press, Oxford

Bavinck M, Gupta J (2014) Pluralism in freshwater and marine governance: a challenge for earth system governance, architecture. Curr Opin Environ Sustain 11:75–85

Broderstad EG, Eythórsson E (2014) Resilient communities? Collapse and recovery of a social-ecological system in Arctic Norway. Ecol Soc 19(3). https://doi.org/10.5751/ES-06533-190301

Cassidy M (2000) Fishing for a language. In: International Collective in Support of Fisheries (ICSF) SAMUDRA Report No. 25, p 18–23

Constitution of the Republic of South Africa (1996) Act 108 of 1996. Government Gazette No 17678

Durette M (2007) Indigenous property rights in commercial fisheries: Canada, New Zealand and Australia compared. Australian National University. Centre for Aboriginal Economic Policy Research, Canberra

FAO (United Nations Food and Agriculture Organization) (2012) Voluntary guidelines on the responsible governance of tenure of land, fisheries and forests in the context of national food security. http://www.fao.org/3/a-i4356e.pdf. Accessed 28 Mar 2018

FAO (2015) Voluntary guidelines for supporting sustainable small-scale fisheries in the context of food security and poverty eradication. http://www.fao.org/3/a-i4356e.pdf. Accessed 28 Mar 2018

Finley-Brook M, Offen K (2009) Bounding the commons: land demarcation in northeastern Nicaragua. Bull Lat Am Res 28(3):342–363

Fleming AE (2015) Improving business investment confidence in culture-aligned indigenous economies in remote Australian communities: a business support framework to better inform government programs. IIPJ 6(3). https://doi.org/10.18584/iipj.2015.6.3.5

Frühling P, González M, Buvollen HP (2007) Etnicidad y ación. El Desarrollo de la Autonomía de la Costa Atlántica de Nicaragua 1987–2007. F&G Editores, Guatemala

Gonzalez M (2017) Beyond the small-scale fisheries guidelines: tenure rights and informed consent in indigenous fisheries in Nicaragua. In: Jentoft S, Chuenpagdee R, Barragán-Paladines MJ et al (eds) The small-scale fisheries guidelines: global implementation. Springer Science, Dordrecht, pp 191–212

Gonzalez M (2018) The embrace of Liwa Mairin: lobster diving and sustainable livelihoods in the Nicaraguan Miskitu Coast. In: Salas S, Barragán-Paladines MJ, Chuenpagdee R (eds) Viability and sustainability of small-scale fisheries in Latin America and the Caribbean. Springer, Dordrecht

Gonzalez M, Jentoft S (2011) MPA in labour. Securing the Pearl Cays of Nicaragua. Environ Manag 47:617–629

ILO (International Labour Organization) (1989) C169 – Indigenous and Tribal Peoples Convention, 1989 No 169

INPESCA (Instituto Nacional de Pesca) (2016) Anuario Pesquero y Acuicola 2015. INPESCA, Managua

Jentoft S (2014) Walking the talk: implementing the international voluntary guidelines for small-scale fisheries. MAST 13:16. https://doi.org/10.1186/s40152-014-0016-3

Jentoft S, Ulfsdatter Søreng S (2017) Securing sustainable Sami small-scale fisheries in Norway: implementing the guidelines. In: Jentoft S, Chuenpagdee R, Barragán-Paladines MJ et al (eds) The small-scale fisheries guidelines: global implementation. Springer Science, Dordrecht, pp 267–290

Jentoft S, Minde H, Nilsen R (eds) (2003) Indigenous peoples: resource management and global rights. Eburon Academic Publishers, Delft

Kooiman J (2003) Governing as governance. Sage Publications, London

Lalancette A (2017) Creeping in? Neoliberalism, indigenous realities and tropical rock lobster (kaiar) management in Torres Strait, Australia. Mar Policy 80:47–59. https://doi.org/10.1016/j.marpol.2016.02.020

Larson AM, Soto MF, Mairena D et al (2016) The challenge of 'territory': weaving the social fabric of indigenous communities in Nicaragua's northern Caribbean Autonomous Region. Bull Lat Am Res 35(3):322–337

Minde H (2005) Assimilation of the Sami – implementation and consequences. Journal of Indigenous peoples rights 3:1–34

Nicaragua National Assembly (1987) Statute of autonomy of the autonomous regions of the Atlantic Coast. La Gaceta 238, Managua

Nicaragua National Assembly (2013) Law 836, Managua

Northern Land Council (2017) Northern land council annual report 2016–2017. Northern Land Council, Darwin.

NOU (Norwegian Official Report (2008) NOU 2008:5, Retten til å fiske i havet utenfor Finnmark. Oslo. Departementenes Servicesenter. Informasjonsforvaltning

Ørebech P (2016) Attaining differentiated responsibility: the hard life of governmental comment in fighting ecological ruin and the triumph of customary law & general principles of law. In: Papaux A, Zurbuchen S (eds) Philosophy, law and environmental crisis. Franz Steiner Verlag, Stuttgart pp, pp 133–153

Paine R (1965) Coast Lapp society Vol II. Universitetsforlaget, Oslo

Pedersen S (2012) The coastal Sámi of Norway and their rights to traditional marine livelihood. Arctic Rev Law Polit 3(1):51–80

PNUD (Programa de las Naciones Unidas para el Desarrollo) (2005) Informe de Dessarrollo Humano 2005. Las Regiones Autonomas de la Costa Caribe, Managua, PNUD

Procuraduría General de la República (2018) Titulación de Territorios Indígenas Costa Caribe, Managua: PGR. http://www.pgr.gob.ni/index.php/areas-de-la-pgr/aa/19-proyectos/34-titulacie-territorios-indnas-costa-caribe. Accessed 28 Mar 2018

Productivity Commission (2016) Marine fisheries and aquaculture. Productivity Commission inquiry report, No 81, Canberra. https://www.pc.gov.au/inquiries/completed/fisheries-aquaculture/report/fisheries-aquaculture.pdf. Accessed 28 Mar 2018

Renwick AR, Robinson CJ, Garnett ST et al (2017) Mapping indigenous land management for threatened species conservation: an Australian case-study. PLoS One 12(3):e0173876. https://doi.org/10.1371/journal.pone.0173876

Sanders T, Xuereb S (2016) Optimising the monitoring of tropical aquatic resources through the development of indigenous scientific capability. Rev Fish Biol Fish 26:727–736. https://doi.org/10.1007/s11160-016-9451-0
Schnierer S, Egan H (2016) Composition of the aboriginal harvest of fisheries resources in coastal New South Wales, Australia. Rev Fish Biol Fisheries 26:693–709. https://doi.org/10.1007/s11160-016-9452-z
Smith C (1990) Om samenes rett til naturressurser – særlig om fiskeriressurser. Lov og rett:507–534
Smyth D (2007) Sea countries of the north-west: literature review on indigenous connection to and uses of the north-west marine region. Department of the Environment and Water Resources, 2007 for the National Oceans Office Branch, Marine Division, Department of the Environment, Water, Heritage and the Arts, Canberra
Stacey N, Steenbergen DJ, Clifton J et al (2017) Impacts of marine protected areas on livelihoods and food security the Bajau as an indigenous migratory people in Maritime Southeast Asia. In: Westland L, Charles A, Garcia S, Sanders J (eds) Marine protected areas: Interactions with fisheries livelihoods and food security, FAO fisheries and aquaculture technical paper, No 603. Rome, FAO, pp 113–126 http://www.fao.org/documents/card/en/c/9ea9ba18-3d8d-4ac4-bebb-a809e59ce70f/
Stahler-Sholk R (1997) Structural adjustment and resistance. In: Prevost GE, Vanden H (eds) The undermining of the Sandinista revolution. St. Martin Press, New York
Sunde J (2014) Customary governance and expressions of living customary law at Dwesa-Cwebe: contributions to small-scale fisheries governance in South Africa. Dissertation, University of Cape Town
Sunde J (2017) Expressions of tenure in South Africa in the context of the voluntary guidelines for securing sustainable small-scale fisheries (SSF Guidelines). In: Jentoft S, Chuenpagdee R, Barragan-Paladines MJ et al (eds) The small-scale fisheries guidelines: global implementation. Springer Science, Dordrecht, pp 139–165
Thurow LC (1981) The zero-sum society: distribution and the possibilities for economic change. Basic Books, New York
Tobin B (2014) Indigenous peoples, customary law and human rights: why living law matters. Routledge, New York
UN (United Nations) (2017) International Day of the World's Indigenous Peoples 9 August. http://www.un.org/en/events/indigenousday/. Accessed 28 Mar 2018
URACCAN (Universidad de las Regiones Autónomas de la Costa Caribe Nicaragüense) (2017) Pueblos Originarios y Afro-descendientes de Nicaragua. Etnografia, Ecosistemas Naturales y Areas Protegidas. IBIS, Managua
Vierros M, Tawake A, Hickey F et al (2010) Traditional marine management areas of the Pacific in the context of national and international law and policy. United Nations University – Traditional Knowledge Initiative, Darwin
von Benda-Beckmann F, von Benda-Beckmann K, Griffiths A (eds) (2009) The power of law in a transnational world. Berghan Press, New York/Oxford
Ween G (2006) Sedvaner og sedvanerett: Oversettelsesproblemer i møte mellom rettsvesen, samer og teknologi. Tidsskrift for menneskerettigheter 24(1):1–34
Wicomb W (2017) The exceptionalism and identity of customary law under the constitution. Constitutional Court Review, Juta, Cape Town, pp 127–146

Chapter 15
Defending the Beach: Transdisciplinary Approaches in Small-Scale Fisheries in Pernambuco, Brazil

Matias John Wojciechowski, Beatriz Mesquita P. Ferreira, Daniele A. Vila-Nova, and Sérgio M. Gomes de Mattos

Abstract At the northern estuaries in the state of Pernambuco, Brazil, small-scale fishing activities are intensively carried out, mostly targeting shellfishes. A compelling number of fishers make this activity their main source of income or as a complementary activity to salaried work, being an important supply of protein and food security for their families. Social inequality remains critical in most rural communities, and ongoing lack of visibility and economic perception affect household income. There are considerable conflicts associated with access to beaches, as well as seas and estuaries, mostly through private land or tourism, aquaculture, agriculture, and urbanization. The highly contested and entangled socio-spatial and institutional contours of small-scale fisheries in the study region turn the story of this fisher population into an emblematic struggle for 'defending the beach'. We describe the wicked nature of the challenges faced by small-scale fisheries in the region, and their struggles to gain land ownership and access to the beach and to coastal fishing areas, such as estuaries.

Keywords Pernambuco · Estuary · Fishing territory · Land ownership · Conflicts

---

M. J. Wojciechowski (✉)
World Fisheries Trust – WFT, Victoria, BC, Canada

B. Mesquita P. Ferreira
Fundação Joaquim Nabuco – FUNDAJ, Recife, Brazil
e-mail: beatriz.mesquita@fundaj.gov.br

D. A. Vila-Nova
Independent researcher, Ecology and Conservation, Recife, Brazil

S. M. G. de Mattos
Ministry of Planning, Recife, Brazil

R. Chuenpagdee, S. Jentoft (eds.), *Transdisciplinarity for Small-Scale Fisheries Governance*, MARE Publication Series 21,
https://doi.org/10.1007/978-3-319-94938-3_15

15.1 Introduction

In Brazil, the coastal zone extends for more than 8600 km over an area of approximately 388,000 km^2, accounting for 398 municipalities distributed from the equatorial north to the temperate south of the country. The Brazilian coastal zone is home to 16 of the 28 metropolitan regions, concentrating around 35 million inhabitants (18% of the total population) (CIRM 2010). The population density on the coast, 88 inhabitants per km^2, is five times higher than the national average (CIRM 2010). These territories suffer constant anthropogenic pressures associated with development vectors that, according to Lopes et al. (2015), threaten ecosystem services humans rely on, particularly the local poor population that depend on these resources for sustaining their livelihood. Among these economic vectors, the fishing sector is experiencing territorial conflicts with tourism (Lopes et al. 2015, 2017), oil and gas (Silva and Walter 2017), port development, chemical, livestock, agriculture, urbanization, and aquaculture. All those development vectors account for 70% of the national GDP (Ribeiro and Coura 2003).

Due to these economic dynamics, the spaces of low demographic density of the Brazilian coastline – historically places of settlement of semi-isolated traditional communities – in the last decades have been incorporated into the market economy, which had tourism and vacation activities as the modern main vector occupation. Real estate speculation increasingly contributed to the displacement of traditional populations of fishers, farmers, and extractivists, often also causing the degradation or even destruction of naturally sensitive areas of the coastal region (CIRM 2010). The expansion of resorts, major hotel projects, and second home areas threatens the integrity of coastal and marine environments. Real estate speculation in the coastal zone tends to cause inadequate land occupation, landscape disfigurement, and the destruction of ecosystems, affecting and increasing conflicts with other activities.

According to Brazil's last national fishing census carried out in 2012 (Brasil 2013), the total fisheries and aquaculture production stabilized at around 1.4 million tons per year. More than 60% of the fisheries total comes from the small-scale fisheries. The same census shows that the Northeast Region of Brazil, the poorest region in the country, where this case study is located, has the highest fishery production with more than 250 thousand tons per year. The region also represents 54.7% of the country's small-scale fishers (approximately 570 thousand), and holds the highest concentration of fishers that depend on estuarine-based fisheries for their livelihood (more than 70% of all estuarine-based fisheries).

Small-scale fisheries play an important role in coastal livelihoods, providing social and economic benefits for a number of local communities and a premium source of regional seafood (Gasalla and Gandini 2016). In the state of Pernambuco, small-scale fisheries comprise the entirety of seafood production, accounting for more than 15,000 tons in 2009 (Brasil 2013), and play an important role in the livelihood of the coastal communities along the littoral, especially on the state's northern coastline, where the Atlantic Rainforest known as Zona da Mata Mesoregion

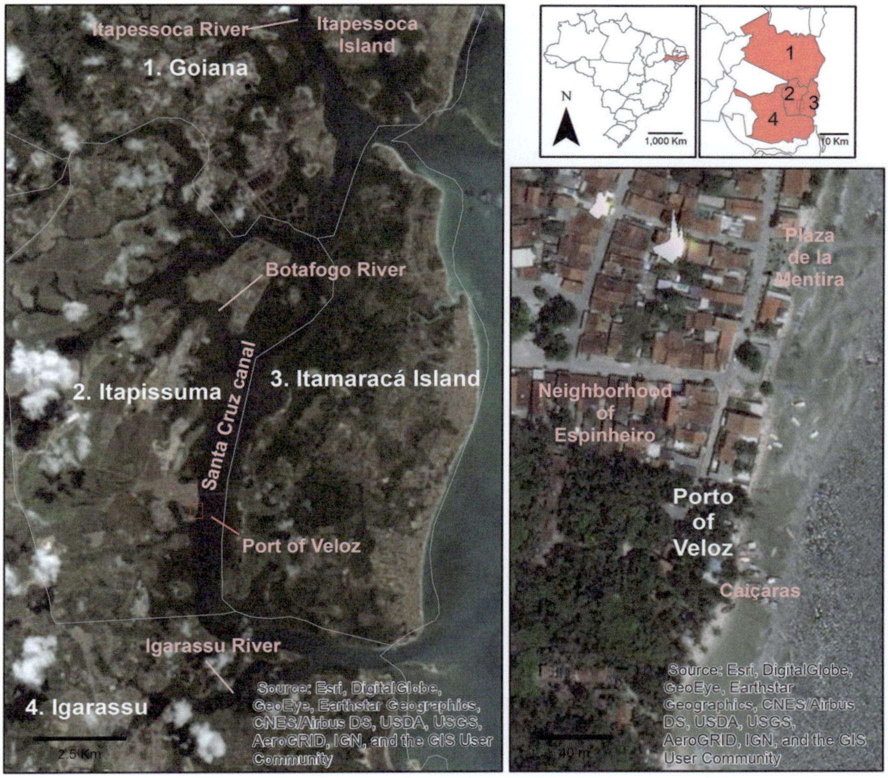

Fig. 15.1 Location of the case study – Santa Cruz Canal Estuarine Complex and the City of Itapissuma showing the four municipalities (*1*. Goiana; *2*. Itapissuma; *3*. Itamaracá Island; and *4*. Igarassu), as well as Porto do Veloz

prevails. The study area is at a major state estuarine complex on the north coast of Pernambuco, encompassing the municipalities of Goiana, Itapissuma, Itamaracá Island, and Igarassu (Fig. 15.1). Mangrove and patches of remaining Atlantic Rainforest cover approximately 30% of these municipalities' area. Small-scale fisheries' predominance in this estuarine region is explained by the fact that this section of the coast of Pernambuco concentrates 61.6% of the total state estuarine areas or 1593.54 km^2 (FIDEM 1987). Therefore, most of the state seafood production comes from Itapissuma and Goiana municipalities (57.3%), where there are several fishing communities (IBAMA 2007).

The historic port called Port of Veloz ('Porto do Veloz' in Portuguese), where we address the transdiciplinary approach in our study, is located in the municipality of Itapissuma, at the Santa Cruz Canal (Fig. 15.1). Due to its proximity to large urban centers, especially the state capital Recife (which is at about 40 km south), the municipality is marked by intense urbanization and population density. In

Fig. 15.2 View of Porto do Veloz from the Santa Cruz Canal. The adjacent land, called Sitio do Canto, can be viewed in the upper region. Itapissuma-PE, Brazil. (Photo credit: Sérgio Mattos)

Fig. 15.3 Small-scale fishing occurs at Santa Cruz Canal, in small wooden boats, mainly targeting shellfishes, at a family basis. (Photo credit: Beatriz Mesquita)

Itapissuma, fishers inhabit such urban areas and meet on the channel benches when on land, whether in the neighbourhood of Espinheiro or in the Praça da Mentira (Fig. 15.1), where they are easily found. Porto do Veloz is an anchoring place for small wooden boats, called *baiteras* and *catraias*, and is a place where fishers keep fishing equipment in small wooden dwellings (Fig. 15.2). Fishing in estuaries and coastal waters or in the inland sea predominates in these fishing communities (Fig. 15.3) (Instituto Oceanográfico 2010).

The highly contested and entangled socio-spatial and institutional contours of small-scale fisheries in the study region turn the story of this estuarine complex fisher population into a representative struggle for 'defending the beach'. This case study shows that the multidimensional and cross-scale predicaments faced by these fishing communities, and the loss of access to their livelihoods, are a function of the historic disconnection between the natural, social, and governing systems. In this regard, this case study attempts to discuss the potential and limits of territorial development as a transdisciplinary approach[1] to target the wicked problems faced by small-scale fisheries.

15.2 Untangling the Wicked Problems of Small-Scale Fisheries

The fisher population of Itapissuma and Goiana municipalities have increased tremendously from the 1970s to the 1990s, in communities originating from the indigenous populations that inhabited the locality (IBGE 2010). In the present day, essentially, the fishing sector in the state of Pernambuco as a whole has found itself at the receiving end of a migration wave caused by a sequence of national development plans and green revolution-driven agrarian tactics. The recurrent 'explosion' of fisher population in the region is embedded in the dehumanizing agribusiness model, which over the years has generated a contingent of farm workers unemployed by the increasingly mechanized agriculture.

Together with this unprecedented growth, came highly speculative and financier-driven urban sprawl, large-scale tourism projects, industrial development, and more recently mangrove-based corporate shrimp aquaculture projects,[2] affecting directly both production and livelihood of small-scale fishers. These anthropic factors, coupled with lack of urban and regional planning and low institutional capacity, at both the state and municipal levels, have contributed to an overall degradation of the estuarine systems (Brasil 2010). Coupled with decades of uncontrolled fishing of these estuarine species by hundreds of people, these factors have been undermining the long term sustainability of fisher livelihoods in the region. Furthermore, the rudimentary processing practices are compromising both working conditions and the quality of the product, resulting in a spiralling race to the bottom in terms of fish size, economic returns, and profit. The result is a serious socioeconomic and environmental situation that needs to be urgently addressed.

---

[1] The transdisciplinary approach referred to herein as well as its categories of analysis including: 'governing systems', 'systems to be governed', and 'governing interactions', as well as their inherent diversity, complexity, dynamics, and scale follow the definition given in Chap. 3 of this book.

[2] In 1999, the establishment of a large-scale 300 ha shrimp farm in the Santa Cruz Canal and 500 ha in between Goiana and Megaó rivers, within the estuarine ecosystem, destroyed several hectares of mangrove swamps in order to occupy them with nurseries.

15.2.1 The Legal Framework of the Small-Scale Fisheries System to Be Governed: Territoriality, Tradition, and Sustainable Use

In this subsection, we attempt to depict the multilayered legal framework that governs the small-scale fisheries systems within the case study. The objective of this section is (1) to relate the specific legal structure of the territory evaluated and (2) to show the challenges faced by fishing communities to access the resources. Despite the current framework's recognition of traditional fisheries' contribution to the region, and the fishing sector in general, challenges are constant, suggesting the presence of territorially-based wicked problems.

Comparing with others, the low economic value of the fisheries means that there are relatively few user conflicts on beaches used by the fisher population. However, there are considerable conflicts associated with access to beaches through private land or near tourist resorts, associated with the social class of fishers, and the concern of fishing in areas with faecal or industrial contamination from adjacent urban areas (Mattos et al. 2017).

Therefore, closer to large urban centres in the state of Pernambuco, interests within the small-scale fisheries sector are more focused on gaining adequate access to the beaches and economic returns than on potential limitations that ensure sustainable use. Nevertheless, local knowledge and rules (Rocha and Lopes 2014; Rodrigues 2015) drive extractive pressure to profitable fishing areas. An analysis provided by Mattos et al. (2017) contextualizes the social system in terms of its level of diversity, complexity, dynamics, and scale, considering the social aspects of an ageing fisher population, gender issues, and family ties that pose many challenges to small-scale fisheries sustainability, including high illiteracy levels, and a daily struggle for survival. These authors indicate that an important starting point for sustainable fisheries management lies in improving the base knowledge required for management, but also appreciating the broader integrated community context, including roles in community development and poverty reduction, and sustaining communities in making a living from fishing, using equitable dialogue, participatory management and innovative, appropriate technology.[3]

In Brazil, fisheries management is generally seasonal and geographically enclosed, with gear restrictions, but legislation does not generally allow for specifically exclusive access rights for either large or small-scale fisheries except for Extractive Reserves (RESEX), defined as a Protected Area in the National Protected Area System (Brasil 2000). The legal designation of a specific territory such as a Marine RESEX defines, specifically, the allowed uses (as per the management plan)

---

[3] From 2008 to 2011 the World Fisheries Trust (WFT), a Canadian-based NGO together with the Ministry of Fisheries and Aquaculture implemented the People of the Tides Project (PoT) – a specific development project to attend to shellfishers in the region. The participatory value-chain mapping exercises carried out within the project among other community development initiatives represent the first attempt by the Ministry of Fisheries to formally recognize the peculiar and unique challenges faced by this fishery.

giving rights to traditional communities to maintain fishing activities and territorial rights. This principle is currently being extended to other co-management approaches such as Fishing Accords (Acordos de Pesca), Terms of Agreement (Termos de Compromisso), and Protected Area Management Councils (Conselhos Gestores) (Araújo et al. 2017) with limited success, especially in highly contested peri-urban settings (Kalikoski et al. 2006; Almeida et al. 2009), such as is the case of Porto do Veloz.

The Brazilian legislation guarantees that the territorial sea, beaches, mangroves, estuarine systems, and rivers with tide influences are "*areas of common use of the Brazilian people*", being, therefore, federal properties, although transitory tenure is possible for traditional communities and others.

Fishing grounds as well as inland areas where fishing community dwellings are established are commonly located at the internal margin of estuarine systems and, therefore, are characterized as marine lands belonging to the Brazilian State, in accordance with the provisions of Article 20 of the Federal Constitution, combined with Decree-Law No. 9760/1946, Title I – Real Property of the Union:

> *Art. 20. The properties of the Union*
> *The marine lands and their added;*
> *II ....*
> *III ....*
> *IV – The rivers, lakes, and islands in the border areas with other countries; The sea beaches; The oceanic and coastal islands, excluded from these, those that contain the headquarters of Municipalities, except those areas affected to the public service and the federal environmental unit, and those referred to in art. 26, II.*

Federal Decree No. 6040/2007 duly characterizes the fishing activity carried out in the study area, under the 'National Policy for the Sustainable Development of Traditional Peoples and Communities.' (PNCPT)[4] The policy defines 'Traditional Peoples and Communities', 'Traditional Territories', and 'Sustainable Development'. The general purpose of this National Policy is:

> *Article 2 – Promote the sustainable development of Traditional Peoples and Communities, with an emphasis on the recognition, strengthening and guarantee of their territorial, social, environmental, economic and cultural rights, with respect and appreciation of their identity, their forms of organization and their institutions.*

The Federal Ordinance No. 89/2010 from the Secretariat of Federal Property grants an "Authorization Term of Sustainable Use – TAUS[5] in favor of traditional communities on a transitional and precarious basis by local authorities of federal marine beaches area and river basins". These areas "*are undoubtedly considered Federal State property, enforced by the federal constitution, and over them any private title is nil*" (§ 1, of Article 2). It is a provisionally granted instrument.

---

[4] Política Nacional de Povos e Comunidades Tradicionais – PNPCT.

[5] Termo de Autorização de Uso Sustentável – TAUS.

In addition to the aforementioned legal instruments, it is important to recognize that the study area also falls under the provisions of the state of Pernambuco Decree No. 32,488/2008. This legislation declares the region (including the Municipalities of Itamaracá, Itapissuma and part of the Municipality of Goiana) as an Environmental Protection Area (EPA), under the name of APA of Santa Cruz. Its Article 2 defines the objectives of the EPA as:

> *Art. 2. …*
> *I – promote sustainable development respecting the environmental support capacity of the ecosystems, enhancing the natural, cultural, artistic, historical and ecotourism vocations of the territory;*
> *II – to protect the estuarine complex of the Santa Cruz Channel and the Itapessoca and Jaguaribe rivers, considered to be of significant environmental importance and great fish potential, in order to conserve its quality, biological diversity and its fishing resources;*
> *III – to protect the rare, endemic, vulnerable and endangered species existing in the 06 (six) Wildlife Refuges and in the 03 (three) Estuarine Areas;*
> *IV – to protect the Beberibe Aquifer, ensuring the conditions of permeability and maintenance of its recharge areas;*
> *V – preserving the local culture of traditional communities with respect to economic activities and local values;*
> *VI – to protect the significant historical, artistic and cultural heritage, reminiscent of the Portuguese colonial period and the Dutch presence in Pernambuco;*
> *VII – to promote the improvement of the quality of life of the region's population.*

In addition, Federal Law 11,959/2009, which deals with the National Policy for the Sustainable Development of Aquaculture and Fisheries, establishes as one of its main objectives the fostering of "*the socioeconomic, cultural and professional development of those who carry out the activity and their communities*" (section IV, Art. 1).

Likewise, State Law 15,590/2015 establishes the Artisanal Fisheries Policy in the state of Pernambuco, in order to "*support the development, promotion and control of small-scale fishing, with the objective of sustainable, socioeconomic, cultural and professional development of those who practice it, their traditional communities, as well as the conservation and recovery of aquatic ecosystems*". It also created the SSF Deliberative State Council (currently in process of being established), one of the few experiences in the country, in a co-management process that includes fishers and civil society organization.

Based on all these principles, the TAUS aims to "*regulate the rational and sustainable use of the natural resources available on the seafront or inland waterway, aimed at the subsistence of traditional populations.*" It is in line with one of the main objectives of Law 11,959/2009, referred to above, regarding "*the sustainable development of fisheries and aquaculture as a source of food, employment, income and leisure, guaranteeing the sustainable use of the fishery resources, as well as the optimization of the resulting economic benefits, in harmony with the preservation and conservation of the environment and biodiversity*"(item I of Article 1).

As can be seen from this complex and multi-level policy framework, a number of challenges arise from such a hierarchical/top-down governance system. In Brazil, fishery policies often fall under a dense and entangled governmental and institutional framework, with structures at the federal, state, and local levels, which together with the existing legal instruments, affect the construction of fishery policies that reflect fishing realities, peculiarities and practices, and at higher level hinders the implementation of the standing fishing policies. Attempts to overcome the prevailing wicked problems found in small-scale fishing communities in this estuarine complex are often faced with limited financial and technical resources and short term political interest. Within this context, it is important to analyze the diversity, complexity, dynamics, and scale of the study area.

Showing all these instruments may give an idea that, besides the shown wicked problems faced by small-scale fisheries, there are others institutional and legal frameworks that worsen the struggle for defending the beach. Although we can assume that a relationship between the constructed and established public policies for small-scale fisheries and the institutional and legal framework exist, it does not suitably sustain fishing communities or provide support to governmental agencies responsible for the implementation of public policies, because communities are usually left behind in decision-making processes. Therefore, the existing management framework that support the involved actors does not generate synergies, thus, weakening the process of implementation of management measures. As a result, the failings of the current system adversely affects the social, economic, technological, and ecological processes characteristic of small-scale fisheries communities in Brazil, with implication on the sectors' sustainability.

15.3 Interactions and Governing System(s) Quality

In this section, we will describe two specific interactions and their respective quality of governing system developed over the last 8 years: (1) the implementation of government 'fishing territories' program in the region; (2) the current struggle to defend access to the beach.

Moreover, these two points on the fishing community timeline are also a testament of resistance and hope that overcomes the theory of neatly aligned governing systems. The messy field of social activism and institutional ruptures created over time, points out to some path-inflections. It is exactly by (re)connecting these events that we attempt to trace a contour line of reflexive governance experiments. It is within this history and geography of unpredictable turning points that the fishing community's rupture with its oppressing path dependence can be evidenced, melding into a path-setting and exciting story of *defending the (local) beach.*

15.3.1 The Northern Coast Fishing Territory Policy and Multi-Stakeholder Council

In the state of Pernambuco, the northern coastal region was granted with a policy called fishing territory designation in 2008.[6] The creation of territory's interinstitutional council in 2009 was coordinated by the Ministry of Fisheries State Superintendence Office and engaged more than 20 stakeholders, including universities and research centers, state development agencies, municipal secretariats, local fishers associations, private sector representatives (mainly aquaculture interest), civil society organizations, and NGOs, among others. The council had a co-management structure and its key role was serving as a regional platform for inter-institutional dialogue, comprehensive planning of the fishing sector and a space for elaborating joint initiatives in favor of small-scale fisheries development. Although the initiative based actions on a multidisciplinary approach to fisheries planning and management, the council's governance faced a number of challenges eventually resulting in its dismemberment.

A study carried out by Joaquim Nabuco Foundation (Silveira and Ferreira 2011) found that after 2 years of implementation the council was not able to pass the creation of the Local Fisheries Development Plan. Besides experiencing difficulties with organizational and operational tasks, the council lacked a clear vision of the problems and strategies needed to strengthen the fishing sector. The main conflict resided between small-scale fisheries representatives and private interests over access to the beach, devastation of the mangroves by large aquaculture projects, and urban pollution. Small-scale fisheries representatives felt abandoned (and in some cases betrayed) by the state's Ministry of Fishery and Aquaculture and its alignment with corporate aquaculture development projects.

The small-scale fisheries and aquaculture conflict was mainly over the presence of large shrimp farms in the region, and their devastating role on the mangroves (key areas for estuarine fisheries)[7] (Fig. 15.4). In addition, lack of implementation of municipal sewage and water treatment programs and historical pollution from the textile industry were seen as the main causes for the deterioration of the estuarine ecosystem. On the other hand, municipalities felt overburdened by their legal obligations due to their limited financial and human resources. Their interests involved

[6] In 2008 the Ministry of Fisheries and Aquaculture identified within Brazil 174 territories designated as 'fishing territories'. The objective of this program was to strengthen the fisheries sector through a territorial development paradigm. The social, economic and institutional stakeholders present in each fishing territory created a development plan to identify and prioritize essential investment to improve fisher's livelihoods. As such the program's pillars included (a) social participation and civic engagement; (b) decentralization of decision-making processes; (c) institutional and organizational capacity building; (d) budgetary efficiency; (e) territorial planning and governance (MPA 2008).

[7] For a more complete discussion on these challenges refer to: Santos et al. (2017) Socioecological assessment for environmental planning in coastal fishery areas: A case study in Brazilian mangroves https://www.sciencedirect.com/science/article/pii/S0964569117300285

Fig. 15.4 Corporate shrimp farms located within Pernambuco northern coast's mangroves

the improvement of fish markets and fish processing plants to improve the fish value chain, however, most of the municipalities in the region did not have the technical expertise, nor the financial means to design, approve, and build these projects with the fishers' engagement generating conflicts of a social and political nature. The role of universities and research centers was mainly perceived as aligned with corporate interests.

Within this highly contested system to be governed, Itapissuma's fisherfolks associations strengthened partnerships with the Fishers Pastoral Council – CPP, local NGOs, some of the more progressive university professors, and civil servants from the state's Fisheries and Aquaculture Superintendence Office. Although the Territorial Council ceased to function in late 2012 (with the Program's end at the national level[8]), the program was fundamental in constructing partnerships that strengthened and, in some cases, reactivated the collaboration of fisher's associations with the CPP and NGOs working in the region.

[8]President Dilma Rouseff and her newly elected Cabinet Ministers negatively evaluated the national program. The general opinion was that the program did not achieve the foreseen socioeconomic goals since its release in 2008. However, a number of academic papers on this topic suggest that the frailty of the territorial program was in large part due to the inability of the local and state institutions to think and act systematically from a territorial perspective, and thus were incapable of integrating the various dimensions of the problem. The program was designed territorial, but it did not fit into the existing institutional structures. At the local level the Council lacked the skill set to coordinate an interinstitutional body. Its inability to define a clear fishery development plan diluted the responsibilities of the council representatives in terms of the implementation of the designed strategies. Furthermore, the council members did not have the skills and institutional capacity to resolve conflicts and negotiate win-win situations.

15.3.2 Fisher's Struggle in Itapissuma to Gain Land Ownership and Access to the Beach

By defending the beach through struggling to gain land ownership and access to fishing territories, many fishers associations' request for granting the Term of Authorization for Sustainable Use – TAUS of Federal property areas within their respective municipalities, issued collectively in favor of each referenced entity. Among them one of the most active is the Fishers Colony Z-10 from Itapissuma, followed by others fishers associations. Some of the contested issues in the "regularization of real right of use of their fishing territories" included: areas historically occupied by fishers for the development of their fishing activities faced privatization and speculation for residential purposes (e.g. condominiums) and tourism (e.g. resorts); construction of fences that prevented fishers from freely circulating in the territory (a right guaranteed by the federal constitution) with their fishing materials and, consequently, impeding the production flow; occupation of coastal and estuarine areas by shrimp farms, nautical structures, etc.; and the need to take the requested issues to others institutional and organizational levels, due to a high level of institutional framework complexity.

In Itapissuma, the TAUS was specifically requested for 'Porto do Veloz' community (Fig. 15.2). In this location, fisher's presence interferes with the owner of the adjacent upper land, and came into conflict with the fishers over a small but crucial enclave of handmade shacks, called *caiçaras*[9] used by fishers to keep their materials and fishing boats, by attempting to prevent the use of the occupied public area (i.e., near the water) by fishers. Previous owners had no conflict with the fishing activities. Nowadays, the current owner charges rent for the use of *caiçaras* (about US $10.00 per month), despite the fact that these are public lands as per the current legislation. When the *caiçaras* structures become obsolete or are in need of renovation the owner takes them apart to expand his coconut tree plantation and tourism business, threatening the entire fishing community. In addition to this onsite conflict, the community struggles with the implementation of a local water and sewage cleaning system by the local government creating malaise among fishers and their visitors/customers.

This is a very important area for the fishers of Itapissuma, harboring about 300 small boats that use rowing or small motors. They are women and men who fish for shrimp, fish and shellfish. Since they live far apart, taking up to 30 min to move to the port, they use the *caiçaras*, not only for material storage, but also for storing unsold product in small freezers. They use shrimp trawl nets, gillnets up to 360 m long, and smaller gear targeting mollusks and other crustaceans. The area has also being used for over 45 years as a boat yard, responsible for manufacturing the vessels used (Fig. 15.5).

---

[9] *Caiçaras* are small fishing warehouse where fishers keep the fishing tackles and boats from theft and against the weather.

Fig. 15.5 Porto do Veloz area, where the fishing community is requesting the TAUS. Upper pictures show where boats are kept during high tide and/or when they are not being used; Bottom pictures show the *caiçaras* with very poor maintenance conditions – older *caiçaras* are being replaced by houses, where the land owner rents them back to the fishermen, or are being replaced by coconut plantations. (Photos credit: Daniele Vila-Nova)

In an attempt to resolve the conflict and provide legal justification for maintaining the access to the beach, the Fishing Colony Z-10, in partnership with the Fishers Pastoral Council, geo-referenced the area. The conducted survey delineated two areas: one with designated traditional usage as per the requested TAUS and a second area with improvements, which would remain within the premises requested by the upper area landowner. The idea was to agree on a division of the conflict area meeting demands from both sides. It is important to mention that the process of geo-referencing the site for granting the TAUS excluded the area delineated by the medium high tide (Limite de Preamar Média – LPM) and the so called 'marine land' (Limite de Terreno de Marinha – LTM).[10] Fishers consider that this 'imaginary' line draws the limits to the area claimed by the landowner, even if it includes the contested 'improved segment'. These lands are regarded by the community as nonnegotiable since they fall within the designation of federal lands.

It is important to note that the territorial conflict illustrated by this case study provides an *in loco* description of issues related to tenure rights of traditional

[10] The 'marine land' is an area of 33 m in width, calculated from the medium high tide measured in 1831.

communities, which are currently at the center of the debate on fisheries sustainability at the international level. In 2012, the Food and Agriculture Organization of the United Nations (FAO) approved the Voluntary Guidelines on Responsible Land of Tenure, Fisheries and Forestry in the Context of National Food Security. The objective is to "*achieving food security and supporting the progressive realization of the right to contribute to the eradication of hunger and poverty and the sustainable use of the environment and to verify the recognition of territorial fisheries rights and, for example, marine protected areas, in order to strengthen the dialogue between fishermen's rights and environmental issues*."

Furthermore, management systems need to be embedded in the community's ecosystem knowledge and need to consider the traditional practices and rules that take place herein. Based on this approach, the management system should be based on principles that seek food security and poverty eradication, supported by the Voluntary Guidelines for Securing Sustainable Small-Scale Fisheries in the Context of Food Security and Poverty Eradication, the SSF Guidelines, also adopted by FAO in 2015. The objective is "*to support responsible fishing and sustainable social and economic development for the benefit of present and future generations, with an emphasis on small-scale fishers, fishworkers and related workers and including vulnerable and marginalized people, promoting a human rights-based approach*."

It is also worth noting that the surveys and studies carried out in the region identify contested fishing territories. In Brazil's context, the historical ineffectiveness of tenure regulation has given way to illegal appropriation and claims to entitlement by speculative real estate developers and other private interests. In this case, the attitude of 'private' interests, denounced by fishers as a threat to the integrity of the community, remains disconnected from the legal realm. These areas are federal lands, enforced by the Constitution. Within this context, any attempt to obtain private entitlement is null. In addition, the cultural diversity of the fishing community is thoroughly shown in the study area, acknowledging the "*way of creating, making, and living*", especially protected by constitutional and legal texts. As such, the indiscriminate advances of tourism enterprises that have dominated land development dynamics in the region represent a real threat to the fishers' livelihood. For this reason, it is the state's duty to act to ensure the sustainable use of the beach area occupied by the fishing community. The lack of institutional capacity to deal with these issues denotes the poor quality of the governing system.

The contested area falls within the designation of used federal area. Historically, the fishers' population use this piece of the beach as an access point to support small-scale fishing activities, to store fishing gear, equipment, and vessels, and to handle and dispose of fishery production, among others. This area is also used for the development of complementary activities, such as planning and management meetings, according to their particular modes of work. Inevitably, the thick layering of challenges can only be addressed through transdisciplinarity. This is a quintessential example of a social conflict, which arises from the interdependence of the social system and the natural system for survival and sustainability. This interdependence is nested within interacting systems that often have conflicting social and economic stakeholders. At the same time, the competent environmental agency and government structures were unable to provide adequate measures to assign fishing areas and to solve the conflicts that have arisen over the disputed beach, aggravating small-scale fisheries' wicked problems and instability.

The organized civil society has been seeking justice, and a greater and better understanding, for territorial ownership rights of traditional communities. In the fishing sector, a number of social organizations have recently launched a Bill of Popular Initiative in defense of the 'Fishing Territory' claiming, among other issues, the "*recognition, protection and guarantee of the right to the territory of traditional fishing communities*", consistent with a historical and anthropological approach. This Bill is an attempt to create tools that allow citizens to present proposals for legitimating norms of life in society (and in relating with the natural system), beyond the capitalist mode of accumulation, for the creation of rights, and for reversal of a situation that causes harm to the community. The Bill is an important step in the defense of a better territorial management system based on traditional fishing communities' knowledge and their balanced livelihoods.

The legal argument of the Bill bases its presumptions on the characterization of fishing communities as traditional or native population, occupants of a traditional territory, corresponding to a transparent process of recognition of self-identification. Amongst its basic principles lie: the sustainable development for the preservation of cultural and historical ways of life, economic activities, and social values; the exploitation of natural resources with traditional low-impact technologies that respect the ecosystem's carrying capacity; and, fostering the protection of this estuarine complex, with a view to preserving its quality, biological diversity, and its fishing resources. Based on the principles of social, economic, environmental, and cultural sustainability in the exploitation of fishing resources, the guarantee of tenure rights will allow fishers to follow management and monitor processes through participatory mechanisms and social control.

The Bill, in defense of fishing territories, assumes that the stories, memories, and fishing livelihoods would be preserved, ensuring that the diversity of peoples and cultures that define the Brazilian society would be maintained. Furthermore, the concept of fishing territories goes beyond the fisheries sector per say. These traditional communities are responsible for guaranteeing the production of healthy food for the Brazilian society and for safeguarding many of the existing ecosystems in the country (Hanazaki et al. 2013; Medeiros et al. 2014; Villasante and Österblom 2015; Gasalla and Castro 2016).

Below we provide a synthesis of the case study through the lens of transdisciplinarity, relating scale and goodness of fit, responsiveness, and the performance assessment of governing system experiences:

Transdisciplinary Experience

- Land tenure rights

Scale and Goodness of Fit

- Institutional and legal framework complexities at national level do not support challenges and diversities at local scale, favouring land grabbling of federal properties;

- Land grabbling historically happened based on economic and political powers at national, regional, and local levels.

Responsiveness

- Access to information based on human rights approach and participatory processes helped by a conjunction of CSO's stakeholders supporting fishing communities took actions struggling for land-ownership rights;
- Need to minimize the legal and institutional bureaucratic system helping small-scale and traditional communities to have access to basic life conditions through recognising territorial rights.

Performance of Governing System

1st Order

- Mapping traditional occupation of federal land properties by small-scale fishing communities;
- Acknowledgement of the social, economic, and cultural system defined in space and time along Brazilian territories;
- Assessment of the institutional and legal framework aiming at responding to local characteristics and way of life.

2nd Order

- Institutional actions are needed at the Federal Government level through Superintendence Office in the State of Pernambuco for mapping small-scale fishing communities' territories;
- Future actions are needed to establish a working group for assessment of the institutional and legal framework to implement territorial rights public policies:

  (a) Operational layer: planning an assessment studies on existing public policies; and
  (b) Strategic layer: involvement of governmental agents and CSO's stakeholders.

3rd Order (Principles)

- Sustainable development for the preservation of cultural and historical ways of life, economic activities, and social values;
- Exploitation of natural resources with traditional low-impact technologies that respect the ecosystem's capacity of support;
- Allowing the protection of this estuarine complex, with a view to preserving its quality, biological diversity, and its fishing resources;
- Bringing together traditional and scientific knowledge on social, economic, and cultural peculiarities of the fishing communities;
- Reinforce management system based on principles that seek food security and poverty eradication;

- Reinforce the rational and sustainable use of the natural resources available on the seafront or inland waterway, aimed at the subsistence of traditional populations;
- Small-scale fisheries communities guaranteed access to the water and land resources necessary to carry out their fishing activities, preserving their livelihoods.

15.4 Final Remarks

The experience described in this case study – of the tenure dispute faced by the fisher population in Itapissuma – portrays the power struggles that are involved in resolving wicked problems that are embedded in oppressive institutional frameworks.

In the case of Itapissuma, the fishers' dispute over the regularization of their fishing territories, where traditions can be found everywhere through an established way of social, political, and natural life, has to deal with traits of neocolonial appropriation of land, often accompanied by the use of brutal violence on behalf of the dominant economic and political players. In this case, observing the social, natural, and governance systems through a transdisciplinary approach shows that '*defending the beach*' is more than just a question to struggle for a neatly nested governing system, but rather a question of messy social activism and its capacity to time the interaction with institutional players according to windows of opportunity that are opened and closed by the ever-shifting institutional landscape and power struggles within the territory.

The transdisciplinary approach challenges, in a certain respect, the dominant narrative that a conservationist or a developmental approach is best fit to resolve the complex challenges of small-scale fisheries. The precipitated and enthusiastic one-model-fits-all narratives are hardly compatible with the social, economic, environmental, institutional, and cultural variants, even when they are focusing on environmental conservation and sustainability. Wicked problems can only be resolved by shining light on the existing power struggles and contested interests over the production and appropriation of fishing territories. Understanding theses power struggles is a first but important step to delineating governance relationships, the governing system, and the system to be governed, its interactions with local fisheries governance, as well as the complexity of the legal and institutional frameworks. This case study shows that in light of the exogenous pressures that mold configuration of the built and natural environment, the point-of-entry for intervention is the territory itself. Fishing territories are not just mere geographically delimitated spaces. They are places where life is born, where knowledge is created and where values are defended based on daily relationships that define its uses and appropriation. We argue that the territoriality can be understood as an intrinsic transdisciplinary concept and method. As such, the territorial struggle portrayed in this case study and the popular quest to guarantee fishing territories can be seen as an important step in defending the beach and the livelihood of traditional fishing communities.

The choice of the 'northern coast' territorial scale allowed us to capture the many facets of the complex relationship between the '*people of the tides*' and the decent access to the beach/sea within a territorial context riddled with exogenous pressures (urban, tourism, speculation, etc.) jeopardizing current natural resources. Moreover, this access depends on the struggle for visibility of the issue and of fisheries as an important sector in regional development. Visibility can thus be understood as the criterion and central objective of intervention in the territory. As such, this case study poses a key question for further reflection: what are the most strategic scales (and scalar linkages) capable of giving visibility to the group of '*tidal workers*' and the '*wicked problems*' that contribute to their invisibility? In this sense, the territorial scale of the north coast becomes the key to '*reading*' the wicked problem, and at the same time a '*gateway*' to transdisciplinarity.

References

Almeida OT, Lorenzen K, McGrath DG (2009) Fishing agreements in the lower Amazon: for gain and restraint. Fish Manag Ecol 16(1):61–67

Araújo LG, Castro F, Freitas RR et al (2017) Struggles for inclusive development in small-scale fisheries in Paraty, Southeastern Coast of Brazil. Ocean Coast Manag 150:24–34

Brasil (2000) Lei n° 9.985, de 18 de Julho de 2000. Regulamenta o art. 225, § 1o, incisos I, II, III e VII da Constituição Federal, institui o Sistema Nacional de Unidades de Conservação da Natureza e dá outras providências

Brasil (2010) Sistema Nacional de Unidades de Conservação – SNUC. Law # 9.985/2000

Brasil (2013) Boletim Estatístico da Pesca e Aquicultura – 2012. Ministério da Pesca e Aquicultura, Brasília, 60 p

Comissão Interministerial Para Os Recursos Do Mar (CIRM) (2010) Plano Nacional de Gerenciamento Costeiro II. Comissão Interministerial de Recursos do Mar Ministério do Meio Ambiente Brasília: 9 p. http://wwwmmagovbr/gestao-territorial/gerenciamento-costeiro/plano-nacional-de-gerenciamento-costeiro#pngc-ii. Accessed 15 Feb 2018

FIDEM (Agência Estadual de Planejamento e Pesquisas de Pernambuco) (1987) Sistema Gestor Metropolitano. Agência Estadual de Planejamento e Pesquisas de Pernambuco. http://www.condepefidem.pe.gov.br/web/condepe-fidem/apresentacao19. Accessed 15 Aug 2017

Gasalla MA, Castro F (2016) Enhancing stewardship in Latin America and Caribbean small-scale fisheries: challenges and opportunities. MAST 15:15. https://doi.org/10.1186/s40152-016-0054-0

Gasalla MA, Gandini FC (2016) The loss of fishing territories in coastal areas: the case of seabob-shrimp small-scale fisheries in São Paulo, Brazil. MAST 15:9. https://doi.org/10.1186/s40152-016-0044-2

Hanazaki N, Berkes F, Seixas CS et al (2013) Livelihood diversity, food security and resilience among the Caiçara of coastal Brazil. Hum Ecol 41:153–164. https://doi.org/10.1007/s10745-012-9553-9

IBAMA (Instituto Brasileiro do Meio Ambiente e dos Recursos Naturais Renováveis) (2007) Estatística da pesca 2005. Brasil Grandes regiões e unidades da federação 147 p. http:// wwwibamagovbr/recursos-pesqueiros/download/25/pdf 2007. Accessed 15 Aug 2017

Instituto Brasileiro de Geografia e Estatística (IBGE) (2010) Censo Estatístico do Brasil. Fundação Instituto Brasileiro de Geografia e Estatística. https://www.ibge.gov.br/. Accessed 15 Feb 2018

Instituto Oceanográfico (2010) Diagnóstico sócio-econômico da pesca artesanal no Estado de Pernambuco. Instituto Oceanário. Governo do Estado de Pernambuco

Kalikoski CD, Almudi T, Seixas CS (2006) O Estado da Arte da Gestão Compartilhada e Gestão Comunitária da Pesca no Brasil. In Revista Jirau, Informativo do Projeto Manejo dos Recursos Naturais da Várzea. Número 15. ProVárzea/Ibama

Lopes PFM, Pacheco S, Clauzet M et al (2015) Fisheries, tourism, and marine protected areas: conflicting or synergistic interactions? Science Direct. doi.org/10.1016/j.ecoser.2014.12.003

Lopes PFM, Mendes L, Fonseca V (2017) Tourism as a driver of conflicts and changes in fisheries value chains in Marine Protected Areas. Science Direct. https://doi.org/10.1016/j.jenvman.2017.05.080

Mattos SMG, Wojciechowski MJ, Macnaughton AE et al (2017) Implementing the small-scale fisheries guidelines: lessons from Brazilian clam fisheries. In: Jentoft S, Chuenpagdee R, Barragán-Paladines MJ et al (eds) The small-scale fisheries guidelines: global implementation, vol 14. Springer, Heidelberg, pp 473–494

Medeiros RP, Serafini TZ, McConney P (2014) Enhancing ecosystem stewardship in small-scale fisheries: prospects for Latin America and the Caribbean. Desenvolvimento e Meio Ambiente 32:181–191

RESEX – Reserva Extrativista (Extractive Reserve); Sistema Nacional de Unidades de Conservação – SNUC: Law # 9.985/2000

Ribeiro MA, Coura MF (2003) A importância da gestão integrada costeira e marinha no controle de impactos socioambientais e seus aportes para o fomento do desenvolvimento sustentável no Brasil. In: Viera PF (ed.), Conservação da diversidade biológica e cultural em zonas costeiras. Enfoques e experiências na América Latina e no Caribe. APED, Florianópolis, pp 265–269

Rocha LM, Lopes PFM (2014) Ecologia humana e mariscagem de Anomalocardia brasiliana no Nordeste brasileiro. In: da Silva GHG, Carolsfeld J, Olivera Gálvez A (eds) GENTE DA MARÉ: Aspectos Ecológicos e Socioeconômicos da Mariscagem no Nordeste Brasileiro. EdUFERSA, Mossoró, pp 157–184

Rodrigues AML (2015) Aspectos da ecologia populacional, salinidade e da pesca do molusco bivalve Anomalocardia brasiliana em região estuarina do semiárido do nordeste brasileiro. Tese de Doutorado. 145f:Il. UFERSA, Mossoró

Silva EP, Walter T (2017) Pesca artesanal e conflitos ambientais: o caso da zona costeira do Rio Grande Do Sul. Revista Caribeña De Ciencias Sociales, V 1:1–10

Silveira PCB, Ferreira BMP (2011) Reservas extrativistas e pesca artesanal: etnografia do campo socioambiental em Pernambuco. Relatório de pesquisa (Research Report), Fundaj/Facepe, Recife

Villasante S, Österblom H (2015) The role of cooperation for improved stewardship of marine social-ecological systems in Latin America. Ecol Soc 20:8. https://doi.org/10.5751/ES-05949-200108

Part VI
Governing the Governance

Chapter 16
Governing Change in Small-Scale Fisheries: Theories and Assumptions

Svein Jentoft

Abstract Small-scale fisheries communities are caught between local necessities and external demands. Globalization brings new challenges, but also opportunities. Climate change is a real threat to future survival, and small-scale fisheries communities are particularly vulnerable to natural hazards. Poverty is an issue, also in the context of climate change, as both are intimately linked. Small-scale fisheries communities are, therefore, undergoing social and ecological change, but change is not the same as progress, and adaptation is not synonymous to development. Change is empirical, something that can be measured and forecasted, whereas progress is ethical and political. With the Voluntary Guidelines for Securing Sustainable Small-Scale Fisheries in the Context of Food Security and Poverty Eradication (SSF Guidelines), we now have an authoritative notion of what progress is supposed to mean in small-scale fisheries globally. However, going from such ideas, via practical interventions, to real development is a stretch. Here, governance matters, but governance can mean several things, is in itself ethical and political, and in need of a constructive process that transcends different disciplines and scales. This chapter discusses what it would take to make change become progress in small-scale fisheries.

Keywords Small-scale fisheries · SSF Guidelines · Poverty eradication · Climate change · Theory of change

> [It] seems a tempting option, not to bother about practical questions of intervention if you are a researcher, and not to worry about theoretical and conceptual hair-splitting if you are a practitioner. However, this is not really an option (Lund 2010, 22).

S. Jentoft (✉)
Norwegian College of Fishery Science, UiT- The Arctic University of Norway, Tromsø, Norway
e-mail: svein.jentoft@uit.no

R. Chuenpagdee, S. Jentoft (eds.), *Transdisciplinarity for Small-Scale Fisheries Governance*, MARE Publication Series 21,
https://doi.org/10.1007/978-3-319-94938-3_16

16.1 Introduction

Any government agency or civil society organizations that aim to be change makers in small-scale fisheries have an idea about what, how, and why change happens, and how they can be part of it. However, this may well be something they have not spent much time reflecting on, as they may be more interested in the practicalities of things. Still, whatever action taken is inevitably a reflection of some perceptions of what change means and what constitutes progress. Without some kind of theoretical road map of the change process, they may be brought to places where they do not want to be. Change makers, be they organizations or individuals, may, therefore, benefit from contemplating their own role, for instance, to what extent they want to assume a lead role or a reactive role, which values they want to promote, which principles they are governed by, and whose interests they want to serve. Kooiman (2003) believes that 'meta-governance', the reflection, deliberation, and decision on such issues, should be an integrated part of governance practice, and not detached from it. The more time and effort spent on meta-governance, the less risk of failure when instigating change at what he calls the 'second' and 'first order governing', i.e. when designing institutions, and when implementing them in the day-to-day governance practice (Kooiman and Jentoft 2009). Thus, meta-governance is essentially about 'governing the governance', which is one of the focuses for the Too Big To Ignore Global Partnership for Small-Scale Fisheries Research (TBTI; toobigtoignore.net).

Small-scale fisheries change makers cannot avoid being implicated in meta-governance. They are likely to be better served by being explicit about how they see themselves in the ongoing change process they are part of, and aim to influence. They should, then, be conscious of how their own ideas of change relates to the theories that already exist. Social scientists have been thinking long and hard about social change, and they may, therefore, have relevant insights to offer. Social science thinking can be abstract, but also draws lessons from observations of how change actually happens, and offers perspectives and tools on what leads to social change and brings progress.

This chapter performs the kind of 'theoretical and conceptual hair-splitting' that Lund (2010) is talking about in the epigram to this chapter. Theories of change consist of a set of concepts, assumptions, and expectations about how change occurs and what it involves. The concepts that build theories also provide us with a language, without which we would not be able to think and speak about change, and in the next instance, act. For thinking deeply about change, and how to steer change in a way that results in something we would associate with progress, we need a sophisticated language. We may even have to invent new concepts, like 'interactive governance' with the 'governing orders' mentioned above (Kooiman 2003; Bavinck et al. 2013).

16.2 Harbingers of Change

Change can be many things, for which we have different words, partly overlapping and at different analytical levels, and which emphasize various forms and dynamics of change. Social systems, like small-scale fisheries systems, are more or less in constant flux, and may transform into a different structure and dynamics over time. Small-scale fisheries are caught in between local needs and external demands. They are undergoing change while they must cope with changes that are happening in their environment. As an interconnected ecological and social system, small-scale fisheries are, from a governance perspective, a moving target, which complicates their governability. Before governors can act on a problem, it may have changed or moved elsewhere.

Climate change and poverty bring existential challenges to fisheries communities. Both affect communities individually, but also simultaneously, as the two are linked (Kalikoski et al. 2018). Not only does climate change undermine the resilience of fisheries communities, who because of poverty are often vulnerable to begin with, poverty also makes communities less effective in coping with climate change related impacts. Living in a poor country also means less support when crisis hits. Poor people are here left to fend for themselves.

The poverty-climate change nexus requires a coordinated effort to address. Poverty eradication is a means of enhancing 'adaptive capacity' to climate change. Building such capacity must involve the mobilization and empowerment of local people, linked with institutions beyond the local community, as communities do not exist in a vacuum. We cannot assume that change at local scales occurs similarly, and follows the same pathways as at larger scales. Still, the same good governance principles, such as transparency, the rule of law, efficiency and effectiveness, ethical behavior, etc., many of which are embedded in the SSF Guidelines, apply at all governance levels.

A theory of change is essential from a governance perspective. Governance is after all a means through which change is directed. Such a theory is deemed to be complex, as it must encompass the conceptual nuances and multiple causes and consequences of change. To be useful for governance, theories must provide hunches as to where to look for change, what exactly to look for, and what to look at. Where change originates, what drives it, and what happens because of change, are all relevant research questions. Crises may, for instance, be a harbinger of change, as they give reasons to challenge the established order. This applies to social conflict, as it requires actors to reconsider and rearrange their positions, relationships, and strategies. Innovations are change by definition, which may trigger social change in ways that leads to progress. With climate change, the need for technical, social, and institutional innovations have become even more imperative (Freduah et al. 2018).

With globalization, the speed and reach of change have increased. Small-scale fisheries are not immune to this influence. The globalization of markets, communication, and culture involves both threats and opportunities, which may in turn trigger change when small-scale fisheries actors respond to them. It is not a given that small-scale fisheries are bound to lose out. It depends on to what extent small-scale fisheries are able to take advantage of the opportunities that globalization provides. Globalization brings new legislation at the international level, driven for instance from within the United Nations (UN) system, which may be helpful to small-scale fisheries. The Voluntary Guidelines for Securing Sustainable Small-Scale Fisheries in the Context of Food Security and Poverty Eradication (SSF Guidelines) (FAO 2015), which the members of the UN Food and Agriculture Organization (FAO) made a priority in 2014 (SSF Guidelines), are such an instrument for change. The Common Fisheries Policy (CFP) of the EU (see Chap. 19 by Arias-Schreiber et al., this volume) is a regional initiative that has yet to prove that they support small-scale fisheries.

Thus, small-scale fisheries are increasingly subject to exogenous change, that is, change generated from the outside, resulting from even higher scales than the nation state. With the Paris Agreement and the UN Sustainable Development Goals (SDGs) (UN 2015), more change at sector level is underway. How these global calls for change will work in small-scale fisheries, whether laudable intentions lead to concrete initiatives at the local level, and to what effect, is a research issue gaining momentum. For TBTI, it is a priority.

The move towards global governance expands the scale of the governing system, which also challenges the transparency of it. For local small-scale fisheries stakeholders, it becomes harder to grasp where change is coming from, and who the change makers are, as they now find themselves at the receiving end of a long chain of decision-making. Exogenous governance also implies the appearance of new stakeholders with ideas and claims in a space previously occupied and controlled from within the small-scale fisheries sector. Sometimes this results in 'ocean grabbing' (Bennett et al. 2015; Bavink et al. 2017). To avoid such an outcome, the tenure rights of small-scale fisheries need legal protection, as emphasized in the SSF Guidelines, which must happen at higher levels than at the local community. Consequently, the institutional premises under which small-scale fisheries stakeholders operate are not only an internal affair. Instead, small-scale fishers and fish workers are relegated into a reactive position. This means less freedom and less self-governance, which have uncertain outcomes, also with regard to poverty eradication and climate change response, given that local fisheries communities must play a proactive role in both. The SSF Guidelines, in this context, display a paradox; as they are calling for states and civil society organizations to intervene, they also emphasize the need for local autonomy and self-determination. How these external actors are involved—in a supportive or in an assertive role- would, therefore, matter. They would need to know the system and the situation into which they step.

16.3 Knowing the System

Small-scale fisheries form complex and dynamic systems of multiple human and nonhuman components, multidimensional relationships and ongoing interactions. Any intervention, small or large, may change the structure and functioning of the system. How it does that, and with which effects, is always associated with a level of uncertainty and risk. It is difficult to predict what a particular intervention will do to the system and to its constituent parts. Understanding what the system is and how it works—for instance, how synergies emerge —would be important. This includes the overview of the components (including the social actors) and relationships that forms the system, as well as the processes that drive it. Those who are inside and hold a position in the system may not always have that overview. Nor do they have the full grasp of how the system may have changed over time. As Marshal McLuhan (1911–1980) said, "I don't know who discovered water, but I'm pretty sure it wasn't a fish." Therefore, the perspective of an outsider may be beneficial. Here is where researchers, with their analytical perspectives, concepts, and methodologies, have a role to play.

For change makers, who have the ambition of system change, the challenge is to identify the buttons to push. They need, in other words, to seek out what Flyvbjerg et al. (2013, 760) call the 'tension points', which they define as "power relations that are particularly susceptible to problematization and thus change." When aiming to change a system, one does not necessarily engage with all system components and relationships at the same time, or at all. You look for the points that can sway other components and relationships, for instance, the bottlenecks. Then, interventions can be selective and targeted. They may also take place sequentially, allowing learning to alter the system structure and its course.

When searching for tension points, one should be open to where in the fisheries governance system these points may be located. One cannot be sure in advance whether tension points are within the system-to-be-governed, the governing system, or the governing interactions. They may well be located in other sectors, like in the education or health sector, or in communication and transport, and/or in the institutions governing these sectors. Tension points may exist locally, in government, or in domestic or international markets. Therefore, in locating the tension points, what constitute these systems and where the boundaries are cannot be determined *ex ante*, but after empirical research has been conducted. System boundaries vary also according to the particular tension point: governance failures may be locally delimited or have larger reach, as the case may be.

16.4 Meta-theories of Change

We all have theories of change, because we live by them. We must relate to the world as it is changing around us. Therefore, we change as well. In order to understand how avenues of change can be redirected, governors must grasp people's own theories of change and determine on what empirical observations such theories are based. Small-scale fisheries are no exception to this rule, also in the context of climate change, given their vulnerability due to their location on the coast.

Academic disciplines have different ideas about change. These ideas have also changed over time. For their own reflections and interventions, small-scale fisheries governors could obviously benefit from them. Their approaches to problem solving are always rooted in some perception of what change is, and why and how it happens, and they would need to think about how their own initiatives are interfering with ongoing change processes. It may also be worth contemplating how theories of change affect actual change. Unlike fish, fishers have theories about themselves, which may even be influenced by science. Social scientists cannot ignore the way fishers, through their own conceptualizations, make sense of their world and the change they experience. Anthony Giddens (1987, 20–21) calls this the 'double hermeneutics' of social science. In this respect, there is a major difference between social science and the single hermeneutics of natural science.

An effort to keep small-scale fisheries in a steady state when everything around them is changing, is hardly possible. Small-scale fisheries have real problems, often stemming from dire working conditions that require change. In order for small-scale fisheries to become more capable of dealing with change, be it globalization, climate related impacts, or the policies imposed on them, small-scale fisheries must themselves undergo change. When TBTI aims to enhance the profile of small-scale fisheries, it is an endeavor to produce the knowledge needed to promote the change that will make them sustainable. This is essentially about the form, direction, pace, and goal of change. The sector must become more robust so that it can withstand change that threatens to undermine it. Small-scale fisheries must also become capable of exploiting existing and new opportunities. TBTI insists that any change that governments or other change makers initiate must be human centered and community focused. The SSF Guidelines talk about the need for institutional reform, like organization building, securing tenure rights, and the need for redistribution of resources from the large- to small-scale sector. Changing power relations, including those of gender, is another major area for change. Most of all, bringing human rights into the equation may require major policy and institutional reform, which the SSF Guidelines want to see.

16.5 Conceptualization of Change

Karl Marx has a strong influence on how social scientists think about change. He famously argued: "Men make their own history, but they do not make it just as they please; they do not make it under circumstances chosen by themselves, but under circumstances directly encountered, given and transmitted from the past." (Marx 1978 [1852]). Inspired by Hegel, Marx saw change not as a linear process or as a repetitive cycle, but as a dialectic, innovative process, that brings society forward. Change, according to this idea, is not simply a series of quantitative shifts that leaves the system intact, but a process where quantitative changes create qualitative, transformative leaps. His focus was on class conflict and asymmetrical power as causes of revolutionary change.

Change can therefore be different things, as can be seen easily by checking for synonyms. Some of these synonyms are at different analytical levels, some are partly overlapping, while some represent qualitatively different forms of change. Change may occur gradually or be abrupt. Change can be marginal or comprehensive, it can be willed, foreseen, or unexpected.

As illustrated in Fig. 16.1, one may think of change concepts forming a gradient, with adaptation on one extreme and revolution on the other, with evolution and transformation in the middle.

As we move along the axis, quantitative change leads to a qualitative shift, involving systemic or holistic change, which is where transformation occurs. Consolidation is change put to rest and equilibrium restored, whereas transformation disturbs the current order in lasting ways. Whatever equilibrium is achieved, it is different from the previous one, and may well contain the inherent impetus for change at a later stage. Thus, the kind of change that we may observe or anticipate, for example, as a result of climate change, may be all these things. In some paragraphs, the SSF Guidelines call for change that is relatively marginal, while in others, like when talking about legislative reform, they suggest major transformation. It is worth noting that climate change discourse, which is also relevant in relation to the SSF Guidelines, tends to operate at the left side of this diagram. The language is about adaptation and adaptive capacity, not so much about social transformation, and hardly ever about revolution. When Naomi Klein (2015), as a clear exception, argues that capitalism is the problem and, therefore, cannot be the solution to climate change, she is talking

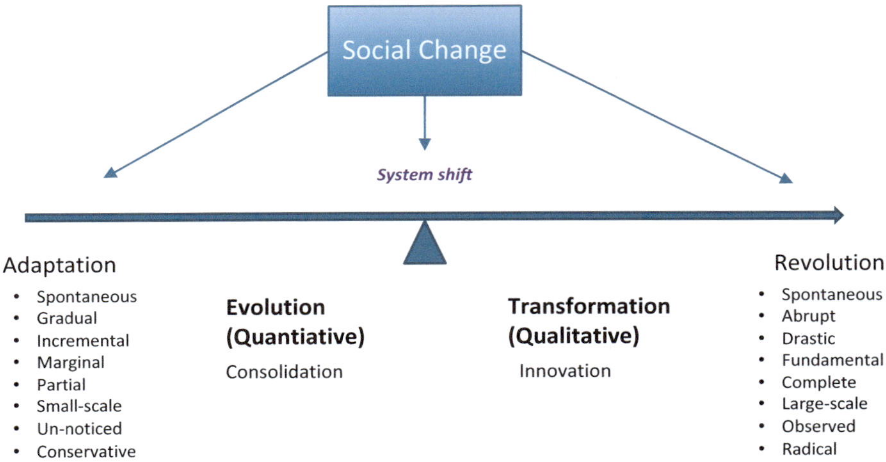

Fig. 16.1 Forms and characteristics of social change

about a system change, and not marginal change. Now is "a time to leap because small steps won't cut it" (Klein 2017, 231). Adaptation is a concept that invites baby steps, whereas revolution involves giant strides. Although the language suggests differently, due to the urgency and seriousness of climate change and poverty, as stressed in the Paris Agreement and the UN SDGs, one would think that the latter would be more appropriate. How small-scale fisheries governance deals with this dissonance is indicative of their governability challenge.

One may wonder why we refrain from using concepts that are more fitting to the challenge, when the urgency and magnitude of the climate change problem suggests that we should be thinking of change on the right side of the diagram. Why 'adaptation', when what is needed is fundamental change? The answer may well be opportunistic and strategic: A softer language is chosen not to startle policy-makers for whom adaptation invokes fewer connotations than, let's say, revolution does. This is also the reason why the SSF Guidelines are voluntary and the language soft. It is up to the states themselves to decide how they will take them on. However, soft language is not the same as neutral language. The adaptation concept has less teeth than revolution, but is no less political. Archbishop Desmond Tutu (2007) views the matter as follows: "Adaptation is becoming a euphemism for social injustice on a global scale." Thus, concepts used have real social consequences because they determine how we perceive the challenge and what actions are needed. Hermann Hesse (2002 [1927], 68) explains this well in his "Steppenwolf" novel:

> Just imagine a garden with hundreds of different trees, thousands of different flowers, hundreds of different fruits and herbs. Now, if the only botanical distinction the gardener knows is that between edible things and weeds, he will not know what to do with nine tenths of his garden. He will uproot the most enchanting flowers, fell the finest trees, or at any rate detest and frown upon them.

We may easily conclude that before this gardener start splitting woods, he should split hair! He is obviously in need of a richer language. Without it, he can neither see nor appreciate what his garden contains, and risks doing irreparable damage to it. In a parallel situation, to eradicate poverty and cope with climate change in small-scale fisheries, a governor is in danger of doing injury, like the gardener. The governor needs a terminology that allows him to recognize small-scale fisheries in all their social and cultural diversity in order to analyze the nexus between poverty and climate change in all its dimensions.

16.6 Resilience Thinking

The way we talk about environmentally related social change now, in adaptive terms, is strongly influenced by so-called 'resilience theory'. Inspired by Gunderson and Holling's 'panarchy' model (2002), this theory has a champion in the Resilience Center at University of Stockholm and 'The Resilience Alliance' (Resilience Alliance 2018). Resilience theory was quite evident in the fourth IPCC report that came out in 2007, from the second working group, which also talked about adaptive social change. The literature has grown enormously since then. "The meteoric rise" in use of the term adaptation (Flood and Schechtman 2014) can be seen in concepts like 'adaptive system', 'adaptive networks', 'adaptive capacity', 'adaptive learning', 'adaptive governance', 'adaptive management', 'adaptive co-management", 'adaptive cycle', and so forth. 'Adaptive' is about to become as popular an adjective as 'sustainable'. Resilience theory is also a good example of how "concepts seep through from the academic, analytical side, to the political, engaged, and operational side" (Lund 2010, 24). This theory has also found its way into FAO's climate-related fisheries discourse (Barange et al. 2018). In this particular context, Sunde (2008) believes that building resilience and creating social transformation are different things.

It would, therefore, be a mistake to believe that conceptual 'hair-splitting' is just of academic interest, with no relevance for, or impact on, the rest of society, including small-scale fisheries. As Kurt Lewin (1999 [1943], 336) noted, "[t]here is nothing as practical as a good theory." Therefore, also in small-scale fisheries governance, theories of change have practical value, as they provide the language needed to identify the problem and see what must be done in order to, among other things, generate resilience or social transformation. There is, however, among social scientists, ambivalence towards resilience theory and the idea of social change as an adaptive, evolutionary process (for example Cote and Nightingale 2012; Flood and Schechtman 2014; Olsson et al. 2015). Regarding climate change adaptation, Mann and Wainswright (2018, 71) state the following:

> None of this is to deny the value of scientific study of nature, the legitimacy of evolutionary theory, or valid uses of the word 'adaptation' in social and political analysis. We are all subjects of ideology. No one can wholly reject one's conceptual inheritance any more than one can wholly refuse the knowledge that it affirms. But grave problems arise when we forget the irrevocably metaphorical quality of all natural and biological concepts that circulate in political life.

In a similar vein, Béné et al. (2014, 616) argue that "if applied uncritically, a resilience-based approach might end up leading us towards abandoning interest in the poor(est) for the sake of strengthening community or even (eco)system-level resilience… it seems that practitioners and development agencies—if they really want resilience to become a true instrument of poverty alleviation—should step back, consider the objectives of their interventions and then assess more carefully and rigorously how resilience can support (or hinder) these objectives."

Not that there have never been social theorists who have been thinking of social change in adaptive - or evolutionary terms, as resilience theory does. Karl Marx and Herbert Spencer (1820–1903) (Stanford Encyclopedia of Philosophy 2002), who both found inspiration from Darwin, did that. For Marx, Darwin provided a basis in natural science for historical class struggle. Spencer is famous for the term 'survival of the fittest', where, like in nature, social evolution rids us of things that are not well adapted. Policy-makers and governors may be tempted to look at small-scale fisheries development in a similar ways. As a natural process, the idea would be that the industrial, large-scale fisheries as a more efficient mode, would unavoidably supplant small-scale fisheries. Any effort to save small-scale fisheries from becoming extinct is, at best, delaying their demise because it would be against the 'the law of nature'. Small-scale fisheries are bound to perish, as an adaptive process tantamount with evolution. The SSF Guidelines (and TBTI), obviously think differently about the fate of small-scale fisheries.

Whereas evolution is a slow process, unnoticeable for those who are part of it; revolutions befall rapidly and loudly. As the most abrupt and drastic form of social change, revolutions start from a more optimistic idea (which may well end with disappointment) of what collective action may accomplish. Revolutionary change is not a quantitative process of aggregation of individual choice outcomes. Neither is it guided by the 'invisible hand' of the market place. Instead, revolutions are a collective enterprise, have a leadership, are 'intelligently designed', run in accordance with a strategy, inspired by an idea of a future state—a 'utopia', as originally coined by Thomas Moore (1478–1535). Revolutions are perhaps rare events, but they do happen from time to time. If, for political reasons, we cannot mention revolution for the actions that are needed in connection with poverty and climate change, we may instead talk about the components of what this concept is made up of, as listed in the figure. If not revolution, we should at least be thinking about social transformation, or 'transformative change'.

16.7 Change as Progress

Bertrand Russell, the philosopher, stated in his Unpopular Essays (1996 [1901]): "Change' is scientific, 'progress' is ethical; Change is indubitable whereas progress is a matter of controversy." In other words, change is something we can study, describe, measure, and predict. One may disagree about the findings, but, then, the remedy would be more research, higher quality data, and better models. Progress, on the other hand, is a concept of change that belongs to a different discursive realm. Progress is normative. It refers to our social values, and is, therefore, political. What is progress for some is not necessarily progress for others, and we would, therefore, not always agree on what it means in concrete situations, like in small-scale fisheries.

In the climate change and poverty eradication discourse, Garrett Hardin's 'tragedy of the commons' (1968) often shows up. This is his most memorable and cited quote:

> Each man is locked into a system that compels him to increase his herd without limit – in a world that is limited. Ruin is the destination toward which all men rush, each pursuing his own best interests in a society that believes in the freedom of the commons. Freedoms in a commons brings ruin to all. (1968, 246)

Climate change is a result of the atmosphere being a common resource that is bound to deteriorate because of the freedom of individuals to pollute. Thus, to address the problem, collective action is needed. We need some external authority, like government, to limit that freedom. However, for some peculiar reason, Hardin's main argument (1968, 1243) seems to be overlooked:

> An implicit and almost universal assumption of discussions published in professional and semipopular scientific journals is that the problem under discussion has a technical solution. A technical solution may be defined as one that requires a change only in the techniques of the natural sciences, demanding little or nothing in the way of change in human values or ideas of morality.

Some problems, like poverty and climate change, do not have a technical fix because they are ultimately social and ethical. Mann and Wainwight (2018, 7) argue: "The political problems we face cannot be fixed by simply delivering sciences to the masses. If good climate data and models were all that were needed to address climate change, we would have seen a political response in the 1980s." Political problems require moral judgement because, as Hardin argues, they challenge values and worldviews. Therefore, the decisions we need to make on such issues require a different process than that of science, one of participation and deliberation in which we all engage. Before we can assess the problem, and subsequently act on it, we need to find a way to agree on how to define it. The problem does not speak for itself; we do. Hence, the formulation of the problem is the problem, as Rittel and Webber (1973, 161) noted in another much cited paper that also talks about poverty. They term such problems as 'wicked'. What makes problems like poverty and climate change 'wicked' is not only that they are difficult to define, but that they are also often part of bigger problems. Neither can we be sure that we have solved them once an for all. People may advance above the poverty line, but they also risk falling back under it. Wicked problems require continued vigilance.

Like Hardin, Rittel and Webber assumed a problem-solving process distinctly different from a scientific one:

> Approaches … should be based on a model of planning as an argumentative process in the course of which an image of the problem and of the solution emerges gradually among the participants, as a product of incessant judgment, subjected to critical argument. (1973,162)

This is a standpoint that 'interactive governance' (Kooiman et al. 2005) would share in a fisheries setting. As Chuenpagdee and Jentoft (2018) demonstrate, in small-scale fisheries this would involve transformative change in governance modes, away from hierarchical, state-driven approaches towards a cooperative mode.

A notable thing about the SSF Guidelines is that they were indeed developed through a participatory process, where a large number of stakeholders around the world were consulted. In the final stage there was a Technical Consultation (in FAO lingo, in which I took part), but rather than being technical, the consultation was largely a deliberation about values, norms, and governance principles involving both state delegates and civil society representatives. By making human rights a basic principle, the SSF Guidelines make a moral statement. Also, when COFI (the Committee of Fisheries in FAO) finally endorsed the SSF Guidelines in June 2014, they did so on the basis of a reached consensus about what constitutes 'progress' in this sector. This, I argue, is precisely what makes SSF Guidelines so powerful.

The SSF Guidelines mention 'change' 25 times, and 'adaptation' 12, but 'transformation' and 'revolution' zero times. Still, it is a rather progressive document, which if implemented, will bring systemic change as far as small-scale fisheries poverty and food security are concerned. It will also help to make small-scale fishers better equipped to cope with climate change. This may well amount to something that could rightfully be defined as 'revolution.'

16.8 Comparative Research

In the book *Angels Fear*, Gregory Bateson (1988) notes that we learn when we observe a difference that, in one way or another, makes a difference to us. For instance, to say that small-scale fisheries in the north are different from small-scale fisheries in countries in the south, is to state the obvious. There is also a wealth of academic literature that tells us how they differ. The important question to ask from a poverty eradication and climate change perspective, and also for a project like TBTI that has a global, comparative focus, is what difference these differences make. Here, the answer is not at all clear.

Many things are similar in small-scale fisheries, regardless of where they are. The concerns are largely the same: the natural environment and ecosystems must be healthy and livelihoods secure. People care for their children and want to live in dignity. Human rights are universal. This is why the SSF Guidelines hold them as basic governance principles. Still, the Guidelines stress the importance of implementing them in a way that is sensitive to cultural context. There is always something unique about a particular place and fishery. The implementation should not work from an assumption that the tools employed will work in the same way, regardless of context. Governors should not let predefined problems and solutions guide their actions, as they may easily end up with the wrong tension points, and, thus, do more harm than good (Jentoft and Eide 2011; Ratner and Allison 2012). As a rule, their theories of change must be put to empirical test, as there is always something to be learned from a new situation. The same applies to the concepts we use; we should not assume that the same word has the same meaning everywhere. Again, Archbishop Desmond Tutu (2007) makes an important point:

> Perhaps the starting point is to reflect on the inadequacy of language. The word 'adaptation' has become part of the standard climate change vocabulary. But what does adaptation mean? The answer to that question is different things in different places.

Change makers must know that laws of nature and those of society are fundamentally different, and that this is a reason why change evolves differently in those two realms. I may drop the pen I hold in my hand, and it will fall to the floor wherever I am in the world, and it does so every time. Physics can explain why this is the case. While the laws of nature are universal, the laws of society are human constructs in a way that is appropriate to context. These 'laws' may differ from country to country, sector to sector, and from community to community. Such 'legal pluralism' is one of the things that make fisheries communities unique, and a reason why some communities prosper while others do not. Therefore, unlike the natural sciences, the social sciences do not deal in universals (Flyvbjerg 2001). Social scientists do not assume, for instance, that a new rule, a particular management tool, or a technical gadget, will perform equally as well everywhere. Social scientists are trained to be skeptical of technical fixes. Until they have investigated the matter empirically, all they have are clues. For this reason, Ostrom et al. (2007) warned against technical fixes or panaceas.

16.9 Change as a Transdisciplinary Project

Academic disciplines provide partial perspectives; what Johan Asplund (1970) called 'aspect vision'. Disciplines also tend to single out one concern, be it conservation (biology), economic efficiency (economics), or community welfare (sociology). Yet, as any fisheries manager would know from experience, fisheries management is about all of these concerns, and more. If they only focus on one concern and are blind to others, they are deemed to fail. To avoid such an outcome in small-scale fisheries, they need to understand the diversity, complexity, and dynamics of this sector, and that the problems they are trying to solve, be it poverty or climate change, are inherently wicked. For this, they need a richer vocabulary than one discipline can offer. They must be prepared for the possibility that a concept may carry different meanings in different contexts, as Desmond Tutu reminds us with regard to 'adaptation'. Onyango and Jentoft (2010) have made the same argument with regard to poverty. Concepts, therefore, need translation before they can be put into action, and that requires empirical work. People's own conceptualizations must be explored (cf. the 'double hermeneutics' of Giddens mentioned above).

Hence, scientists must not only engage with people in other disciplines, but also with people outside academia, in an effort of transdisciplinarity (Wainwright 2010; see Chap. 21 by Msomphora and Jentoft, this volume; Chap. 22 by Said et al., this volume). Albert Einstein (2009 [1949]) argued: "We should be on our guard not to overestimate science and scientific methods when it is a question of human problems; and we should not assume that experts are the only ones who have a right to express themselves on questions affecting the organization of society." Knowledge, including the practical, experience-based knowledge that stakeholders have, must be sought wherever it can be found. This is especially important when confronted with 'wicked problems', which are too complex and too intertwined with other issues for a single discipline or a single actor to handle alone (Kooiman 2003).

There is, however, something to be said for the ‘etic’—or the outsider view (Harris 1976)—perspective of academics, if we can concur with Marshall McLuhan cited earlier. Theory provides the observer with the mental distance that is needed to get an idea of the system as a whole. Poverty is a social issue, with structural causes and implications, also with regard to climate change (Jentoft and Eide 2011). It, therefore, benefits from the scientific perspective. But scientific perspectives must be calibrated with the empirical reality of those who live it. Poverty is a lived, personal experience. What it really means to be poor in times of climate change requires an ‘emic’ perspective (Harris 1976)—the insider’s view—which only the poor and vulnerable themselves have. Therefore, the transdisciplinary process must not only be pro-poor, but also inclusive of the poor. The process must be interactive and communicative, because it requires both the ‘emic’ and the ‘etic’ lens. The etic perceptive of the scientist does not have to be the starting point. Theory can explain empirical facts, like poverty, but so can experience, as David Hume argued (cf. Hume’s ‘fork’) (DeMichele 2016). Thus, rather than theory driving empirical research, empirical facts inspire theory development. Glaser and Strauss’s (1967) concept of ‘grounded theory’ captures this latter approach.

The SSF Guidelines advocate for a ‘holistic’ approach, drawing on both scientific and local knowledge. They also call on a broad range of actors, not just fisheries authorities but also other government agencies, civil society, and academia, in an effort that may be called ‘transdisciplinary’. In small-scale fisheries, we cannot do without the natural sciences and their knowledge about issues that are universal, like ecosystem dynamics. But we also need the type of knowledge that Aristotle called ‘*phronesis*’, which is sometimes translated as prudence or experienced-based knowledge (cf. Jentoft 2006). This is the deep understanding of the difference that context and experience make about what it means to be ethical. Without it, we are less capable of creating change that small-scale fishing people would associate with ‘progress’.

Climate change makes poverty an even more pressing challenge, both because it may aggravate poverty, and by doing so, reduces the capacity of the poor and vulnerable to cope with it. Change may therefore occur as a vicious cycle. Progress can only occur if this cycle is broken. Then, climate change adaptation and poverty eradication must go hand in hand. As a human rights issue, poverty eradication is important in itself. Poverty eradication is also a means to make poor people in small-scale fisheries less vulnerable to climate change. In both instances, it requires a sound ethical underpinning, as the solution is not a technical one.

Acknowledgements This chapter was written while on sabbatical leave with FAO, Rome, from January to July 2018. I am grateful for the support and constructive comments by Daniela Kalikoski, Ratana Chuenpagdee, and two anonymous reviewers.

References

Asplund J (1970) Om undran inför samhället. Argos, Lund

Bateson G, Bateson MC (1988) Angels fear: towards an epistemology of the sacred. The University of Chicago Press, Chicago

Barange M, Bahri T, Beveridge M, Cochrane K, Funge-Smith S, Poulain F (eds) (2018) Impacts of Climate Change on fisheries and aquaculture: Synthesis of current knowledge, adaptation and mitigation options. FAO Fisheries Technical Paper 627

Bavinck M, Chuenpagdee R, Jentoft S et al (eds) (2013) Governability in fisheries and aquaculture: theory and applications. Springer Science, Dordrecht

Bavink M, Berkes F, Charles A et al (2017) The impact of coastal grabbing on community conservation – a global reconnaissance. MAST. https://doi.org/10.1186/s40152-017-0062-8

Béné C, Neshhan A, Davies M et al (2014) Resilience, poverty and development. J Int Dev. https://doi.org/10.1002/jid.2992

Bennett NJ, Gowan H, Satterfield T (2015) Ocean grabbing. Mar Policy 57:61–68

Chuenpagdee R, Jentoft S (2018) Transforming the governance of small-scale fisheries. Mar Stud. https://doi.org/10.1007/s40152-018-0087-7

Cote M, Nightingale AJ (2012) Resilience thinking meets social theory: Situating social change in socioecological systems (SES) research. Prog Hum Geogr 36(4):475–489

DeMichele T (2016) Hume's fork explained. http://factmythcom/humes-fork-explained/. Accessed 26 Mar 2018

Einstein A (2009) Why socialism? Mon Rev: Indep Soc Mag. 61(1). https://monthlyreview.org/2009/05/01/why-socialism/. Accessed 26 Mar 2018

FAO (United Nations Food and Agriculture Organization) (2015) Voluntary guidelines for securing sustainable small-scale fisheries in the context of food security and poverty eradication. http://www.fao.org/3/a-i4356e.pdf. Accessed 26 Mar 2018

Flood S, Schechtman J (2014) The rise of resilience: evolution of a new concept in coastal planning in Ireland and the US. Ocean Coast Manag 12(A):19–31. https://doi.org/10.1016/j.ocecoaman.2014.08.015

Flyvbjerg B (2001) Making social science matter: why social inquiry fails and how it can succeed again. Cambridge University Press, Cambridge

Flyvbjerg B, Landman T, Schram S (2013) Tension points in real social science: a response. Br J Sociol 64(4):758–762. https://doi.org/10.1111/1468-4446.12047_4

Freduah G, Fidelman P, Smith TF (2018) Mobilising adaptive capacity to multiple stressors: insights from small-scale coastal fisheries in the Western Region of Ghana. Geoforum 91:61–72

Giddens A (1987) Social theory and modern sociology. Polity Press, Cambridge

Glaser BG, Strauss AL (1967) The discovery of grounded theory: strategies for qualitative research. Aldine, Chicago

Gunderson LH, Holling CS (eds) (2002) Panarchy: understanding transformations in human and natural systems. Island Press, Washington, DC

Hardin G (1968) The tragedy of the commons. Science 162(3859):1243–1248

Harris M (1976) History and significance of the emic/etic distinction. Annu Rev Anthropol 5:329–350

Hesse H (2002) Steppenwolf. A novel. Picador USA, New York

Jentoft S (2006) Beyond fisheries management: the phronetic dimension. Mar Policy 30:671–680

Jentoft S, Eide A (eds) (2011) Poverty mosaics: realities and prospects in small-scale fisheries. Springer, Dordrecht

Klein N (2015) This changes everything: capitalism vs. the climate. Simon & Schuster, New York

Klein N (2017) No is not enough: defeating the new shock politics. Allen Lane, London

Kooiman J (2003) Governing as governance. SAGE, London

Kooiman J, Jentoft S (2009) Meta-governance: values, norms and principles, and the making of hard choices. Public Adm 87(4):818–836

Kooiman J, Bavinck M, Jentoft S et al (eds) (2005) Fish for life – interactive governance for fisheries. Amsterdam University Press, Amsterdam
Kritsonis A (2004–2005) Comparison of change theories. Int J Sch Acad Intellect Divers 8(1):1–7
Lewin K (1999) Psychology and the process of group living. In: Gold M (ed) The complete social scientist: a Kurt Lewin reader. American Psychological Association, Washington, DC, pp 330–345
Lund C (2010) Approaching development: an opinionated review. Prog Dev Stud 10(1):19–34
Mann G, Wainwright J (2018) Climate Leviathan: a political theory of our planetary future. Verso, London
Marx, K (1978 [1852]) The eighteenth Brumaire of Louis Bonaparte. 1st edn. Foreign Language Press, Peking. http://www.marx2mao.com/M&E/EBLB52.html. Accessed 26 Mar 2018
Olsson L, Jerneck A, Thoren H et al (2015) Why resilience is unappealing to social science: theoretical and empirical investigations of the scientific use of resilience. Sci Adv 1(4):e1400217. https://doi.org/10.1126/sciadv.1400217
Onyango P, Jentoft S (2010) Assessing poverty in small-scale fisheries in Lake Victoria, Tanzania. Fish Fish 11:250–263
Ostrom E, Janssen MA, Anderies JM (2007) Going beyond panacea. PNAS 104(39):15176–15178. https://doi.org/10.1073/pnas.0701886104
Ratner BD, Allison EH (2012) Wealth, rights, and resilience: an agenda for governance reform in small-scale fisheries. Dev Policy Rev 30(4):371–398
Resilience Alliance (2018) Home page. https://www.resalliance.org/. Accessed 26 Mar 2018
Rittel HWJ, Webber MM (1973) Dilemmas in a general theory of planning. Policy Sci 4:155–169
Russell B (1996) Unpopular essays. Routledge, London
Stanford Encyclopedia of Philosophy (2002) Herbert Spencer. https://plato.stanford.edu/entries/spencer/#Bib. Accessed 26 Mar 2018
Sunde J (2008) Building resilience of transformation? SAMUDRA Rep 51:20–24
Tutu D (2007) We do not need climate change apartheid in adaptation. United Nation Development Programme – Human Developments Reports. http://hdr.undp.org/en/content/we-do-not-need-climate-change-apartheid-adaptation. Accessed 26 Mar 2018
UN (United Nations) (2015) Sustainable Development Goals: 17 goals to transform our world. http://www.un.org/sustainabledevelopment/sustainable-development-goals/. Accessed 26 Mar 2018
Wainwright J (2010) Climate change, capitalism, and the challenge of transdisciplinarity. Ann Am Assoc Geogr 100(4):983–991. https://doi.org/10.1080/00045608.2010.502439

Chapter 17
Transdisciplinary Engagement to Address Transboundary Challenges for Small-Scale Fishers

Joeri Scholtens, Andrew M. Song, Johny Stephen, Catalina García Chavez, Maarten Bavinck, and Merle Sowman

Abstract Small-scale fisheries and their governance are increasingly affected by natural, social, and political issues that originate outside their immediate control and locality. This chapter explores how researchers, non-governmental organizations (NGOs), and fisher organizations can collaborate in the pursuit of empowerment small-scale fisheries vis-à-vis such 'external' dynamics, with a focus on maritime boundaries. To do so, we first analyse how transboundary maritime issues complicate the operation and welfare of small-scale fishers and may further their marginalization. Second, we explore how transdisciplinary engagement can be key to better understanding and addressing such transboundary challenges. Taking an action-oriented approach, we analyse the opportunities and pitfalls of transdisciplinary collaboration to empower small-scale fishers through five types of

J. Scholtens (✉) · M. Bavinck
Amsterdam Institute for Social Science Research, University of Amsterdam, Amsterdam, The Netherlands
e-mail: j.scholtens@uva.nl; j.m.bavinck@uva.nl

A. M. Song
Australian Research Council Centre of Excellence for Coral Reef Studies, James Cook University, Townsville, Australia

WorldFish, Honiara, Solomon Islands
e-mail: andrew.song@jcu.edu.au

J. Stephen
Tata Institute of Social Science, Hyderabad, India

C. G. Chavez
Erigaie Foundation, Amsterdam Institute for Social Science Research, University of Amsterdam, Amsterdam, The Netherlands
e-mail: m.c.garciachaves@uva.nl

M. Sowman
Department of Environmental and Geographical Science, University of Cape Town, Cape Town, South Africa
e-mail: merle.sowman@uct.ac.za

R. Chuenpagdee, S. Jentoft (eds.), *Transdisciplinarity for Small-Scale Fisheries Governance*, MARE Publication Series 21,
https://doi.org/10.1007/978-3-319-94938-3_17

intervention strategies: *capacity building* (strengthening fisher organizations), *institution building* (building bridges between disparate actors), *discourse* (reframing the nature of the problem), *law* (appealing to national or international courts), and *mobilization* (mobilizing fishers to confront power). Each type of intervention is illustrated with a case study from various parts of the world. We argue that despite potentially conflicting incentives, interests, and accountabilities, transdisciplinary engagement can be both a meaningful and effective practice to empower small-scale fishers vis-à-vis transboundary challenges.

Keywords Transboundary · Transdisciplinary · Maritime boundaries · Empowerment · Interventions · Small-scale fishers

17.1 Introduction

Small-scale fisheries and their governance are increasingly connected to natural, social, and political trends that originate outside their immediate control and locality. To better understand small-scale fishers' influence on such 'external' dynamics, or the lack thereof, we must explore cross-sector, cross-scale, and cross-boundary linkages. In this chapter we take on this challenge, with 'boundaries' as the focal point of our discussion.

The seas are replete with boundaries, mostly indiscernible in the physical sense but drawn on paper in the course of recent decades with aims to improve maritime governance (e.g. Exclusive Economic Zones, Large Marine Ecosystems, marine spatial planning, and Marine Protected Areas - MPAs). Song et al. (2017) have highlighted the paradoxical implications of boundary proliferation at sea: because marine ecosystems and fishers, due to their fluid and mobile nature, do not mould themselves easily to physical borders, more boundaries in the water have actually exposed more *transboundary* fishing practices. Drawing boundaries thus inevitably leads to more transboundary movement. Concomitantly, this phenomenon has also given rise to a vast body of transboundary fisheries research, as well as to efforts of state- and non-state actors to govern such transboundary movements (see Song et al. 2017 for details).

The purpose of the chapter is twofold. First, we aim to understand how transboundary maritime issues complicate the operation and welfare of small-scale fishers and ask whether and how transboundary dynamics may further the marginalization of small-scale fishers across the world. For example, there are disputes over international maritime boundaries or with transboundary fishing fleets, which can work to restrict fishing operations. Small-scale fisheries may also be negatively impacted by newly established MPA boundaries, or—more subtly—by emergent international conventions, guidelines, discourses, or regulations.

Second, this chapter aims to explore the relevance of transdisciplinary collaboration between academics and non-academics for understanding and addressing transboundary issues. Such collaboration can occur in various stages of knowledge-action production, including agenda-setting, knowledge coproduction, and collaborative interventions. In this chapter, we focus on the latter and take an action-orientated approach, focusing on collaborations between researchers, NGOs, practitioners, and fisher organizations in an effort to empower small-scale fisheries through various intervention strategies. In doing so, we hope to provide a reflection on the possibilities for and the necessity of such transdisciplinary engagement in addressing some urgent transboundary fisheries problems. Our argument is also that transdisciplinary collaboration can provide useful insights for improving transboundary interactions of small-scale fisheries. Discipline is a type of boundary after all. As knowledge (and experience) domains that pertain to small-scale fisheries are in reality fluid and plural, involving multiple disciplinary perspectives and bridging between academics and non-academics, using principles such as cooperation, multiscalar linkages, empathy, reflexivity, and attention to power (Lang et al. 2012; Polk 2015), constitutes a significant transboundary endeavour.

This chapter is structured as follows: we commence with an overview of transboundary fisheries issues, and identify a range of challenges that arise for small-scale fisheries. We follow with a discussion about the potential of transdisciplinary work to empower small-scale fishers. We subsequently synergize these two discussions to highlight five types of transdisciplinary interventions that can help address transboundary challenges encountered by small-scale fishers, and illustrate those with brief case examples.

17.2 Transboundary Challenges for Small-Scale Fishers

Transboundary issues abound in contemporary fisheries. Even if we limit our definition of 'boundary' to the spatial domain (that is, lines on a map depicting division of water surfaces and columns for various governing purposes), we observe many kinds of boundaries in oceans and inland waters alike. For example, boundaries are used to delineate marine protected areas, exclusive economic zones, and large marine ecosystems. Subsequently, we also identify many inadvertent and intentional crossings of these boundaries by numerous social, biological, and oceanographic components such as fishers, boats, currents, and fish themselves. 'Transboundary' has, therefore, become a salient topic these days for research and practice (Song et al. 2017). We define a transboundary perspective to be an approach that examines the effects of boundary-setting and provides ways to reconcile or transcend the limitations of static and rigid spatial demarcation for fisheries management (Song et al. 2017).

Transboundary events in fisheries can intersect with the modus operandi of small-scale fisheries in, at least, four ways. First, because target resources move across boundaries, such boundaries may create intensified competition and conflicts between fisher groups positioned on alternate sides. Determining optimal and cooperative harvesting strategies, as well as conservation measures becomes an important consideration. Secondly, many fishing strategies are of a mobile nature, and, therefore, it is common to see fishers and boats travelling across boundaries. Migrating, temporary, and visiting fishers (whether legal or illegal) interact with local ones, which may push those with more primitive gear or less established political connections to the margins. Because of their strategic position and movement near boundaries, small-scale fishers have also been implicated in the geopolitical wrangling of high-level boundary disputes among states (Song 2015; Roszko 2015). Thirdly, as sale of fish and seafood to other jurisdictions is common, trade is often transboundary. Many small-scale fishers engage in the inter-country and even inter-continent trade of their catch. They are, therefore, increasingly subject to, and become vulnerable to, sometimes forceful trade rules, tariff restrictions, price fluctuations, and import food safety standards (Purcell et al. 2017). Lastly, small-scale fishers may be invited as participants in the governance discussions or negotiations about how to more effectively and cooperatively manage transboundary fisheries. This is a positive step that could increase their visibility and enable a fairer treatment. However, there is also a danger that they are used as a bargaining chip or become a token presence in order to achieve gains in other sectors (Scholtens 2016b).

From this general description of the different avenues through which transboundary issues may impact small-scale fishers and small-scale fisheries, we further examine three specific challenges that draw upon geographical concepts: scale, politics, and sovereignty.

17.2.1 Scalar Mismatches and Institutional Fragmentation

Very common in the governance of transboundary fisheries is a mismatch "between the geographic scale of ecosystem functioning and the spatial extent of the institutional arrangements managing such a system" (Duraiappah et al. 2014). According to Berkes (2010, 236) this "gross misfit of [...] scales is one of the fundamental reasons why management often fails." Institutional scale may disagree not only with ecosystem scales, but in the case of transboundary fisheries, also with the spatial range of the fisheries activity (Scholtens and Bavinck 2013). Van Tatenhove (2013, 300) employs the term 'institutional ambiguity' to describe such "mismatch between the institutional settings and the specific territorial locations ... where [fishers] operate." Berkes (2010, 236) argues that seeking an exact fit between the two systems may often not be very realistic given the highly dynamic and mobile nature of both fish stocks and fishermen. Rather, mismatches need to be addressed by having appropriate interactions, both horizontally (at a single level) and vertically (between levels), as only then can various overlapping jurisdictions coordinate efforts (Scholtens and Bavinck 2014).

While scalar mismatches are thus common, the added problem with transboundary fisheries issues is the involvement of multiple governments (including defence, fisheries, and foreign affairs ministries), transnational private sectors companies, multiple fisher groups, as well as international organizations, generating a highly fragmented institutional and legal arena. Institutional arrangements may include both international laws and guidelines, bilateral agreements, state law, and community law, as well as customary norms, none of which enjoy exclusive authority over fishing grounds (Scholtens 2016b). This not only challenges governability, or the capacity for governance, but the resulting legal ambiguities may also work to disadvantage small-scale fishers (even though occasionally they can also provide an advantage).

17.2.2 Sovereignty and Geopolitical Issue Linkage

Maritime boundaries are inevitably tied to vital state preoccupations such as sovereignty, security, and defence. When small-scale fishers operate across boundaries, or are affected by other transboundary fishing operations or regulations, they become easily encapsulated in political calculations that range far beyond fisheries. In such cases, small-scale fisheries governance thus becomes entangled in national or higher level political processes that have little to do with fisheries per se. Constituting just one of the many issues in international relations, transboundary fisheries conflicts have been frequently used as a pawn in regional geopolitics (Mitchell 1976; Dupont and Baker 2014). In such a situation, transboundary fisheries become subject to issue linkage, in other words, "the simultaneous discussion of two or more issues for joint settlement" (Poast 2012, 278), in which fisheries concerns end up being linked with unrelated bilateral issues into a package deal. Examples in the literature abound. The Sri Lankan government has repeatedly released Indian trawlers that were caught poaching in its waters to appease its big neighbour, despite opposition of small-scale fishers whose operations were devastated by those trawlers (Scholtens and Bavinck 2014). Chinese fishers have been used as subsidized pawns to display territorial claims of the Chinese state, to the frustration of Vietnamese fishers (Roszko 2015). Song (2015) demonstrates how fishers are used by the South Korean state to represent its geopolitical agenda and reinforce its objectives of boundary legitimization. Such examples show that small-scale fishers can be employed for an instrumental or ideological gain in a larger game of geopolitical swagger (Roszko 2015; Song 2015).

It must be said that issue linkage is not necessarily a negative phenomenon and may sometimes be used to solve stalemates in bilateral negotiations (Warner 2016). However, with small-scale fishers typically having a poor representation in those negotiations, their interests are easily compromised for national interests that are often better protected and conveyed.

17.3 Empowering Fishers Through Transdisciplinary Collaboration[1]

The question, now, is how can these transboundary challenges be linked to the empowerment—or the increase of collective influence—of small-scale fishers, and what can we gain from transdisciplinary engagement? That is, what types of interventions are required to address these transboundary challenges faced by small-scale fishers, and what is the role of transdisciplinary collaboration therein? To address these questions we first need to clarify what we mean by empowerment.

Empowerment can be understood as a process or an outcome, and as a goal in itself, or rather valued instrumentally for achieving other ends (Jentoft 2005). In this chapter, we understand empowerment as the process of improving resource access of marginal users, with access defined as the ability to benefit from resources (Ribot and Peluso 2003). To understand the nature of empowerment processes, Mohan and Stokke (2000) draw a useful distinction between two approaches labelled revisionist neoliberal and post-Marxist. From the first perspective, empowerment entails a harmonious process of building bottom-up organizational structures. Here, empowerment has the potential of a win-win situation, which can be achieved through multi-stakeholder deliberations, with the assumption that bringing about positive change for marginal groups can take place within existing power structures (Mohan and Stokke 2000). The second perspective understands empowerment as social mobilization that challenges hegemonic interests within the state and the market (Mohan and Stokke 2000, 249). From this perspective, empowerment is a conflictive and potentially zero-sum game that cannot occur without disempowerment; "those who are being empowered are doing so at someone's expense" (Jentoft 2005, 2). This perspective stresses that marginalization is produced by often contentious socio-political interactions and that empowerment is therefore a relational phenomenon.

In terms of empowering small-scale fishers for gaining resource access, the difference between these two approaches is significant. A harmonious approach to empowerment may entail building institutional capacity among fisher organizations and facilitating the negotiation of collective outcomes with competing resource users. An antagonistic approach may rather entail confronting exclusion, inviting public protests and challenging rights and discourses that privilege dominant resource users (Fowler and Biekart 2013).

Each type of empowerment provides different opportunities and challenges for transdisciplinary collaboration. We define transdisciplinary collaboration here as a form of collaborative action between practitioners, NGOs, fisher organizations, academics, and possibly even governments, in different possible constellations. As explained in the introductory chapters of this book, transdisciplinary collaboration is appealing for many reasons, and few people would be outright opposed to it. Yet, it has a range of potential vulnerabilities and limitations and should not be

[1] The arguments made in this section build upon Scholtens and Bavinck (forthcoming).

romanticized. For example, interests, incentives, accountabilities, values, and epistemologies can differ markedly between scholars working for universities and professionals working as activists and non-activists for NGOs or representative organizations. Stereotypically, academics are primarily interested in understanding change, whereas practitioners and policy-makers rather want to enact change. “In a caricatured way, both groups [researchers and practitioners] are prone to develop images of each other: to the researcher, the operator may be vain, and naïve as the target set out in plans is missed; to the operator, on the other hand, the researcher may well be independent but equally insignificant and of little proper use” (Lund 2010, 22).

From a researcher’s perspective, collaboration with civil society organizations and fisher representatives may be valuable to formulate more demand-driven questions, gain access to valuable insider information, engage in action research, ensure impact of one’s research, possibly add new or better questions, and perhaps, most importantly, add a significant element of accountability. Yet, scientists may also be concerned with losing scientific integrity by taking position and engaging in action. Transdisciplinary collaboration may also lack the necessary alignment with scholars’ theoretical research interests, and their questions may be considered irrelevant from a practitioner’s point of view. For practitioners, collaboration with researchers may be helpful to acquire data to inform, support, or legitimize an action plan. Cooperation with universities may also be useful to have access to a certain form of authority and networks. In addition, academic research may be useful to reflect on and improve intervention strategies, and allow to challenge assumptions and scrutinize NGO practices. Yet, collaboration with academics can also pose major frustrations in terms of having different time horizons, publication requirements, and obsession with ‘neutrality’ and ‘objectivity’.

17.4 Addressing Transboundary Challenges Through Transdisciplinary Collaboration

Below we elaborate on a variety of strategies that transdisciplinary consortia can engage in to tackle small-scale fisheries challenges of transboundary nature. For the purpose of this chapter, we identify five possible types of intervention strategies (which are not necessarily exhaustive) on the basis of the authors’ own experience with transdisciplinary work. They are capacity building interventions (e.g. strengthening fisher organizations), institutional interventions (e.g. building bridges between disparate actors), legal interventions (e.g. appealing to national or international courts), discursive interventions (e.g. reframing the nature of the problem), and mobilization interventions (e.g. mobilizing fishers to confront power). Each of these strategies has its own strengths and weaknesses, and obviously different contexts require their own mix of approaches. Different types of organizations also tend to favour different types of interventions; while some may consider building social movements the only way to confront social exclusion, others may consider such interventions harmful and blocking potential avenues for constructive engagement.

17.4.1 Capacity Building Interventions – Let's Organize

Transboundary negotiations can be overly complicated and generally not very transparent. Laden with political interests, any meaningful involvement might seem discouraging and time-consuming to small-scale fishers. Capacity building of fishers is a widely articulated strategy of promoting empowerment and reducing marginalization, seen as necessary for promoting their participation in transdisciplinary dialogues for finding innovative solutions. It is arguably the core assertion of the systematic, harmonious view that aligns with the mainstream practices and theories. Many small-scale fishers do not have the capacity or continued interest to participate in transdisciplinary collaboration. Engaging across (international or management) boundaries can also be particularly challenging and may require an added set of capacities to be able to negotiate with those positioned on the 'other' side of boundaries (see Box 17.1). Hence, appropriate training and necessary support that encompass awareness of potential regulatory, cultural, and ecological differences across boundaries, as well as use of positive, empathic language would be desired. At the same time, the transdisciplinary fisher-scientist collaborations themselves can further enhance confidence, networking, lobbying, decision making, and adaptive capability of the fishers. In addition to capacity building at the level of an individual, there is also empowerment at the level of collectives and institutions that must happen. For that, fishers need to re-structure themselves into an organized group to better represent their interests. Coping with the unpredictability of transboundary fisheries, striving to locate mutually-agreeable solutions through transdisciplinary initiatives such as co-management (Jentoft 2005), and dealing with politics of scale to productively engage with distant and high-level political and economic forces would all require, as the first step, organized and enhanced capacity on the part of the fishers and others involved.

17.4.2 Institutional Interventions – Let's Build Bridges and Linkages

Institutional interventions aim to improve the quality and capacity of institutions dealing with or affecting small-scale fishers, in particular the coordination and functional linkages between institutions. As mentioned above, in transboundary contexts there are frequent institutional mismatches, for example, resulting in the inability of a fisheries authority to control a fishery in its jurisdiction because of its highly mobile transboundary nature. Local or national institutions may also be unable to respond to rapidly changing international fish markets or fishing techniques (Berkes 2010). Improving coordination between formal and informal institutions at different levels of scale, mobilizing different forms of knowledge, and involving both state- and non-state actors are considered important from this point of view (Kooiman et al. 2005; Fanning et al. 2007; Berkes 2010).

In the context of transboundary conflicts, in cases where conventional inter-state diplomacy is hampered or has come to a standstill, NGOs may engage in more

Box 17.1 Building Capacity of Fishers to Look Beyond Boundaries

Many tropical island nations face concerns of food insecurity with a high proportion of rural households relying on fisheries for food and income. However, the nearshore reefs and lagoons that surround many of these islands accessed by small-scale fishers are considered to be heavily fished. Resources traditionally harvested in these relatively shallow and close-to-shore areas, including trochus, reef fish and sea cucumber, can no longer be consistently relied upon to support livelihoods, necessitating a relieved fishing pressure and a time for a stock recovery (e.g. through closed seasons or protected areas). As part of securing alternative fishing options for small-scale fishers, NGOs and researchers are working together with national fisheries departments and regional organizations, such as the Pacific Community, to operationalize offshore fish aggregating devices (FADs) that give coastal communities affordable and safe access to oceanic, pelagic species, such as tuna and mackerel.

For example, in Timor-Leste, where fishers typically use outrigger canoes and small motor boats, they are limited by the distance they can travel. In Atauro Island, nearshore fringing reefs drop off quickly to great depths of up to 3000 meters, posing an operational boundary for these fishers. Importantly, in facilitating to go beyond these tacit boundaries, the technical transfer of FADs is being promoted under the auspices of community-based management (or traditional management system called *tara bandu*), which encourage direct involvement of local communities and building their technical and organizational capacity. Similarly, in Vanuatu, FAD development is part of the community's overall fisheries management plan. In collaboration with the Vanuatu Fisheries Department, a FAD Management Committee is set up in the target villages to implement the community FAD guidelines on various aspects, including FAD access rules and fees, gear restrictions, catch-and-effort monitoring, and maintenance of the FADs. Fishers are also trained on using vertical longlines for catching pelagic fish and squid, and five new trolling gear types. In this way, FADs provide a vehicle for capacity building and community organization as well as integration of different groups and disciplines in order to expand fishing possibilities beyond nearshore environments.

Sources:

http://blog.worldfishcenter.org/2017/06/against-the-tide-a-fad-fit-for-timor-lestes-artisanal-fishers/

https://www.worldfishcenter.org/content/communities-tackle-coral-reef-sustainability-timor-leste

http://www.spc.int/fame/fr/projets/devfish2/193-nearshore-fad-support-to-atauro-island-timor-leste

Amos et al. (2014).

informal and possibly creative forms of transboundary collaboration or conflict resolution. This so-called 'track-II diplomacy' may cut through the red tape of conventional diplomacy and be more creative, being less bound by formal procedures. "Track II offers a 'bottom up' […] approach in contrast to the top down, mediated solutions often touted in the past. In this process, NGOs who have become deeply embedded in societies can act as facilitators and conduits, that is as agents who make productive exchange among players possible and who provide 'ground truth' to governments and other interested parties'" (Kelleher and Taulbee 2017, no page number). In other words, transboundary conflicts may well be served by collaboration between fisher organizations and broader civil society actors, possibly supported by researchers, for example, to benefit from the legitimacy they may enjoy with state authorities (see Box 17.2).

17.4.3 Discursive Interventions – Let's Reframe

Scientists and policy-makers tell each other stories of what fisheries are, what problems they face, and what issues are real and deserving. These narratives typically frame victims and villains, problems and solutions, and causes and effects. They are also continuously challenged and never permanently fixed. Fisheries management jargon is rich in such narratives, and some of them implicitly undermine the viability and relevance of small-scale fishers. For example, the narrative of Malthusian overfishing suggests there would be too many fishers out there chasing too few fish (Pauly 1994; Finkbeiner et al. 2017); the tragedy of the commons narrative frames fishers as non-communicative resource exploiters in need for stringent regulation, while narratives of 'wealth based fisheries' and ITQs give implicit preference to economic efficiency and 'fisheries rationalization' over socially equitable outcomes (Cunningham et al. 2009; Pinkerton 2017). Over the course of past decades, with ample technical (development) cooperation to train southern scholars and prospective managers in 'modern fisheries management', these narratives have, in the course of globalization, moved across boundaries and resulted in a global expert consensus (Thorpe and Bennett 2001).

These mainstream narratives can, however, be effectively deconstructed by scholars and practitioners in order to reveal how they were constructed in the first place and what interests they serve. While dominant discourses typically represent vested interests and may, therefore, be highly resilient to change, alternative- or counter-narratives can be and are developed as well. Such deconstructing, challenging, and reframing of dominant narratives that misrepresent small-scale fishers' interest lend themselves particularly well to transdisciplinary collaboration. Challenging discourses requires persistence and strong collaboration with media to affect not only policy but also shape wider public opinion. Another key stumbling block to changing the discourse is how wedded scientists are to their particular discipline and affiliated methodological approaches. Transdisciplinarity requires us to go beyond collaboration across disciplines and also involves societal groups in the framing of the research problems, the co-production of knowledge, and the framing of solutions (see Box 17.3).

Box 17.2 Transboundary Dialogues for Change

Despite the existence of a maritime boundary between India and Sri Lanka, the overcapitalized trawler fleet of Tamil Nadu has a long history of targeting the rich Sri Lankan fishing grounds. The grounds had become particularly attractive as north Sri Lankan fisheries had collapsed as a result of the devastating civil war raging in the country from 1983 to 2009. From 2010–2016 a project called REINCORPFISH worked with a group of academics, NGOs, and fisher organizations from both sides, aiming to transform this transboundary fisheries conflict.

The transdisciplinary consortium took a bottom-up approach. It reasoned that since fishers from both sides shared a common language and a Tamil ethno-political identity, there may be fertile ground for a transboundary dialogue between fisher representatives from both countries. In 2010, the consortium organized so-called fisher-to-fisher dialogues, which were initially met with cautious support from both states. After a week of intense dialogue and debate, the two groups of fishers reached a promising agreement, stipulating that Indian trawl fishers would terminate trawling in Sri Lankan fishing grounds after a transitioning period of 1 year; during the transition period, trawlers would be allowed to continue fishing in a limited section of Sri Lankan waters for a total of 70 fixed days.

The Sri Lankan government, however, soon rejected the fishers' proposal and subsequently started to oppose the dialogue process, arguing that the transboundary fishing conflict was an issue for the two governments to solve. It accused the NGO for hand-picking fisher delegates to suit its own political needs, and for buying into the logic of 'sharing waters' with India, thus undermining Sri Lanka's territorial integrity. While the organizers of the dialogue viewed the issue mostly from a livelihoods and fisheries management perspective, for the Sri Lankan government sovereignty and security were the primary framing. The backlash also reflected a larger tension between the post-war authoritarian Sri Lankan regime and rights-based NGOs in Sri Lanka. The lack of government support on both sides allowed Indian fishers not to adhere to the agreement, and to continue trawling in Sri Lankan fishing grounds. What became clear from this initiative is that collaboration between academics, NGOs, and fishers can in principle be fruitful to address transboundary challenges by facilitating creative (interim) solutions, but also that such initiatives are fragile as they rely on endorsement and support by respective governments to bear fruit. This is particularly difficult to achieve since transboundary issues can easily flame a range of state anxieties.

Sources: Stephen et al. 2013; Scholtens and Bavinck forthcoming.

Box 17.3 Reframing Transboundary Fishing as IUU Fishing

Transboundary fishing activities can be framed in various ways. While fleets operating illegally or semi-legally in distant or foreign waters may be variously called pirate fishers or roving bandits, they are increasingly termed with the relatively new name of 'IUU fishing'. The phrase Illegal Unregulated and Unreported (IUU) fishing was introduced by the FAO in 2001 and gained momentum in 2010 when the EU adopted a far-reaching policy to combat IUU fishing practices across the world. This policy includes the possibility to unilaterally ban seafood exports to the EU by countries that it identified as non-cooperative in the fight against IUU fishing.

In the Palk Bay case (as introduced in Box 17.2), the transboundary operations of the Indian trawler fleet in Sri Lankan waters have for many years been subject to intense and enduring struggles by Sri Lankan fishers and various supporters from broader civil society. How to frame this conflict between fishers in two countries has always been an important and contested issue in these struggles. Framings used by Sri Lankan actors varied widely, but revolved around 'an intrusion to Sri Lanka's sovereignty', 'an injustice incurred on Sri Lanka's fishers', or 'an unsustainable act of destruction to the fragile Palk Bay ecosystem', while competing with the Indian framing that the entire Palk Bay belonged to its traditional fishing grounds and the maritime boundary line was illegitimate in the first place.

Interestingly, in 2012 a group of activists in Sri Lanka mobilized lawyers, fisher leaders, and the media to again tackle the same transboundary intrusion, but this time reframing it as an act of 'IUU fishing'. The transboundary trawler intrusions indeed qualify by all means for the nomer IUU fishing, being both unregulated and illegal going by Indo-Sri Lankan bilateral agreements. This new framing not only approached the problem from a different angle, but also opened up a new political arena by bringing new actors—most notably the Indian seafood export sector and the European Union—into the equation. The framing of India being engaged in 'IUU fishing' in the Palk Bay effectively internationalized an otherwise bilateral issue, alluding to a politics of scale where civil society actors bypassed the Sri Lankan State and appealed to the EU, thus putting indirect pressure on India by exposing its illegal fishing activities, and potentially making Indian fish exporters nervous (with a threat of a ban on Indian fish exports to the EU). While the EU responded that it wished not to interfere in this 'bilateral issue', the new framing had been quite effective in steering up the debate and putting a novel form of pressure on India. The use of the IUU framing, however, come with its own perils. Not only is the internationalization of domestic or bilateral issues a risky affair that can take unpredictable turns, the action taken by activists in Sri Lanka can also be read as endorsing a western hegemony in setting international fishing standards.

Source: Scholtens 2016b; Stephen et al. 2013.

Box 17.4 Demanding Justice from the Court in the Colombia-Nicaragua Boundary Dispute

Within the international framework of UNCLOS, the resolution of maritime boundary conflicts is supposed to take into consideration equity concerns, the interest and needs of the states as well as the conservation of the marine environment (UNCLOS, Art. 59). Based on these principles, in 2012, the International Court of Justice (ICJ) announced its decision regarding the maritime border conflict between Nicaragua and Colombia, which redrew the maritime boundary such that it now came to be positioned 200 nautical miles from the Nicaraguan coastline. This implied an abrupt loss for Colombia of a 75,000 km^2 sea area. This resolution subsequently spurred a range of unforeseen conflicts. These included hostile treatment of Colombian vessels and fishermen by Nicaraguan authorities, a lawsuit against Colombia for not abiding by ICJ's decision and violating Nicaraguan sovereign rights, alerts from environmental NGOs and activists regarding the risks for the Seaflower Biosphere Reserve, and claims from the island's inhabitants who objected to how the decision had taken away their ancestral territory and infringed their livelihood rights.

In response to the multiple unexpected effects of the newly drawn maritime boundary, the Native National Authority (NNA, a group of Colombian community leaders from San Andres and Old Providence Islands elected to represent and defend the wellbeing and the future of the *raizal* ethnic group) devised a two-pronged strategy. First, they tried to informally negotiate a joint solution with governmental representatives and fishermen from Nicaragua in order to preserve their traditional activities in the contested area by establishing common fishing zones and trade route agreements. Second, in 2014 the NNA community leaders started collaborating with the University of Medellín to take the issue to court. According to Estrada Vélez, the professor and lawyer who joined the initiative of the community leaders, this unique collaboration followed from his conviction that "academics must be a means of social transformation and human rights defense."

They jointly sent a note of protest to the ICJ, and presented a lawsuit against those articles of the Colombian Constitution that dictate the government to abide by ICJ rulings. These actions challenged the constitutionality of the ICJ judgment, arguing that, first, it "suffers from a structural deficiency by not allowing voice, participation or recognition of indigenous people" (Gaviria Liévano 2014, 182), and, second, the two states had been too narrowly concerned with defining their respective sovereign rights and delimiting what they considered to be their territories, while violating the rights to ancestral property and food security, and producing harmful economic, ecological, and social effects on the archipelago (Centro Colombiano de Estudios Constitucionales 2017).

(continued)

Box 17.4 (continued)

While at the time of writing the ICJ has not yet taken a final decision, the joint action undertaken by the fisher community leaders, academics, and lawyers resulted in a partial victory in late 2017. The ICJ announced that it accepted the counterclaim concerning Nicaragua's infringement of the customary rights of the San Andrés Archipelago inhabitants to access and exploit their traditional fishing grounds.

Sources: Gaviria Lievano (2014); Centro Colombiano de Estudios Constitucionales (2017); UNCLOS (1982).

Flyvbjerg et al. (2012) brings the argument one step further, by arguing that this work of reframing is in fact one of the key roles for social scientists. He suggests to focus on 'tension points', in other words, issues in which power relations "are particularly susceptible to problematisation and thus to change, because they are fraught with dubious practices, contestable knowledge and potential conflict. Thus, even a small challenge – like problematisation from scholars – may tip the scales and trigger change…" (Flyvbjerg et al. 2012, 288).

17.4.4 Legal Interventions – Let's Appeal

Transboundary fisheries movements of fishers, as well as the stipulation of international maritime boundaries themselves, are typically regulated in one way or another by national, bilateral, or international laws and agreements. These may be domestic regulations prohibiting or permitting foreign fishing fleets to operate in one's EEZ, high-level guidance such as the 'Voluntary Guidelines for Sustaining Small-Scale Fisheries' as adopted by the FAO member states in 2015 (FAO 2015), or the UNCLOS providing a measure of multilateral consensus for the equitable sharing and effective management of oceanic resources. Many ocean spaces are subjected to a multiplicity of potentially conflictive rule systems, leading to situations of legal pluralism, where stakeholders may end up 'forum-shopping' for the solution that fits their interests best. Legal interventions on behalf of small-scale fisheries become relevant when the problem at stake stems from an unwillingness or inability on the part of the authorities to enforce existing regulations, agreements and guidelines that are potentially protective of small-scale fisheries.

Small-scale fishers' activities have been frequently obstructed by illegal or semi-legal interventions that include displacements, dispossession of land or resource access, or the ignorance of historical or ancestral rights. Sometimes these processes have been called ocean or coastal grabbing (Bennett et al. 2015; Bavinck et al. 2017), even though such involuntary transfers of ownership are not necessarily illegal. In such scenarios, when objections, attempts at consultations and other administrative procedures fail, appealing to the court can be a strategy of last resort. Dealing with the court requires its own type of transdisciplinary collaboration between lawyers, fisher organizations, and possibly also academics. Especially when international courts become involved, fishers may require high-level national or international expertise to be able to frame their claims effectively (see Box 17.4).

Taking the legal route, however, has its limitations, also in cases where small-scale fishers clearly have the law on their side. While in an ideal world, legal systems operate independent of political systems, in many contexts and countries the courts and legal systems do not operate in a political vacuum. As Ribot and Peluso argue (2003), possessing the appropriate legal right to access fish resources does not always mean that one is *actually* able to benefit from those resources. In other cases, the law may itself be considered the source of disempowerment designed to disproportionally benefit (trans)national elites, rather than an avenue for obtaining and protecting the rights of vulnerable user-groups. In such cases, this would make the law itself a target of protest rather than a means of empowerment.

17.4.5 Social Mobilization – Let's Fight

When governments are implicated in infringements on the rights of small-scale fishers, closing their eyes to misbehaving third parties, or are clearly in no mood to respond to legitimate concerns, collective mobilization can be a powerful strategy. For example, governments may ignore historical fishing rights, engage in shady joint ventures selling oceanic resources, facilitate or ignore coastal developments undermining small-scale fishers' resource access, or simply displacing them altogether. In such cases, change may not be effectively pursued through dialogue and collaboration, but rather requires collective mobilization to challenge the status quo, and the building up of pressure in ways that authority holders feel compelled to respond.

There is a burgeoning literature that helps to understand under what conditions social movements may arise, fail, or succeed, the diversity of repertoires and strategies employed across time and space, the importance of leadership, resource mobilization, networks and alliances, as well as the vitality of timing in relation to political opportunities (e.g. Tilly 1993; Tarrow 2011). While this literature is rarely applied to small-scale fishers and their organizations, they certainly have a rich history of collective mobilization, whether organized at the local, national, or international level. For example, the Indian National Fisheries Forum booked a major success in getting foreign fishing fleets banned from Indian waters in the 1990s (Sinha 2012). Mobilization of fishers in South Africa against their exclusion from fishing rights led to significant change (Isaacs 2011). Fishers in Sri Lanka successfully mobilized against the introduction of sea planes (Kumara 2014), and less successfully against the illegal intrusion of Indian trawlers into their waters (Scholtens 2016a). The International Collective for the Support of Fishworkers (ICSF) and the

World Fisheries Forum (WFF) have played key roles internationally in pursuing this path of change to challenge injustices incurred on small-scale fishers.

Academics may not be the most obvious partner when it comes to social mobilization. They may rather shy away from open engagement with activism, fearing that it could undermine their respected positions as 'objective' and 'neutral' observers. This may complicate transdisciplinary collaboration and create distance and mistrust between scientists on the one hand and fisher activists on the other. Yet, a wide range of scholars—sometimes called activist-scholars or scholar-activists—consider their practice of science as inherently political and the problematisation of and mobilization against the status quo a key element of their *raison d'être*. In such cases, collaboration with scholars can be useful not only to provide a degree of authority and legitimacy to a cause, but also because they can provide and organize relevant knowledge. Beyond the individual scholar, some universities also have a rich history of functioning as platforms supporting popular mobilization and for 'speaking truth to power'. In other cases, the role of academics may rather be more a supportive one of providing information, developing arguments, doing action research with fishers, and building an alternative narrative (see Box 17.5).

17.5 Discussion and Conclusion

This chapter has explored the potential for transdisciplinary collaboration between academics, NGOs, and fisher organizations in the pursuit of empowering small-scale fishers to deal with transboundary issues. In an increasingly globalized world, it is not an option to confine analysis of small-scale fishers to local affairs, as their viability is increasingly affected, if not undermined by forces originating at higher levels of scale. This chapter has thus focused on how transboundary issues have complicated the operation and welfare of small-scale fishers, and explored how different modalities of transdisciplinary engagement can be instrumental in addressing these challenges.

While transdisciplinary work is usually thought of as collaborative knowledge creation, this chapter has argued that it can also be understood as jointly engaging in action. We highlighted five types of interventions relevant for empowering small-scale fishers to deal with transboundary challenges. The type of intervention required is partly dependent on the nature of the transboundary problem at hand. Problems of institutional mismatch may be better served by governance interventions, while geopolitical issue linkages may rather require legal interventions. When fishers' marginalization is the result of institutional ignorance or inadequate sharing of information, capacity building of fishers or state authorities, or advocacy work may be more relevant. When fisher rights are deliberately undermined, social mobilization or legal action may be required; in such situations empowerment has become a zero-sum game.

In reality, however, the distinction between the different types of interventions may be less clear-cut; they may well be combined and mutually reinforce one

Box 17.5 Partnerships Mobilize Fishers for Change in South Africa

The advent of democracy in South Africa in 1994, resulted in transformation of state institutions and society at all levels. Yet, despite a progressive Constitution, the interests of the fishing industry and the conservation community took precedent over addressing the rights of small-scale fishers. At about this time, fishers from Ebenhaeser and Papendorp, two villages located on the Olifants estuary on the west coast, approached researchers at the University of Cape Town (UCT) for assistance in ascertaining the reasons for the declining fish catches and challenging governments plans to significantly reduce fishing activities on the estuary. These fishers, who had been forced off their lands in the early 1920s and settled adjacent to the estuary, have been fishing in these waters for nearly 100 years. They regard the estuary as their traditional fishing grounds. Since 1993, a partnership between researchers at UCT and fishers of the Olifants estuary developed that sought to support fishers in their efforts to claim their rights, implement a community-based monitoring system, facilitate greater involvement of fishers in decision making, and oppose a government plan to phase out gillnetting, which would undermine their rights, livelihoods, and way of life (Sowman 2017).

However, in 2008, fishers at the Olifants estuary were confronted with a government proposal to declare a 'no-take' Marine Protected Area in the Olifants estuary. Building on the partnership with researchers at UCT, and two NGOs – Masifundise and the Legal Resources Centre – fishers were able to organize and mobilize around this threat to close the estuary. Fishers wrote letters to the Minister, threatened legal action, participated in protests, held workshops, and gathered data to challenge this proposal. One of the main obstacles to finding a resolution was the scientists' conviction that a 'no-take' MPA was the only approach to address conservation concerns. After five years of community activism, research, and engagement with other estuary stakeholders, the fishers reached an agreement with all stakeholders that ensured the rights of fishers were respected but that conservation interests would also be addressed. This transformation process involved engaging in action research and dialogue amongst fishers, researchers, and other estuary stakeholders to determine estuary management proposals that were supported by the fishing community. Incorporating local perspectives and ecological knowledge in developing management proposals for the estuary were central in moving the narrative forward from a state centred, science-based approach to one that respected fisher rights and the role they play in stewardship of resources. The key factors that led to this outcome were (1) an enabling policy and legal environment that requires respect for human rights; (2) a robust partnership between fishers and civil society; (3) providing a 'space' that was conducive to discussion and deliberation amongst diverse governance actors; and (4) the presence of a respected process leader that recognized the importance of gaining input from all partners, and acknowledged the diversity of interests and knowledge as a resource, rather than an obstacle. This social learning process, although conflictual at first, led to transformations at many different levels and an outcome that was broadly supported.

Sources: Sowman 2009, 2017.

another. When small-scale fishers need to assert themselves vis-à-vis the state or corporate interests, a combination of institutional strengthening, reframing the problem by engaging with dominant public or scientific discourses, reaching out for dialogue to show a willingness to explore synergistic opportunities, resorting to legal action as well as outright mobilization and confrontation may be quite necessary. In the South African case, a combination of capacity building, dialogue, working with social partners, protest, and the threat of legal action proved crucial for gaining the necessary attention for the struggle of small-scale fishers and follow up action by the state. However, in other cases combining different approaches through one-and-the-same partnership can also be counterproductive. For example, in the case of the Palk Bay fisheries, the same partnership involved in facilitating the transboundary dialogue process also tried to mobilize fishers against the Sri Lankan government. The latter approach failed: the authoritarian government was keen to repress the NGO involved in organizing the protests, which subsequently also compromised the efforts of the same NGO to build capacity among fisher organizations and facilitate the dialogue process with Indian trawler operators.

The discussed intervention strategies are not exclusively relevant for addressing boundary challenges of course, nor are they exclusively applicable to fisheries. Yet, in this chapter we have tried to make a strong argument that the multiple transboundary challenges faced by small-scale fisheries are of such a character that they require broad based interventions, involving both academics, practitioners, NGOs, and fishers. Working in isolation will not work, for challenges are simply too complicated and demanding in their scope. This does not mean that collaboration between researchers and practitioners is always required for pursuing social and environmental justice in fisheries. The possible contribution of academics is highly dependent on context, the nature of the challenge at hand, and, simply, the courage or commitment of concerned individuals, and ranges from just a supportive one of providing information, to full-fledged collaboration with fishers to challenge status quo. Productive engagement between academics, fishers, and other civil society organizations, however, cannot be taken for granted; it needs to be carefully crafted, continuously fine-tuned in terms of mutual expectations and possibly conflicting priorities, incentives and modes of working. Just like any partnership, building trusting relationships and going beyond ad-hoc problem solving is key. And perhaps most challengingly, the accountability of all partners involved—fisher representatives, NGOs, and researchers—needs to be continuously tuned downwards towards actual fishers rather than to donors, research organizations, and bureaucrats.

References

Amos G, Nimoho G, Fujii M et al (2014) New FAD development approach strengthens community-based fisheries management in Vanuatu. SPC Fish Newslett 144:40–47

Bavinck M, Berkes F, Charles A et al (2017) The impact of coastal grabbing on community conservation–a global reconnaissance. Mar Stud 16(1):8

Bennett NJ, Govan H, Satterfield T (2015) Ocean grabbing. Mar Policy 57:61–68

Berkes F (2010) Linkages and multilevel systems for matching governance and ecology: lessons from roving bandits. Bull Mar Sci 86:235–250

Centro Colombiano de Estudios Constitucionales (2017) Acción contra el Pacto de Bogotá. https://www.cecec.co/documentos-litigio-pacto-de-bogota. Accessed 25 Mar 2018

Cunningham S, Neiland AE, Arbuckle M et al (2009) Wealth-based fisheries management: using fisheries wealth to orchestrate sound fisheries policy in practice. Mar Res Econ 24(3):271–287

Dupont A, Baker CG (2014) East Asia's maritime disputes: fishing in troubled waters. Wash Q 37:79–98. https://doi.org/10.1080/0163660X.2014.893174

Duraiappah AK, Asah ST, Brondizio ES et al (2014) Managing the mismatches to provide ecosystem services for human well-being: a conceptual framework for understanding the New Commons. Curr Opin Environ Sustain 7:94–100

Fanning L, Mahon R, McConney P et al (2007) A large marine ecosystem governance framework. Mar Policy 31:434–443. https://doi.org/10.1016/j.marpol.2007.01.003

FAO (Food and Agriculture Organization of the United Nations) (2015) Voluntary guidelines for securing sustainable small-scale fisheries. http://www.fao.org/fishery/ssf/guidelines/en. Accessed 20 Mar 2018

Finkbeiner EM, Bennett NJ, Frawley TH et al (2017) Reconstructing overfishing: moving beyond Malthus for effective and equitable solutions. Fish Fish 18(6):1180–1191. https://doi.org/10.1111/faf.12245

Flyvbjerg B, Landman T, Schram S (2012) Real social science. Cambridge University Press, Cambridge

Fowler A, Biekart K (2013) Relocating civil society in a politics of civic-driven change. Dev Policy Rev 31(4):463–483

Gaviria Lievano E (2014) La desintegración del Archipiélago de San Andrés y el Fallo de la Corte Internacional de la Haya. Temis, Bogotá

Isaacs M (2011) Creating action space: small-scale fisheries policy reform in South Africa. In: Jentoft S, Eide A (eds) Poverty mosaics: realities and prospects in small-scale fisheries. Springer, Dordrecht, pp 359–382

Jentoft S (2005) Fisheries co-management as empowerment. Mar Policy 29:1–7. https://doi.org/10.1016/j.marpol.2004.01.003

Kelleher A, Taulbee JL (2017) Chapter 5: building peace Norwegian style: studies in track 1,5 diplomacy. In: Carey HF (ed) Subcontracting peace: the challenges of NGO peacebuilding. Routledge, London, pp 69–84

Kooiman J et al (eds) (2005) Fish for life: interactive governance for fisheries and aquaculture. Springer, Dordrecht

Kumara H (2014) Struggle over the Negombo Lagoon. NAFSO Publication N. 37. NAFSO, Negombo. https://www.scribd.com/doc/238637669/Struggle-over-the-Negombo-Lagoon. Accessed 13 July 2017

Lang DJ, Wiek A, Bergmann M et al (2012) Transdisciplinary research in sustainability science: practice, principles and challenges. Sust Sci 7(1):25–43

Lund C (2010) Approaching development: an opinionated review. Prog Dev Stud 10:18–34. https://doi.org/10.1177/146499340901000102

Mitchell B (1976) Politics, fish, and international resource management: the British-Icelandic cod war. Geogr Rev 66:127–138

Mohan G, Stokke K (2000) Participatory development and empowerment: the dangers of localism. Third World Q 21:247–268. https://doi.org/10.1080/01436590050004346

Pauly D (1994) From growth to Malthusian overfishing: stages of fisheries resources misuse. Tradit Mar Resour Manag Knowl Inf Bull 3:7–14

Pinkerton E (2017) Hegemony and resistance: disturbing patterns and hopeful signs in the impact of neoliberal policies on small-scale fisheries around the world. Mar Policy 80:1–9

Poast P (2012) Does issue linkage work? Evidence from European alliance negotiations, 1860 to 1945. Int Organ 66:277–310. https://doi.org/10.1017/S0020818312000069

Polk M (2015) Transdisciplinary co-production: designing and testing a transdisciplinary research framework for societal problem solving. Futures 65:110–122

Purcell SW, Crona BI, Lalavanua W et al (2017) Distribution of economic returns in small-scale fisheries for international markets: a value-chain analysis. Mar Policy 86:9–16

Ribot JC, Peluso NL (2003) A theory of access. Rural Sociol 68:153–181

Roszko E (2015) Maritime territorialisation as performance of sovereignty and nationhood in the South China Sea. Nations and Nationalism 21(2):230–249

Scholtens J (2016a) The elusive quest for access and collective action: North Sri Lankan fishers' thwarted struggles against a foreign trawler fleet. Int J Commons 10(2):929–952

Scholtens J (2016b) Fishing in the margins. North Sri Lankan fisher's struggle for access in transboundary waters. Dissertation, University of Amsterdam

Scholtens J, Bavinck M (2013) South Indian trawl fisheries – assessing their governability. In: Bavinck M et al (eds) Governability of fisheries and aquaculture: theory and practice. Springer, Dordrecht, pp 177–199

Scholtens J, Bavinck M (2014) Lessons for legal pluralism: investigating the challenges of transboundary fisheries governance. Curr Opin Environ Sustain 11:10–18. https://doi.org/10.1016/j.cosust.2014.09.017

Scholtens J, M Bavinck (forthcoming) Transforming conflicts from the bottom-up? Reflections on civil society efforts to empower marginalized fishers in postwar Sri Lanka. Ecology and Society, forthcoming

Sinha S (2012) Transnationality and the Indian Fishworkers' Movement, 1960s–2000. J Agric Change 12(2–3):364–389

Song AM (2015) Pawns, pirates or peacemakers: fishing boats in the inter-Korean maritime boundary dispute and ambivalent governmentality. Pol Geogr 48:60–71

Song AM, Scholtens J, Stephen J et al (2017) Transboundary research in fisheries. Mar Policy 76:8–18. https://doi.org/10.1016/j.marpol.2016.10.023

Sowman M (2009) An evolving partnership: collaboration between 'experts' and a net-fishery. Gateways, Int J Commun Res Engag 2:119–143

Sowman M (2017) Turning the tide: transforming attitudes and behaviour to conservation in South African coastal areas: a case study at the Olifants Estuary. In: Armitage D, Charles T, Berkes F (eds) Governing the coastal commons. Communities, resilience and transformation. Earthscan from Routledge, New York

Stephen J, Menon A, Scholtens J et al (2013) Transboundary dialogues and the politics of scale in Palk Bay fisheries: brothers at sea? South Asia Res 33:141–161. https://doi.org/10.1177/0262728013487633

Tarrow SG (2011) Power in movement: social movements and contentious politics. Cambridge University Press, Cambridge

Thorpe A, Bennett E (2001) Globalisation and the sustainability of world fisheries: a view from Latin America. Mar Res Econ 16(2):143–164

Tilly C (1993) Contentious repertoires in Great Britain 1758–1834. Soc Sci Hist 17(2):253–280

United Nations Convention on the Law of the Seas (1982) UN General Assembly. http://www.un.org/depts/los/convention_agreements/texts/unclos/unclos_e.pdf. Accessed 20 Mar 2018

van Tatenhove JPM (2013) How to turn the tide: developing legitimate marine governance arrangements at the level of the regional seas. Ocean Coast Manag 71:296–304. https://doi.org/10.1016/j.ocecoaman.2012.11.004

Warner JF (2016) Of river linkage and issue linkage: transboundary conflict and cooperation on the River Meuse. Globalizations 13(6):741–766

Chapter 18
Using Transdisciplinary Research Solutions to Support Governance in Inland Fisheries

Shannon D. Bower, Andrew M. Song, Paul Onyango, Steven J. Cooke, and Jeppe Kolding

Abstract The diverse nature of internal and external threats and fishery attributes in inland fisheries indicates that the development of long-term solutions to governance issues will require interaction among multiple disciplines and actors. Pollution, habitat alteration, invasive species, and hydropower development are widespread problems that are often external to threats imposed by inland fisheries, but greatly impact fishery productivity. Within inland fisheries, challenges of overfishing, equitable access, conflict, and an overall lack of political will to sustain inland fisheries at the regional, national, and international policy levels serve to pressure the sector further. Power dynamics, governance systems, and regulations play a role in determining the perspectives from which solutions to these issues are viewed, and thereby the perspectives from which they are defined as a success. Promoting transdisciplinary research in inland fisheries can support development of successful governance solutions by providing relevant insights to identify and inform these perspectives. Here, we offer examples of redefined governance problems and potential strategies for addressing them using

S. D. Bower (✉) · S. J. Cooke
Fish Ecology and Conservation Physiology Laboratory, Department of Biology, Carleton University, Ottawa, ON, Canada
e-mail: Shannon.Bower@carleton.ca; Steven.Cooke@carleton.ca

A. M. Song
Australian Research Council Centre of Excellence for Coral Reef Studies, James Cook University, Townsville, Australia

WorldFish, Honiara, Solomon Islands

P. Onyango
Department of Aquatic Sciences and Fisheries, University of Dar es Salaam, Dar es Salaam, Tanzania
e-mail: onyango_paul@udsm.ac.tz

J. Kolding
Department of Biology, University of Bergen, Bergen, Norway
e-mail: Jeppe.Kolding@uib.no

R. Chuenpagdee, S. Jentoft (eds.), *Transdisciplinarity for Small-Scale Fisheries Governance*, MARE Publication Series 21,
https://doi.org/10.1007/978-3-319-94938-3_18

transdisciplinary research approaches. We conclude by offering suggestions for improving transdisciplinary research in inland fisheries.

Keywords Freshwater fisheries · Inland fisheries · Sustainability · Transdisciplinary research

18.1 Introduction

Inland fisheries are often overlooked in the predominantly marine-focused global discussions of fisheries sustainability, yet they can be crucial to nutrition and livelihoods, particularly in developing countries (Béné 2006; Andrew et al. 2007; Kolding et al. 2016; Lynch et al. 2017). Inland fisheries face similar governance challenges to their coastal and marine counterparts, including complex economic relationships and value-chains (e.g. welfare vs wealth-based; Béné et al. 2010; Kolding et al. 2014), inequitable governance relationships (Chuenpagdee and Song 2012; Jentoft and Chuenpagdee 2015) and unsustainable fishing practices (Voluntary Guidelines for Securing Sustainable Small-Scale Fisheries, FAO 2015). Inland fisheries are also challenged by additional factors that are not experienced by their coastal and marine counterparts, including sharing fishing areas with multiple users from other sectors and industries (Beard et al. 2011; Lynch et al. 2016, 2017; Song et al. 2017a) and their reliance on external drivers such as rainfall and land development. These competing demands on water from other sectors include irrigation, hydropower, drinking water, and flood controls. The high degree of connectivity in inland systems means that inland fisheries in one part of a watershed may be challenged by situations occurring in another part of the watershed, outside of their sphere of governance and influence (Nguyen et al. 2016). Thus, pollution, habitat loss or alteration, invasive species, hydropower development, and lack of policy supports are all key issues that have a broader governance dimension that affects global inland fisheries governance (Beard et al. 2011; Welcomme et al. 2010; Lynch et al. 2016).

The complexity and interlinked nature of governance challenges in inland fisheries suggest that these constraints likely act as ‘wicked problems’, or issues that are characterized by high uncertainty, complex linkages to other issues, and lack clear solutions (Rittel and Webber 1973). Wicked problems are complex, persistent, or re-occurring (Khan and Neis 2010), and it is not clear when they are solved, as no right or wrong solution can be proved scientifically (Jentoft and Chuenpagdee 2009). The tightly coupled nature of inland fisheries and freshwater ecosystems and the resulting complexity of issues faced within the sector (as wicked problems), means that successful solutions will require interaction among multiple disciplines and perspectives.

Transdisciplinary research is defined as “research that addresses questions of broad societal interest and fosters integration not only among researchers from different disciplines but also with individuals and organizations from outside academia” (Blythe et al. 2017, 114). This indicates that outcomes developed through transdisciplinary research will not be based in a single field of interest or study but will encompass aspects of each in a mutual learning scenario (Roux et al. 2017). To ensure adequate representation from all fields, transdisciplinary research approaches for inland fisheries should include participation from multiple arenas, including: fishers, community representatives, physical and natural sciences, social sciences, policy, management, and other stakeholders. Transdisciplinarity in inland fisheries will include multiple jurisdictions and broad geographical issues, thereby requiring similar participation considerations across boundaries (Song et al. 2017b).

18.2 Transdisciplinary Research Teams

Transdisciplinary research has been used to solve governance problems in multiple areas of study, such as sustainability science (Brandt et al. 2013) and public health (Hall et al. 2012) and has been applied in coastal and marine fisheries scenarios to support ecosystem-based management (e.g. Paterson et al. 2010). However, few examples of genuine transdisciplinary research processes have been applied to inland fisheries issues. In other fields and examples, transdisciplinary research teams feature processes such as team development, issue identification, analysis, implementation, and translation (Hall et al. 2012; Jahn et al. 2012; Brandt et al. 2013); though, it should be noted that a generalized framework for transdisciplinary team formation and function is lacking (Brandt et al. 2013).

Given the need to include perspectives from fishers and community members when considering inland fisheries issues, it is essential to consider power dynamics in transdisciplinary team development. Power dynamics are an important consideration in terms of defining success if marginalized communities are not present during the solution development process. Power asymmetry can act as a barrier to governance during the implementation and translation process (from team outcomes to governance body) (Crona and Bodin 2010). Power dynamics can also hinder adaptation to potential solutions (Sathyapalan and George 2015). Thus, we recommend ensuring that fishers and community members are actively involved and empowered from the start of the transdisciplinary process to collaborate on identifying issues and to define success on an issue by issue basis, such that all group members can strive to agree on which proposed solution constitutes a success. Mechanisms for identifying and actively collaborating with essential partners described in the literature are introductory workshops, formation of stakeholder advisory committees, and regularly scheduled workshops and meetings, for instance (DeLorme et al. 2016).

During the issue identification and analysis process, various tools are recommended to create opportunities for finding common ground to improve the success of transdisciplinary approaches. Blythe et al. (2017) focus on having participants identify interrelated components within the system as a bridge to finding common ground. Roux et al. (2017) suggest that 'bridging agents' are helpful for knowledge sharing. 'Bridging agents' (also known as 'boundary agents' and 'knowledge brokers') are individuals whose skill sets include strong social networks and social capital who can effectively interface between different perspectives and knowledge types (Roux et al. 2017). Blythe et al. (2017) also identify the need to embrace complexity and contrast in perspective and approach when weighing options. Roux et al. (2017) suggest that using boundary objects, such as models or maps, and encouraging discussion in novel and neutral locations are effective strategies for dealing effectively with transdisciplinary complexity.

The processes described above for the formation and function of transdisciplinary teams are unlikely to be universal and may vary on an issue by issue basis, potentially hampering reproducibility (Jahn et al. 2012). Indeed, outcomes and implementation for similar issues may be very different for transdisciplinary teams, depending on the team members involved and the issues and solutions defined.

Implementation of research results and translation into governance processes are considered integral components of transdisciplinary research (Brandt et al. 2013). In other words, the transdisciplinary process is not complete until outcomes have been implemented, which involves governance and management processes. However, translating research results into policy decisions is challenging as translation can be negatively impacted by power asymmetry in the transdisciplinary process (as described above) and occurs over different timelines and under different influences than research processes (e.g. political influences; Simon and Schiemer 2015). Ideally, these issues can be mitigated by ensuring that stakeholders and policy representatives are involved from initial stages through to, and including, translation.

Following these recommendations for the function of transdisciplinary teams, it follows that modes of interactive governance are a suitable lens for translating transdisciplinary research team outcomes. Kooiman et al. (2005, 17) defined interactive governance as "the whole of interactions taken to solve societal problems and create societal opportunities, including the formulation and application of principles guiding those interactions and care for the institutions that enable them." Ideally, by involving stakeholders in an ongoing, solution-driven research process, all team members become familiar with each point of view and possible solution. This sense of ownership of the process would be more likely than other approaches to encourage long-term awareness of and participation in governance (Kapoor 2001; Bulkeley and Mol 2003). Thus, transdisciplinarity can provide a specialized microcosm that reflects the macrocosm displayed by the issue at hand, as a transdisciplinary team would ideally include representatives of all viewpoints and specialties. The movement from macrocosm (fishery) to specialized microcosm (fishery governance bodies) allows for the complexity of the issues to be managed more effectively. This in turn suggests that applying transdisciplinarity via interactive governance processes that move from disagreement (on problems) to agreement (on solutions) can result in longer term success.

At first glance, using a transdisciplinary research approach to support this idealized governance process in inland fisheries appears as complex and challenging as the wicked problems the approach attempts to solve. However, a key feature of transdisciplinary approaches is that there is no *a priori* requirement to agree on the problem. Indeed, it can be expected that most parties involved in a transdisciplinary approach will not agree on the source or focus of the issue (Kahane 2017). However, in transdisciplinary approaches, while it is not necessary to agree on the source of the problem, it is necessary to agree on the solution (Jentoft and Chuenpagdee 2009; Kahane 2017).

18.3 Using Transdisciplinary Research Approaches to Address Governance Issues

The use of transdisciplinary research approaches in any field requires a more complex definition of a 'successful' solution. Since success in solving wicked problems eludes universal definition, it is better to say that we create effective solutions to existing problems in their current form, on the understanding that new forms of the same problem may arise (Rittel and Webber 1973; Game et al. 2014). Defining 'success' in solutions is subjective, but can we draw conclusions about an enabling environment that supports governance solutions that are broadly viewed as successful? What role does transdisciplinary research play in supporting the development of this enabling environment? Research examining attributes in fisheries management and decision making show stakeholder participation in policy and legislation development, defined boundaries and local leadership are part of what supports successful implementation of co-management initiatives in marine and coastal small-scale fisheries in Asia (Pomeroy et al. 2001). When success was defined as increases in stock abundance and unit price, a similarly oriented study by Gutiérrez et al. (2011) identified strong leadership, quotas (individual/community), social cohesion, and protected areas as the most important components of success. In the latter example, it is noteworthy that these criteria were established by the authors and based on a large-scale commercial paradigm. However, in both cases, success was more likely when multiple positive attributes (such as those identified above) were present in a fishery. It is important to recognize that numerous possible solutions could be viewed as successful and that no one solution is optimal, but the presence of similar 'enabling' attributes can be encouraged, and the likelihood of successful solutions (as defined by transdisciplinary teams) can be enhanced by encouraging enabling environments in transdisciplinary research processes. If implemented according to system- and issue-specific needs during team processes, translation stages can be used to connect optimum external (defined boundaries, legislation, external partners) and internal (local leadership, stakeholder participation) management attributes to promote an enabling environment of success (e.g. success of multiple attributes) in interactive governance.

Unfortunately, there are few examples of transdisciplinary research-based solutions to wicked problems in inland fisheries literature through which to identify attributes that could support an enabling environment for interactive governance in this sector. However, there are numerous case study examples that can help identify some of the most common scenarios in inland fisheries issues that could help guide the identification of these attributes. Here, we explore some examples of problem situations in inland fisheries to illustrate how these issues are transdisciplinary in nature, and discuss whether attributes similar to those that contributed to fisheries management successes could support successful governance solutions through transdisciplinary research. We discuss the ways in which transdisciplinary research has played, or could play, a role in developing solutions, and offer suggestions of a way forward towards improving transdisciplinary research through redefining successful solutions to wicked problems in inland fisheries. The first two examples (Laurentian Great Lakes, Fraser River basin) are compiled from external sources. The final example is from Too Big to Ignore's Inter-Sectoral Governance of Inland Fisheries Ebook, with chapter authors indicated in the case study title (Mhlanga and Nyikahadzoi 2017).

18.4 Case Study Examples

18.4.1 Pacific Salmon Science and Management in the Fraser Basin, British Columbia

Pacific salmon are socioeconomically, culturally, and ecologically valuable, and the Fraser Basin of British Columbia is the most important watershed for Pacific salmon in Canada. It is for this reason that the Pacific salmon fisheries in the Fraser are among the most intensively managed in the world. The governance system is complex and includes real time in-season management, co-management efforts between government and stakeholders, joint management between the US and Canada, and First Nation treaty rights. The fisheries are also complex, with three fishing sectors targeting adult Pacific salmon during the freshwater phase of their spawning migration: First Nations, commercial, and recreational. All three fishing sectors are managed by the Canadian Department of Fisheries and Oceans with a subset of fishing activities (those that are trans-boundary between Canada and USA at some point during migration) subject to the international Pacific Salmon Treaty. Given the complexity of the actors and institutions involved, it is not surprising that there has been occasional conflict. Nonetheless, collective concern for the state of the Pacific salmon resources has led to all parties embracing a science-based approach to management. Beyond the routine stock assessment needed to inform fisheries management, there have also been extensive efforts to engage in transdisciplinary research to support decision making. To that end, teams of researchers spanning the natural and social sciences have been assembled to tackle some of the more pressing challenges. For example, ongoing concern for understanding the fate of salmon released as bycatch led to a research program that included field studies that engaged various fishing sectors, laboratory studies to understand mechanisms, and parallel social science studies to understand fisher and manager perspectives on salmon science and management. Despite the coproduction of the research program, it took over 5 years and the accumulation of much evidence before there was opportunity to validate and operationalize this science into formal science advice via the Canadian Science Advisory Secretariat process (Patterson et al. 2017). The social science findings regarding barriers to knowledge mobilization were particularly important for understanding the factors that influence whether new knowledge will be accepted or rejected by managers and stakeholders (Young et al. 2016). The burden of evidence and long-standing (and ongoing) stakeholder/manager engagement were critical for generating meaningful changes in fisher behaviour and policy.

This case study shows several features described by the fisheries management example as supporting an enabling environment. Stakeholders representing different fisheries, scientists from multiple disciplines, managers, and policy-makers are involved in an ongoing research program and have a say in its development. This indicates that stakeholders participate on equal levels in the process, and the agreement to prioritize a science-based approach to management indicates a degree of social cohesion.

While this case study represents the most coherent transdisciplinary approach to inland fisheries research of those presented, it is important to note that it was not originally intended to be a transdisciplinary process. Indeed, this example shows that the benefits of a transdisciplinary approach become evident over time and accumulation of scientific evidence. This case study also illustrates the potentially lengthy timeline between research outcomes and implementation and translation processes of outcomes (into formal science advice, in this case).

18.4.2 The North American (Laurentian) Great Lakes

The North American (Laurentian) Great Lakes is a large-scale freshwater system that connects five major lakes (Lake Superior, Lake Michigan, Lake Huron, Lake Erie, and Lake Ontario) and smaller tributaries with a complex mosaic of human settlements and activities that stretch over 650 jurisdictional units ranging from municipal to bi-national (McCrimmon 2002). The Great Lakes fishery experienced several major developments throughout its history, including the decline of a commercial sector, the rising socioeconomic importance of a recreational fishery, as well as the persistent political struggle of Indigenous fishers in asserting their resource rights (Hudson and Ziegler 2014). Similar to the Fraser River example, the governance system for the Great Lakes fishery involves multiple groups, including joint management between US and Canada and advisory co-management through the Great Lakes Fishery Commission which includes representatives from stakeholder states and provinces. Supported by an ecosystem-based management approach (Minns 2013; Jackson 2015), the Great Lakes fisheries have over time presented several key examples of effective multi-jurisdictional collaboration, including the control of sea lamprey (*Petromyzon marinus*; Gaden et al. 2013). However, active collaboration among different stakeholder groups has proven a challenge. In particular, a lack of consideration of fishers' experiential knowledge among fisheries managers, as well as the historical failure to recognize fishing rights of Indigenous peoples have been lasting sources of contestation that have hindered progress towards mutually agreeable management successes (Norman 2015). Multidisciplinary research collaboration in the Great Lakes does occur, though Indigenous groups have, indeed, been the weakest partner (i.e. small and blue; Fig. 18.1; Song et al. 2016), suggesting this collaboration has not strongly informed management processes.

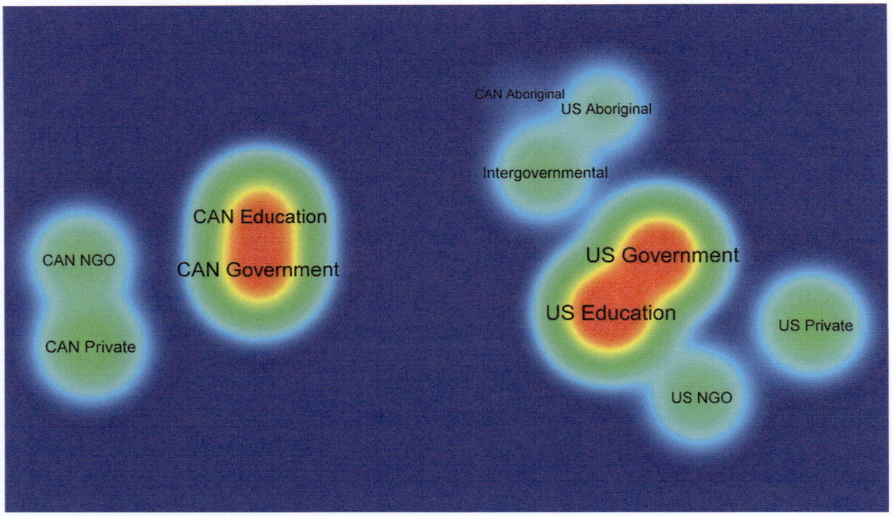

Fig. 18.1 Density visualization ('heat map') showing the pattern of research collaboration in the North American Great Lakes-based organizations. The Canadian Indigenous group (shown as CAN Aboriginal) is nearly invisible; detached from other Canadian groups and only closely supported by US Indigenous counterparts (shown as US Aboriginal) and the Great Lakes Fishery Commission. Node color indicates the level of collaboration each organizational type engages in. Distance between nodes denotes the intensity of collaboration (Source: Song et al. 2016)

Though many similar elements appear in the Great Lakes case study that can be seen in the Fraser River case study (stakeholder participation, research collaboration), the presence of ongoing resource-access conflict and negative perceptions surrounding Indigenous peoples' access to research collaboration indicate that this approach has not been as successful. The issue identified by Song et al. (2016) suggests that adopting a more idealized transdisciplinary approach, beginning with concrete strategies for ongoing active collaboration with Indigenous communities may help to narrow the gap identified in the research process and contribute to more successful long-term solutions to the issue. Further, the issue described in this case study highlights a critical point: transdisciplinary team formation and processes are not external to existing problems or conflict, but to function as intended will need to account for them and actively include conflicting viewpoints.

18.4.3 Competing Resource Claims on Lake Kariba, Mhlanga and Nyikahadzoi 2017

Lake Kariba on the Zambezi River is the world's largest man-made lake by volume, designed for hydropower generation and shared by Zambia and Zimbabwe. Fishing activities play a central role in the governance of Lake Kariba and include offshore commercial operations and inshore small-scale artisanal producers. Governance structures featuring transdisciplinary collaboration are not formalized at the national level, but there is an underfunded, bilateral committee set up between Zimbabwe and Zambia to discuss joint fisheries management issues. Despite massive support to enable governance-level collaboration (Jul-Larsen et al. 1998), fishing regulations and management differ between the two countries (Kolding et al. 2003, 2015), for example, with respect to allowable mesh sizes, and conflict occurs between gill net fishers and fisheries managers in both countries. The combination of historical racial segregation in Zimbabwe, large numbers of fishers in different fishing sectors (artisanal, offshore, recreational) in two countries sharing the same body of water has led to a series of conflicts and tensions between artisanal and offshore (kapenta) fishers over wealth and opportunity. For historical political reasons, artisanal/gillnet fishers are typically from the Black community and less wealthy, while kapenta fishers are typically from the White community and more wealthy. Between 1988 and 1997, a large bilateral and transdisciplinary management project was conducted to enable and facilitate collaborative processes (Jul-Larsen et al. 1998), which initially had good results. However, the deteriorating political situation in Zimbabwe caused isolation and retracted donor support resulted in a collapse of most activities. A permit redistribution exercise by the Zimbabwean government reallocated permits to Black community members, such that 80% of kapenta licenses are now owned by Black fishers, but the conflict between the two sectors remains a recurring conflict.

To develop long term successful solutions, decision-makers will have to address resource status (natural science), regulatory measures (law), historical racial tensions (social science), and the links between these issues and fishery activity (social-ecological systems approach). Solutions to these problems appear to be nascent, but a transdisciplinary research approach could be used at multiple scales here: both to address the historical tensions among fishers, and to address the different management methods between countries. Thus, the Kariba case illustrates that if the political landscape is not conducive, or external constraints not identified, then even the best intentions may not flourish. Cross-boundary collaboration at the governance level would be required for translation of relevant research outcomes, suggesting that, in this case, transdisciplinary research may require development of interactive governance systems prior to successful initiation.

18.5 Patterns of Governance Problems and Solutions in Case Studies

Collectively, the case studies presented here illustrate the 'wicked' nature of governance problems in inland fisheries. Competition for resources among inland system users, unequal power dynamics, shared boundaries, and a lack of communication among parties are featured in the examples provided, and showcase the additional complexity of inland fisheries problems. The case studies also illustrate complex management and conflict situations occurring at different scales (local to international) that governance systems are required to address. Attributes of successful solutions that were identified for fisheries management also appear in these inland case studies. For example, social cohesion and stakeholder participation evidently played a role in the successful development of formal scientific advice in the Fraser River case study, as all parties agreed to embrace a science-based approach to management. In contrast, the North American Great Lakes management example showed that Canadian Indigenous fishers are insufficiently recognized in research collaboration, which indicates that their essential perspectives are not adequately incorporated into the process of developing solutions.

The case studies also serve to highlight the differences between the idealized processes in transdisciplinary research and the real-world processes likely to constrain development or execution of transdisciplinary processes. For example, in the Great Lakes case study, we see how failure to collaborate with stakeholder groups effectively during research phases can lead to negative stakeholder perceptions of solutions. Additionally, while we argued in the earlier sections that transdisciplinary research teams can support interactive governance processes, the Lake Kariba example illustrated the uneasy relationship between research and management, where deep, unresolved political tensions prevented willingness to implement interactive governance processes.

In addition to the challenges posed by potentially inter-dependent research and governance processes, a further challenge in highly complex systems is that solutions to wicked problems can have unintended consequences in one or more of the system attributes (Game et al. 2014), resulting in unforeseen tradeoffs. Social and economic tradeoffs often compromise the ecological system, which results in instances where the ecological system is harmed by both the conflict and the solution, which may result in increased risk to the social-ecological system over time (e.g. Nayak et al. 2016). The harm in the tendency to favour tradeoffs that negatively impact the longer-term integrity of the ecosystem is compounded by the challenge of measuring current system performance (Game et al. 2014). This makes it harder to quantify negative effects on the ecosystem. The use of a transdisciplinary research teams for developing management solutions can alleviate this risk, particularly if representatives from all viewpoints participate in the solution creation process, as these consequences are more likely to be foreseen. However, it is likely that unforeseen consequences of management solutions will occur, and indeed, solutions developed through transdisciplinary processes may also lead to unforeseen consequences.

18.6 Enabling Environments in Transdisciplinary Approaches

In much the same way that an enabling environment of success has been determined for fisheries management, we see here that a similar enabling environment for success needs to be identified for transdisciplinary research approaches to solution development in inland fisheries. A crucial point is that these enabling environments are likely synonymous. Blythe et al. (2017) note that transdisciplinarity is challenged by favouring dominant paradigms (such as favouring natural sciences over social sciences; Visser 2004), in much the same way that fisheries management is challenged by lack of inclusion of marginalized communities or inadequate representation (Pomeroy et al. 2001). If one discipline or paradigm dominates the dialogue process, outcomes will be hampered, and essential inputs lost (Kolding and van Zwieten 2011; Blythe et al. 2017). The case study examples illustrate how difficult this problem can be to overcome.

Pomeroy et al. (2001) noted that clearly defined boundaries supported positive fisheries management outcomes, and Roux et al. (2017) suggest that exploring boundaries through various mechanisms can support transdisciplinary approaches by clearly defining perspectives and finding common ground. It was difficult to discern the nature of boundary exploration beyond geographical boundaries and delineation among representative groups (such as stakeholders and scientists) in the case studies. Indeed, the connectivity of inland systems and the inherent complexity of inland fisheries governance structures suggest that clearly identifying and exploring boundaries and external constraints early in a transdisciplinary research process may be of particular use in solving problems in inland fisheries systems.

Some challenges to adopting transdisciplinary research approaches may be more likely to occur for inland fisheries than in other sectors. It is reasonable to posit that a similar enabling environment for inland fisheries governance (through transdisciplinary research) to that of fisheries management will also involve consideration and inclusion of multiple factors (e.g. representation, knowledge brokers, transparency, neutral locations). However, the number and diversity of stakeholders in inland fisheries issues suggests that the need for both representative and epistemological diversity will be a key feature in successful inland fisheries governance.

18.7 Conclusions/Recommendations

Although transdisciplinary research is not necessarily new, the value of a transdisciplinary research approach for developing long-term, successful solutions to wicked problems is increasingly relevant to inland fisheries issues. In this chapter, we have explored the ways in which inland fisheries governance issues can be viewed as wicked problems that would benefit from a transdisciplinary research approach. We have explored the enabling environment of success surrounding long-term solutions in fisheries, examined case studies to see whether this enabling environment was present, and how transdisciplinary research played, or could have played, a role in fostering positive outcomes. It is noteworthy that none of the case study examples offer 'completed' processes or permanent solutions, but rather reflect on ongoing and dynamic nature of inland fisheries governance. We ended by examining some key challenges in adopting transdisciplinary approaches and considered how these challenges would apply to inland fisheries scenarios. We conclude by summarizing chapter outcomes to support transdisciplinarity in problem solving for inland fisheries governance:

- The case studies of inland fisheries examined in this paper highlight the wicked nature of inland fisheries governance problems and the realized or potential value of transdisciplinary approaches to solving them;
- Issue-specific solutions may be well-served by use of processes known to solve wicked problems that include features involving ongoing dialogue, interactivity, and iterative decision making: all are necessary components of transdisciplinary research approaches;
- Certain factors applying to representation (e.g. marginalized or unrepresented interest groups), boundary definition, and paradigm dominance are likely to be common in inland fisheries transdisciplinary approaches due to the number of competing sectors/stakeholder types and should thus be planned for and prioritized;
- Identifying attributes that foster an enabling environment for successful translation of transdisciplinary research outcomes to governance processes should be a research priority for inland fisheries.

Acknowledgements We would like to thank the TBTI Inland Fisheries Cluster members for their participation in the Inland Fisheries Governance Ebook, which contributed greatly to this chapter. We would also like to extend our sincere thanks to our reviewers, whose careful attention to this manuscript was immensely helpful. We hope this chapter is of value to anyone wishing to explore transdisciplinarity in inland fisheries research.

References

Andrew NL, Béné C, Ratner BD et al (2007) Diagnosis and management of small-scale fisheries in developing countries. Fish Fish 8(3):227–240

Beard TD, Arlinghaus R, Cowx IG et al (2011) Ecosystem approach to inland fisheries: research needs and implementation strategies. Biol Lett 23:481–483

Béné C (2006) Small-scale fisheries: assessing their contribution to rural livelihoods in developing countries. FAO, Rome

Béné C, Hersoug B, Allison EH (2010) Not by rent alone: analyzing the pro-poor functions of small-scale fisheries in developing countries. Dev Policy Rev 28(3):325–358

Blythe J, Nash K, Cumming G et al (2017) Feedbacks as a bridging concept for advancing transdisciplinary sustainability research. Curr Opin Environ Sustain 26:114–119

Brandt P, Ernst A, von Wehrden H et al (2013) A review of transdisciplinary research in sustainability science. Ecol Econ 92:1–15

Bulkeley H, Mol AP (2003) Participation and environmental governance: consensus, ambivalence and debate. Environ Value 12(2):143–154

Chuenpagdee R, Song AM (2012) Institutional thinking in fisheries governance: broadening perspectives. Curr Opin Environ Sustain 4(3):309–315

Crona B, Bodin Ö (2010) Power asymmetries in small-scale fisheries: a barrier to governance transformability. Ecol Soc 15(4):32

DeLorme DE, Kidwell D, Hagen SC et al (2016) Developing and managing transdisciplinary and transformative research on the coastal dynamics of sea level rise: experiences and lessons learned. Earth's Future 4:194–209. https://doi.org/10.1002/2015EF000346

FAO (United Nations Food and Agriculture Organization) (2015) Voluntary guidelines for securing sustainable small-scale fisheries in the context of food security and poverty eradication. FAO, Rome

Gaden M, Goddard CI, Read J (2013) A history of multi-jurisdictional management of the shared Great Lakes fishery: transcending conflict and diffuse political authority. In: Taylor W, Lynch A, Leonard N (eds) Great Lakes fishery management and policy. Michigan University Press, East Lansing, pp 305–337

Game ET, Meijaard E, Sheil D et al (2014) Conservation in a wicked complex world; challenges and solutions. Conserv Lett 7(3):271–277

Gutiérrez NL, Hilborn R, Defeo O (2011) Leadership, social capital and incentives promote successful fisheries. Nature 470(7334):386–389

Hall KL, Vogel AL, Stipelman BA et al (2012) A four-phase model of transdisciplinary team-based research: goals, team processes, and strategies. Transl Behav Med 2(4):415–430

Hudson JC, Ziegler SS (2014) Environment, culture, and the Great Lakes fisheries. Geogr Rev 104(4):391–413

Jackson SA (2015) A fresh perspective for marine fisheries: ecosystem based fisheries management in the Laurentian Great Lakes. Dissertation, University of Washington

Jahn T, Bergmann M, Keil F (2012) Transdisciplinarity: between mainstreaming and marginalization. Ecol Econ 79:1–10

Jentoft S, Chuenpagdee R (2009) Fisheries and coastal governance as a wicked problem. Mar Policy 33(4):553–560

Jentoft S, Chuenpagdee R (2015) Interactive governance for small-scale fisheries: global reflections. Springer International Publishing, Dordrecht

Jul-Larsen E, Graça FB, Nielsen JR et al (1998) Research and Fisheries Management; the Uneasy Relationship. Review of the Zambia-Zimbabwe SADC Fisheries Project. Chr. Michelsen Institute, R1998:1, Bergen Norway. https://brage.bibsys.no/xmlui/handle/11250/2435726. Accessed 15 Apr 2018

Kahane A (2017) Collaborating with the enemy: how to work with people you don't agree with or like or trust. Berrett-Koehler Publishers, Oakland

Kapoor I (2001) Towards participatory environmental management? J Environ Manage 63(3):269–279

Khan AS, Neis B (2010) The rebuilding imperative in fisheries: clumsy solutions for a wicked problem. Prog Oceanogr 87(1):347–356

Kolding J, van Zwieten PA (2011) The tragedy of our legacy: how do global management discourses affect small scale fisheries in the south? In Forum for development studies 38(3). Routledge, pp 267–297

Kolding J, Musando B, Songore N (2003) Inshore fisheries and fish population changes in Lake Kariba. In: Jul-Larsen E, Kolding J, Nielsen JR et al (eds) Management, co-management or no management? Major dilemmas in southern African freshwater fisheries. Part 2: case studies, FAO fisheries technical paper 426/2. FAO, Rome

Kolding J, Béné C, Bavinck M (2014) Small-scale fisheries – importance, vulnerability, and deficient knowledge. In: Garcia S, Rice J, Charles A (eds) Governance for marine fisheries and biodiversity conservation interaction and coevolution. Wiley-Blackwell, Oxford, pp 317–331

Kolding J, Jacobsen NS, Andersen KH et al (2015) Maximizing fisheries yields while maintaining community structure. Can J Fish Aquat Sci 73(4):644–655

Kolding J, van Zwieten PAM, Mosepele K (2016) Where there is water there is fish – small-scale inland fisheries in Africa: dynamics and importance. In: Tvedt T, Oestigaard T (eds) A history of water 3(3) Water and food: from hunter-gatherers to global production in Africa. I.B. Tauris, London

Kooiman J, Bavinck M, Jentoft S et al (2005) Fish for life. Amsterdam University Press, Amsterdam

Lynch AJ, Cooke SJ, Beard TD Jr et al (2016) The social, economic, and environmental importance of inland fish and fisheries. Environ Rev 24:115–121

Lynch AJ, Cowx IG, Youn S et al (2017) Inland fisheries – invisible but integral to the UN Sustainable Development Agenda for ending poverty by 2030. Global Environ Chang 47:167–173

McCrimmon DA Jr (2002) Sustainable fisheries management in the Great Lakes: scientific and operational challenges. Lake Reserv Manage 7(3):241–254

Mhlanga W, Nyikahadzoi K (2017) Competing claims in a multipurpose lake: mapping resource conflicts on Lake Kariba. In: Song AM, Bower SD, Chuenpagdee R et al (eds) Inter-sectoral governance of inland fisheries, TBTI publication series, E-01/2017, Too Big to Ignore. WorldFish, St John's, pp 71–83

Minns CK (2013) The science of ecosystem-based management on a global scale: the Laurentian Great Lakes, Lake Ontario, and the Bay of Quinte as a nested case study. Aquat Ecosyst Health 16(3):229–239

Nayak PK, Armitage D, Andrachuk M (2016) Power and politics of social–ecological regime shifts in the Chilika lagoon, India and Tam Giang lagoon, Vietnam. Reg Environ Chang 16:325–339

Nguyen VM, Lynch AJ, Cooke SJ et al (2016) To manage inland fisheries is to manage at the social-ecological watershed scale. J Environ Manage 181:312–325

Norman ES (2015) Governing transboundary waters: Canada, the United States, and indigenous communities. Routledge, New York

Paterson B, Isaacs M, Hara M et al (2010) Transdisciplinary co-operation for an ecosystem approach to fisheries: a case study from the South African sardine fishery. Mar Policy 34(4):782–794

Patterson DA, Robinson KA, Cooke SJ et al (2017) Guidance to derive and update fishing-related incidental mortality rates for Pacific Salmon. DFO Canadian Science Advisory Secretariat. Doc. 2017/011. vii + 56 p

Pomeroy RS, Katon BM, Harkes I (2001) Conditions affecting the success of fisheries co-management: lessons from Asia. Mar Policy 25(3):197–208

Rittel HWJ, Webber MM (1973) Dilemmas in a general theory of planning. Policy Sci 4:155–169

Roux DJ, Nel JL, Fabricius C et al (2017) Transdisciplinary research for systemic change: who to learn with, what to learn about and how to learn. Sustain Sci 12:1–16

Sathyapalan J, George S (2015) Governability challenges in sustaining small-scale fisheries in an urban context: a study of Cochin Backwaters, India. In: Jentoft S, Chuenpagdee R (eds) Interactive governance for small-scale fisheries. Springer International Publishing, Dordrecht, pp 85–99

Simon D, Schiemer F (2015) Crossing boundaries: complex systems, transdisciplinarity and applied impact agendas. Curr Opin Environ Sustain 12:6–11

Song AM, Hickey GM, Krantzberg G et al (2016) Assessing transboundary scientific collaboration in the Great Lakes of North America. J Great Lakes Res 42(1):156–161

Song AM, Bower SD, Chuenpagdee R et al (2017a) Inter-sectoral governance of inland fisheries: research needs and foci. In: Song AM, Bower SD, Chuenpagdee R et al (eds) Inter-sectoral governance of inland fisheries, TBTI publication series, E-01/2017, Too Big to Ignore. WorldFish, St John's, pp 1–17

Song AM, Scholtens J, Chuenpagdee R et al (2017b) Transboundary research in fisheries. Mar Pol 76:8–18
Visser LE (2004) Reflections on transdisciplinarity, integrated coastal development, and governance. In: Visser L (ed) Challenging coasts. Transdisciplinary excursions into integrated coastal zone development, Mare publication series 1. Amsterdam University Press, Amsterdam, pp 23–47
Welcomme RL, Cowx IG, Lorenzen K (2010) Inland capture fisheries. Philos Trans R Soc B 365(1554):2881–2896
Young N, Corriveau M, Hinch SG et al (2016) How do potential knowledge users evaluate new claims about a contested resource? problems of power and politics in knowledge exchange and mobilization. J Environ Manage 184:380–388

Chapter 19
Governing the Governance: Small-Scale Fisheries in Europe with Focus on the Baltic Sea

Milena Arias-Schreiber, Sebastian Linke, Alyne E. Delaney, and Svein Jentoft

Abstract 'Governing the governance' refers to the overarching values and interests on which governing institutions are built and operate relative to small-scale fisheries. Who are these values representing and how consistent are institutional settings with respect to them? How effective are existing institutions for accommodating diverse values in their governing practices? This chapter explores the governance system and experiences of small-scale fisheries in the Baltic Sea, particularly in the context of the European Union (EU) Common Fisheries Policy (CFP). We focus on the EU governing system and its relation with the values and interests of small-scale fisheries and the organizations representing them, and evaluate the scope for the governability of small-scale fisheries in the Baltic Sea. Our chapter use the case of fishers' representation in the EU Baltic Sea Advisory Council for this purpose and discusses the consequences of a missing link between small-scale coastal fisheries' values and the overarching EU fisheries governance system regulated by the CFP. Finally, we explore how this missing linkage might change under current regional and local initiatives to support small-scale fisheries in Europe, and conclude on the governability and transdisciplinary challenges this shift may entail.

M. Arias-Schreiber (✉)
Swedish Institute for the Marine Environment, Gothenburg University, Gothenburg, Sweden
e-mail: milena.schreiber@gu.se

S. Linke
School of Global Studies, Gothenburg University, Gothenburg, Sweden
e-mail: sebastian.linke@gu.se

A. E. Delaney
Innovative Fisheries Management, Aalborg University, Aalborg, Denmark
e-mail: ad@ifm.aau.dk

S. Jentoft
Norwegian College of Fishery Science, UiT-The Artic University of Norway, Tromsø, Norway
e-mail: svein.jentoft@uit.no

R. Chuenpagdee, S. Jentoft (eds.), *Transdisciplinarity for Small-Scale Fisheries Governance*, MARE Publication Series 21,
https://doi.org/10.1007/978-3-319-94938-3_19

Keywords Small-scale fisheries · Baltic Sea · Common Fisheries Policy · Stakeholder participation · Values · Governability

19.1 Introduction

Fish and fishing in the Baltic Sea is more than a source of food, employment, and income to support livelihoods. For centuries, fishing has represented a component of the way of life, place identity, and distinctiveness for inhabitants along the Baltic coast (Højrup 2011; Järvik et al. 2012; Kulmala et al. 2013). For this reason, Baltic Sea small-scale fishers continue to apply many traditional fishing techniques and associated knowledge, resisting industrialization and vast technological developments (e.g., Højrup 2011). During the last decades, these small-scale fishers are no longer thriving and their numbers have decreased significantly (Björkvik 2013; Papaioannou et al. 2014).[1]

Many Baltic small-scale fishers have abandoned their fishing occupation and the recruitment of new fishers is nowadays low or non-existent (Krogseng 2016; STECF 2017). Consequently, small-scale and coastal fisheries are vanishing around the Baltic Sea, and with them, cultural heritage, fishers' contributions to coastal communities, provision of fresh local food, traditional knowledge, and tourist attractions (Strehlow 2010; Björkvik 2013). Maladapted or ambiguous policy and governance approaches for small-scale fishers and their communities play a crucial role in these developments (Johnsen and Vik 2013; Hentati-Sundberg et al. 2014). A key feature of this problematic governance situation pertains to the 'wickedness' of the problem, in other words, the multifaceted features of small-scale coastal fisheries and the challenge to define it as a single problem, making possible solutions difficult to identify and develop (Jentoft and Chuenpagdee 2009).

As conceptualized by interactive governance theory (Kooiman and Bavinck 2005), a fishery can be divided into three main building blocks: the *governing system* (social and natural), the *system-to-be-governed*, and the *governing interactions*. Governance is, at the same time, a result of the interactions between governing institutions (the *governing system*) and the targeted social-natural system (*system-to-be-governed*) (Kooiman 2008). Despite institutional inertia, path dependency, and socio-ecological resilience, these systems and their interactions are not static and alterations over time are inevitable. In the Baltic Sea, academic research on changes in the *system-to-be-governed* has mostly focused on the systematic assessment of the ecological system and has guided science-based policy and decision making (Österblom et al. 2010; Gilek et al. 2016). Social changes in the fisheries *system-to-be-governed* and particularly related to coastal and small-scale fisheries

[1] In the Baltic states, the number of fishers, including small-scale fishers, increased after independence in 1991, when Soviet restrictions for going out to the sea were abolished, market regulations relaxed, and fishing became uncontrollable. In Estonia, the number of fishers dropped again over the present century caused by reduced fish stocks (see Raid and Järvik 2008; Järvik et al. 2012).

have been seldom addressed despite significant changes in the *governing system* due to the expanding role of the European Union (EU) and its Common Fisheries Policy (CFP) (Sellke et al. 2016). Our chapter addresses this gap by exploring the extent to which small-scale fisheries and their values are considered in the Baltic Sea governance system, particularly under the new CFP. While this policy's aims to improve the status of fish stocks have been, recently, to some extent acknowledged (Gascuel et al. 2016; STECF 2017), our chapter claims that small-scale coastal fisheries sustainability has been a neglected issue that deserves more attention.

The chapter is divided into four sections. This introduction is followed by an overview of small-scale fisheries' positioning under the umbrella of the EU's CFP, emphasizing the Baltic small-scale fisheries representation in the governance system. In the second section, using published interviews with Baltic Sea small-scale fishers, a range of values expressed by these fishers are identified and analyzed in terms of the quality dimension of their fishery's governability. The third section introduces two recently established initiatives (one regional and one local) that might bring possible transitions to enhance the governability of the Baltic small-scale fisheries and change the current situation. The paper concludes with the prospects of small-scale fisheries in the Baltic Sea and the coastal communities that still rely on them.

19.2 Governing Small–Scale Fisheries Under the Common Fisheries Policy (CFP)

Since its establishment in 1983, the CFP serves as the overarching governing framework for fisheries management in EU member states. Through this policy, the EU legislates at a supranational level imposing directives and regulations for fisheries management. The CFP grants each member state a 12 nautical mile zone around their coastlines, where states decide upon fisheries regulations and management. This area is alleged to be the edge to which small-scale fisheries fleets generally operate. With this approach, the CFP aims at handing over the management of the small-scale fisheries sector to its member states. Yet, splitting the administration of fisheries in coastal and offshore waters did not enhance the coastal small-scale fisheries governance in the EU. Even though the 12 nautical mile separation seems to empower the member states with the governance of small-scale fisheries, as we will discuss below, overarching regulations and laws imposed through the CFP directly impacts small-scale fishing fleet operations.

Over the years, the primary objective of the CFP has been to improve or maintain the yields of commercial fish stocks and balance the fleet's fishing capacity with the size of targeted fish stocks. To maintain fish stocks, fisheries are managed through a system of quota setting and other technical measures informed by scientific advice. The EU small-scale fisheries sector is directly affected by the quota system, since they target many of the regulated commercial fish stocks. Additionally, measures to reduce fishing fleets, so called ‘structural policy’, have also involved smaller vessels (Guyader et al. 2013). Therefore, the CFP approach to small-scale fisheries can be depicted as *laissez faire*, with, at best, the sloppy assumption that a focus that targets large-scale fisheries will not have detrimental impacts on small-scale fisheries. The problem is exacerbated with the difficulties in EU policies in defining small-scale fisheries, which is based on the size of the vessel disregarding, for example, their impact on the marine ecosystem or operational radios. This resulted in a neglect, or at least an underestimation, of EU small-scale fisheries’ ecological and societal contributions (Guyader et al. 2013; Lloret et al. 2016). Moreover, member states fisheries policies pay little attention to the small-scale fisheries sector by directing their fishing policies towards ecological and economic sustainability regardless of the scale of the fleet (De Vos and Kraan 2015; Pita et al. 2015). The establishment of Marine Protected Areas without small-scale fishers’ consultation finally resulted in the loss of their traditional fishing grounds, as have been reported in Spain (Gómez et al. 2006) and Malta (Said et al. 2017). National policies have, in some cases, clearly favored more profitable large-scale fisheries interests. This is, for example, the case of the introduction of Individual Transferable Quotas in the Danish and the pelagic Swedish fishery, a policy that fits well with fisheries based on mobility, large investments, and economies of scale to achieve profitability, in contrast to small-scale fishers who lack financial capital, and normally avoid large investments, considering them too risky (Høst 2015; Boonstra and Hentati-Sundberg 2016). There is a shortage of attention to coastal and small-scale fisheries from the member states in the Baltic Sea, for example, in Denmark (Høst 2015), Germany (Papaioannou et al. 2014), and Sweden (Boonstra and Hentati-Sundberg 2016; Arias-Schreiber et al. 2017).

As mentioned above, commercial fish stocks are managed in the EU through the system known as Total Allowable Catch and quotas. The national allocation of these quotas among fishing operators of different fleets is the decision of each member state. The most common allocation system is based on ‘catch history’, where quota allocation is proportional to the calculation of fisher’s historical landings as a percentage of the fishery’s total landings. This system has adversely transformed the EU small-scale fisheries sector, which is constantly struggling to get more equitable shares based on criteria like environmental and social benefits (i.e. employment) of the operators. As explained by a small-scale fisher during a meeting for the Baltic Sea Advisory Council:

> When the introduction of the EU national quota system was a rumor many years ago, the small-scale fishers didn’t believe that the EU was going to be able to control our catches. We have been fishing for centuries with no such controls. When we ask now the EU people and our authorities for larger quotas, they reply asking as: why didn’t you ask for them from the very beginning? – retired Baltic Sea coastal fisher, interviewed in Hamburg 2017.

19.2.1 The Baltic Sea Small–Scale Fisheries

The Baltic coastline stretches along nine countries including Finland and Russia, the Baltic countries (Estonia, Latvia, and Lithuania), Denmark, Sweden, Poland, and Germany. All these countries own fishing fleets and, with the exception of Russia, are currently member states of the EU.

Although comprehensive statistics are missing, research on the small-scale fisheries in the Baltic Sea shows a decrease in the number of vessels and fishers over the last century (Delaney 2007; Björkvik 2013; Guyader et al. 2013; Papaioannou et al. 2014; Salmi 2015), and the first decade of the twenty-first century. Countries like Lithuania, Poland, and Sweden decreased their number of small-scale vessels by more than 30% between 2000 and 2010 (Lloret et al. 2016). In 2015, the size of the small-scale fleet was estimated at 5626 vessels, which is equivalent to 91% of the total Baltic fleet and constitutes around 9% of the total small-scale fisheries fleet in the EU (Lloret et al. 2016; STECF 2017). Almost 7000 fishers are estimated to be involved in this fishery.

Internationally, the Baltic Sea is perceived as an area of proactive political development with a new history of collaboration between countries since the end of the European east-west division in 1992 (Kern and Löffelsend 2008). However, these attributes for good governance did not overcome the problem of overfishing, with its associated ecological, economic, and social impacts (Gilek et al. 2016). Prior to 2004 and the insertion of Poland and the Baltic countries (Estonia, Lithuania and Latvia) in the EU, the states bordering the Baltic Sea managed transnational fisheries issues multilaterally via the *International Baltic Sea Fisheries Commission*. With the EU enlargement, the commission became redundant and ceased its activities in 2005. Currently, the two remaining parties, the EU and Russia, arrange bilateral fisheries agreements, approved by EU Council Regulations. Thus, over the last decades, fisheries governance in the Baltic Sea region has witnessed a process of significant 'Europeanisation' (Gilek et al. 2016).

As for the whole of the small-scale fisheries sector in the EU, Baltic Sea small-scale fisheries are governed by the CFP by a multitude of measures and regulations, including: (1) the allocation of quotas based on Total Allowable Catch recommendations for commercially important species (cod, herring, sprat, plaice, and salmon); (2) gear limitations; (3) multi-annual plans for the management of certain fish stocks; and (4) spatial and seasonal fishery closures (Papaioannou et al. 2014). Baltic Sea small-scale fisheries are also integrated within the framework of inter-regional institutions such as the Helsinki Commission (HELCOM) from 2000, the Baltic Sea Advisory Council (BSAC) from 2006, and the Baltic Fisheries Forum BALTFISH established in 2010 (Fig. 19.1). Regarding fisheries, HELCOM aims at ensuring the *ecological* sustainability of the Baltic Sea, while recognizing the economic and cultural contributions of fisheries. For instance, HELCOM's Baltic Sea Action Plan adopted in 2007 states that *"coastal fisheries are of great importance to the society from both a socio-economic and a cultural point of view*" (HELCOM 2007, 21). Using the case of the BSAC, we show below how these organizations contributed little to support the governance quality of small-scale fisheries in the Baltic.

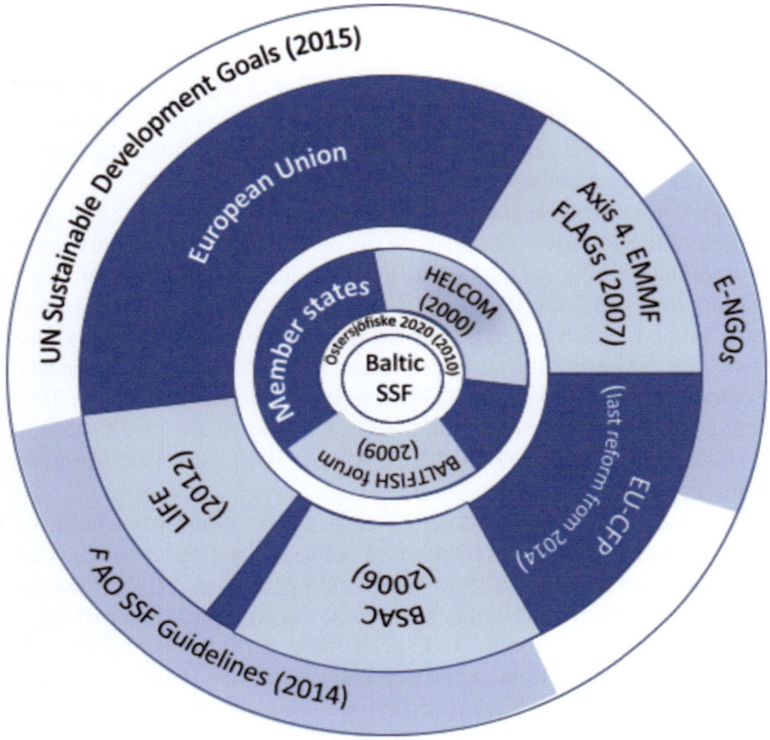

Fig. 19.1 Complex constellation of institutional actors interacting in the small-scale Baltic Sea governing system

19.2.2 The Scope for Small-Scale Fisheries Representation in the CFP

This section discusses challenges for stakeholder representation from small-scale fisheries in the overarching regional governance framework of the Baltic Sea. As part of the CFP's regionalization (Raakjær and Hegland 2012), and reforms for stakeholder participation, so-called *Regional Advisory Councils* were established between 2004 and 2008 (now just *Advisory Councils*, ACs). The ACs are comprised of 60% representatives of fisheries interests (mainly large-scale national fishing organizations) and 40% so-called 'other interests', which are dominated by concerns of environmental non-governmental organizations (ENGOs). The Baltic AC was founded in 2006 and has been working thoroughly since then as the central forum to connect different stakeholder interests around the Baltic Sea and providing recommendations to the EU Commission on fisheries matters. ACs also became an object for fisheries social science, analyzing the scope and procedures for stakeholder participation (see Ounanian and Hegland 2012 for the BASC, Linke and Jentoft 2013, 2014, 2016), knowledge exchange and inclusion (see Long 2010;

Griffin 2013; Hatchard and Gray 2014), as well as the overall contribution of ACs to fisheries governance in the EU. While the academic literature also discusses ambiguous features of the ACs, with regard to their capacities for meaningful participation, prospects for transdisciplinary knowledge inclusion, and democratic issues of representation, the EU Commission has, after a few years, drawn positive conclusions: "Despite the difficulties encountered in the start-up phase" regional ACs have "*already made a positive contribution to the development of the CFP*" (COM 2008). As expressed by Raakjær and Hegland (2012, 7), they might be understood most realistically as an "*interim institutional stage … facilitating better information sharing and cultivating stakeholder relationships*". However, questions pertaining to the representation of small-scale fishers and community interests have not been discussed to a significant degree. As mentioned above, local fishing communities and the small-scale fisheries sector experience the consequences of decision making at the EU level (CFP), which is informed by regional-level inputs from the ACs. However, their interests are not automatically considered when large-scale national fishing organizations and ENGOs negotiate issues at AC meetings. The gap in EU fisheries stakeholder representation resulting from this mismatch has been increasingly voiced since the most recent CFP reform in 2013. This reform process took place parallel to the development of the Voluntary Guidelines for Securing Sustainable Small-Scale Fisheries in the Context of Food Security and Poverty Eradication (Jentoft 2014; Arias-Schreiber et al. 2017). It might, hence, be no sheer coincidence that the issue of small-scale fishers' representation in the ACs is featured prominently in the new basic regulation of the CFP stating that "*Member States should endeavor to give preferential access for small-scale, artisanal or coastal fishermen*" (EC 2013, 24). The emphasis on addressing the small-scale fisheries sector as a politically relevant stakeholder is also chiefly reflected in a letter from the Fisheries Commissioner at the time of the CFP reform, Maria Damanaki, to the ACs (dated 4 October 2013):

> In order to be able to carry out this role effectively ["proactive advisory role"], it is essential that the ACs have the necessary representativeness, i.e. that all legitimate stakeholders have a fair opportunity to participate and express their view. In this context, it is one of my priorities to ensure that small-scale fishermen, as well as all other legitimate stakeholders, have a real impact in the decision-making process through their effective participation in the Advisory Councils. It is particularly important for small-sale fishermen to organize themselves and become truly actively involved in the activities of the Advisory Councils … so that small-scale fishermen have a voice in Europe, and I am confident that progress can be made on this issue.

While the North Sea AC responded to the letter (NSAC 2013), the issue has to our knowledge not been seriously addressed for all ACs. As Damanaki's letter calls for a new realization of small-scale fisheries interests in the ACs, perhaps other types of organizations might be more relevant for representing small-scale and local community stakeholders at national, regional, or international levels.

One such type that the EU has established is the *Fisheries Local Action Groups* (FLAGs) . FLAGs were the key instruments of the Axis 4 of the *European Maritime Fisheries Fund* to support fishing communities throughout Europe (Linke and

Bruckmeier 2015; Phillipson and Symes 2015).[2] While FLAGs are highly relevant institutions for supporting local fishers and their communities, they have no formal link in the EU fisheries management system or any established form of cooperation between these groups and the ACs. Questions regarding whether such a connection should be established have been raised, for instance, when the Bornholm FLAG applied to become a member of the BSAC in 2012. The BSAC administration thereupon sent a letter to the EU Commission, which triggered a response (letter of 2012 copied to all ACs; BSAC 2012) highlighting general issues of local community fishing interest's representation. Although the EU Commission admitted that some FLAGs may represent legitimate stakeholders, the response letter stated clearly that this was not generally the case. It is noted that the large number of FLAGs (ca. 150 around the Baltic at that time) would generate practical problems for their representation in ACs.[3] However, it is also stated in the letter that FLAGs may still become 'observers' of AC meetings, though this is everyone's right, as meetings are open. In conclusion, the Commission's letter states that "RAC membership does not seem a suitable nor workable solution for FLAGs, but participation in meeting as observers, on a case by case basis, and coordination and contacts between FLAGs and RACs who so wish, could be alternative solutions" (BSAC 2012). Regardless of these complications involved when including small-scale operators via FLAGs in their current form in EU fisheries policy and decision making, these interests are by and large excluded from informing decision making. Small-scale fishers and local fishing communities are presently not at all represented at the regional or EU level through any coherent or bridging institution, as no organizations at local or higher scales exist to perform their representative function. This is a gap, which the more recently established organization LIFE (*Low Impact Fishers of Europe*) tries to address, for example, through their recent initiative 'LIFE in the Advisory Councils' (LIFE 2017, see below).

19.2.3 Baltic Sea Small–Scale Fisheries Values and the CFP

Like interests and power, collective values also underpin any practical decision regarding fisheries governance. These values shape the understanding of how the fisheries system works and how to act in particular situations (Song et al. 2013; Young et al. 2016). They also determine the perspective of stakeholders when they argue on fisheries management matters. Values and interests do not only affect the decision-making process but also the institutional frameworks where decisions are taken and implemented (Kooiman and Jentoft 2009). Thus, when people argue about representation and participation in fisheries, they are likely to attribute values

[2] FLAGs are now an instrument of the Community-led Local Development program of the European Maritime Fisheries Fund. See www.farnet.eu

[3] The BSAC for example has only 26 seats in its Executive Committee and 40 members of the General Assembly (www.bsac.dk).

to these aspects of governance, which may not have been specific to fisheries but to governance as a democratic process.

Values, defined here as *means of what is important* for a fisher, have been reported in small-scale fisheries literature. Besides fishing as a mean of supporting livelihoods, it has been reported that small-scale fishers value fishing for representing a specific *way of life* with a meaning in itself where profit is as valuable as other social commitments (Sønvisen 2014; Onyango 2015; Krogseng 2016). Fishing is also valued because it provides small-scale fishers with a communal and personal identity (van Ginkel 2001) or a sense of independence, self-reliance, and freedom (Delaney 2003; Björkvik 2013). Similarly, Song and Chuenpagdee (2015) showed that in Korea, small-scale fishers valued the health of the ecosystem and the equal distribution of resources among fishers relatively to their level of effort, diligence or/and investment. These fishers, much like the 'lobster gangs' of Maine (Acheson 1988), considered the adjacency principle (i.e. access to use a fishing ground should be first granted to those who live near it) as a central norm to guide decisions (Song and Chuenpagdee 2015). Thus, small-scale fisheries are known to possess values that adhere to social justice and ecological sustainability (Johnson 2006; Chuenpagdee and Jentoft 2015).

Baltic small-scale fishers do value their occupation where profit is as valuable as other personal fulfillments. As expressed by a Baltic coastal fisher:

> You have to love the sea and fishing to be a fisher. The financial side is not that important and that's why there are no outsiders, nobody not from Kuźnica aboard our boats. We know that our earnings aren't that high, but we love the sea and our work as fishers – coastal Polish fisher 2007, interview by B. Marciniak and E. Kuzebski published in Delaney (2007, 80).

In the Swedish Baltic eel fishery, small-scale fishers regard as important the persistence of their ways of fishing and claim that "*fishing should be done as it always has been done*" (Björkvik 2013, 26). This seems to be associated with a need to maintain their identities and a related discomfort with change. Fishers' knowledge assets and learning from experience and practice are also important components of this small-scale fishery (Björkvik 2013). Furthermore, social equity and equality are important values for Baltic Sea small-scale fishers (Ignatius and Haapasaari 2018) and are considered neglected by existing fisheries policies as compared to economic efficiency. This paradox of economic efficiency is nicely illustrated by Boonstra and Hentati-Sundberg (2016, 88) quote from a Baltic coastal fisher, asserting: "*Isn't it better to have 10 boats that fish instead of 1 boat that fishes more than these 10 altogether?*" Small-scale fishers in Estonia, Sweden, and Denmark are also aware of the advantage of small-scale fisheries methods versus trawlers in terms of fish stock and ecological sustainability and feel they should be given greater access to quotas, while trawlers should be kept away from their local waters. Baltic Sea small-scale fishers also value their knowledge, the sense of freedom that coastal fisheries provide, and the importance of fisheries for their communities. These values are revealed during interviews, as shown here:

> The most important thing is that the government is listening to the fishermen, and that they don't just sit with calculations; because the fishermen understand more about the fishing industry than can be reflected in numbers. There can be streams or algae for example…the fish are flexible, the water temperature: it's too cold or it's too warm. But it's me as a professional fisherman who understands these things in order to be able to catch a lot of fish. When you go through these systems [EU, bureaucratic, etc.) there's no real relevance in real life. The decisions that the decision-makers are making are not from the fishermen. They come from somewhere else – Retired Baltic Sea fisher, interview by E. Krogseng published in Krogseng (2016, 30).
>
> The lifestyle to be a fisherman, the freedom: that's what I like. I like to go out and fish and not care what week it is or what month it is the freedom that I really enjoy to be a fisherman. I worked two years in a professional industry and I will never do that again. I promised myself that I would never again work in a place with four walls and a roof. I did that from 1991 to 1993 and it's after that that I became a professional, independent fisherman – Salmon Baltic Sea fisher 2012, interview by E. Krogseng published in Krogseng (2016, 30).
>
> I think that if it were only possible to earn a living fishing that even the highly educated ones will return. The desire to work on the sea and a longing for freedom will draw them back – coastal Polish fisher 2007, interview by B. Marciniak and E. Kuzebski published in Delaney (2007, 78).

Fishers and fish workers see the close relationship between a sustainable small-scale fishery, on the one hand, and the place where they live on the other. They also bring 'added' value because they provide a base for other industries and activities that would not exist if not for the fisheries activities, as highlighted by these interviews:

> If we don't catch a sufficient amount of cod, the existence of our families will be threatened. Only in a few families is there no fisher, so it's clear that the fate of our entire village depends on cod catches – cod Polish fisher 2007 interview by B. Marciniak and E. Kuzebski published in Delaney (2007, 82).
>
> The whole society around here would die out. This area is really famous for salmon and eel. And now when it's forbidden with fishing salmon, and soon it will be prohibited to fish eel. Hörvik is world famous for their fishing and when it disappears, everything else will disappear. There won't be any more tourists around here – Eel Baltic Sea fisher 2012, interview by E. Krogseng published in Krogseng (2016, 30).

The limited participation of the small-scale fisheries sector in EU fisheries decision making under the CFP (see above) resulted in Baltic Sea small-scale fishers' values being ignored in EU fisheries policies. As a consequence, sense of freedom, independence, and the satisfaction to use experienced-based knowledge to challenge the intrinsic challenges associated with fisheries have been flattened with the multitude of fisheries regulations currently applied in the Baltic Sea. Fisheries technical regulations are so numerous (see Hentati-Sundberg et al. 2014) that in 2006 a new regulation (EC No 2187/2005) was enacted to adopt a 'single legislative text' to help familiarize fishers with the vast number of rules (Baltic Sea Technical Rules 2005). Similarly, the link between fisheries and the central role that coastal fishers play for rural fishing communities' sustainability has been neglected. Thus, EU funds to support EU Rural Development policies were detached from fisheries policies objectives (as in the case of FLAGs). Finally, equity is not mentioned as one of the objectives of the CFP where ecological and economic objectives prevail. Consequently, the CFP is not considered to be an instrument of 'social' policy (Stobberup et al. 2017) despite aiming at ecological, social, and economic goals.

For the FLAGs, it seems early still to determine their ultimate effectiveness and permanency of the established networks or measure the degree of complementarity with other local development initiatives (Phillipson and Symes 2015). After one decade of funding and more than 300 FLAGs projects currently established, there is still little evidence to date of if and how FLAGs will play a role in higher fisheries governance levels, for example, at regional, EU, or Member State level (ibid).

19.3 Current Transitions – Solutions at Regional or Local Levels?

Fisheries management in the EU has historically been dominated by ecological and economic objectives through its targets for maximum sustainable yield for commercial fish stocks. While this is, to some extent, understandable, the inadequacy of grounding fishery governance on bio-economic models for resource stewardship is increasingly recognized, especially when other sources of insecurity among fishers exist. The mistake becomes more visible when small-scale fishers are the focus of attention. EU ACs in their current performance might not substantially change these fisheries management paradigms as our analyses suggest. Subsequently, when small-scale fisheries and fishing communities have no representation, their particular values, concerns, and knowledge will have to seek other venues, such as the FLAGS, grassroots associations, fisher's organizations like LIFE, the media, or political parties.

In the meantime, the impacts of ignoring the small-scale fisheries sector have become evident, and calls for change start to resonate. Recent supranational policy instruments that demand changes in the governance of coastal and small-scale fisheries have emerged. These new policies include the "Voluntary Guidelines for Securing Sustainable Small-Scale Fisheries in the Context of Food Security and Poverty Eradication" (Jentoft 2014), the UN Sustainable Development Goal and its Target 17, "*ensure property rights to small-scale fishers*", and the latest reform of the EU CFP that came into effect on January 2014. While these policy instruments differ in their scope, the first two being voluntary and the latter binding, they all aim at supporting changes in the governance of small-scale fisheries. For instance, current CFP rules for the functioning of ACs recognize the need to ensure the participation of the small-scale fisheries sector in these organizations:

> Based on designations from the sector organizations and from the other interest groups for the seats allotted to them respectively, the general assembly appoints an executive committee of up to 25 members. After consultation of the Commission, the general assembly may decide to appoint an executive committee of up to 30 members to ensure appropriate representation of small scale fleets (EC 2013).

Such statements claim to give more weight to small-scale fisheries representation in the CFP, and also demonstrate what constructive policy initiatives targeting small-scale fisheries and coastal communities may entail. In the meantime, more promising and forceful initiatives seem to emerge and foster at regional and local levels, as shown in the next section.

19.3.1 The 'Low Impact Fishers of Europe' (LIFE) Organization – A Regional (EU) Level Initiative

LIFE is a fisheries organization, launched in November 2012, to support the participation and representation of small-scale fishers at EU, regional, and member states institutions. The organization describes itself as an overarching initiative of fishers with the aim of providing a clear and coherent voice representing the EU small-scale fisheries sector. (LIFE n.d.). As explained on its website,

> LIFE represents those fishermen who tend to be smaller scale and have a low environmental impact but who at the same time have a high social and economic value within the communities they support and who support them. The owner usually works aboard, undertakes day trips, has low greenhouse gas emissions per kilo of fish landed and fishes in a sustainable manner. On this basis LIFE has developed a Mission Statement that clearly illustrates the ethos, aims and aspirations of the organization (LIFE n.d.).[4]

Around 30 EU fisher associations are currently members of LIFE, including four that operate in the Baltic Sea. LIFE supports small-scale fishers in organizing themselves for representation and participatory purposes. Over the last years, LIFE has provided information and capacity building to Baltic Sea fishers and has achieved the official membership of a representative of the small-scale fisheries sector in the Executive Committee of the BSAC. LIFE members accompany small-scale representatives to ACs meetings, guide them, and ensure that they have a voice in this kind of EU fora.

19.3.2 'Östesjöfiske 2020' in Simrishamn, Sweden – A Local Initiative

'Östersjofiske 2020', in the coastal municipality of Simrishamn in South Sweden, is an initiative to empower local coastal fishers and their communities in the Baltic Sea. Supported by local authorities, 'Östersjöfiske 2020' was established in 2010 and serves as a focal point where diverse activities, including research and entrepreneurship, merge to support coastal fisheries. In 2017, the leaders of the initiative managed to collect signatures from 17 Swedish coastal municipalities and sent a letter to the central government criticizing the fishing quota system and claiming for sustainable fisheries strategies based on ecological, social, and economic goals. The letter gained widespread attention and media coverage, creating public awareness on the impacts of current policies and the deterioration of the traditional coastal small-scale fisheries sector in Sweden. Among other activities, 'Östersjöfiske 2020', organizes annual conferences, which connect regional managers, policy-makers,

[4]The four Baltic fisheries organizations that are members of LIFE are: Fischerieschutzverband Schleswig-Holstein (Germany), Foreningen for Skånsomt Kystfiskeri (Denmark), Srodkowopomorska (Polen) and *Sveriges Yrkesfisakres Ekonomiska Förening* (Sweden).

scientists, local fishers, and fisheries representatives for evaluating the initiative's progress, and presenting and planning future developments. The initiative is regarded increasingly successful in addressing problems and creating change for the future of coastal fishers and their communities. It focuses on three main areas: fisheries, research, and environment. This initiative recognizes that fisheries are vital to Simrishamns inhabitant's identity, attractiveness, and to the local business environment, and serves as a driving force for rural development.

19.4 Concluding Remarks

We argue that EU policies over the last decades have negatively impacted the small-scale fishery sector in the Baltic Sea. The prioritization of ecological and economic objectives above social goals in fisheries management, the disadvantages in resources competition with large-scale (and possibly also recreational fisheries), and the exclusion from the policy realm and decision-making process have resulted in small-scale fishers' marginalization. Granting member states the freedom to regulate within their 12 nautical miles from the coastline has not prevented small-scale fisheries from being impacted by EU policies. This situation is not surprising since fish *do not respect borders* and the EU multilevel governance system does not allow national regulations to overrule EU legislation.

In the terms of interactive governance theory and the governability of small-scale fisheries, as the overall 'quality' of the social-political entity, the Baltic Sea governance system clearly does not entail all the attributes that one would associate with the concept of 'good governance'. The governing system has undergone substantial change with the reformed CFP, from 2004, with the establishment of the ACs. This has made the governance system more open to stakeholder participation and sensitive to regional contexts. However, the system is still not truly representative for all relevant stakeholders since small-scale fisheries are not included. Thus, without representation and a voice, small-scale fisheries values, concerns, and interests are filtered out. One may, therefore, question whether the CFP, while moving towards a more open system, which in itself would be in accordance with 'good governance' principles, has left small-scale fisheries worse off than before. In fact, the way ACs are structured, reducing small-scale fisheries participants to observers, seems to have tilted the playing field and disadvantage them. Obviously, this is a paradox and a risk. Interactive governing systems risk exacerbating existing inequities and marginalities if they do not provide space for all legitimate stakeholders. Different fishing styles and interests have not been considered with negative consequences for the small-scale fishers. More progressive approaches and understanding of the fisheries sector as a whole might have prevented these developments.

Small-scale fisheries have legitimate and urgent concerns in the Baltic Sea, but the current governing system does not allow them to enter the conversation. Hence, we find small-scale fisheries in a highly ambiguous situation from a governability perspective—entrenched between top-down EU governance and some dispersed local and regional institutions. Yet, with the pressures now coming from within the small-scale fisheries sector, from coastal communities, and indeed from central EU levels as shown above, things may change. Whether small-scale fisheries will be able to get a foot in the door in the Baltic Sea case, remains to be seen, because also representation, in other words, who are allowed to sit at the table, is a political issue, as interests differ and are often in conflict. It is unclear whether other stakeholders, who currently dominate, will welcome small-scale fisheries to the table, which is why such an institutional reform would need not just the 'backing' from the EU level but also the initiative. As one small-scale fisheries representative in the BSAC put it:

> Membership is not enough. Without good preparation, the study of documents and proposals before the meetings as well as personal attendance we can even with higher numbers of members not achieve anything and will remain a negligible minority. This really means a tough job – retired fisher, LIFE Hamburg Workshop 2017.

Ignoring the values and interests of small-scale fisheries in the governing system of the Baltic Sea has contributed to their present-day invisibility and in many cases near extinction. Fisheries governance permanently reflects some meta-order elements, like values, images, and beliefs, whether explicit or not, which guide policies and actions. If the values of small-scale fishers are not taken into consideration, priorities of the values of other stakeholders, like those of large-scale fishers, managers, and scientists, will take precedence. Although there exists a growing consensus on the need to promote the small-scale fisheries sector in the EU and worldwide, as manifested by the increased attention of international and regional institutions (for instance the Small-scale Fisheries Guidelines; the UN Sustainable Development Goals, the lately CFP reform, and LIFE, or local initiatives), the sector has yet to overcome decades of an unfavorable management regime (Papaioannou et al. 2014; Salmi 2015).

In contrast to the strong top-down influence of the CFP, we find a plethora of dispersed governing organizations and initiatives that may help in addressing small-scale fisheries and community interests for future developments and their policy-influence. LIFE, and the 'Östersjöfiske 2020' initiative in Simrishamn are examples of these organizations. Although this seems to bring the small-scale fisheries to a new place in policy and management discussions, the crucial question is whether it might already be too little, too late to rescue it in many places, such as Denmark and Sweden. LIFE works at the regional and EU level, while 'Östersjöfiske 2020' works at the local and member state level. Regional organizations such as HELCOM or BALTFISH could play a role in providing strategic partnerships with these initiatives for addressing various interests connected to the small-scale and community interests. It is clear that LIFE is gaining trust among small-scale fishers and, albeit slowly, is fostering a better representation at the EU level in places like the Baltic Sea where the current invisibility in the decision-making process jeopardizes the future of this fisheries sector. For local initiatives, how their interests and goals will reproduce and be integrated in national and regional political spheres or how governments will promote a good policy environment for them to persist remain big challenges for their success.

Small-scale fisheries governance in the Baltic Sea reassembles well with Chhotray and Stoker's definition of governance as "*being about the rules of collective decision-making in settings where there are a plurality of actors and organizations and where no formal control system can dictate the terms of the relationship between these actors and organizations*" (Chhotray and Stoker 2009 cited in Jouanneau and Raakjær 2014, 332). This diversity of well-established actors and organizations poses challenges for the governability of this fishery. The complex interactions within the governing system have apparently contributed to the small-scale fisheries sector's incapacity to share their values and interests in policies and regulations, where ecological and market-based economic values and concerns have prevailed. New partnerships and networks and governance structures will have to oppose path dependency in the Baltic governing system where "things seem to be more frozen" (Jouanneau and Raakjær 2014, 337) and EU governance has, over the years, gained supremacy over local contexts voices and needs. Representation and participation of the small-scale fisheries in the co-creation of knowledge—to make their societal contribution visible to decision-makers—and the co-development of policies are vital for the future survival of this fishing sector in the Baltic Sea.

Acknowledgments This research was partially funded by the main author's research grant from the Centre for Sea and Society at Gothenburg University. The authors thank two anonymous reviewers for their valuable comments and recommendations to an early version of our manuscript.

References

Acheson JM (1988) The lobster gangs of Maine. University Press of New England, Hannover

Arias-Schreiber M, Säwe F, Hultman J et al (2017) Addressing social sustainability for small-scale fisheries in Sweden: institutional barriers for implementing the Small-Scale Fisheries Guidelines. In: Jentoft S, Chuenpagdee R, Barragán-Paladines MJ et al (eds) The Small-Scale Fisheries Guidelines: global implementation. Springer International Publishing, Cham, pp 717–736

Baltic Sea Technical Rules (2005) Presentation of the main points of Regulation (EC) 2187/2005. https://ec.europa.eu/fisheries/sites/fisheries/files/docs/body/baltic-sea-technical-rules_en.pdf. Accessed 30 Apr 2018

Björkvik E (2013) Explaining the decline in Swedish Baltic Sea small-scale fisheries: a historical analysis of fishers in their social and ecological context. Stockholm Resilience Centre, Stockholm University, Stockholm

Boonstra WJ, Hentati-Sundberg J (2016) Classifying fishers' behaviour. An invitation to fishing styles. Fish Res 17:78–100

BSAC (Baltic Sea Advisory Council) (2012) Letter from Ernesto Penas Lado. European Commission, DG MARE to Baltic Sea AC on "Applications for membership by Bornholm FLAG and PAPF". https://www.bsac.dk. Accessed 30 Apr 2018

COM (2008) Communication from the Commission to the Council and the European Parliament. Review of the functioning of the Regional Advisory Councils. Brussels (COM/2008/364)

Chuenpagdee R, Jentoft S (2015) Exploring challenges in small-scale fisheries governance. In: Jentoft S, Chuenpagdee R (eds) Interactive governance for small-scale fisheries: global reflections. Springer International Publishing, Cham, pp 3–16

De Vos B, Kraan M (2015) To define or not to define; implications for the governability of small-scale coastal fisheries in the Netherlands. In: Jentoft S, Chuenpagdee R (eds) Interactive governance for small-scale fisheries: global reflections. Springer International Publishing, Cham, pp 629–647

Delaney A (2003) Setting nets on troubled waters: environment, economics, and autonomy among nori cultivating households in a Japanese fishing cooperative. Dissertation, University of Pittsburgh

Delaney A (2007) Profiling of small-scale fishing communities in the Baltic Sea: a study requested by the European Commission. IFM-Innovative Fisheries Management, Aalborg University (Call FISH/2005/10)

EC (2013) Regulation (EU) No 1380/2013 of the European Parliament and of the council of 11 December 2013. European Parliament, Strasbourg

Gascuel D, Coll M, Fox C et al (2016) Fishing impact and environmental status in European seas: a diagnosis from stock assessments and ecosystem indicators. Fish Fish 17:31–55

Gilek M, Karlsson M, Linke S et al (2016) Environmental governance of the Baltic Sea: identifying key challenges, research topics and analytical approaches. In: Gilek M, Karlsson M, Linke S, Smolarz K (eds) Environmental governance of the Baltic Sea. Springer International Publishing, Cham, pp 1–17

Gómez S, Lloret J, Demestre M et al (2006) The decline of the artisanal fisheries in Mediterranean coastal areas: the case of Cap de Creus (Cape Creus). Coast Manag 34:217–232

Griffin L (2013) Good governance, scale and power a case study of North Sea fisheries. Routledge, New York

Guyader O, Berthou P, Koutsikopoulos C et al (2013) Small scale fisheries in Europe: a comparative analysis based on a selection of case studies. Fish Res 140:1–13

Hatchard JL, Gray TS (2014) From RACs to advisory councils: lessons from North Sea discourse for the 2014 reform of the European Common Fisheries Policy. Mar Policy 47:87–93

HELCOM (Baltic Marine Environment Protection Commission – Helsinki Commission) (2007) Baltic Sea action plan. In: HELCOM ministerial meeting. http://www.helcom.fi/Documents/Baltic%20sea%20action%20plan/BSAP_Final.pdf. Krakow, Poland, 15 November 2007

Hentati-Sundberg J, Hjelm J, Österblom H (2014) Does fisheries management incentivize non-compliance? estimated misreporting in the Swedish Baltic Sea pelagic fishery based on commercial fishing effort. ICES J Mar Sci 71:1846–1853

Højrup T (2011) The need for common goods for coastal communities: common community quotas for sustainable life-modes in coastal fisheries: the common fisheries policy has to recognize: the alternative to privatization of fishing rights in the home waters of Europe. Narayana Press, Gylling

Høst J (2015) Governing through markets: societal objectives, private property rights and small-scale fisheries in Denmark. In: Jentoft S, Chuenpagdee R (eds) Interactive governance for small-scale fisheries: global reflections. Springer International Publishing, Cham, pp 319–336

Ignatius S, Haapasaari P (2018) Justification theory for the analysis of the socio-cultural value of fish and fisheries: the case of Baltic salmon. Mar Policy 88:167–173

Järvik A, Raid T, Sadul J-V, Järv L (2012) Overexploitation of Fish Resources and Small-Scale Fisheries in the Northeastern Baltic Sea: Social Aspects of Ecosystem-Based Fisheries Management in the West Estonian Archipelago. In: Kruse GH, Browman HI, Cochrane KL, Evans D, Jamieson GS, Livingston PA, Woodby D, Zhang CI (eds) Global Progress in Ecosystem-Based Fisheries Management. University of Alaska Fairbanks, Alaska Sea Grant. https://doi.org/10.4027/gpebfm.2012.013

Jentoft S (2014) Walking the talk: implementing the international voluntary guidelines for securing sustainable small-scale fisheries. MAST 13:16

Jentoft S, Chuenpagdee R (2009) Fisheries and coastal governance as a wicked problem. Mar Policy 33:553–560
Johnsen JP, Vik J (2013) Pushed or pulled? understanding fishery exit in a welfare society context. MAST 12:4
Johnson DS (2006) Category, narrative, and value in the governance of small-scale fisheries. Mar Policy 30:747–756
Jouanneau C, Raakjær J (2014) 'The hare and the tortoise': lessons from Baltic Sea and Mediterranean Sea governance. Mar Policy 50:331–338
Kern K, Löffelsend T (2008) Governance beyond the nation-state: transnationalization and europeanisation of the Baltic Sea Region. In: Joas M, Jahn D, Kern K (eds) Governing a common sea: environmental policies in the Baltic Sea region. Earthscan, London
Kooiman J (2008) Exploring the concept of governability. J Comp Policy Anal: Res Pract 10:171–190
Kooiman J, Bavinck M (2005) The governance perspective. In: Kooiman J, Bavinck M, Jentoft S et al (eds) Fish for life. Amsterdam University Press, Amsterdam, pp 11–24
Kooiman J, Jentoft S (2009) Meta-governance: values, norms and principles, and the making of hard choices. Public Adm 87(4):818–836
Krogseng K (2016) The decline of small-scale fisheries: a case study in Blekinge. In: NM009 Rural Development and Natural Resource Management – Master's Programme 120 HEC
Kulmala S, Levontin P, Lindroos M et al (2013) Atlantic salmon fishery in the Baltic Sea — a case of trivial cooperation? Strateg Behav Environ 3:121–147
LIFE (2017) LIFE in the advisory councils: foot in the door, but more seats at the table needed. http://lifeplatform.eu/life-advisory-councils/. Accessed 30 Apr 2018
LIFE (Low Impact Fishers of Europe) (n.d.) Welcome to life. http://lifeplatform.eu/. Accessed 30 Apr 2018
Linke S, Bruckmeier K (2015) Co-management in fisheries – experiences and changing approaches in Europe. Ocean Coast Manag 104:170–181
Linke S, Jentoft S (2013) A communicative turnaround: shifting the burden of proof in European fisheries governance. Mar Policy 38:337–345
Linke S, Jentoft S (2014) Exploring the phronetic dimension of stakeholders' knowledge in EU fisheries governance. Mar Policy 47:153–161
Linke S, Jentoft S (2016) Ideals, realities and paradoxes of stakeholder participation in EU fisheries governance. Environ Sociol 2:144–154
Lloret J, Cowx IG, Cabral H et al (2016) Small-scale coastal fisheries in European Seas are not what they were: ecological, social and economic changes. Mar Policy. https://doi.org/10.1016/j.marpol.2016.11.007
Long R (2010) The role of regional advisory councils in the European Common Fisheries Policy: legal constraints and future options. Int J Mar Coast Law 25:289–346
NSAC (2013) Representation of small scale fishers. Letter from NSRAC Secretariat to Lowri Evans, Director General DG Mare
Onyango PO (2015) Governability, self-governance and co-governance in the context of Lake Victoria fisheries, Tanzania. In: Jentoft S, Chuenpagdee R (eds) Interactive governance for small-scale fisheries: global reflections. Springer International Publishing, Cham, pp 179–198
Österblom H, Gårdmark A, Bergström L et al (2010) Making the ecosystem approach operational—can regime shifts in ecological- and governance systems facilitate the transition? Mar Policy 34:1290–1299
Ounanian K, Hegland TJ (2012) The regional advisory councils' current capacities and unforeseen benefits. MAST 11:10
Papaioannou EA, Vafeidis AT, Quaas MF et al (2014) Using indicators based on primary fisheries' data for assessing the development of the German Baltic small-scale fishery and reviewing its adaptation potential to changes in resource abundance and management during 2000–09. Ocean Coast Manag 98:38–50

Phillipson J, Symes D (2015) Finding a middle way to develop Europe's fisheries dependent areas: the role of fisheries local action groups. Sociol Rural 55:343–359

Pita C, Pereira J, Lourenço S et al (2015) The traditional small-scale octopus fishery in Portugal: framing its governability. In: Jentoft S, Chuenpagdee R (eds) Interactive governance for small-scale fisheries: global reflections. Springer International Publishing, Cham, pp 117–132

Raakjær J, Hegland TJ (2012) Introduction: regionalising the common fisheries policy. MAST 11:5

Raid T, Järvik A (2008) Current situation and possible ways of sustainable development of small-scale fishery in Estonian coastal sea. ICES CM 2008/O:21:14

Said A, Macmillan D, Schembri M et al (2017) Fishing in a congested sea: what do marine protected areas imply for the future of the Maltese artisanal fleet? Appl Geogr 87:245–255

Salmi P (2015) Constraints and opportunities for small-scale fishing livelihoods in a post-productivist coastal setting. Sociol Rural 55:258–274

Sellke P, Dreyer M, Linke S (2016) Fisheries: a case study of Baltic Sea environmental governance. In: Gilek M, Karlsson M, Linke S, Smolarz K (eds) Environmental governance of the Baltic Sea. Springer International Publishing, Cham, pp 45–71

Song AM, Chuenpagdee R (2015) Eliciting values and principles of fishery stakeholders in South Korea: a methodological exploration. Soc Nat Resour 28:1075–1091

Song AM, Chuenpagdee R, Jentoft S (2013) Values, images, and principles: what they represent and how they may improve fisheries governance. Mar Policy 40:167–175

Sønvisen SA (2014) Contemporary fisher images: ideologies, policies and diversity. J Rural Stud 34:193–203

STECF (Scientific, Technical and Economic Committee for Fisheries) (2017) The 2017 Annual Economic Report on the EU fishing fleet (STECF-17-12). Scientific, technical and economic committee for Fisheries Luxembourg

Stobberup K, Garza Gil MD, Stirnemann-Relot A et al (2017) Research for PECH committee – small-scale fisheries and "Blue Growth" in the EU. European Parliament, Policy Department for Structural and Cohesion Policies, Brussels

Strehlow HV (2010) The multiannual management plan for cod in the Baltic Sea: reactions and sentiments in two German fishing communities. ICES J Mar Sci 67:1963–1971

Van Ginkel R (2001) Inshore fishermen: cultural dimensions of a maritime occupation. In: Symes D, Phillipson J (eds) Inshore fisheries management. Springer Netherlands, Dordrecht, pp 177–193

Young ML, Foale S, Bellwood DR (2016) Why do fishers fish? a cross-cultural examination of the motivations for fishing. Mar Policy 66:114–123

Part VII
Towards Transdisciplinarity in Fisheries Governance

Chapter 20
Beyond the Basics: Improving Information About Small-Scale Fisheries

Melinda Agapito, Ratana Chuenpagdee, Rodolphe Devillers, Jennifer Gee, Andrew F. Johnson, Graham J. Pierce, and Brice Trouillet

Abstract Small-scale fisheries require knowledge in decision-making, particularly as they face many of the same issues affecting large-scale, industrial fisheries, such as declining fish stocks, marine habitats degradation, resource use competition, and climate change. There are other characteristics of small-scale fisheries, however, that cause additional challenges. For example, small-scale fisheries target many species that are not usually exploited by their large-scale counterpart. The export values from small-scale fisheries catches are also generally lower than those from large-scale, due partly to the subsistence nature of small-scale fisheries and the relatively high proportion of use directly for household consumption in local communities. Systematic data collection and information systems drawing data from multiple sources, through on-board sampling and market sampling, have been focused mostly on large-scale fisheries and the economic contribution of small-scale fisheries may not justify investment in the 'machinery' of modern fishery science. In this chapter, we briefly review approaches and models used in assessing large-scale fisheries as part of the modern fisheries management and discuss their applicability to small-scale fisheries. Next, we present four examples of initiatives that aim to

M. Agapito (✉) · R. Chuenpagdee · R. Devillers
Department of Geography, Memorial University of Newfoundland, St. John's, NL, Canada
e-mail: magapito@mun.ca; ratanac@mun.ca; rdeville@mun.ca

J. Gee
Food and Agriculture Organisation (FAO) of the United Nations, Rome, Italy
e-mail: Jennifer.Gee@fao.org

A. F. Johnson
Marine Biology Research Division, Scripps Institution of Oceanography, San Diego, CA, USA
e-mail: anj016@ucsd.edu

G. J. Pierce
Departamento de Ecología y Recursos Marinos, Instituto de Investigaciones Marinas (CSIC), Vigo, Spain
e-mail: g.j.pierce@abdn.ac.uk

B. Trouillet
Université de Nantes, CNRS, UMR LETG, Nantes, France
e-mail: brice.trouillet@univ-nantes.fr

R. Chuenpagdee, S. Jentoft (eds.), *Transdisciplinarity for Small-Scale Fisheries Governance*, MARE Publication Series 21,
https://doi.org/10.1007/978-3-319-94938-3_20

improve information about small-scale fisheries through: (i) national-level fisheries statistics; (ii) information crowdsourcing; (iii) effort estimation; and, (iv) integration of fishers' knowledge in marine spatial planning. The chapter concludes with recommendations about ways forward.

Keywords Information system · ISSF · FAO statistics · Crowdsourcing · Fishing effort · Transdisciplinarity

20.1 Introduction

Knowledge is important in making decisions. However, before knowledge can be obtained, one must generate information out of sets of symbols or data (Ackoff 1989). The Food and Agriculture Organization of the United Nations (FAO-UN), for instance, recognizes that a systematic system to collect and analyse data is important in generating useful information products and publications that could assist member-countries in implementing responsible fishery practices (FAO-UN 2018). Thus, recently there is increased effort to collect, organize, store, and communicate data to build an information system. For example, the ecosystem approach to fisheries (EAF) projects of FAO-UN in various regions of the world have dedicated a considerable amount of effort to generate and make use of biological and socioeconomic data (FAO-UN 2018). Recently, international agreements such as the Sustainable Development Goals (SDGs), particularly Goal 14, rely on information related to sustainable use of marine resources to measure progress that directly and indirectly impacts other goals, such as poverty alleviation (SDG 1) and zero hunger (SDG 2) (United Nations 2015). The Voluntary Guidelines for Securing Sustainable Small-Scale Fisheries (SSF Guidelines), another international instrument, encourage countries to establish a system to collect biological, social, cultural, and economic data, and to develop platforms that can facilitate the exchange of such information (FAO 2015a).

Data about small-scale fisheries are particularly important because fisheries management is facing a multitude of challenges including those related to overexploitation, illegal, unregulated, and unreported (IUU) fishing, and climate change. Data are also needed to mitigate the uncertainties arising from lack of comprehensive monitoring program for small-scale fisheries. Relative to small-scale fisheries, many large-scale fisheries are data-rich when it comes to understanding the dynamics of the fished stocks, yet may still be data-poor in relation to their wider environmental, ecological, economic, and social impacts, and in relation to the degree of compliance with regulations. In short, data paucity is particularly acute in small-scale fisheries due to the diversity, complexity, dynamics, and scale issues in the social system along the fish chain (Chuenpagdee et al. 2017). Consequently small-scale fisheries would require specific strategies concerning their monitoring and resource assessment.

The industrial sector often uses quantitative data and modeling exercises for optimal management at least when the optimum is narrowly defined to consider only the target stock. Although complexities such as the interconnectivity and changing ocean systems may apply to large-scale fisheries to some extent, the use of traditional stock assessment and fishery management in the large-scale sector have essentially ignored these complexities. The small-scale fisheries, however, readily acknowledged much of the complexity that impacts the sector that is neither easily quantified nor expressed as monetary values. For instance, rather than relying on monetary income from fishing as the sole measure of poverty, determining wellbeing and livelihood sustainability of small-scale fishing communities recognizes other vital factors. These include food security, access to health, education, political, and geographical marginalization, as well as ownership structure of land and boats (Béné and Friend 2009). These factors can help better understand what could hamper or support small-scale fisheries communities to become sustainable. Therefore, the required datasets and analyses will have to depend also on qualitative data, and approaches that could help bring out the nuances embedded in the small-scale fisheries sector that are not readily observed or quantified.

Another reason data on small-scale fisheries are needed is because their characteristics are quite diverse. Globally, there are about 120 million full-time and part-time workers who are directly or indirectly dependent on small-scale fisheries, and around 95% of small-scale landings are destined for local consumption (World Bank et al. 2012). Unlike a large-scale fishery, which often targets a single species in a single trip, a small-scale fishery can target multiple species of fish, shellfish, and other marine organisms, employing various types of techniques and gears. Small-scale fishing can be seasonal and, while it is often considered the primary occupation, many small-scale fishers are also engaged in other livelihood activities. Fishing can also be part of an informal economy, with family members participating in post-harvest activities, including home-based traditional processing, direct selling of fish at local markets, and trading with retailers and wholesalers. These characteristics of the small-scale fishing sector suggest that information about small-scale fisheries cannot always be acquired using the same approaches used to collect data about large-scale fisheries.

Despite the differences between large-scale fisheries and small-scale fisheries, the use of quantitative fisheries assessment in large-scale fisheries has influenced how small-scale fisheries data have been collected and analyzed. An example is the work by Thompson and FAO (1988) and more recent research (Pauly 2006; Chuenpagdee and Pauly 2004; World Bank et al. (2012), that offer comparative quantitative estimates of benefits from small-scale fisheries in comparison to large-scale fisheries. These studies have offered several metrics to measure the social and environmental benefits from fisheries, including the number of people employed in the fisheries, the proportion of the catch used for human consumption, and the annual fuel consumption, as well as the amount of discards. Regarding sustainability, small-scale fisheries, in most accounts, fare much better than large-scale fisheries, suggesting 'misplaced priorities' in fisheries development, which has mostly focused on economic efficiency involving large-scale fisheries, thus, requiring a serious reconsideration of global fisheries governance, especially as the interest in blue growth and blue economy is escalating (Pauly 2018). Together, these efforts to compare large-scale and small-scale fisheries helped the research community better understand small-scale fisheries.

Still, more can be done to develop appropriate approaches, methods, and tools that can capture both quantitative and qualitative information about the small-scale fisheries sector. This chapter aims to advance the discussion about innovative ways of improving the information system about small-scale fisheries. It begins with a quick overview of the approaches, models, and tools used in present fisheries management and discusses the applicability to small-scale fisheries. Next, it presents four specific examples of recent efforts to collect, build, and use data about small-scale fisheries in accord with their nature and characteristics. This chapter highlights examples including the new data template that the FAO-UN uses to collect small-scale fisheries data and the Information System on Small-Scale Fisheries (ISSF) developed by Too Big To Ignore (TBTI),[1] the Global Partnership for Small-Scale Fisheries Research. Also important are innovative ways to generate information using data that can be made available as in the case of estimating small-scale fishing effort. Finally, is the critical value of making use of quantitative and qualitative data obtained through a participatory process such as community mapping and citizen science (de la Torre-Castro et al. 2014; Cigliano et al. 2015; Fairclough et al. 2015). As an example, this chapter highlights precautions about the use of marine spatial planning (MSP), a participatory approach that is gaining popularity, but less is known as to how it can effectively integrate information about small-scale fisheries. This chapter concludes with suggestions about ways forward framed around the call in the SSF Guidelines to employ a transdisciplinary perspective to improving data about small-scale fisheries sector to advance its welfare and to promote fisheries sustainability.

20.2 Modern Fisheries Management and the Applicability to Small-Scale Fisheries

Large-scale fisheries in national waters are generally monitored, assessed, and managed under a governance system, with the aim of ensuring sustainability. Sustainability, in its narrowest sense can be measured according to the setting and compliance with quotas to allow for a Maximum Sustainable Yield (MSY), or one of its variants, of target species, or in a broader sense accounting for ecological, economic, and social objectives. However, there has been limited success in achieving sustainability under this regime in large-scale fisheries, so one may justifiably question whether it could work for small-scale fisheries. Fifty-four years after co-authoring a book (Beverton and Holt 1957) that laid the foundations of modern fishery science, Holt (2011) considered MSY to be the worst idea in fishery management, "a perfect example of pseudo-science with little empirical or sound

[1] Too Big To Ignore (http://toobigtoignore.net/)

theoretical basis. As a target for management of fisheries, or even as the anchor for so-called 'reference points', it is inadequate and its pursuit increases the likely unprofitability, and even collapse, of fisheries." Finley (2011) notes that MSY has also failed fishers, leading to falling income and job losses. Finley and Oreskes (2013), argue that MSY was a political tool, with overfished stocks and unemployed former fishers effectively being collateral damage. Hughes (2011) concludes that deterioration of marine ecosystems can be blamed squarely on poor governance. Even if the modern-day concept of MSY has a more solid scientific foundation, EEA (2015) reported that only 76% of EU Atlantic fisheries and 14% of EU Mediterranean fisheries met at least one of two sustainability criteria (fishing effort and stock reproductive potential consistent with MSY) and, overall, only 12% of EU marine fisheries met both criteria.

Compared to large-scale fisheries, the impact of small-scale fisheries on exploited stocks and the wider environment, as well as the economic gain from the fishery, may be relatively small, but the social importance of small-scale fisheries to coastal communities tends to be comparatively high. It may be assumed that small-scale fisheries are more environmentally friendly, for example with a low discarding rates. While many small-scale fishing gears are low impact (ICSF 2016), some do incur adverse impacts (e.g. Alfaro-Shigueto et al. 2010) and Kelleher (2005) estimates that small-scale fisheries accounted for only 11% of global fishery discards, but this result needs to be reassessed as complete data on discard rates were only available for fewer than half of the small-scale fisheries considered and, for the others, anecdotal evidence was used to assign a discard rate of 1% or less. Also, some small-scale fisheries are not being managed and others managed independently of large-scale fisheries, thus potentially compromising long-term sustainability. In the European Union (EU), even if small-scale fisheries vessels are subject to Common Fisheries Policy (CFP) regulations on fleet capacities and technical measures, their activity is mainly governed by local or regional authorities (e.g. various small-scale cephalopod fisheries) (e.g. Pierce et al. 2010; Pita et al. 2015). According to Guyader et al. (2013), in policy terms small-scale fisheries have largely been ignored by Europe, which may have left them exposed to competition from within the sector and from other sectors, such as large-scale fleets. In addition, small-scale fisheries catches are not always well-documented (e.g. total fishing mortality may be underestimated, increasing the risk of overexploitation of stocks).

In the EU, the majority of small-scale fisheries occur in nearshore waters, however, both sectors are mostly fishing within the exclusive economic zone and in many cases they exploit the same stocks. On account of this similarity and many differences, small-scale fisheries are neither exactly the same as nor totally different from large-scale fisheries. Arguably they mainly fall at opposite ends of a continuum but with considerable overlap, for example in terms of functional groups of marine animals which are exploited (Fig. 20.1). Thus, it is worth exploring the extent to which existing approaches can be transferred from large-scale to small-scale fisheries and vice versa to get improved information about small-scale fisheries. This transfer of approaches follows the notion that although large-scale fisheries

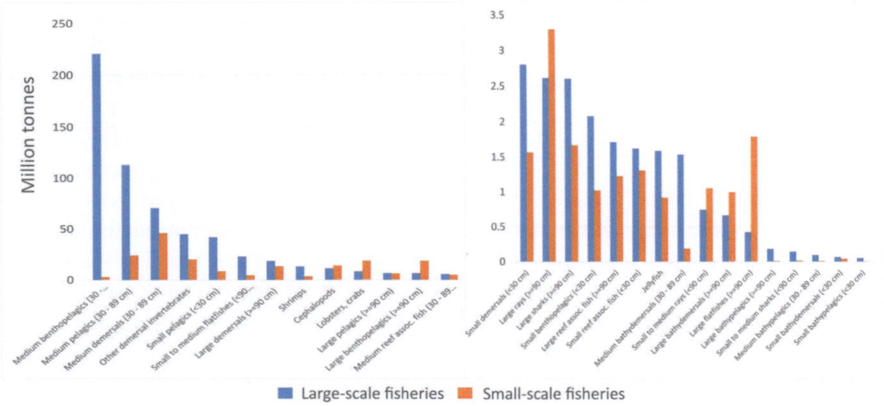

Fig. 20.1 Various marine functional groups, with catches in millions of tonnes, caught by both small-scale and large-scale fishing sectors at the global level. (Source: Sea Around Us Project)

are far from being sustainable, the sector has more assessment methods and accumulated data compared to small-scale fisheries.

20.2.1 Barriers to Applying Large-Scale Fisheries Data Collection Methodologies

Monitoring small-scale fisheries presents challenges both in collecting the data needed for fishery assessment and in monitoring the compliance with regulations. Many countries, especially in the north, obtain information on fishing effort, catches, bycatch, landings, and discards of large-scale fleets from a combination of logbook data, onboard monitoring, and market records. In small-scale fisheries, small boats have limited capacity to carry observers, they target a range of species with a range of gears (often several in the same boat), and they do not always land their catches at regular fish markets. Thus, it can be difficult to collect data both generally and in relation to the exploitation of individual species and many stocks exploited by small-scale fisheries are 'data poor'. In some instances however, data limitations flow directly from choices made by responsible authorities (e.g. exemption of small boats from the need to complete logbooks in the EU).

Alternative data sources, including, data from remote and automatic surveillance, like Global Positioning System (GPS) and Vessel Monitoring System (VMS), onboard camera systems, drones (Stop Illegal Fishing 2016), and satellite imagery (e.g. Waluda et al. 2002) have been employed in some fisheries. For example, Bartholomew et al. (2018) report a study of a small-scale elasmobranch gillnet fishery in Peru, which suggests that cameras can be an effective alternative to onboard observers. However, such approaches can be expensive and can raise legal and ethical questions, for example, under data protection and privacy laws. In addition, much useful data can be obtained from the fishers, for example, to populate alternative models such as the Gomez-Muñoz model to estimate fishing effort (Gomez-Muñoz 1990; Rocha et al. 2006).

20.2.2 Examples of Large-Scale Fisheries Stock Assessment Methods Applied to Small-Scale Fisheries

Approaches to stock assessment developed for long-lived finfish taken by large-scale fisheries can be adapted to short-lived species and small-scale fisheries (e.g. Pierce and Guerra 1994). Stock assessment models typically have two components, a model describing population dynamics and one linking the parameters of the first to information that can be obtained from a fishery (Hilborn and Walters 1992). As such, small-scale fisheries target species can present problems (i) related to species population ecology (e.g. when small-scale fisheries target populations of short-lived species like octopus, which comprise of a single cohort, for which age determination is difficult and which have highly variable growth rates so that length is a poor proxy for age), and (ii) many small-scale fisheries are data-poor, especially in developing countries (e.g. Ramírez et al. 2017).

Production models have long been applied to large-scale cephalopod fisheries on the Saharan Bank (e.g. Bravo de Laguna 1989) and were recently used in the octopus fishery in the Gulf of Cadiz, Spain (ICES 2014), a fishery in which both trawl and artisanal fleets participate (Silva et al. 2002). A key assumption of this approach is that environmental carrying capacity is a constant. However, in the short-lived, rapidly growing cephalopods, a high sensitivity to environmental variation (e.g. Pierce et al. 2008; Rodhouse et al. 2014) means that carrying capacity effectively changes from year to year. Thus, the successful application of production models may reflect the higher resilience of cephalopod populations to overexploitation, compared to longer-lived species (see Caddy 1983), with the variable carrying capacity simply introducing noise, but no bias into the available data.

In relation to age-structured cohort-based assessment models, nowadays the most common type of model in assessment of large-scale fisheries, application to small-scale fisheries can be difficult, because age determination is difficult in many small-scale fishery target species (e.g. crabs, sea urchins, goose barnacles, and the small squid *Alloteuthis subulata*). However, in some of these species, length can be used as a substitute for age – although not in cephalopods, which have very variable growth rates. In addition, cohort-based models can be recast to follow 'micro-cohorts' (i.e. groupings of animals with similar hatching dates) from month to month, something which has been achieved for cephalopods (e.g. Royer et al. 2006; Gras et al. 2014), the short lifespans of which (often only 1 year) preclude the traditional approach to following annual cohorts through time.

Empirical models linking abundance to environmental conditions can also work for short-lived species, with promising results achieved for octopus (e.g. Sobrino et al. 2002; Otero et al. 2008). In principle, environmental predictors can be added to traditional stock assessment models, such as production models, as in the case of octopus in the Gulf of Cadiz (ICES 2014). Finally, even where it is impossible to apply formal assessment models, relevant information on stock status and exploitation rate of small-scale fishery target species can be obtained based on fishery monitoring (e.g. ICES 2010).

20.3 Examples of Methods and Tools to Tailor Information About Small-Scale Fisheries

The SSF Guidelines have provided an excellent justification of how, given the unique characteristics and contributions of small-scale fisheries, they require different governance approaches. The need for an improved understanding and better data about small-scale fisheries is part of the call to FAO member states. Similar to how the SSF Guidelines require plans and strategies for a successful implementation, some directions about ways to address data paucity in small-scale fisheries is necessary to encourage member states to take action. We present four examples of existing initiatives that could provide a basis for further discussion about moving forward.

20.3.1 FAO Fisheries Statistics in the Context of Small-Scale Fisheries

The UN FAO has under one of its mandates the responsibility to collect, analyze, and report data on fisheries and aquaculture from member countries (FAO 2015b). The earliest reported statistics date back to 1950 for capture and aquaculture production. Recently, FAO has come up with new presentations of datasets and statistics specific to small-scale fisheries, as discussed below.

As part of its statistical work, FAO actively supports and fosters the development of internationally agreed methods, definitions, and tools for data compilation. To this end, it has cooperated through mechanisms like the Coordinating Working Party on Fisheries Statistics (CWP) for the past 50 years. In 1995, the definitions applied in the national reporting questionnaires were brought in line with the International Labour Organization (ILO) International Standard Classification of Occupations (ISCO-88), including sex-disaggregated reporting, as recommended by the UN Statistical Commission. Before this change in the questionnaire, the first occurrence of sex-disaggregated reporting was provided by Japan in 1970 (Gee and Bacher 2017). The application of method specific to socioeconomic data collection has resulted in a comparable dataset of particular relevance to small-scale fisheries.

This data is now available through the FAO EastMed project (FAO-EASTMED 2018) with the method being disseminated in a practical handbook (Pinello et al. 2017).

Since its inception, the FAO Fisheries and Aquaculture Department has created statistical databases that currently include information on capture production, fishing fleets, aquaculture production, employment, commodities, and food balance sheets. Data reporting is scheduled as an annual activity with member states transmitting their annual figures, for the previous calendar year, after the midpoint of the year; when appropriate, reports may include estimates. Fundamentally, the reliability of the data analysis rests with the quality of the data reported. The number of reporting countries and level of detail provided in the data fluctuates between years.

The definition of small-scale fisheries relies on the specific context in which the fisheries are situated (FAO 2015a). Matrix-based definitions, such as that proposed in the workshop proceedings 'Improving our knowledge on small-scale fisheries: data needs and methodologies' (FAO 2017), can serve as a guide for selecting the most relevant statistics for small-scale fisheries in a particular context. The flexibility of these definitions to best match country needs is in line with the SSF Guidelines. However, as there is no single, universal small-scale fisheries definition, it is not possible to have a specific small-scale fisheries category identified within the global FAO statistics.

Although the data on capture fisheries and aquaculture production, commodities, and food balance sheets all cover fundamental areas of the sectors, at this time there are no means to distinguish small-scale fisheries-relevant components of these datasets. In this light, the areas of particular relevance to small-scale fisheries are the data collected on employment and fleets. The collection of employment data uses a questionnaire that captures the number of people engaged disaggregated by gender for both occupational category and working time. In the employment data, a separate category included subsistence fishing, as well as inland and coastal fisheries. Although small-scale fisheries are not necessarily subsistence fisheries, they are certainly one component of small-scale fisheries.

The questionnaire on national fishing fleets is broken down into two categories: decked and undecked vessels. The smallest length categories are defined differently between decked and undecked categories, with 0–6 m and 6–12 m for undecked vessels, and an aggregated class of <12 m length overall (LOA) for decked vessels. Again, at the furthest limits of these categories, there is some overlap between small-scale and not small-scale, and the exact definition of these categories rests with the countries and the contexts in which the fisheries operate. Along with the other FAO fisheries statistical datasets, fleet and employment data have been shared through the FAO Yearbook of Fishery and Aquaculture Statistics publications, and the full complement of reported sex-disaggregated employment statistics available to-date is planned for a release.

20.3.2 Crowdsourcing Small-Scale Fisheries Data: The ISSF Platform

Datasets for small-scale fisheries are often inconsistent and incomplete due to complex patterns of harvest and post-harvest activities that take place in diverse areas, often remote, and in the absence of effective reporting mechanisms. This problem leads to imperfect information and to large knowledge gaps that hamper effective decision-making. Current global databases capture a limited type of information, such as species names (e.g. OBIS database), catch, and weight (e.g. Sea Around Us project), failing to capture data about the social, economic, or political contexts in which those fisheries operate. Such additional data can, however, be reached in local studies, and when put together, could contribute to a broader understanding of small-scale fisheries.

In the absence of mechanisms that allow for a consistent collection of detailed small-scale fisheries data, citizen science and crowdsourcing open innovative means to collect and compile data in support of policy and decision-making. Crowdsourcing systems have been increasingly popular in the past decade, supporting the creation of large global databases, such as the Wikipedia dictionary, the OpenStreetMap mapping project, or projects like eBird and FishBase. Those projects mobilize large communities of contributors that help fill knowledge gaps as well as improve the quality of compiled data.

Following these examples, TBTI adopted a crowd-sourcing approach to create a small-scale fisheries database, resulting in the Information System on Small-scale Fisheries (ISSF)[2] (Chuenpagdee et al. 2017). Arguably the largest existing global database on small-scale fisheries, ISSF is an open-source, open-data web portal that allows anyone with access to the Internet to share information describing small-scale fisheries. This crowdsourcing platform aims to collect fragments of knowledge accumulated by people all around the world and put it together to help answer important questions about small-scale fisheries that could not be answered by looking at individual data pieces. ISSF allows sharing different types of information, such as about the people studying small-scale fisheries, organizations working in this field, publications about small-scale fisheries, descriptions of specific small-scale fisheries from around the world, and even videos and other multimedia material related to this sector (e.g. interview with a small-scale fisher). Those descriptions can be done at different geographic levels, ranging from a specific place to the global level.

As a crowdsourcing platform, the ISSF database is constantly growing, currently consisting of data from about 400 people, 350 organizations, 1800 publications, and includes detailed records/profiles of small-scale fisheries from about 200 locations. Those profiles record 20 core characteristics about the sector (e.g. gear types, species, gender information, governance mode) for each location, with the core characteristics identified through several years of consultation with the small-scale

[2] http://issf.toobigtoignore.net

fisheries experts. The profile layer, allows a systematic description of small-scale fisheries, resulting in a detailed comparison of fishing communities from around the world and offering a *de facto* standard that can be used to collect data in the future.

ISSF provides a platform for sharing small-scale fisheries data but also makes those data accessible to anyone through the use of various tools. Specifically, data are reported both in the form of text (i.e. tabular format) and on a map. Information is searchable using simple search terms (e.g. species names, locations) and types of themes/issues or characteristics (e.g. data related to gender questions, labor migration). Preliminary analyses of the data collected using ISSF were published and presented as a literature overview (Rocklin 2016a) and the people involved in small-scale fisheries (Rocklin 2016b).

The TBTI-ISSF project demonstrates that crowdsourcing methods are beneficial for gathering knowledge about small-scale fisheries. Such systems have incredible potential but are limited by the number of contributors entering data onto the system, hence some places/counties may have more data than others. Combining such approach with mobile data collection tools (e.g. apps) could help source data more rapidly to generate the information required to characterize, in detail, regions or countries.

20.3.3 Estimating Small-Scale Fishing Efforts

Consistent signs of fish stock over-exploitation, increases in global fishing effort, declines in landed fish biomass (Pauly and Zeller 2016) and growths in human populations along many of the world's coastlines all highlight the importance of understanding where fish are caught and how much effort is required to catch them. However, in addition to the challenges related to data collection discussed above, methods to analyze small-scale fisheries data, especially of the social dimension is also scarce and can hinder understanding not only of the social systems, but also of the ecological systems surrounding fisheries. For example, to fully understand the impacts of fishing, knowledge of where fishing boats operate and how much they fish is critical.

In addition to the amount and value of fish caught, the amount of fishing (i.e. the effort) and the area over which that fishing takes place (i.e. the geographic range or spatial extent) are key measures for the successful implementation of fisheries management strategies and can assist in forecasting the ecological consequences of current and future fishing activities (Hinz et al. 2013). They can also help elucidate some of the social factors driving fisheries exploitations and fishing behaviors (Guillemot et al. 2009). The lack of data associated with many small-scale fisheries makes the calculation of fishing effort difficult.

To address the problem, Johnson et al. (2017) designed a method, 'Predicted Fishing Effort' (PFE), for estimating the amount of fishing activity based on the number of boats present in an area, the average distance of fishing operations from shore, and the local human coastal populations around the Gulf of California,

Mexico. Also, by correlating the PFE metric to spatially explicit local fisheries landings, the method was also able to predict the amount of fish biomass extracted from areas of previously unknown fishing activity, evaluate the potential future growth in the local fishery, and provide an estimate of the most cost-effective number of fishing boats per area without a fishery experiencing diminishing returns (i.e. adding more boats, but returning fewer landed fish). The authors validated the method using fine resolution boat tracking data from the same study region, but such data are not commonly available for small-scale fishing fleets, particularly in developing countries. Therefore, in such cases, additional validation methods should be sought, such as fisher interviews and onboard observer estimates.

Across a broader spatial scale, Stewart et al. (2010) compiled a comprehensive database of fishing effort metrics and the spatial limits of fisheries to map fishing effort density across six ocean regions. Their method also considers the utility of using socioeconomic variables as indicators of fishing pressure at the national level including the Human Development Index (HDI), Gross Domestic Product (GDP), and population size. The Fishing Effort Envelope Tool (FEET), which combines distance from shore and port, and the depth of the fishery, was used to delimit the potential area in which fishers may operate, to distribute fishing effort across this potentially fished area in 1 km^2 cells. Their results show that fishing density increased as a function of population size, similar to the results of Johnson et al. (2017), and decreased as a function of coastline length. Methods to calculate fishing effort such as the PFE and FEET are by no means the final hurdle in estimating the fishing effort of small-scale fishing vessels that do not employ vessel tracking technologies. However, such innovative methods will reward by helping to inform managers and policy-makers.

20.3.4 The Case of Spatial Information in Marine Planning

Marine spatial planning (MSP) is becoming popular to define the conditions for sharing the maritime space between various uses (e.g. marine renewable energy, mining, oil and gas, submarine cables), while taking into consideration environmental and biodiversity conservation (e.g. marine protected areas). However, this process creates a new challenge for the small-scale fisheries sector.

MSP is rooted in the principles of ecosystem-based management (EBM) (Degnbol and Wilson 2008; Douvere 2008). To date, around 60 MSP processes have been documented worldwide (IOC-UNESCO and EC-DGMARE 2017). Also, one may perceive MSP as a friendly approach that can help achieve fisheries sustainability through economic growth (Qiu and Jones 2013). In practice, however, MSP takes a wide range of forms and continues to struggle with finding balance, representing itself as prescriptive, spatial, and competitive or environmentally sensitive, which could, in some cases, present more risks than opportunities for fisheries (Jentoft and Knol 2014). Hence, MSP can be far from neutral. From a fisheries perspective, MSP may struggle with inconsistent and uncoordinated decisions (Kannen

2014) or can potentially contribute to some form of 'ocean grabbing' (Bennett et al. 2015) and 'sea sparing' (Wolff 2015).

Past and ongoing MSP experiences show that there are questions about data, notably *spatial* data. For instance, the United Kingdom authorities concluded in the East Inshore and East Offshore Marine Plans that "the lack of uniformity and stakeholder consensus regarding fisheries data combined with the difficulties in predicting the future of fisheries, makes formulating prescriptive marine plan policies for this sector a challenge" (HM Government 2014, 162). Mapping or zoning fisheries presents challenges due to a high spatial and temporal variability and a large spatial distribution of fisheries activities. Accordingly, most existing marine plans failed to integrate fisheries, such as the cases in the Netherlands, Germany, and Israel, whose fisheries have apparently disappeared from the maps. Such a situation could happen, in particular, to small-scale fisheries for three reasons: (i) high dependency on the inshore zone, where competition for space is the strongest; (ii) low spatial adaptability; and, (iii) data-poor (e.g. VMS or AIS data only existing for ships up to 12 or 15 m). Therefore, planners must carefully apply MSP when it involves small-scale fisheries.

Additionally, issues about spatial data for marine planning go beyond their availability and accuracy. Combining the narratives of managerial approaches with technical fixes (Swyngedouw 2009), a 'spatially-focused MSP' rather than a strategic one, can drift towards being post-political (Tafon 2018), satisfying the idea that spatial planning can be a ploy: "The goal of those who promote planning is to reduce managers' power over strategy making" (Mintzberg 1994, 479). As a result, MSP can be reduced to a techno-managerial data gathering and map-making exercise (Boucquey et al. 2016), a process that is not impartial, but rather strongly political. Considering that "power defines what counts as rationality and knowledge and thereby what counts as reality" (Flyvberg 1998, 227), improving data could also mean looking into the power attached to, and promoting realities other than, generally, the ones created by science and technologies (Whatmore 2009). Therefore, to carefully consider small-scale fisheries in MSP, one must reconsider the commonly used metrics (as well as the wordings). For instance, typical metrics (e.g. fishing effort, landed weightor values) are not enough to represent the importance of small-scale fisheries.

While the need to integrate non-scientific knowledge is widely accepted in fisheries management (Haggan et al. 2007; Hind 2015), it is rarely applied to spatial planning. Therefore, supporting the use of non-scientific knowledge, by helping fishers translate their knowledge into geospatial technologies, could allow them to be empowered. On this basis, it would form a 'socio-technical agencement'(Akrich et al. 2006; Callon 2006), making fishers' representatives and technology a 'hybrid collective' of human and non-humans, able to (inter)act as a whole in MSP processes. A partnership between fishers and scientists developed in France since 2010 (see Trouillet et al. forthcoming) has experimented with this process. If such experience may not be easily replicable, it calls nevertheless to pay greater attention to small-scale fisheries in MSP.

20.4 Moving Forward: Improving Information About Small-Scale Fisheries

Humans use knowledge and information on a daily basis to make simple decisions, such as what to eat, or complex ones like what fisheries management tools and approaches to implement. The role of data and information is crucial in making these decisions. In most cases, a decision-maker faces a situation where he/she needs to generate data to come up with the best possible and informed choice.

It is important to note that past and current efforts have already led to possibilities and milestones that have, to some extent, motivated the research community to improve data and the data gathering methods. Information that shows the global importance of small-scale fisheries in comparison to large-scale fisheries indicates how small-scale fisheries are too big to ignore. Numerous local case studies describe the various contexts in which small-scale fisheries operate and how different they are from the large-scale fisheries sector (e.g. Jentoft and Chuenpagdee 2015). These 'thick descriptions' about small-scale fisheries provide evidence that better data can improve governance. Prospects may also lie in the applicability of quantitative tools and methods used in large-scale fisheries to small-scale fisheries data. Innovative techniques, such as the PFE (Johnson et al. 2017), add hopes to mapping the small-scale fishing effort on a broader scale. Popular methods such as crowdsourcing, like ISSF, and citizen science also show positive contributions to addressing data gaps in the small-scale fisheries sector, particularly those that are hard to obtain using formal monitoring and assessment methods.

So what else can we do to make the knowledge-generation of and about small-scale fisheries help with the sustainability of this sector? One of the options lies in the transdisciplinary approach, a process that requires a collective understanding of the issue, not only of the problems in fisheries, but also of other issues impacting the fisheries sector. Transdisciplinarity is different from multidisciplinarity whereby understanding is drawn from various disciplines but that the knowledge remains within the discipline (Choi and Pak 2006).

A transdisciplinary approach can help researchers and managers by moving the issue beyond managing the symptoms of the problem and first understand what the problems may be and what information can express them. In doing so, researchers and decision-makers alike would need to embrace various avenues of inquiry under the qualitative and quantitative categories (e.g. stakeholder views, expert judgment, and scientific method). Often, this requires going out of the norms and being creative about the best way to use the existing information and tap into new methods and data as innovative knowledge sources.

This need for a transdisciplinary approach is fundamentally recognized within the SSF Guidelines under its guiding principle that calls for more holistic and integrated solutions for small-scale fisheries sustainability. As one would expect, a transdisciplinary approach for improving knowledge about small-scale fisheries would require a mix of different types of information and a wealth of knowledge, encompassing ecological, social, economic, and governance. This wealth of information can be used to measure desired outcomes (or indicators), such as the percentage of fishers under poverty threshold that would be useful to track down goals such as, for instance, SDGs (FAO 2017). However, indicators are like the tip of the iceberg, since what goes on underwater is far deeper and broader than what is visible from above, and that is where the transdisciplinarity process would be the most useful. Hence, to manage small-scale fisheries, managers, researchers, and other stakeholders will need to explore holistic and integrated perspectives that begin with identifying the appropriate information.

For example, since small-scale fishers experience a higher degree of marginalization (economically, geographically, and politically) compared to large-scale fisheries and other sectors, careful consideration should be given to how best small-scale fishers are represented through sets of information. As has been noted, the different forms of information (e.g. model scenarios, maps, and narratives) do not necessarily represent small-scale fisheries stakeholders equitably at the decision-making table (Jentoft and Knol 2014). As a result, reaching an agreement as to which data to use and how to present data could be a challenge (HM Government 2014). Hence, there are more opportunities to think more carefully and collectively about the reasoning behind what and how information about small-scale fisheries are obtained and represented alongside known challenges in particular with the recognition that more data specific to small-scale fisheries are needed (see FAO 2017).

Applying transdisciplinary perspectives in identifying and obtaining small-scale fisheries information, as well as in making decisions, may offer new insights. For example, it is worthwhile to consider how small-scale fisheries information might look like if it were to fully take advantage of the broad range of methods (Brandt et al. 2013) and mixed-methods research (Mingers 2015) that pay equal attention to the value of both the quantitative and qualitative data. One can also consider how different small-scale fisheries knowledge would be if it were to come from research that is co-designed, co-produced, and co-disseminated by multiple actors involved (Mauser et al. 2013), which are the underlying principles for an 'open' transdisciplinarity (Brandt et al. 2013). Therefore, when thinking of ways to improve small-scale fisheries information, we could consider the following: i) whether the information speaks to the real problems that concern the small-scale fisheries sector; ii) the extent to which the small-scale fisheries knowledge types are inclusive (natural and social), appropriate, and practical; and, iii) the level of involvement of practitioners that is suitable in generating and making use of the small-scale fisheries information. In moving forward, transdisciplinarity will involve more work and coordinated actions, so how can this be initiated and implemented are essential considerations. In Chap. 22 (this volume), Said et al..... offer thoughts in this regard, as well as the challenges involved.

Moreover, the objectives stated in the SSF Guidelines provide a roadmap on what matters to small-scale fisheries (FAO 2015a). These objectives are also beneficial for identifying indicators and actors. Hence, the SSF Guidelines could be a useful starting point to determine priorities, including the associated data, through a transdisciplinary approach. Success in pursuing unconventional means to obtain greater and improved knowledge may not be as far as it seems, since it could build on the advantages already held by small-scale fisheries sector, in other words, the stronger sense of ownership that leads to a more responsible stewardship and collaborative management of natural resources.

References

Ackoff R (1989) From data to wisdom. J Applies Syst Anal 16:3–9
Akrich M, Callon M, Latour B (eds) (2006) Sociologie de la traduction. Textes fondateurs. In: Sociologie de la traduction. Presses des Mines, Paris
Alfaro-Shigueto J, Mangel JC, Pajuelo M et al (2010) Where small can have a large impact: structure and characterization of small-scale fisheries in Peru. Fish Res 106:8–17. https://doi.org/10.1016/J.FISHRES.2010.06.004
Bartholomew DC, Mangel JC, Alfaro-Shigueto J et al (2018) Remote electronic monitoring as a potential alternative to on-board observers in small-scale fisheries. Biol Conserv 219:35–45. https://doi.org/10.1016/J.BIOCON.2018.01.003
Béné C, Friend RM (2009) Water, poverty and inland fisheries: lessons from Africa and Asia. Water Int 34:47–61. https://doi.org/10.1080/02508060802677838
Bennett NJ, Govan H, Satterfield T (2015) Ocean grabbing. Mar Policy 57:61–68. https://doi.org/10.1016/J.MARPOL.2015.03.026
Beverton RJH, Holt SJ (1957) On the dynamics of exploited fish populations. Fisheries investigations series 2 marine fisheries no.19. Ministry of Agriculture, Fisheries and Food. In Status of marine fish stocks. European Environment Agency, Copenhagen, fisheries
Boucquey N, Fairbanks L, St. Martin K et al (2016) The ontological politics of marine spatial planning: assembling the ocean and shaping the capacities of "community" and "environment". Geoforum 75:1–11. https://doi.org/10.1016/J.GEOFORUM.2016.06.014
Brandt P, Ernst A, Gralla F et al (2013) A review of transdisciplinary research in sustainability science. Ecol Econ 92:1–15. https://doi.org/10.1016/J.ECOLECON.2013.04.008
Bravo de Laguna J (1989) Managing an international multispecies fishery: the Saharan trawl fishery for cephalopods. In: Caddy JF (ed) Marine invertebrate fisheries: their assessment and management. Wiley, New York, pp 591–612
Caddy J (1983) The cephalopod: factors relevant to their population dynamics and to the assessment and management of stock. In: Advances in assessment of world cephalopod resources. Fisheries technical paper no. 231. FAO, Rome, Italy, pp 416–452
Callon M (2006) What does it mean to say that economics is performative? CSI working papers series 005
Choi B, Pak A (2006) Multidisciplinary, interdisciplinary and transdisciplinarity in health research, services, education and policy: 1. Definitions, objectives, and evidence of effectiveness. Clin Invest Med 29:351–364. https://doi.org/10.1016/j.jaac.2010.08.010
Chuenpagdee R, Pauly D (2004) Small is beautiful? A database approach for global assessment of small-scale fisheries. In: Proceedings of the 4th World Fisheries Congress, May 2–6, 2004. Vancouver, British Columbia, Canada
Chuenpagdee R, Rocklin D, Bishop D et al (2017) The global information system on small-scale fisheries (ISSF): a crowdsourced knowledge platform. Mar Policy. https://doi.org/10.1016/J.MARPOL.2017.06.018
Cigliano JA, Meyer R, Ballard HL et al (2015) Making marine and coastal citizen science matter. Ocean Coast Manag 115:77–87. https://doi.org/10.1016/J.OCECOAMAN.2015.06.012
de la Torre-Castro M, Di Carlo G, Jiddawi NS (2014) Seagrass importance for a small-scale fishery in the tropics: the need for seascape management. Mar Pollut Bull 83:398–407. https://doi.org/10.1016/J.MARPOLBUL.2014.03.034

Degnbol D, Wilson DC (2008) Spatial planning on the North Sea: a case of cross-scale linkages. Mar Policy 32:189–200. https://doi.org/10.1016/J.MARPOL.2007.09.006
Douvere F (2008) The importance of marine spatial planning in advancing ecosystem-based sea use management. Mar Policy 32:762–771. https://doi.org/10.1016/J.MARPOL.2008.03.021
EEA (European Environment Agency) (2015) Status of marine fish stocks. EEA, Copenhagen
Fairclough DV, Brown JI, Carlish BJ et al (2015) Breathing life into fisheries stock assessments with citizen science. Sci Rep 4:7249. https://doi.org/10.1038/srep07249
FAO (Food and Agriculture Organization of the United Nations) (2015a) Voluntary guidelines for securing sustainable small-scale fisheries. FAO, Rome
FAO (2015b) Fisheries and Aquaculture topics. Fisheries statistics and information. In: Fisheries and Aquaculture Department [online]. Rome, Italy
FAO (2017) Improving our knowledge on small-scale fisheries: data needs and methodologies – Workshop proceedings. FAO Fisheries and Aquaculture Proceedings 56. FAO, Rome
FAO-EASTMED (2018) Scientific and institutional cooperation to support responsible fisheries in the Eastern Mediterranean. http://www.faoeastmed.org/. Accessed 3 Mar 2018
FAO-UN (2018) Ecosystem Approach to Fisheries (EAF). http://www.fao.org/fishery/topic/16034/en. Accessed 28 Feb 2018
Finley C (2011) Fish unlimited: how maximum sustained yield failed fishermen. Solut J 2:77–81
Finley C, Oreskes N (2013) Maximum sustained yield: a policy disguised as science. ICES J Mar Sci 70:245–250. https://doi.org/10.1093/icesjms/fss192
Flyvberg B (1998) Rationality and power: democracy in practice. The University of Chicago Press, Chicago
Gee J, Bacher K (2017) Gender in aquaculture and fisheries: engendering security in fisheries and aquaculture. Asian Fish Sci Spec Issue 30S:277–290
Gomez-Muñoz VM (1990) A model to estimate catches from a short fishery statistics survey. Bull Mar Sci 46:719–722
Gras M, Roel BA, Coppin F et al (2014) A two-stage biomass model to assess the English Channel cuttlefish (Sepia officinalis L.) stock. ICES J Mar Sci 71:2457–2468. https://doi.org/10.1093/icesjms/fsu081
Guillemot N, Léopold M, Cuif M et al (2009) Characterization and management of informal fisheries confronted with socio-economic changes in New Caledonia (South Pacific). Fish Res 98:51–61. https://doi.org/10.1016/J.FISHRES.2009.03.013
Guyader O, Berthou P, Koutsikopoulos C et al (2013) Small scale fisheries in Europe: a comparative analysis based on a selection of case studies. Fish Res 140:1–13. https://doi.org/10.1016/J.FISHRES.2012.11.008
Haggan N, Neis B, Baird I (eds) (2007) Fishers' knowledge in fisheries science and management. UNESCO Publishing, Paris
Hilborn R, Walters CJ (1992) Quantitative fisheries stock assessment: choice, dynamics and uncertainty. Chapman and Hall, New York
Hind EJ (2015) A review of the past, the present, and the future of fishers' knowledge research: a challenge to established fisheries science. ICES J Mar Sci 72:341–358. https://doi.org/10.1093/icesjms/fsu169
Hinz H, Murray LG, Lambert GI et al (2013) Confidentiality over fishing effort data threatens science and management progress. Fish Fish 14:110–117. https://doi.org/10.1111/j.1467-2979.2012.00475.x
HM Government (2014) East inshore and east offshore marine plans. London, UK
Holt S (2011) Maximum sustainable yield: the worst idea in fisheries management. In: Breaching blue. https://breachingtheblue.com/2011/10/03/maximum-sustainable-yield-the-worst-idea-in-fisheries-management/
Hughes T (2011) The future of marine governance. Solut J 2:18–20
ICES (International Council for the Exploration of the Sea) (2010) Report of the Working Group on Cephalopod Fisheries and Life History (WGCEPH), 9–11 March 2010, Sukarrieta, Spain. ICES CM 2010/SSGEF:09. ICES, Copenhagen

ICES (2014) Report of the Working Group on Cephalopod Fisheries and Life History (WGCEPH), 16–19 June 2014, Lisbon, Portugal.ICES CM 2014/SSGEF:02. Copenhagen
ICSF (2016) New, but long overdue. Samudara Report No. 72
IOC-UNESCO, EC-DGMARE (2017) 2nd international conference on marine/maritime spatial planning, 15–17 March 2017, Paris, France; IOC. Workshop report
Jentoft S, Chuenpagdee R (eds) (2015) Interactive governance for small-scale fisheries: global reflections. Springer, Cham
Jentoft S, Knol M (2014) Marine spatial planning: risk or opportunity for fisheries in the North Sea? Marit Stud 2014 13:1–16. doi:https://doi.org/10.1186/2212-9790-13-1
Johnson AF, Moreno-Báez M, Giron-Nava A et al (2017) A spatial method to calculate small-scale fisheries effort in data poor scenarios. PLoS One 12:4. https://doi.org/10.1371/journal.pone.0174064
Kannen A (2014) Challenges for marine spatial planning in the context of multiple sea uses, policy arenas and actors based on experiences from the German North Sea. Reg Environ Chang 14:2139–2150. https://doi.org/10.1007/s10113-012-0349-7
Kelleher K (2005) Discards in the world's marine fisheries an update. United Nations Food and Agriculture Organization fisheries technical paper 470. FAO, Rome
Mauser W, Klepper G, Rice M et al (2013) Transdisciplinary global change research: the co-creation of knowledge for sustainability. Curr Opin Environ Sustain 5:420–431. https://doi.org/10.1016/j.cosust.2013.07.001
Mingers J (2015) Helping business schools engage with real problems: the contribution of critical realism and systems thinking. Eur J Oper Res 242:316–331. https://doi.org/10.1016/J.EJOR.2014.10.058
Mintzberg H (1994) The rise and fall of strategic planning. Harv Bus Rev:107–114
Otero J, Álvarez–Salgado XA, González AF et al (2008) Bottom-up control of Octopus vulgaris abundance in a wind-driven upwelling ecosystem (NE Atlantic). Mar Ecol Prog Ser 362:181–192
Pauly D (2006) Major trends in small-scale fisheries, with emphasis on developing countries and some implications for the social sciences. Mari Stud 4:7–22
Pauly D (2018) A vision for marine fisheries in a global blue economy. Mar Policy 87:371–374. https://doi.org/10.1016/J.MARPOL.2017.11.010
Pauly D, Zeller D (2016) Catch reconstructions reveal that global marine fisheries catches are higher than reported and declining. Nat Commun 7:10244. https://doi.org/10.1038/ncomms10244
Pierce GJ, Guerra A (1994) Stock assessment methods used for cephalopod fisheries. Fish Res 21:255–285. https://doi.org/10.1016/0165-7836(94)90108-2
Pierce GJ, Valavanis VD, Guerra A et al (2008) A review of cephalopod – environment interactions in European Seas. Hydrobiologia 612:49–70. https://doi.org/10.1007/978-1-4020-9141-4_5
Pierce GJ, Allcock L, Bruno I et al (2010) Cephalopod biology and fisheries in Europe. Cooperative research report 303. International council for the exploration of the Sea
Pinello D, Gee J, Dimech M (2017) Handbook for fisheries socio-economic sample survey – principles and practice. FAO fisheries and aquaculture technical paper no. 613. FAO, Rome
Pita CB, Pereira J, Lourenço S et al (2015) The traditional small-scale octopus fishery in Portugal: framing its governability. In: Jentoft S, Chuenpagdee R (eds) Interactive governance for small-scale fisheries: global reflections. Springer, Cham, pp 117–134
Qiu W, Jones PJS (2013) The emerging policy landscape for marine spatial planning in Europe. Mar Policy 39:182–190. https://doi.org/10.1016/J.MARPOL.2012.10.010
Ramírez JG, Lleonart J, Coll M et al (2017) Improving stock assessment and management advice for data-poor small-scale fisheries through participatory monitoring. Fish Res 190:71–83. https://doi.org/10.1016/J.FISHRES.2017.01.015
Rocha F, Otero J, Outeiral R et al (2006) Modelling small-scale coastal fisheries of Galicia (NW Spain) based on data obtained from fisheries: the case of Sepia officinalis. Sci Mar 70:593–601. https://doi.org/10.3989/scimar.2006.70n4593

Rocklin D (2016a) Small-scale fisheries of the world. In: Chuenpagdee R, Rocklin D (eds) State-of-the-Art in small-scale fisheries. TBTI Publication Series, St. John's, p 8

Rocklin D (2016b) Who's who in small-scale fisheries. In: Chuenpagdee R, Rocklin D (eds) Small-scale fisheries of the world. TBTI Publication Series, St. John's, p 8

Rodhouse PGK, Pierce GJ, Nichols OC et al (2014) Environmental effects on cephalopod population dynamics. Adv Mar Biol 67:99–233. https://doi.org/10.1016/B978-0-12-800287-2.00002-0

Royer J, Pierce GJ, Foucher E, Robin JP (2006) The English Channel stock of Sepia officinalis: modelling variability in abundance and impact of the fishery. Fish Res 78:96–106. https://doi.org/10.1016/J.FISHRES.2005.12.004

Silva L, Sobrino I, Ramos F (2002) Reproductive biology of the common octopus, Octopus vulgaris Cuvier, 1797 (Cephalopoda: Octopodidae) in the Gulf of Cádiz (SW Spain). Bull Mar Sci 71:837–850

Sobrino I, Silva L, Bellido J, Ramos F (2002) Rainfall, river discharges and sea temperature as factors affecting abundance of two coastal benthic cephalopod species in the Gulf of Cádiz (SW Spain). Bull Mar Sci 71:851–865

Stewart KR, Lewison RL, Dunn DC et al (2010) Characterizing fishing effort and spatial extent of coastal fisheries. PLoS One 5:e14451. https://doi.org/10.1371/journal.pone.0014451

Stop Illegal Fishing (2016) How drones are being used in the fight against poachers. https://stopillegalfishing.com/press-links/drones-used-fight-poachers/. Accessed 14 Apr 2018

Swyngedouw E (2009) The antinomies of the postpolitical city: in search of a democratic politics of environmental production. Int J Urban Reg Res 33:601–620. https://doi.org/10.1111/j.1468-2427.2009.00859.x

Tafon RV (2018) Taking power to sea: towards a post-structuralist discourse theoretical critique of marine spatial planning. Environ Plan C Polit Space 36:258–273. https://doi.org/10.1177/2399654417707527

Thompson D, FAO (1988) The world's two marine fishing industries – how they compare. Naga, the ICLARM Quarterly 11:17

United Nations (2015) Rethinking poverty: report on the world social situation 2010 social perspective on development. https://www.un.org/development/desa/socialperspectiveondevelopment/2015/08/20/rethinking-poverty-report-on-the-world-social-situation-2010/. Accessed 17 Jan 2018

Waluda CM, Trathan PN, Elvidge CD et al (2002) Throwing light on straddling stocks of Illex argentinus: assessing fishing intensity with satellite imagery. Can J Fish Aquat Sci 59:592–596. https://doi.org/10.1139/f02-049

Whatmore SJ (2009) Mapping knowledge controversies: science, democracy and the redistribution of expertise. Prog Hum Geogr 33:587–598. https://doi.org/10.1177/0309132509339841

Wolff M (2015) From sea sharing to sea sparing – is there a paradigm shift in ocean management? Ocean Coast Manag 116:58–63. https://doi.org/10.1016/J.OCECOAMAN.2015.07.004

World Bank, FAO, WorldFish Center (2012) Hidden harvest: the global contribution of capture fisheries. Report No. 66469-GLB. Washington, DC

Chapter 21
Transdisciplinary Science for Small-Scale Fisheries

Mbachi Ruth Msomphora and Svein Jentoft

Abstract The move from what is called Mode 1 to Mode 2 science involves a shift from the ethos and practice of academic, curiosity driven, and disciplinary oriented research to a science more targeted for policy-making. In Mode 2 fisheries science, this would imply a problem-oriented focus, such as poverty alleviation within small-scale fisheries, but driven mainly by industry's interests. Transdisciplinary research, which seeks the middle ground between Mode 1 and 2, takes the process a step further by recognizing the relevance of nonscientific, experience-based knowledge, and by involving those who 'own' the problem and who have stakes in the policy-making process. This is a research process in which the voices of small-scale fisheries people are not only heard, but where fishers and fish workers have an active role in defining the problem to be investigated and in generating the knowledge needed to effectively address it. However, ideal as it may seem from a governance and governability perspective, transdisciplinary research is not without challenges. For instance, what does the move from Mode 1 to Mode 2 and beyond mean for reliability, credibility, and quality assurance of science? What is gained and what is lost? This chapter discusses what such a transformation may imply to the scientific community and what consequences it might have for the way fisheries science is practiced and perceived in a governance context. It also lays out a strategy for transdisciplinary fisheries science that is both scientifically qualified and applicable for enhancing the governability of small-scale fisheries in the real world.

Keywords Fisheries management · Research · SSF Guidelines · Mode 1 · Mode 2 · Mode 1.5 · Science · Knowledge production · Quality assessment · Peer review

---

M. R. Msomphora (✉)
Research and Publishing Support - Department of Library services, Science and Health Library, UiT – The Arctic University of Norway, Tromsø, Norway
e-mail: mbachi.msomphora@uit.no

S. Jentoft
Norwegian College of Fishery Science, UiT-The Arctic University of Norway, Tromsø, Norway
e-mail: svein.jentoft@uit.no

R. Chuenpagdee, S. Jentoft (eds.), *Transdisciplinarity for Small-Scale Fisheries Governance*, MARE Publication Series 21,
https://doi.org/10.1007/978-3-319-94938-3_21

21.1 Introduction[1]

Small-scale fisheries provide food and employment to millions of people around the world. They are also a keystone for coastal communities, whose survival depends on them. For these reasons, small-scale fisheries are 'too big to ignore and too important to fail'. However, and puzzling as it may seem, small-scale fisheries figure low on the political agenda, government support circumvents them, and more often than not, fisheries development efforts miss them. As a result, many small-scale fisheries people find themselves marginalized and impoverished, without a voice. Moreover, their potential contribution to food security, economic growth, employment, and community development is consequently undermined (Jentoft and Eide 2011). For small-scale fisheries to thrive, good management and governance of fisheries resources is the basic condition – but only one of many – that needs to be brought in order. Indeed, as advanced by the Voluntary Guidelines for Securing Sustainable Small-Scale Fisheries (SSF Guidelines), human rights principles should underpin governance efforts for this sector (Jentoft 2014). Efforts to make small-scale fisheries sustainable must also include research on a broad range of issues, as outlined in Section 11 of the SSF Guidelines. For instance, Paragraph 11.1 reads:

> States should establish systems of collecting fisheries data, including bio ecological, social, cultural and economic data relevant for decision-making on sustainable management of small-scale fisheries with a view to ensuring sustainability of ecosystems, including fish stocks, in a transparent manner. Efforts should be made to also produce gender-disaggregated data in official statistics, as well as data allowing for an improved understanding and visibility of the importance of small-scale fisheries and its different components, including socio-economic aspects.

The SSF Guidelines also emphasize the need to recognize that small-scale fisheries communities are "holders, providers and receivers of knowledge", which is a research topic in itself. Small-scale fisheries research should be broad and applied in the range of issues investigated, and be inclusive by integrating small-scale fisheries people's knowledge into a research process that moves beyond traditional academic, disciplinary research. Paragraph 11.9 underscores the importance of integrating this research knowledge into governmental decision-making processes.

In their view of the need for knowledge development and research, the SSF Guidelines are in line with current trends in scientific approaches. Traditionally, scientific knowledge was acquired through basic research, which in the next instance led to applied research, then development and, finally, policy application (Silbey 2006; Sismondo 2008; Turner 2008). From being largely curiosity-driven within academic institutions, scientific research is increasingly becoming oriented towards resolving societal problems by providing knowledge to inform strategic policy goals (Irvine and Martin 1984; Amsterdamska 2008). In a more problem-oriented approach, science becomes multidisciplinary and transdisciplinary, where the

[1] This chapter builds on Mbachi Ruth Msomphora: The role of science in fisheries management in Europe: from Mode 1 to Mode 2. Maritime Studies (2016, 15:3. DOI https://doi.org/10.1186/s40152-016-0042-4).

research community is mobilized to provide knowledge that is relevant for policy implementation. Research is, therefore, perceived to be socially constructed and intertwined with other spheres of society (e.g. policy-making) through different, but overlapping social practices and values (Walt 1994; Silbey 2006). This also seems to be the tendency of direction and practice in marine and fisheries science (Wilson 2009; Linke and Jentoft 2013).

According to Gibbons et al. (1994) and Ziman (2000a), the transition described above is categorized as a move from 'Mode 1 science' (traditional academic science) to 'Mode 2 science' (modern post-academic science), which would seem as a move in the right direction from a governance perspective, including as a means of making research more relevant for small-scale fisheries. However, the move from Mode 1 to Mode 2 science is not without challenges, as it brings out, and puts into question, scientific virtues that should not be lost in the process, such as reliability, credibility, trustworthiness, and quality assurance.

This chapter discusses what a transformation from Mode 1 to Mode 2 science may imply for the scientific community and what consequences it might have for the perception and practice of research in the context of small-scale fisheries governance. It also lays out a strategy for transdisciplinary fisheries science that is both scientifically qualified, and applicable for enhancing the governability of small-scale fisheries in the real world. Of particular interest is how transdisciplinary science for small-scale fisheries may maintain the qualities of both Mode 1 and Mode 2 science in the effort of making science matter in the implementation of the SSF Guidelines. How should a transdisciplinary approach relate to the ethos of science and be relevant and constructive for small-scale fisheries governance at the same time? Thus, more precisely, in our conceptualization, transdisciplinary science, with respect to what Msomphora (2016a) calls Mode 1.5, is an ideal type that combines the qualities of both Mode 1 and Mode 2.

In the following section, we explore in more detail the ideas that guide Mode 1 and Mode 2 science. The section also describes the ideals of transition from Mode 1 to Mode 2 science. The section 'Theoretical Context' outlines the information theory used to perform the analysis about the shift from Mode 1 to Mode 2 in fisheries science. The section that follows describe the essentials of an ideal middle ground mode approach (Mode 1.5 Science) that retains the best qualities of both Mode 1 and Mode 2 science, while avoiding their shortcomings. The next section contains a discussion of the relationship between science and policy-making in relation to issues concerning fisheries research. Lastly, the concluding section draws the link between the modes of science and transdisciplinary research in small-scale fisheries.

21.2 Mode 1 and Mode 2 Science

As Ziman (2000b) and others note, Mode 1 refers to traditional academic (university) science or basic science, whereas Mode 2 characteristics have primarily been important in applied science. Both science models have existed side by side for a long time. However, applied science (with Mode 2 features) currently seems to outgrow basic science. In this process, as Ziman (2000b) observes, the normative ideals of applied science also seem to be influencing the way the traditional academic science is practiced and justified. Fishery science is, according to Holm (2003), an example of what Jasanoff calls regulatory science. As such, we would expect that it has important characteristics of Mode 2 science. Nevertheless, as Holm (2003) notes, Mode 1 ideals have also been important to justify fishery science. This does not mean that the general transition from Mode 1 to Mode 2 as a generalized ideal is irrelevant for fishery science. For example, in line with Mode 2 science, the International Council for the Exploration of the Sea (ICES) has opened up and made a number of its processes transparent in response to certain concerns that are evident in the EU's Common Fisheries Policy (CFP) reforms (ICES 2018). This has in effect moved its science in the direction of Mode 2.

While this transition of science attracts considerable support, it has not been universally endorsed (Gibbons et al. 1994; Limoges 1996; Hackett et al. 2008; Jasanoff 2009). What consequences might the shift have on the way science is perceived, practiced, legitimized, and converted into policy? Should this move be criticized or supported? What works with the Mode 2 conceptualization in the context of small-sale fisheries, and what does not? Is there a way to reframe Mode 1 and Mode 2 science for what is practically happening on the ground that prevents the latter from draining the virtues of the former?

As an applied, regulatory science, fishery science may perhaps never approach the Mode 1 ideal of disinterestedness and autonomy in the first place, but this does not mean that Mode 1 science ideals are irrelevant for fisheries governance. In fact, since there is a tendency of blurring boundaries between science and politics within fisheries management, "Mode 1 ideals became all the more important" (Holm 2003, 6). The chapter, therefore, explores the dilemmas in maintaining the reputation of science, while avoiding a dilution of the values that are associated with the move from Mode 1 to Mode 2 science. There is hardly any reason why a science targeting small-scale fisheries should be exempted from the general criteria and ideals of what constitutes good science, particularly if there is no necessary conflict between a science adhering the standards that are associated with Mode 1 and Mode 2 science that is practical and relevant for small-scale fisheries. Transdisciplinary science, what we suggest in Mode 1.5, must be able to assume the best qualities of the two modes while avoiding the lesser ones.

21.3 Theoretical Context

Mode 1 and Mode 2 are 'ideal types' of science, not strict empirical representations (cf. Weber's concept) (Burger 1987; Msomphora 2016a). They are, in other words, perfect forms or states (types) of science, used as yardsticks for comparison (and for phrasing research questions) when the perceived ideals meet empirical reality. As ideal types, Mode 1 and Mode 2 science concepts enable the demonstration of the implications that their knowledge production may have on science used for small-scale fisheries policy-making and management, for instance, in the context of the SSF Guidelines and their implementation. In the following, the key features of Mode 1 and Mode 2 science are described and contrasted.

21.3.1 Mode 1 – CUDOS

Mode 1 science is what we think of as traditional academic science. This science model is associated with Merton's 'ethos of science' (1996), a science that is perceived as 'pure', and separate from policy-making (Gibbons et al. 1994; Merton 1996). In Mode 1, science is institutionalized within disciplines in universities, aimed at the extension of certified knowledge (Merton 1996; Ziman 2000a). The standards and recipes for how to conduct research should originate from the scientific community. The scientific community defines which research methods qualify as scientific research. Merton (1996) pointed out that Mode 1 procedures must provide the relevant definition of knowledge, one that is empirically confirmed and consists of logically consistent statements about regularities. More specifically, in Merton's conceptualization, the five sets of institutional norms govern scientific research: communalism, universalism, disinterestedness, originality, and skepticism (CUDOS). Although, 'originality' is not on the list of the Mertonian norms, it is included in the modified definition of CUDOS, which, as Ziman (2000b) observes, is most widely used in the contemporary science debates.

The first norm, communalism (C), requires that the fruits of academic science be regarded as 'public knowledge' (Ziman 2000a). It covers the practices involved in the communication of research outcomes to other scientists, students, and to the society. The second norm, universality (U), requires that contributions of science not be exclusive due to social status, religion, nationality, or other criteria. The third norm, disinterestedness (D), alludes to the principle of scientific objectivity: scientists must repress their natural enthusiasm for their own ideas and undertake 'pure' research without commercial applications as motivation (Merton 1996; Ziman 2000a). This particular norm speaks to the philosophical idea of scientific objectivity. The fourth norm is originality (O), which is essential for energizing the scientific enterprise (Merton 1996; Ziman 2000a). Lastly, skepticism (S) permits carefully controlled critical perspective and peer review. This norm, as Ziman (2000a, 137) points out, "merely stresses the constructive role of refutation as the natural partner of conjecture in the production of reliable knowledge."

Mode 1 scientific research is meant to bring new, reliable knowledge to society, which may, or may not, be used for policy implementation (Gibbons et al. 1994; Croissant and Smith-Doerr 2008). In this case, policy implementation is not necessarily a primary driver, but may well be an outcome. For Mode 1, research-based knowledge flows in one direction, from scientists to industries or policy-makers. However, there is also a reciprocal relationship, where society provides funds and general support to the scientific community. Nevertheless, in Mode 1 science there is a clear demarcation between science and policy-making, and between scientists and those who provide funding (Gibbons et al. 1994; Ziman 2000a; Croissant and Smith-Doerr 2008).

Mode 1's perception of a definite, linear, and predicable scientific method is now considered highly questionable (Funtowicz and Ravetz 1992; Ziman 2000a; Weinberg 2005; Lam 2007; Styhre and Lind 2010; Mavi 2014; Funtowicz and Ravetz 2015). Its imaginary, autonomous, self-knowing ideals are not perceived to be realistic and not sufficiently productive for policy-making. Mode 1 science is losing its hegemony (Ziman 1996; Nowotny 1999; Funtowicz and Ravetz 2003; Carolan 2006; Styhre and Lind 2010).

21.3.2 Mode 2 – PLACE

In Mode 2, the boundary between science and policy-making is fuzzy and permeable. It denies the need of general rules to judge the validity of new ideas, or stable categories in which to place new data (Ziman 1996, 2000a). Mode 2 science includes a broader notion of 'excellence' than Mode 1 science. It favors wide definitions of knowledge and decentered diversity of science (Gibbons 2000; Hessels and Van Lente 2008).

In contrast to Mode 1 science, Mode 2 is problem-oriented and not easily reducible to any of the participating academic disciplines (Kraak 2000). Knowledge is constructed in accord with commercial, political, or other government and industry interests (Cole 1992; Funtowicz and Ravetz 1992; Jasanoff 2004; Weinberg 2005; Styhre and Lind 2010; Mavi 2014). Mode 2 science, therefore, is not just a 'new mode of knowledge production' but a formula for a potential new research culture (Ziman 2000a; Jasanoff 2004). Mode 2 science is first initiated by industry or policy-makers, then the scientists are commissioned to undertake the research (Wilson 2009). The knowledge produced is then distributed back to the contractor. Therefore, knowledge flow is not linear, as in Mode 1 science, but intertwined and circular. In Mode 2, the needs of the clients predominantly drive science, rather than the curiosity of the scientist. The two parties may have very different ideas of what constitutes adequate science.

Similarly to CUDOS, Ziman (2000b), among others, has described Mode 2 science as proprietary (P), local (L), authoritarian (A), commissioned (C), and dependent on expert skills (E) for a particular problem; forming the acronym PLACE. It is 'proprietary' as research results are considered as 'intellectual property' and can be withheld from the public by remaining unpublished, for instance, for commercial reasons. Mode 2 science may lack the ambition of producing solutions that are universal, as it concentrates on providing solutions to problems that are inherently 'local' or context-specific.

The conflation of science and society (producer and consumer) characteristics of Mode 2 science, and the fact that the 'government science policy' nexus (especially in the developed world) tends to support the privatization of public funded research, has led to the commercialization of the production of scientific knowledge (Nowotny et al. 2001; Sismondo 2011). With Mode 2 research, government commitments are shifted; research is often privately funded, and knowledge produced becomes private, intellectual, and patented property in the contractors' control. This 'authoritarian' feature of Mode 2 science has resulted in commercialization and privatization of knowledge (Nowotny et al. 2003; Sismondo 2011).

Contrary to Mode 1, Mode 2 science is perceived as cost effective, since the government and/or industry funding the research puts strict financial ceilings on their patronage, and requires value for their money (Ziman 1996). As 'commissioners' of the research, governments and industries insist that scientists are accountable and responsive to their needs and that they are explicitly concerned about the relevance and impact of their findings. Thus, in Mode 2, science and society mutually penetrate each other (Lidskog 2008).

Mode 2 science often involves multidisciplinary 'expert' teamwork and inter-institutional networking (Ziman 2000a). Ideally, their different disciplinary perspectives of the problem give a comprehensive knowledge base for the solution advanced. The risk, however, is that some of the grounds gained, in respective to disciplinary sciences, may get lost (Elam and Glimel 2004); what is gained by a broader multidisciplinary approach is lost in disciplinary depth. Another problem is the lack of the quality control that peer review involves (Irwin 2004, 2008; Hessels and Van Lente 2008). This also allows values that are controlled by interests outside the scientific community to penetrate the scientific process, where success is confused with getting funding (Guggenheim 2006).

21.3.3 From CUDOS to PLACE

While closing the gap between basic and applied science, the move from Mode 1 to Mode 2 is drawing traditional science into a new sphere of activity and culture. Not only does this transition raise issues of funding, intellectual property rights, disciplinary identity, criteria of excellence, career aspirations, and institutional management, just to mention a few, but also seriously threatens the most central value of academic science (Mode 1), namely, its objectivity (Ziman 1996). Objectivity makes science valuable to society because it is meant to give the public reliable, disinterested knowledge. The transition to Mode 2 science also turns public knowledge into intellectual property. Research will mainly be directed with industrial or political interests (Ziman 2000a). Academic tenure may no longer protect researchers from commercial influences incurred by the major research donors (e.g. industries and politicians). Hence, scientists' work will usually deal with matters where some values are highly prioritized, for example, safety, profitability, and efficacy, while neglecting others, like those associated with equity and justice (Jasanoff 2004, 2009; Weinberg 2005; Irwin 2008; Styhre and Lind 2010; Mavi 2014; Funtowicz and Ravetz 2015). In this context, the transition from Mode 1 to Mode 2 science would add to the marginalization of small-scale fisheries, since their interests are often underrepresented (Blasiak and Wabnitzc 2018).

Also, as Ziman (1996, 2000a) warns, systematic intellectual criticism (cf. Merton's 'organized skepticism'), which is the key to the validity of academic Mode 1 science, will be downplayed. However, as Weingart (2011) points out, as a response to the lack of clear demarcation between science and policy-making, scientists are encouraged to become aware of the public interest. It may influence the policy-makers to adapt their way of thinking to scientific knowledge. Thus, while politicians will try to influence science for policy development, Mode 2 may also make policy development more science-based (Lidskog 2008).

21.4 Mode 1.5 – CULAVS

There is, therefore, a need for an approach (Mode of science) that retain the best qualities of both Mode 1 and Mode 2 science, while avoiding their shortcomings. This, we argue, is what transdisciplinary approaches should aim for. This would involve a 'Mode 1.5 science' so to speak, with the ethos of the science acronym: CULAVS (Msomphora 2016a) (Fig. 21.1). This ideally implies that science should be communal (C) – the fruits of the scientific research should be regarded as 'public knowledge'. Knowledge produced should be both universal (U) – implying that scientific propositions should be general enough to apply in any cultural context – and local (L) – also open to problem solving oriented research in particular contexts.

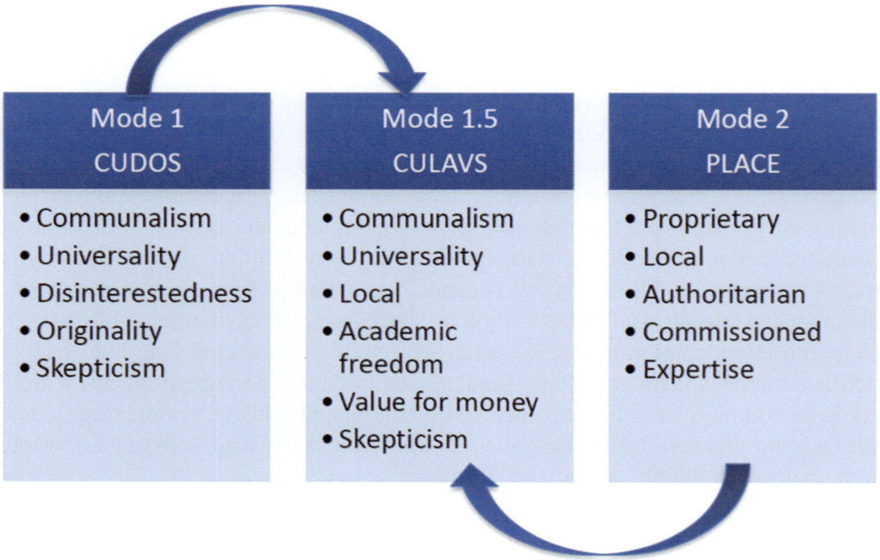

Fig. 21.1 Key features of sciences in different modes

Additionally, there should be academic freedom in research (A), where scientists are self-winding in the choice of research problems and techniques, and their own curiosity can still drive them. However, since research requires resources to be conducted, which the SSF Guidelines say should be provided, researchers should become more accountable and responsible for quality and impact of their work, hence assurance of value for money (V). To allow for skepticism (S), research should be transparent, meaning that secrecy should be avoided in order to permit carefully controlled critical analysis of the knowledge produced.

This is not just a matter of principle but also a means through which the quality of research is enhanced. Hence, the knowledge produced can be of great value to the society, improving the policy-making in small-scale fisheries, while avoiding private biases and the dilution of science values and standards. However, if reliability in Mode 2 science is really reduced to what works, then it means that science is shifting from 'searching the truth' to a more problem solving oriented research. Without the former, the latter would easily lead astray. A critical question for Mode 2 science is, therefore, whether it serves certain political interests and agendas. The move from Mode 1 to Mode 2 science makes it difficult for research that contributes to policy-making to be accepted within the academic community, especially if such research breaks with established standards and perceptions about what science is and what its purpose is. It may also jeopardize the standing of research within the view of the general public, because its values and objectives may be criticized.

However, both CUDOS (Mode 1) and PLACE (Mode 2) science have problems with transparency. Whereas Mode 1 has a tendency of retreating to the 'ivory tower', denying the question of the public and what is going on there; Mode 2 has a problem of becoming exclusive to the contractors—those who can pay for it. In both instances, science becomes elitist rather than common property. Therefore, we herein argue that transdisciplinary science, with the basic features of Mode 1.5, should, of course, secure transparency if the transition of scientific mode is really happening.

21.5 Discussion

According to Amsterdamska (2008, 206), the shift from Mode 1 to Mode 2 science changes the focus to "questions about the use of instruments or ideas in a particular location and situation, rather than universal knowledge, in production and intervention rather than representation, and in science as a mode of working and doing things in and to the world, rather than as a system of propositions arranged into theories." However, as Etzkowitz and Webster (1995) notice, Mode 2 science increases 'capitalization of knowledge' (the awarding of science with financial assets) across science as a whole; although it is argued that this issue is no more than an extension of Mode 1. The transition to Mode 2 signals the implication that scientists no longer associate fundamentally with their specialties and disciplines, but instead engage in a variety of interactions with a heterogeneous group of actors, from individuals to policy-makers to funding agencies (Ziman 2000a; Hackett et al. 2008). Again, one must caution power relations and dynamics among actors, which may influence decisions about research funding. The meager funding to social science research, which is the core of research focusing on small-scale fisheries, relative to natural science research funding supporting the interests of industrial, large-scale fisheries, is a good illustration of negative consequences of the transition from Mode 1 to Mode 2 science.

Mode 2 science moves away from the search for general knowledge driven by curiosity and a need to understand basic processes in nature and society, to answering specific practical questions on demand (Guston 2001). Thus, there is a possibility that Mode 2 research is predefined by the contracting clients. Therefore, as Edwards and Wajcman (2005) argue, this has implications not only for participation in decision-making processes, but also for the quality of knowledge produced and its justification, public access to produced information (Hilgartner 1998), or greater secrecy (Wright and Wallace 2000), and for the structure of authority in producing scientific knowledge. It could also undermine the assurance that all researchers' work is according to the same quality standards with regard to reliability and trustworthiness of research outcomes.

Mode 1 science is not without flaws either. It does not always live up to its ideals. Despite the rule of knowledge sharing, withholding of data still happens (Committee on Science et al. 1995; Savage and Vickers 2009). Data may be withheld for legitimate reasons like anonymity. The status of the scientists may dictate the quality of knowledge produced. In addition, scientists are subject to norms of social responsibility, such as accountability, and contribution to the demands and risks confronting humanity. The move towards Mode 2 brings scientists down from the ivory tower and into the real world, where their roles and contributions are taken seriously. This move is not necessarily a betrayal of traditional norms and values of science, albeit there are obvious risks involved of throwing out the baby with the bath water. These risks must be taken seriously. The questions to be addressed are, therefore: Do we really need to abandon all the values and principles of Mode 1 for Mode 2 science? Can we maintain in Mode 2 science, what is good in Mode 1, and vice versa? Is there a middle ground, such as Mode 1.5 (Msomphora 2016a)? This is clearly an issue that must also be addressed within the context of the SSF Guidelines, which emphasize the constructive role of science and research for the implementation to take hold.

Gibbons (2000, 47) points out that what is required "are not policies which aim to orient research in a means-end relationship with same social objective, but rather policies which promote the generation of a range of contexts of application, that draw on the resources of the socially distributed knowledge production system, and that take advantage of the self-organizing nature of Mode 2 research." Policy-makers, government agencies, industries, and private organizations should, according to this view, act more as brokers than fenders, despite the fact that some level of public funding may still be vital (Gibbons 2000). The SSF Guidelines do right in emphasizing the need for research funding, which is a particular challenge for the transdisciplinary approaches that they are promoting. This will require breaking new grounds with respect to Mode 1.5 characteristics, as transdisciplinary research currently does not enjoy much support of the wider academic community, for mixed reasons, such as lack of a universally established definition and lack of secure institutional support, including funding (Jahn et al. 2012). Nor is there a common glossary, a focused communication platform, or a commonly shared research network (Brandt et al. 2013). Transdisciplinary science is thus hindered in acquiring the status that Mode 1, and increasingly Mode 2 science, enjoy. However, we argue that this is what is to be expected given the embedded nature of transdisciplinary approaches in particular societal problems and contexts, such as small-scale fisheries.

Thus, the transdisciplinary approach (CULAVS), which seeks the middle ground between Mode 1 and Mode 2, is still in the making, as a legitimate and *bona fide* scientific endeavor within small-scale fisheries; and much remains to be done to make it so. Given that, transdisciplinary approaches are for problems that need to be addressed with some urgency. As they keep coming back, similar to fisheries problems, it would require a process with no finish line. As an approach, it is also questionable whether a unified glossary is needed (Brandt et al. 2013). Still, quality standards need to be upheld. Contracting clients may set the structure that can ensure quality, but not take on the burden of performing assessments themselves. The latter is where the research community has particular expertise.

Mode 2 and Mode 1.5 science both allows stakeholders to take part in knowledge production. This is a step towards a more democratic and transparent system, which empowers stakeholders in a way peer review does not (Brandt et al. 2013). Stakeholder involvement in the research arena results in transaction costs where findings are disputed, and researchers must spend more time and effort explaining their research methodology to stakeholders that are research illiterate (Wilson 2009). This may create conflicts; but conflicts, according to Coser (1956), may serve as an interactive arena for creation and modification of measures, as well as the development of new institutional networks to facilitate learning, information exchange, and transparency in the management process, while enhancing compliance to the measures (Msomphora 2016b).

21.6 Concluding Remarks

Mode 1 and Mode 2 science are 'ideal types' of science, which allow the authors to demonstrate the implications 'in the extreme'. In reality, the modes tend to be more mixed. Comparing the ideal types with practice helps to clarify potential areas of improvement, but also help to explain the importance of context, which needs to be taken account. Not all situations and research projects are identical. The distance between the ideal type and the practical research design and process begs for reflection and justification. Why is there such a gap? Is the gap justified? Their might be good reasons for being pragmatic. Still, Mode 1 to Mode 2 have pitfalls that need to be addressed. In the struggle to become more applied and relevant, the quality standards of science risk erosion. It is important "to avoid discouraging or disadvantaging, complex-system science… that integrates scientific rigor with policy/practice that is increasingly necessary in today's world" (Simon and Schiemer 2015, 9).

Science for policy-making must remain reliable, credible, and trustworthy. Transdisciplinary is no excuse for sloppiness. We argue that CULAVS (Mode 1.5), which seeks a middle ground between Mode 1 and Mode 2 science, would be more appropriate. By seeking the middle ground, we may know that transdisciplinary knowledge will be of value to the users and provide for evidence-based policies, without compromising the ethos of science as an institution, be it within small-scale fisheries or in other sectors of society.

References

Amsterdamska O (2008) Plactices, people, and places. In: Hackett EJ, Amsterdamska O, Lynch M, Wajcman J (eds) The handbook of science and technology studies, 3rd edn. The MIT Press, Cambridge, MA, pp 205–209

Blasiak R, Wabnitzc CCC (2018) Aligning fisheries aid with international development targets and goals. Mar Policy 88:86–92

Brandt P, Ernst A, Gralla F et al (2013) A review of transdisciplinary research in sustainability science. Ecol Econ 92:1–15

Burger T (1987) Max Weber's theory of concept formation: history, laws, and ideal types. Duke University Press, Durham

Carolan MS (2006) Science, expertise, and the democratization of the decision-making process. Soc Nat Resour 19(7):661–668

Cole S (1992) Making science: between nature and society. Harvard University Press, Cambridge, MA

Committee on Science, Engineering and Public Policy, National Academy of Sciences, National Academy of Engineering, Institute of Medicine (1995) On being a scientist: responsible conduct in research. National Academy Press, Washington, DC

Coser LA (1956) The functions of social conflict. The Free Press, New York

Croissant JL, Smith-Doerr L (2008) 27 organizational contexts of science: boundaries and relationships between university and industry. In: Hackett EJ, Amsterdamska O, Lynch M, Wajcman J (eds) The handbook of science and technology studies, 3rd edn. The MIT Press, Cambridge, MA

Edwards P, Wajcman J (2005) The politics of working life. Oxford University Press, Oxford

Elam M, Glimel LH (2004) 'Knowledge society as the republic of science enlarged: the case of Sweden'. Case study report, Science Technology and Governance in Europe, Discussion Paper 26

Etzkowitz H, Webster A (1995) Science as intellectual property. In: Jasanoff S, Markle GE, Peters JC, Pinch T (eds) Handbook of science and technology studies. Sage, Thousand Oaks

Funtowicz SO, Ravetz JR (1992) Three types of risk assessment and the emergence of post-normal science. In: Krimsky S, Golding D (eds) Social theories of risk. Praeger, New York, pp 230–251

Funtowicz SO, Ravetz JR (2003) Post-normal science. International Society for Ecological Economics (ed), Online Encyclopedia of Ecological Economics. http://www.ecoeco.org/publica/encyc.htm. Accessed 15 Apr 2018

Funtowicz SO, Ravetz JR (2015) Peer review and quality control. In: Wright JD (ed) International encyclopedia of the social & behavioral sciences, 2nd edn. Elsevier, Oxford, pp 680–684

Gibbons M (2000) Universities and the new production of knowledge: some policy implications for government. In: Kraak A (ed) Changing modes: new knowledge production and its implica-

tions for higher education in South Africa. Human Science Research Council Press, Pretoria, pp 38–55
Gibbons M, Limoges C, Nowotny H et al (1994) The new production of knowledge: the dynamics of science and research in contemporary societies. Sage, London
Guggenheim M (2006) Undisciplined research: the proceduralisation of quality control in transdisciplinary projects. Sci Public Policy 33(6):411–421
Guston DH (2001) Integrity, responsibility, and democracy in science. Sci Policy: J Sci Health Policy 1(2):168–189
Hackett EJ, Amsterdamska O, Lynch M et al (2008) The handbook of science and technology studies, 3rd edn. MIT Press, Cambridge, MA
Hessels LK, Van Lente H (2008) Re-thinking new knowledge production: a literature review and a research agenda. Res Policy 37(4):740–760
Hilgartner S (1998) research. In: A. Thackray (ed.) Private science: biotechnology and the rise of the molecular sciences. Philadelphia: University of Pennsylvania Press, pp. 202-218
Holm P (2003) Crossing the border: on the relationship between science and fishermen's knowledge in a resource management context. MAST 2(1):5–33
ICES (International Council for the Exploration of the Sea) (2018) ICES homepage. http://www.ices.dk/Pages/default.aspx. Accessed 15 April 2018
Irvine JM, Martin BR (1984) Foresight in science: picking the winners. Frances Pinter, London
Irwin A (2004) Expertise and experience in the governance of science: what is public participation for? In: Edmond G (ed) Expertise in regulation and law. Ashgate, Aldershot, pp 32–50
Irwin A (2008) STS perspectives on scientific governance. In: Hackett EJ, Amsterdamska O, Lynch M, Wajcman J (eds) The handbook of science and technology studies, 3rd edn. MIT Press, Cambridge, MA, pp 583–607
Jahn T, Bergmann M, Keil F (2012) Transdisciplinarity: between mainstreaming and marginalization. Ecol Econ 79:1–10
Jasanoff S (2004) The idiom of co-production. In: Jasanoff S (ed) States of knowledge: the co-production of science and the social order. Routledge, London, pp 1–12
Jasanoff S (2009) Science at the bar: law, science, and technology in America. Harvard University Press, Cambridge, MA/London
Jentoft S (2014) Where there is a will: the Norwegian model of fisheries governance, via the Norwegian Raw Fish Act and fish sales organizations, is worth examining. Samudra 2014(68):22–25
Jentoft S, Eide A (2011) Poverty mosaics: realities and prospects in small-scale fisheries. Springer Netherlands, Dordrecht
Kraak A (2000) Changing modes: a brief overview of the mode 2 knowledge debate and its impact on South African Knowledge formation. In: Kraak A (ed) Changing modes: new knowledge production and its implications for higher education in South Africa. Human Science Research Council Press, Pretoria, pp 1–37
Lam A (2007) Knowledge networks and careers: academic scientists in industry–university links. J Manag Stud 44(6):993–1016
Lidskog R (2008) Scientised citizens and democratised science. Re-assessing the expert-lay divide. J Risk Res 11(1–2):69–86
Limoges C (1996) L'université à la croisée des chemins: une mission à affirmer, une gestion à réformer. Paper presented at the Colloque Le lien formation-recherche à l'université: Les pratiques aujourd'hui. Quebec, Gouvernement du Québec Ministère de l' Éducation
Linke S, Jentoft S (2013) Exploring the phronetic dimension of stakeholders' knowledge in EU fisheries governance. Mar Policy 47(C):153–161. https://doi.org/10.1016/j.marpol.2013.10.010i
Mavi RK (2014) Indicators of entrepreneurial university: Fuzzy AHP and Fuzzy TOPSIS Approach. J Knowl Econ 5(2):370–387
Merton RK (1996) The ethos of science. In: Sztompka P (ed) On social structure and science. University of Chicago Press, Chicago/London, pp 267–276

Msomphora MR (2016a) The role of science in fisheries management in Europe: from mode 1 to mode 2. MAST 15(3):01–23
Msomphora MR (2016b) Conflict resolution and the delegation of authority in fisheries management: the case of Outer Hebrides Inshore Fisheries Group in Scotland. Mar Policy 73:263–275. https://doi.org/10.1016/j.marpol.2016.08.006
Nowotny H (1999) The need for socially robust knowledge. TA-Datenbank-Nachrichten 3(4):12–16
Nowotny H, Scott P, Gibbons M (2001) Re-thinking science: knowledge production in an age of uncertainty. Polity Press in association with Blackwell, Cambridge
Nowotny H, Scott P, Gibbons M (2003) Mode 2 revisited: the new production of knowledge. Minerva 41(Special issue):179–194
Savage CJ, Vickers AJ (2009) Empirical study of data sharing by authors publishing in PLoS journals. PloS one [Online] 4(9). https://doi.org/10.1371/journal.pone.0007078
Silbey S (2006) Science and technology studies. In: Turner BS (ed) The Cambridge dictionary of sociology. Cambridge university press, New York, pp 536–540
Simon D, Schiemer F (2015) Crossing boundaries: complex systems, transdisciplinarity and applied impact agendas. Curr Opin Environ Sustain 12:6–11
Sismondo S (2008) Science and technology studies and an engaged program. In: Hackett EJ, Amsterdamska O, Lynch M, Wajcman J (eds) The handbook of science and technology studies, 3rd edn. The MIT Press, Cambridge, MA, pp 13–32
Sismondo S (ed) (2011) An introduction to science and technology studies, 2nd edn. John Wiley & Sons, Chichester
Styhre A, Lind F (2010) The softening bureaucracy: accommodating new research opportunities in the entrepreneurial university. Scand J Manag 26(2):107–120
Turner S (2008) The social study of science before Kuhn. In: Hackett EJ, Amsterdamska O, Lynch M, Wajcman J (eds) The handbook of science and technology studies, 3rd edn. The MIT Press, Cambridge, MA, pp 33–62
Walt G (1994) How far does research influence policy? Eur J Pub Health 4(4):233–235
Weinberg AM (2005) Science and its limits: the regulator's dilemma. Philos Soc Action 31(2):71
Weingart P (2011) Science, the public and the media – views from everywhere. In: Carrier M, Nordmann A (eds) Science in the context of application, Boston studies in the philosophy of science. Springer, Dordrecht, pp 337–348
Wilson DC (2009) The paradoxes of transparency: science and the ecosystem approach to fisheries management in Europe, vol 5. Amsterdam University Press Manchester University Press distributor, Amsterdam
Wright S, Wallace DA (2000) Varieties of secrets and secret varieties: the case of biotechnology. Politics Life Sci 19:45–57
Ziman J (1996) Is science losing its objectivity? Nature 382(6594):751–754
Ziman J (2000a) Postacademic science: constructing knowledge with networks and norms. In: Segerstråle U (ed) Beyond the science wars: the missing discourse about science and society. State University of New York Press, Albany, pp 135–154
Ziman J (2000b) Real science: what it is and what it means. Cambridge University Press, Cambridge

Chapter 22
The Principles of Transdisciplinary Research in Small-Scale Fisheries

Alicia Said, Ratana Chuenpagdee, Alfonso Aguilar-Perera, Minerva Arce-Ibarra, Tek Bahadur Gurung, Bonnie Bishop, Marc Léopold, Ana Isabel Márquez Pérez, Sérgio M. Gomes de Mattos, Graham J. Pierce, Prateep K. Nayak, and Svein Jentoft

Abstract The diverse characteristics, values, and importance of small-scale fisheries imply at least two key considerations. First, there is no tailor-made, one-size-fits-all solution to the problems and challenges facing small-scale fisheries; thus, policy and governance must be sensitive to the contexts. Second, the close relationship and interactivity between the natural and the social dimensions of small-scale fisheries suggests that knowledge and understanding about small-scale fisheries may need to

A. Said (✉)
Too Big To Ignore Global Partnership for Small-Scale Fisheries Research, Memorial University of Newfoundland, St. John's, NL, Canada
e-mail: alicia.said@mun.ca

R. Chuenpagdee
Department of Geography, Memorial University of Newfoundland, St. John's, NL, Canada
e-mail: ratanac@mun.ca

A. Aguilar-Perera
Departamento de Biología Marina, Universidad Autónoma de Yucatán, Mérida, Mexico

M. Arce-Ibarra
Departmento de Sistemática y Ecología Acuática de El Colegio de la Frontera Sur, Chetumal, Mexico
e-mail: aarce@ecosur.mx

T. B. Gurung
Nepal Agricultural Research Council, Kathmandu, Nepal

B. Bishop
Education Coordinator, Vancouver, BC, Canada
e-mail: bbish@mun.ca

M. Léopold
Institut de Recherche pour le développement (IRD) c/o Institut Halieutique et des Sciences Marines (IH.SM), Université de Toliara, Toliara, Madagascar
e-mail: marc.leopold@ird.fr

R. Chuenpagdee, S. Jentoft (eds.), *Transdisciplinarity for Small-Scale Fisheries Governance*, MARE Publication Series 21,
https://doi.org/10.1007/978-3-319-94938-3_22

transcend the boundaries of academic disciplines. These are the premises for research and activities conducted in the Too Big To Ignore (TBTI) – Global Partnership for Small-Scale Fisheries Research. Taking a transdisciplinary approach to research, training, and learning about small-scale fisheries can help address real-world problems and reveal opportunities to move towards pragmatic solutions. In this chapter, we discuss what transdisciplinarity involves, what the underlying principles are, and what makes it distinct from other perspectives. We argue that transdisciplinarity in small-scale fisheries requires institutional and academic innovation at local and national scales that facilitates interactive and transformative learning.

Keywords Research techniques · Transdisciplinary lens · Knowledge integration · Trust

22.1 Why a Transdisciplinary Approach

Small-scale fisheries are diverse and operate within different political, economic, social, and cultural structures. These fisheries target a range of marine resources through different means of capture within inland, coastal, and marine environments, and exhibit various levels of adaptation and resilience as they engage in pre-harvest, harvest, and post-harvest activities. Although inherently heterogeneous, small-scale fisheries worldwide display important values that make them a significant component for sustainable fisheries (Johnsen 2018), playing a key role in providing food security and livelihoods to millions of people worldwide (FAO 2015), contributing to community cohesion and social networks (Jentoft 2000), and displaying strong stewardship ethics towards the protection of species and habitats (Nayak and Berkes 2011).

A. I. Márquez Pérez
Universidad Nacional de Colombia, San Andres, Colombia
e-mail: aimarquezpe@unal.edu.co

S. M. G. de Mattos
Ministry of Planning, Recife, Brazil
e-mail: sergio.mattos@planejamento.gov.br

G. J. Pierce
Departamento de Ecología y Recursos Marinos, Instituto de Investigaciones Marinas (CSIC), Vigo, Spain
e-mail: g.j.pierce@iim.csic.es

P. K. Nayak
School of Environment, Enterprise and Development, University of Waterloo, Waterloo, ON, Canada
e-mail: pnayak@uwaterloo.ca

S. Jentoft
Norwegian College of Fishery Science, UiT- The Arctic University of Norway, Tromsø, Norway
e-mail: svein.jentoft@uit.no

Notwithstanding their fundamental roles in societies worldwide, the majority of small-scale fisheries are facing major challenges that affect their viability and sustainability. These include declining fisheries resources, competition with large-scale fisheries, restricted access to rights and markets, increased exposure to competition from global and corporate markets, and anticipated effects of climate change (Chuenpagdee 2011). The complexity of problems in small-scale fisheries makes them susceptible to ecological, social, political, and economic fluctuations, most of which are unpredictable, and, thus, communities are, most of the time, ill-equipped to deal with them.

Following Rittel and Webber (1973) and Jentoft and Chuenpagdee (2009), major problems in small-scale fisheries are considered wicked because they are not easily identified, agreed upon or solved. The problems are interconnected and interdependent and have no boundaries as they encompass social, natural, and governance systems. These characteristics pose a great challenge for fisheries governance and solving fisheries problems needs to go beyond studying fish stocks and conducting stock assessments, to understanding the socioeconomic realities of fishing communities, the political economy within which fisheries management is implemented, as well as the governing and regulatory framework that defines the context of small-scale fisheries (Bundy et al. 2008). If, for instance, engendering small-scale fisheries sustainability is focused only on the maximum sustainable yield of fish stocks, and does not cater for the social and economic realities of the sector, governance efforts are likely to fail (Symes and Hoefnagel 2010; Urquhart et al. 2011).

The need to broaden the perspective about the underlying causes of the problems is imperative when dealing with difficult issues such as marginalization, restriction of access to fisheries resources, and violation of rights that many small-scale fishers around the world experience. Some of these could have resulted from long histories of exclusion, such as the cases of indigenous communities (McCormack 2017). Marginalization could also be a consequence of neoliberal political economies that do not sufficiently cater for the regeneration of the small-scale fisheries (Pinkerton 2017). In essence, this neoliberal current could be infused by a country's political economy, and fishers might lack the capacity to shield themselves and protect their rights within an era of burgeoning corporate growth (Pinkerton and Davis 2015; Lalancette 2016; Said et al. 2016). Resolving the marginalization issue would require a process in which all these possible links and interconnections are mapped out in a way that gives a holistic understanding of the situation.

This is a very challenging task due to the heterogeneity of the sector, the scale of the problem, and the dynamics that are reinforcing the situation. Essentially, issues and challenges of otherwise similar small-scale fisheries in two different settings would be distinct, and thus both would necessitate diverse knowledge and core expertise that articulate their context-driven realities. Some small-scale fisheries in developing countries, for example, have problems of declining fishery yields due to large foreign fleets fishing in their waters (Antonova 2016), whereas some fishing communities in developed countries are facing problems due to increased competition by uncontrolled recreational fisheries (Tunca et al. 2016). Addressing the fishers' resource access needs in developing countries would require a global inquiry that encompasses an understanding of, for example, the EU fisheries partnership agreements in place (Antonova 2016), while the issue of developed countries is more contextualized within their national political territory. What is similar in both cases is the need for different types of knowledge emanating from both the scientific and non-scientific spheres, including experts, members of the community, fishers, and policy-makers.

This way, the research becomes sufficiently extensive to give a detailed account of the situation including social, economic, and governance realities within which the problems are happening and, thus, the policies can be tailored accordingly. If policies do not fit the complexities of the fisheries systems, they might address the symptoms but not solve the root of the problem (Degnbol and Mccay 2007). As a relatively new approach that tries to facilitate the understanding of the complexity of small-scale fisheries problems, transdisciplinary research provides us with tools and methods to look at the concerns and to inform policies through a more holistic lens. In its full-fledged state, transdisciplinarity can also lead to transformations in institutions and their functions, and can result in innovation in the way they govern small-scale fisheries systems.

In this chapter, we define both the need for and the role of transdisciplinary research in small-scale fisheries sustainability, explore the principles and visions that underpin the transdisciplinary approach, and present a framework on how to integrate multiple knowledge types and values to address complex small-scale fisheries problems. We show how, by integrating theory and practice, and academic and non-academic knowledge, a transdisciplinary approach would: (1) introduce a new perspective and a way to understand the problem that can be easily grasped by multiple audiences; (2) open pathways for ongoing problem solving, where opinions about solutions are overtly discussed; (3) facilitate transitions and transformations with the ultimate goal of supporting the sustainability of small-scale fisheries and fishing communities; (4) provide an avenue to reengage in the problem-solving loop based on lessons learned; and (5) help to craft management and policy solutions that are tailored to the specific and agreed-upon diagnoses (McClanahan et al. 2009).

The following section explains how transdisciplinarity is an approach that goes between, across, and beyond different knowledge types to understand small-scale fisheries´ diversity, complexity, and dynamics. The subsequent section describes the various principles of transdisciplinary research, and the application of a transdisciplinary lens that integrates the social, ecological, and governance domains. The penultimate section discusses transdisciplinary challenges and mechanisms to address them, and the conclusion section summarizes the role of transdisciplinarity and its contribution to propelling societal transformation for sustainable small-scale fisheries.

22.2 An Integrated Approach Through a Transdisciplinary 'Lens'

Bringing together different and distinct disciplines and integrating them to investigate a research problem has been applied in small-scale fisheries research and studies. Choi and Pak (2006) refer to this concept as 'multiple disciplinary' research, while others use more common terms like multidisciplinarity and interdisciplinarity to describe their research. Examples of these are found in bio-economic theory (Cuervo-Sánchez et al. 2018), illegal, unreported, and unregulated fisheries (Gezelius 2008); poverty and segregation studies (Bavinck 2014); food and fishing systems (Nelson et al. 2013); inter-sectoral conflicts involving fisheries at sea (DuBois and Zografos 2012); human-rights approach to fisheries (Charles 2011); legal pluralism in fisheries governance (Jentoft and Bavinck 2014); the wellbeing approach (Weeratunge et al. 2013), and the interactive governance theory (Kooiman 2005). In different ways and to varying degrees, these approaches have applied the concept of a transdisciplinary approach to identify problems in small-scale fisheries (Box 22.1).

In this chapter, we focus our 'lens' in a way that captures these multiple approaches and to explain why investigating small-scale fisheries in their context necessitates a transdisciplinary approach, which not only looks in from different stances, but also creates holistic and integrated knowledge about small-scale fisheries. We submit that the transdisciplinary research approach enables the knowledge generation to go across, beyond, and over disciplinary boundaries (Kerne 2005), to produce a new form of knowledge that does not depend on any specific discipline (Choi and Pak 2006) (Box 22.1). In other words, through a transdisciplinary approach, different experts including local fishers, scientists, and policy-makers share and acquire skills from one another, and recombine the information to create

Box 22.1 The Main Differences Between Multidisciplinary, Interdisciplinary, and Transdisciplinary Research (Adapted from Choi and Pak 2006)

Multidisciplinarity – a process where knowledge is produced by the combination of different disciplines. However, the knowledge remains within the boundaries of those fields and is drawn separately to address a problem.

Interdisciplinarity – a process where knowledge is produced by analyzing, synthesizing, and linking many disciplines into a coordinated and coherent whole. Through interdisciplinarity, new approaches and new methods may emerge as possible outcomes.

Transdisciplinarity – a process where knowledge is produced by bringing together the understanding of the natural, social, and governance domains, in a way that transcends each of their traditional boundaries. Innovation and transformation, also of the way we do sciences, may emerge as an outcome.

new knowledge that does not belong to any particular discipline. This means that the group will create a set of knowledge about a fishery system by looking at the social, natural, and governance systems, at their relationships and interactions, and how they connect with the processes and the institutions that work at different temporal, geographical, and organizational scales.

Specifically, this chapter speaks directly on the need to embark on a transdisciplinary journey to address sustainability in small-scale fisheries, using as a starting point and following Kooiman (2005), the understanding of their diversity, complexity, dynamics, and scale. We look at transdisciplinary research as a mechanism that provides the context and environment for different experts to define and resolve real-world sustainability problems (Lang et al. 2012). The approach can begin from a multidisciplinary endeavour in which experts from various disciplines, such as ecology, biology, economics, and sociology, come together to discuss a fishery problem. Some of the experts can be interdisciplinary 'by discipline', such as experts in ecological economics or human ecology who undertake a multiple disciplinary approach in their own research and practice. Other experts would include people from local communities who bring forth knowledge based on their traditional and practical wisdom gained from life experiences (Jentoft 2006), policy-makers who have the capacity to speak about policy processes and are responsible for decision making, non-governmental organizations (NGOs), civil society organizations (CSOs), and activist groups who aim to influence decisions in the fishery, and the informed and concerned public who care about what goes on in the fishery. While getting together, these different experts share their knowledge, and work towards identifying the problem, and by the end of the research investigation, they will be able to share "a conceptual model of the problem that integrates and transcends each of their separate disciplinary perspectives" (Choi and Pak 2006, 355).

In other words, the transdisciplinary approach urges the assembly of disciplines in recombining information in a way that "draws the mind to puzzle about connection between information elements" (Kerne 2005). An example could be identifying the different links and connections leading to marginalization in resource access and the rippling consequences thereto, including their impacts on the local communities and the local economies. Social learning is an important component of the whole transdisciplinary research journey, because through it, different team members are exposed to the plethora of ideologies, theories, perspectives, and experiences of their participating counterparts. Through this social learning – which is different from the formal education learning – team members become better positioned and confident to interact, and also are capable of developing the respect and empathy needed to work in a heterogeneous team (Pohl and Hadorn 2008). The process of social learning could be developed through the collection of what Clifford Geertz refers to as 'thick descriptions' (Geertz 1973), in this case, about small-scale fisheries.

The 'thickness' can be derived from different data sources including real-life histories, stock assessments, monographs, narratives, and other inductive qualitative and quantitative data analysis (Brown 2010). By collectively providing to the development of the 'thickness' and highlighting the different links of the problem, a transdisciplinary group would be in a better position to create new and shared knowledge about the problem and discuss innovative and imaginative ways of addressing it (Brown 2010). Agreement and consensus-building could become challenging throughout the process of knowledge sharing and integration, especially if the problem veers into different interconnected systems. As will be discussed in the next section, it is the role of the transdisciplinary team to ensure that the participants can collectively contribute to both the identification and the solution of the problem through the journey (Carew and Wickson 2010).

22.3 The Principles of a Transdisciplinary Approach

Like any other form of research, trandisciplinarity needs to follow principles to ensure that the process of investigation runs smoothly and that it retains its focus on resolving problems in small-scale fisheries sustainability. The following principles, and the process of implementing the transdisciplinary research described in the subsequent section, are derived from a process undertaken as part of TBTI research cluster on the topic (see Aguilar-Perera et al. 2017 for details). The principles for transdisciplinary research are related to three aspects: (i) the 'approach' including the attitude of open-mindedness in knowledge integration; (ii) the 'personal traits' of the team members necessary to foster the interactive and transdisciplinary learning; and (iii) the 'process' of engaging in the transdisciplinary endeavour. Given that transdisciplinary research involves a process that goes beyond the normative practice of inquiry, these principles need to be given careful attention. Failure to adhere with the fundamental elements of, for example, stakeholder integration, would dilute the very foundation that makes the research different from other forms of multiple disciplinary attempts.

Approach Principles Transdisciplinary research is an 'approach' both in the framing of the problems and in the processes and tools required to address them. As part of problem framing, key principles include, but are not limited to, holistic, multiscale, and systemic. In problem solving, principles include questioning the status quo, challenging stereotypes, thinking outside-of-the box, and building the creative and imaginative capacity of participants to conceive effective solutions. In this respect, transdisciplinarity should facilitate the flow of information necessary to provide timely and effective responses to local and urgent situations, strengthening local organizations, with an aim to enhancing quality of life and social wellbeing. Since addressing fisheries problems is not a task of a single person or any one discipline, skill sharing is needed for ongoing problem solving through collaborative, adaptive, interactive, and context-dependent processes. This may involve examining and assessing methods for knowledge production, integration, and communication (both among and beyond disciplines). Moreover, if research indicates that more stakeholders need to be included to provide detailed technical information, then identifying and involving the stakeholders have to follow specific criteria to minimize conflict of interest. Stakeholder analysis may need to be performed, as suggested by Mitchell et al. (1997). Involving new team members would need re-sharing of the skills set, so everyone would be able to approach the problem in a level playing field. Trial and error is expected and therefore it is advisable for the transdisciplinary approach to start simple and expand gradually. Participants must be patient; anticipating and acknowledging that problem solving might take time, even years, depending on how successfully the tools are implemented in the different contexts.

Personal Trait Principles Human personality traits are important for transdisciplinary research. Team members need to develop an orientation towards working in groups, with people from different background, experiences, and disciplines. Being empathic is clearly indispensable since individuals need to be receptive and able to see problems from multiple perspectives and value the different opinions that develop through the process. Also, a positive attitude towards teaching, learning, and transforming is necessary, especially since transdisciplinary research ventures into complex and dynamic contexts that are difficult to pin down and can create frustration. As much as possible, individuals need to give an equal chance for others to participate. It is very common for individuals who are more vociferous than others to dominate the process in team-based research, hence, moderating these dynamics is a necessary personal trait skill that one might need to develop. This is a very important trait, which if not well developed, will make it difficult for team members to listen with sincerity and learn from each other.

Becoming open to new forms of knowledge is another important personal trait that might not feel natural, and one might need to adopt the right attitude to be more receptive. In some cases, opening up might require diligence, especially for those who find it difficult to think 'outside their discipline'. One must also acknowledge the restrictions imposed by the boundaries of their own disciplines. Bateson (1987) metaphorically explains how a discipline not only conditions its members to think within a box, but also to think that the box is not a box, and, thus, makes it difficult for an individual to conceptualize any other perspective of the box.

In some cases, specific and unique conditions pertaining to the sociocultural dimensions of the communities need to be catered for. When it comes to communication, for example, the electronic transmission of information might not be possible amongst those who are not computer literate. Moreover, the provision of narratives and thick descriptions by particular individuals might not be achievable if they do not have the tools to describe their story. In these circumstances, it might be necessary to provide specific platforms to elicit community participation, especially when the community itself is unable to value and access its own knowledge, or, due to everyday matters, is unable to reflect and write its story on the same level as the academic members within the transdisciplinary team (Hendricks 2010). Bringing in community-based experts or consultants facilitates the deliberative processes, and helps community participants to draw on their own wisdom. Ultimately, both the form of collaboration and the tools of integration determine the intensity of exchange between the different team members (Pohl and Hadorn 2008). All these human traits and adoption of tools to improve social learning and collaboration are indispensable, especially for the plurality of inquiry designs and various ways of knowing (Russell 2010).

Process Principles Transdisciplinary research should be a flexible approach that emphasizes social learning, experimentation, and an attitude of 'learning by doing' in complex systems (Griffith 2010). For example, fisheries management techniques should be developed and contextualized, through an iterative and adaptive process, to correspond with the characteristics of the small-scale fisheries. There is no fixed mechanism to attain the best measures for fisheries management, but several key process principles can be identified. These principles embrace inclusion of diverse viewpoints and knowledge and the need to recognize power relations at all steps (e.g. stakeholder identification, and knowledge access, production, sharing, and use) to ensure that potentially marginalized groups are meaningfully included. Moreover, assuring that these different knowledges are meaningfully integrated during the transdisciplinary approach is also critical.

Another important component is the development of trust amongst the different participants. Trust is earned over time and should improve when participants gain more experience of working and collaborating together (Harris and Lyon 2013). The development of a common 'language' is also an important principle that should be part of the social learning process, for it facilitates interactions and reconfigures our social discourse, especially when tackling complex issues. For example, discipline-bounded jargon can inhibit effective dialogue between different parties. Scientific jargon used in the academic disciplines could be difficult for participants from the fishing communities to understand, while local knowledge expressed in local dialects and languages of fishing communities could be challenging for nonlocals to comprehend (Jentoft 2006). The intensification of communication could be complemented by a social networking interface, such as through social media, that may be used to maintain an open forum in which different members can occasionally communicate and develop a solid relationship that goes beyond the actual 'formal' research network.

22.4 Implementing the Transdisciplinary Approach

The initiation of the transdisciplinary approach can vary but the entry point is usually an event or crisis identified by researchers, local stakeholders, policy-makers, and/or society. For example, resource depletion could be an issue raised by the fishers who experience a decline in yield, or it could be identified by scientists through stock assessments, or brought up by NGOs or CSOs. Once the problem is identified, the different stakeholders including researchers, community members, fishers, and policy-makers engage together to investigate the roots of the problem, how it is linked to the social, natural, and governance systems, and the scale within which the problem is happening. Thus, transdisciplinary research can be initiated either by academics following a scientific project, or by the government agencies, which need to follow up on an issue raised by fishers and require new knowledge to revert the problem and establish a new policy, or by the community itself.

Following this starting point, the next step is the assembling of the research team and the collective design. The team members normally do not come together on their own. There are always one or two process initiators, who will need to identify the team members, based on specific criteria, which could include direct experience and knowledge of the particular fisheries or a broader understanding of the region. The team would then establish the scale of the problem, and start formulating ideas towards addressing the problem, keeping in mind that the solutions may lie outside of the context where the problem is situated. In other words, the problems of small-scale fisheries are rarely about what goes on in the sector, partly because the fisheries are not a closed system. Instead, fisheries problems are interconnected with other activities happening either in the same space or nearby. This implies that the team members from different disciplines and sectors would bring their most relevant knowledge and the broadest experience to look at the problem and its interconnectedness with the social, political and ecological realms.

Arriving to a consensus about the problem may seem unlikely since each participant will bring forth his or her own definition of what the problem is and what causes it. Most likely, a small-scale fisheries sector characterized by stock depletion, or overfishing, could be defined as a 'market failure' by economists, a 'regulatory failure' by fisheries managers, and a 'community failure' by anthropologists. A fisheries ecologist would see 'data gaps' for stock assessment, a policy-maker would perceive the need for 'better monitoring and enforcement,' and a fisher would see a 'livelihood issue' that determines his/her individual future. While all these definitions and perceptions are dependent on one's knowledge and life experience, collectively they can provide a coherent insight of the problem. The different individuals can deliberate on whether stock depletion actually exists, and if it does, what could be leading to it, and with what consequences. Whilst sharing their knowledge, fishers might explain that they have seen a reduction in the catches in the past, which could then prompt the discussion about the historical trend and other types of changes that may have occurred.

At this point, the team members, taking this timeline into account, would start questioning why stock depletion is actually happening: Is it resulting from excessive effort from the small-scale sector, the large-scale sector, the recreational sector, or the combination? What has changed in the fishing patterns? Could it be a result of climate change? How is the market operating? What are the impacts on the formal market and on the community resilience? Is it a local or national problem? What is the national political situation around it? Is there a political will to address the situation? And ultimately, what should be done to reduce stock depletion? What will happen if the fishing effort of small-scale fisheries is reduced? How will small-scale fishers be affected if effort-reduction policy is implemented? What will be the livelihood impacts? What are the alternatives? Essentially, these questions can indicate why stock depletion is in fact a wicked problem, and answering it needs multi-scale and multi-stakeholder approaches, depending on the context in which the approach is developed.

Throughout the transdisciplinary research process, the group would benefit from an exercise of switching the 'lens' we use to examine the issues. Figure 22.1 depicts three knowledge domains within which the primary research questions arise, i.e., natural science, social science, and governance. Questions would, thus, start from a single domain and then, through knowledge sharing and deliberation, the team would start making the links and make the question a transdisciplinary one. There are different ways in which this can be achieved. First, a disciplinary question in any given domain can move towards the transdisciplinary domain (e.g. a single arrow from one domain to the middle overlapping transdisciplinary area) and turn into a transdisciplinary question. Second, similar or related questions from more than one disciplinary domain can move to the transdisciplinary domain and synergistically form one transdisciplinary question (e.g. meeting of multiple arrows from different disciplinary domains in the transdisciplinary domain). Once the transdisciplinary question is arrived at, the transdisciplinary team is in a better position to deliberate on possible methods, strategies, and processes to move towards a transdisciplinary solution.

In practice, for example, the questions could be 'what happened to the fish stock? (natural domain)', leading to 'how are fishing communities dealing with declining catches? (social domain)', and 'what management techniques can be implemented to resolve overfishing (governance domain)?' Whilst belonging to different knowledge systems, evidently these questions are linked, and the approach to understand them must take into account this interconnectedness. The problem of the 'lack of monitoring and weak enforcement', for example, is an element that belongs to the governance domain, while the 'incompliance' would be in between the social and governance domains. Information about the 'fish stocks' would classify within the natural science domain. However, the 'impact of declining stocks on the community resilience' would feature in between the social and natural science domains.

By engaging in this approach, the transdisciplinary group would be able to dig deep within the problem to understand the root cause of why, for example, overfishing is happening and how to potentially reduce it. In this case, overfishing could be prevented through an allocation system designed specifically for small-scale fisheries, in a way that protects them from the recreational and industrial sectors. In sum, a transdisciplinary team must initially engage with domain-specific questions to identify connections and overlaps between them, and turn these questions into complex, but solvable, transdisciplinary questions. Through this process, the team collectively applies their expertise and knowledge systems, discusses the problem and arrives to potential solutions that could work.

22.5 Challenges in Transdisciplinary Research

As any other research design, transdisciplinary research can bring forth a number of challenges during the different stages of implementation. Difficulties can ensue during the process of problem identification and framing, perhaps due to political and conflicting issues that could affect the level of trust gained amongst the transdisciplinary team members. Linked to this could be the lack of integration across the different knowledge types, and other challenges including uncertainties and funding obstacles that could hinder the achievement of a real transdisciplinary research. In what follows, we describe some of the challenges that are common in transdisciplinary research and ways in which we can address them.

22.5.1 The Process of Problem Identification and Framing

We have shown in Fig. 22.1 how the transdisciplinary 'lens' can help identify the nature of the problem through the different social, natural, and governance domains. Identifying the nucleus of the problem and agreeing on the solutions to resolve the problem can be problematic. Some might not agree on how to frame the problem or

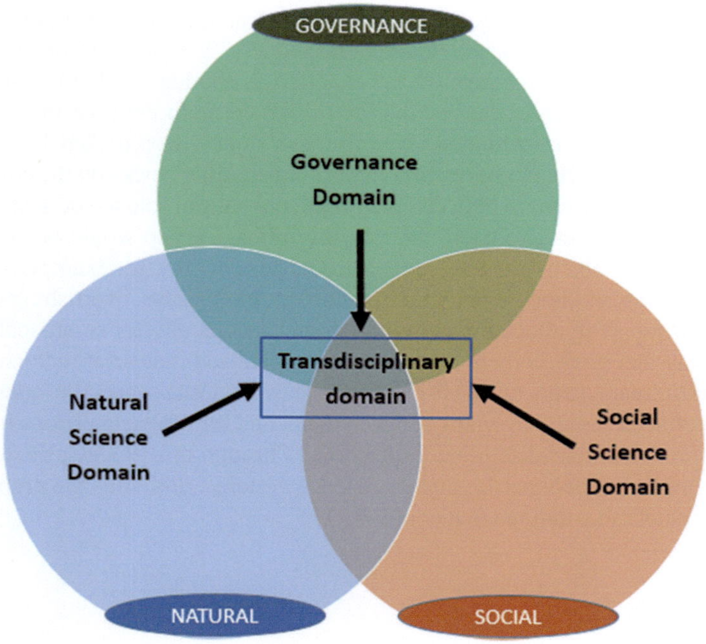

Fig. 22.1 Approaching a problem through a transdisciplinary lens. The transdisciplinary domain is where the three disciplinary domains overlap

what the real driver behind it is. Others might contest the need for particular management measures as a solution to the problem. Disputes are a normal outcome of a transdisciplinarity process, and should be considered as part of a collective change and not a reason to give up (Msomphora 2016), as indicated in the researchers' experiences in Box 22.2.

Box 22.2 Researchers' Experience with Transdisciplinary Journey

Collaborative research carried out in Mexico's Lowland Maya area – Minerva Arce-Ibarra

In 1998, I was invited to join a research team addressing the sustainability of the Maya rainforest. As a former fishery biologist, my task was to address fishing pursued by Maya peasants at "cenotes" (sinkholes) interspersed in the forest. The team was multidisciplinary in nature, with a geographer, ethno-botanist, forest engineer, agriculture engineer, and a fishery biologist. During the second year of the project, some conflicts arose because of contestation of the leadership of the project and epistemic problems with the methodology. Furthermore, Maya authorities were hesitant in granting consent to enter their lands. The following year three team members left the project and a teacher and a lawyer joined, the latter mainly to address indigenous customary law. During this year, a Maya student joined the team, which facilitated the relationship with the Maya authorities for doing the research. Eventually, trust increased among the team members and they wanted to pursue more integrative work. This included, among others, hundreds of hours of discussions to address the use of systemic thinking, using 'mind maps' to connect different fields of knowledge as well as using interdisciplinary concepts, such as 'territory', to address the sustainability of Maya forest and its resources. Since 2008, due to this collaborative research and my Ph.D. in Interdisciplinary studies (2007), I am able to address both the sustainability of Maya fisheries and the multiple productive cycles that Maya peasants carry out through the years.

Community-based research is a transdisciplinary venture – Prateep K. Nayak

In 2007, I started my doctoral fieldwork in Chilika Lagoon, India to understand the marginalization of small-scale fisheries. It is interesting to recall how my methodological approach was developed in different ways to respond to proposals from fishing community partners who advised me to implement open-ended questionnaires in a way that elicited the fisher's voice. Following a successful implementation of the first two questionnaires amongst 30 selected households in two villages, I was further challenged by community leaders about the effectiveness of focusing only on two villages, which was somewhat exclusionary of the rest. With the hope that the results of this study could bring change, some fishing communities expressed their wish that the collection of their voices was as important and volunteered to be active col-

(continued)

Box 22.2: (continued)

laborators and contributors. At this point, I was intellectually flattered by the collaborative gesture from my fisher friends and their seriousness about the results, and effectively we covered a total of 150 villages. The questionnaire for this large census was drafted alongside the fishers, and was implemented with every fisher of the villages, through committee or village meetings. The scale at which this survey was conducted required extensive collaboration and commitment, and the overwhelming response was a true evidence of this. This prompted me to think about concrete plans to make these results public, and along with fishers of Chilika, we planned a state level policy workshop to present the findings and to elevate them through media coverage. This experience remained with me as a powerful image of transdisciplinary research that transcends artificial boundaries set by orthodox science.

From social sciences to transdisciplinarity in small-scale fisheries – Svein Jentoft

Transdisciplinarity is interaction where you are learning from people coming from disciplines other than your own, people with different backgrounds and experiences from what you have. When working in a transdisciplinary research project, you communicate with colleagues who often do not always share your worldview about what constitutes good science, what approach to follow, and what questions to ask. When working with other people from other disciplines, you are often caught by surprise, which may well feel intimidating; you may come to realize that your ideas have limitations, that your assumptions are erroneous. You are likely even to become a better scientist not within a different discipline, but with the discipline in which you belong. This only sounds like a paradox until you have had the transdisciplinary experience. This is what I learned from the FishGovFood project that Jan Kooiman initiated in early 2000, which led to the Fish for Life book (Kooiman 2005). The FishGovFood project, which involved many co-authors from different disciplines (social and natural sciences) engaging in discussions about concepts and arguments, was one of the best universities I ever attended. However, it did not stop with that. This way of working continued with the Povfish project, (Poverty Mosaics: Realities and Prospects in Small-Scale Fisheries; (Jentoft et al. 2011)), the TBTI project, and a series of volumes. Thus this book, which sums up what we think about the merits of transdisciplinary approaches for small-scale fisheries, has a transdisciplinary history. This is how small-scale fisheries research should proceed. The current volume is just the end of the beginning.

The team members need to understand that working towards reaching agreement and a complete solution might be a long and contentious process. It is perfectly normal that, at this point, members may start experiencing fatigue and resentment to the process. Linke and Jentoft (2014) explain how fishers have resented their

Box 22.3 Indicators to Evaluate the Process of the Transdisciplinary Approach

- The level of transparency in the communication of knowledge
- Diverse membership in the transdisciplinary group
- The degree of active roles of the different members
- Dissemination of results to the various members
- Willingness of fishers to participate in research
- Knowledge integration at every step of the process
- Reflection and improved self-awareness

participatory role while engaging in knowledge-integration attempts since the policy-making individuals did not consider the fishers' knowledge as relevant and did not take their 'phronesis' (wisdom-based knowledge) seriously, simply because it did not fit the 'scientific' domain of the maximum sustainable yield discussions. Feeling discouraged and frustrated, fishers felt that they were wasting their time and eventually left the process (Linke and Jentoft 2014).

Periodic monitoring of the process can mitigate these problems. Co-reflection on the participative process by the transdisciplinary team would be ideal (Roux et al. 2010). This could be done through a performance assessment such as an indicator-criteria system (Box 22.3) Participants can also assess the potential challenges encountered and provide a good guidance on possible reconsiderations on future pathways to ascertain that efforts and time invested are not in vain. Having said that, if the challenges persist and the problems that one ought to solve are likely to expand (e.g. other stocks start to deplete), imminent intervention through, for example, a 'temporary fishery closure' policy would need to be implemented despite the lack of perfect understanding (Dyball 2010).

22.5.2 Trust and Legitimacy Issues in Co-design of Solutions

A transdisciplinary team must develop sufficient trust and mutual confidence to transcend disciplinary boundaries and adopt a more holistic approach to recognize problems and co-design solutions (Harris and Lyon 2013). If trust is not gained, then the transdisciplinarity process to reach an ultimate goal might not be fulfilled. Problems of small-scale fisheries sustainability bring forth a number of trust and legitimacy challenges especially since the research team may hold different (possibly contradictory) agendas, and has to deal with complex situations that may require immediate actions to resolve the crisis. Trust issues could relate to questions about the validity of the knowledge-base or the management/policy options proposed. This can be an issue where there is distrust between the scientific community and the fishers due to a history of conflicting views about fishery stock abundance (e.g. Rochet et al. 2008). Other forms of trust issues can result from instances involving lack of transparency in communicating information, or even worse, due to speculations about one's personal interests related to, for example, elite capture in the allocation of fisheries resources (Sundström 2015). Signs of corruption amongst fishers' representatives or policy-makers can hamper the legitimacy of the process, and can generate a culture of suspicion amongst fishers who may then refuse to join any other transdisciplinary project in the future. Moreover, if fishing communities do not believe or trust the process, they might not support the policies that are enacted, leading, therefore, to resistance and incompliance (Goti-Aralucea et al. 2018).

22.5.3 *Lack of Integration Across Knowledge Types*

A common challenge reported in transdisciplinary research is the integration of various knowledge types, both of academic and non-academic nature (Polk 2014). This mainly happens because participants are either reluctant or unable to leave their respective 'comfort zone' to engage in a truly joint endeavour (Bergmann and Jahn 2008). Resistance in scientists has been associated with the apprehension of them to adopt different methodologies and terminology that do not fit within the norms and expectation of their own discipline, making it difficult for them to fully engage in the social learning process (Cornell et al. 2013). Challenges have been recorded amongst individuals trained in the rigorous natural sciences who find the transdisciplinary process cumbersome (Aslin and Blackstock 2010). Problems of integration can also be a consequence of power relations related to social hierarchies and the hegemony of western scientific knowledge that does not easily accommodate indigenous and other forms of knowledge.

Moreover, lack of integration could result from the feeling that non-academics have less of a say in the process, or are only allowed to speak when they are asked to contribute. This is likely to happen when the discussions are more focused on the ecological sustainability and technical jargon deprives them from fully interacting in the process. When the focus is too much on the fish and the empirical evidence provided by scientists, and when the knowledge of the people is simply considered as anecdotal notes that do not penetrate the process, then knowledge integration becomes difficult (Linke and Jentoft 2014). Similarly, focusing too much on the people and forgetting about the fish can cause the same problem of knowledge imbalance and lack of fusion.

Problems of integration could also be deriving from the fact that members feel the approach does not give them sufficient space. Some might feel their knowledge is being subjugated by more vocal participants, who are perceived as more powerful and/or capable to get their story across and incorporated within the general narrative during the problem and solution identification processes. In some cases, capacity-building and training, within a realistic timeframe, could improve one's ability to participate more actively. Moreover, different media can be used to enhance interaction and communication amongst members from different social segments. Rich picturing and board games, for example, have been considered as useful to probe into knowledge systems that could be dormant or unutilized. In a particular case in the Philippines, board games enabled participants to engage in discussions on seascape management relating to long-ingrained conflict concerning the community (Wyborn and Clealand 2010). Simultaneously, self-reflection and questioning can be also effective to help scientists to acknowledge the existence of other kinds of knowledge. Ultimately, when the scientist becomes mentally prepared to orient him/herself towards transdisciplinary research, and is open to learn through new ways and methodologies from distinctive types of knowledge cultures, the process becomes easier, as described in the researchers' experiences (Box 22.2).

22.5.4 Dilemmas for Transdisciplinary Researchers

Although the transdisciplinary process provides new forms of knowledge, and recently has been lauded as an ideal research tool to resolve complex sustainability issues (Brandt et al. 2013), there still remain major obstacles that restrain it from becoming a fully recognized and institutionalized practice. Most of the transdisciplinary work that is conducted is funded by projects that are in place for a few years, but with no assurance that further funding will be obtained.[1] Moreover, academic institutions are still very much conventional with different faculties operating within disciplinary boundaries, making it difficult for multiple disciplinary research to take root (Cornell et al. 2013), although some disciplinary synergies have recently started sprouting in some universities.[2]

Scientists investing in interdisciplinary research have also found it challenging to publish their work (Wise et al. 2017). Coupled with lack of funding towards these endeavours, it has been difficult for scientists to work towards developing a career that is more interdisciplinary. Given this backdrop, the future challenge is to retain the momentum of transdisciplinary research by training more individuals to be more 'prepared' to engage in transdisciplinary research ventures. If the future can be composed of more transdisciplinary research, then scientists investing in this future would have a less degree of risk in their careers, and would feel more fulfilled by their cross-disciplinary expertise (Aslin and Blackstock 2010).

[1] This mostly is a challenge to researchers that work in research and academia, as fishers and members of the community who engage in the transdisciplinary process still retain their livelihood from the sea and not from the research itself.

[2] e.g. School of Anthropology and Conservation in University of Kent, UK and Interdisciplinary Research Excellence in the University of Southampton, UK.

22.6 Conclusion

The transdisciplinary approach contributes to the global discourse on small-scale fisheries sustainability, as promoted in the SSF Guidelines. It does this not only through provisioning research and knowledge, but also as a mechanism that seeks social transformation by providing the adequate tools to address complex societal problems inherent to natural resource governance. As both a framework and pedagogy, the transdisciplinary approach allows scientists, practitioners, fishers, and many other stakeholders to identify the problems and opportunities through an interactive and iterative process. This is why, through this chapter, we acknowledge the urge for integration of information from different academic and non-academic disciplines, and demonstrate how the transdisciplinary research needs to be accomplished in line with specific principles that engender a process of communication, facilitating problem identification and solutions. While there is no guarantee that transdisciplinarity would work or lead to better outcomes, embarking on a transdisciplinary journey is the first step in realizing that complex, wicked problems demand an approach which not only integrates multiple disciplines, but also transcends their boundaries and transforms the perspectives of the participants.

This chapter demonstrates why the transdisciplinary approach is challenging and requires effort, commitment, and relational skills. We reiterate on the need to involve various stakeholders, provide an enabling environment and sufficient space to engage, and make every team member an owner, rather than a guest, in the process. This approach helps reveal the multifaceted realities of small-scale fisheries and generates insights into the value of successful small-scale fisheries governance. This way, the team can generate the necessary knowledge-infused imagination needed to resolve pressing sustainability crises deeply entrenched in communities. On a final note, it is fundamental to acknowledge that transdisciplinarity requires both progressive effort and ongoing reflection, and must be sufficiently iterative to address potential gaps or failures that might be evident throughout any transdisciplinary endeavour.

References

Aguilar-Perera A, Arce-Ibarra AM, Bishop B et al (2017) Towards sustainable small-scale fisheries: key considerations for transdisciplinary teaching and training. Too big to ignore research report number R-04/2017. http://toobigtoignore.net/report-transdisciplinary-approach-for-ssf/. Accessed 2 May 2018

Antonova AS (2016) The rhetoric of "responsible fishing": notions of human rights and sustainability in the European Union's bilateral fishing agreements with developing states. Mar Pol 70:77–84. https://doi.org/10.1016/j.marpol.2016.04.008

Aslin HJ, Blackstock KL (2010) "Now I"m not an expert in anything': challenges in undertaking transdisciplinary inquiries across the social and biophysical sciences. In: Brown V, Harris J, Russell J (eds) Tackling wicked problems through transdisciplinary imagination. Earthscan, London, pp 117–129

Bateson G (1987) Steps to an ecology of mind. Jason Aronson Inc, London
Bavinck M (2014) Investigating poverty through the Lens of Riches – immigration and segregation in Indian capture fisheries. Dev Pol Rev 32:33–52. https://doi.org/10.1111/dpr.12042
Bergmann M, Jahn T (2008) CITY:mobil: a model for integration in sustainability research. In: Hirsch Hadorn G, Hoffmann-Riem H, Biber-Klemm S et al (eds) Handbook of transdisciplinary research. Springer, Berlin/Germany, pp 89–102
Brandt P, Ernst A, Gralla F et al (2013) A review of transdisciplinary research in sustainability science. Ecol Econ 92:1–15. https://doi.org/10.1016/j.ecolecon.2013.04.008
Brown V (2010) Collective inquiry and its wicked problems. In: Brown V, Harris J, Russell J (eds) Tackling wicked problems through transdisciplinary imagination. Earthscan, London, pp 61–83
Bundy A, Chuenpagdee R, Jentoft S, Mahon R (2008) If science is not the answer, what is? An alternative governance model for the world's fisheries. Front Ecol Environ 6:152–155. https://doi.org/10.1890/060112
Carew AL, Wickson F (2010) The TD wheel: a heuristic to shape, support and evaluate transdisciplinary research. Futures 42:1146–1155. https://doi.org/10.1016/j.futures.2010.04.025
Charles A (2011) Small-scale fisheries: on rights. Trade Subsid Marit Stud 10:85–94
Choi B, Pak A (2006) Multidisciplinary, interdisciplinary and transdisciplinarity in health research, services, education and policy: 1. Definitions, objectives, and evidence of effectiveness. Clin Invest Med 29:351–364. https://doi.org/10.1016/j.jaac.2010.08.010
Chuenpagdee R (2011) World small-scale fisheries: contemporary visions. Eburon, Delft
Cornell S, Berkhout F, Tuinstra W et al (2013) Opening up knowledge systems for better responses to global environmental change. Environ Sci Pol 28:60–70. https://doi.org/10.1016/j.envsci.2012.11.008
Cuervo-Sánchez R, Maldonado JH, Rueda M (2018) Spillover from marine protected areas on the pacific coast in Colombia: a bioeconomic modelling approach for shrimp fisheries. Mar Pol 88:182–188. https://doi.org/10.1016/J.MARPOL.2017.10.036
Degnbol P, Mccay BJ (2007) Unintended and perverse consequences of ignoring linkages in fisheries systems. ICES J Mar Sci 64(4):793–797
DuBois C, Zografos C (2012) Conflicts at sea between artisanal and industrial fishers: intersectoral interactions and dispute resolution in Senegal. Mar Pol 36:1211–1220. https://doi.org/10.1016/j.marpol.2012.03.007
Dyball R (2010) Human ecology and open transdisciplinary inquiry. In: Brown V, Harris J, Russell J (eds) Tackling wicked problems through transdisciplinary imagination, first. Earthscan, London, pp 273–284
FAO (Food and Agriculture Organization of the United Nations) (2015) Voluntary guidelines for securing sustainable small-scale fisheries. FAO, Rome
Geertz C (1973) Thick description: toward an interpretive theory of culture. Fontana Press, New York
Gezelius SS (2008) The problem of implementing policies for sustainable fishing. In: Gezelius SS, Raakjær J (eds) Making fisheries management work. Implementation policies for sustainable fishing. Springer, Heldelberg, pp 1–25
Goti-Aralucea L, Fitzpatrick M, Döring R et al (2018) Overarching sustainability objectives overcome incompatible directions in the common fisheries policy. Mar Pol 91:49–57. https://doi.org/10.1016/j.marpol.2018.02.006
Griffith R (2010) Rethinking change. In: Brown V, Harris J, Russell J (eds) Tackling wicked problems through transdisciplinary imagination, first. Earthscan, London, pp 251–259
Harris F, Lyon F (2013) Transdisciplinary environmental research: building trust across professional cultures. Environ Sci Pol 31:109–119. https://doi.org/10.1016/j.envsci.2013.02.006
Hendricks C (2010) Inclusive governance for sustainability. In: Brown V, Harris J, Russell J (eds) Tackling wicked problems through transdisciplinary imagination. Earthscan, London, pp 150–160
Jentoft S (2000) The community: a missing link of fisheries management. Mar Pol 24:53–59
Jentoft S (2006) Beyond fisheries management: the Phronetic dimension. Mar Pol 30:671–680

Jentoft S, Bavinck M (2014) Interactive governance for sustainable fisheries: dealing with legal pluralism. Curr Opin Environ Sustain 11:71–77. https://doi.org/10.1016/j.cosust.2014.10.005
Jentoft S, Chuenpagdee R (2009) Fisheries and coastal governance as a wicked problem. Mar Pol 33:553–560
Jentoft S, Eide AH, Gunnarsdottir M-V (2011) Poverty mosaics: realities and prospects in small-scale fisheries. Springer, Dordrecht
Johnsen JP (2018) Small-scale fisheries governance in Norway: hierarchy, institutions and markets University of Norway. In: Pascual Fernández JJ, Pita C, Bavinck M (eds) Small-scale fisheries in Europe: status, resilience and governance. Springer, Dordrecht
Kerne A (2005) Doing interface ecology: the practice of metadisciplinary. Proceedings of the USA. URL: In: ACM SIGGRAPH 05 electronic art and animation catalog, August 1–4, 2005, Los Angeles, California,
Kooiman J (2005) Fish for life: interactive governance for fisheries, 3rd edn. Amsterdam University Press, Amsterdam
Lalancette A (2016) Creeping in? Neoliberalism, indigenous realities and tropical rock lobster (kaiar) management in Torres Strait. Aust Mar Pol 80:47–59. https://doi.org/10.1016/j.marpol.2016.02.020
Lang DJ, Wiek A, Bergmann M et al (2012) Transdisciplinary research in sustainability science: practice, principles, and challenges. Sustain Sci 7:25–43. https://doi.org/10.1007/s11625-011-0149-x
Linke S, Jentoft S (2014) Exploring the phronetic dimension of stakeholders' knowledge in EU fisheries governance. Mar Pol 47:153–161
McClanahan TR, Castilla JC, White AT, Defeo O (2009) Healing small-scale fisheries by facilitating complex socio-ecological systems. Rev Fish Biol Fish 19:33–47. https://doi.org/10.1007/s11160-008-9088-8
McCormack F (2017) Sustainability in New Zealand's quota management system: a convenient story. Mar Pol 80:35–46. https://doi.org/10.1016/j.marpol.2016.06.022
Mitchell RK, Wood DJ, Agle B (1997) Toward a theory of stakeholder identification and salience: defining the principle of who and what really counts authors (s): Ronald K Mitchell, Bradley R Agle, Donna J Wood. Source: The Academy of Management Review, vol 22, no 4. (Acad Manag Rev 22:853–886, October). https://doi.org/10.5465/AMR.1997.9711022105
Msomphora MR (2016) Conflict resolution and the delegation of authority in fisheries management: the case of outer Hebrides Inshore Fisheries Group in Scotland. Mar Pol 73:263–275. https://doi.org/10.1016/j.marpol.2016.08.006
Nayak P, Berkes F (2011) Commonisation and decommonisation: understanding the processes of change in the Chilika Lagoon, India. Conserv Soc 9:132. https://doi.org/10.4103/0972-4923.83723
Nelson CH, Lowitt K, Nagy M et al (2013) Future research approaches to encourage small-scale fisheries in the local food movement. J Agric Food Syst Community Dev 3:1–5
Pinkerton E (2017) Neoliberalism and global small-scale fisheries (Special issue). Mar Pol 80:1–176
Pinkerton E, Davis R (2015) Neoliberalism and the politics of enclosure in North American small-scale fisheries. Mar Pol. 1–10. https://doi.org/10.1016/j.marpol.2015.03.025
Pohl C, Hadorn GH (2008) Article methodological challenges of transdisciplinary research. Natures Sci Sociétés 16:111–121. https://doi.org/10.1051/nss
Polk M (2014) Achieving the promise of transdisciplinarity: a critical exploration of the relationship between transdisciplinary research and societal problem solving. Sustain Sci 9:439–451. https://doi.org/10.1007/s11625-014-0247-7
Rittel HWJ, Webber MM (1973) Dilemmas in a general theory of planning. Pol Sci 4:155–169. https://doi.org/10.1007/BF01405730
Rochet MJ, Prigent M, Bertrand JA et al (2008) Ecosystem trends: evidence for agreement between fishers' perceptions and scientific information. ICES J Mar Sci 65:1057–1068. https://doi.org/10.1093/icesjms/fsn062

Roux DJ, Stirzaker RJ, Breen CM et al (2010) Framework for participative reflection on the accomplishment of transdisciplinary research programs. Environ Sci Pol 13:733–741. https://doi.org/10.1016/j.envsci.2010.08.002

Russell J (2010) A philosophical framework for an open and critical transdisciplinary inquiry. In: Brown V, Harris J, Russell J (eds) Tackling wicked problems through transdisciplinary imagination, first. Earthscan, London, pp 31–60

Said A, Tzanopoulos J, MacMillan D (2016) Bluefin tuna fishery policy in Malta: the plight of artisanal fishermen caught in the capitalist net. Mar Pol 73:27–34. https://doi.org/10.1016/j.marpol.2016.07.025

Sundström A (2015) Covenants with broken swords: corruption and law enforcement in governance of the commons. Glob Environ Chang 31:253–262. https://doi.org/10.1016/j.gloenvcha.2015.02.002

Symes D, Hoefnagel E (2010) Fisheries policy , research and the social sciences in Europe: challenges for the 21st century. Mar Pol 34:268–275. https://doi.org/10.1016/j.marpol.2009.07.006

Tunca S, Ünal V, Miran B et al (2016) Biosocioeconomic analysis of marine recreational fisheries: a comparative case study from the Eastern Mediterranean, Turkey. Fish Res 174:270–279. https://doi.org/10.1016/j.fishres.2015.10.025

Urquhart J, Acott T, Reed M et al (2011) Setting an agenda for social science research in fisheries policy in Northern Europe. Fish Res 108:240–247. https://doi.org/10.1016/j.fishres.2010.12.026

Weeratunge N, Béné C, Siriwardane R et al (2013) Small-scale fisheries through the wellbeing lens. Fish Fish 15:255–279. https://doi.org/10.1111/faf.12016

Wise PH, Shaw GM, Druzin ML et al (2017) Risky business: meeting the structural needs of transdisciplinary science. J Pediatr 191:255–258. https://doi.org/10.1016/j.jpeds.2017.08.072

Wyborn C, Clealand D (2010) Fences and windows: using visual methods to explore conflicts in land and seascape management. In: Brown V, Harris J, Russell J (eds) Tackling wicked problems through transdisciplinary imagination, first. Earthscan, London, pp 161–170

Chapter 23
Transcending Fisheries Knowledge: From Theory to Integration

Ratana Chuenpagdee and Svein Jentoft

Abstract The multifaceted nature of small-scale fisheries issues and concerns calls for nuanced approaches to governance. Traditionally, government departments are responsible for all three orders of governance, i.e. day-to-day problem solving, institutional design and arrangements, and setting guiding principles. With the increased recognition of the wicked nature of fisheries problems, changes in governing mode from top-down centralized system to co-management and community-based management is now common. Efforts to engage public and involve stakeholders in resource management are growing around the world. The Voluntary Guidelines for Securing Sustainable Small-Scale Fisheries (SSF Guidelines) call for more, however. Grounded in human rights principles and with an emphasis on poverty eradication and food security, the SSF Guidelines stipulate that governments, civil society organizations and research communities should broaden their perspectives in the examination of the problems and in the consideration of possible solutions. This transdisciplinary approach to fisheries governance cannot emerge without conscious efforts to build a solid knowledge base from all relevant stakeholders, and to implement education and training programs that promote transdisciplinary learning and practice. This chapter presents a novel platform for such capacity development, created by the Too Big To Ignore Global Partnership for Small-Scale Fisheries Research.

Keywords Governance orders · Governing modes · Co-management · Transdisciplinary training · Wicked problems · Fisheries systems · Value chain

R. Chuenpagdee (✉)
Department of Geography, Memorial University of Newfoundland, St. John's, NL, Canada
e-mail: ratana.chuenpagdee@mun.ca

S. Jentoft
Norwegian College of Fishery Science, UiT-The Arctic University of Norway, Tromsø, Norway
e-mail: svein.jentoft@uit.no

R. Chuenpagdee, S. Jentoft (eds.), *Transdisciplinarity for Small-Scale Fisheries Governance*, MARE Publication Series 21,
https://doi.org/10.1007/978-3-319-94938-3_23

23.1 The Unique Opportunity

The continued degradation and overexploitation of fisheries resources observed around the world sends an important signal for governance. It suggests that the existing management approaches and the current emphases in science and research are either insufficient or misplaced. Given the diversity, complexity and dynamics associated with small-scale fisheries, it is likely that the governance capacity is also inadequate to deal with the challenges and the 'wicked' nature of the problems that these fisheries bring (Jentoft and Chuenpagdee 2009). Governing actors at all levels are often struggling with competing and shifting priorities, due to numerous interests from a wide range of stakeholders, coupled with changes in environment, society, culture, economics and politics. At most, they may be able to keep up with the governance demands of daily routine, addressing present-day problems that they are confronting.

Governing actors are also limited by the methods and approaches that they have in their toolboxes, which are mostly equipped by advocates of certain ideology and perspectives. They draw from these tools to address a wide range of problems, irrespective of whether the problems need a brush or a hammer (Degnbol et al. 2005). Instead, the methods and potential solutions need to be looked at differently by stakeholders, as suggested in the transdisciplinarity literature (see for instance Lang et al. 2012 and Said et al., Chap. 22 in this volume). Marine protected areas (MPAs) are a classic example of how a good idea can fall short of expectations and create numerous unintended consequences when applied indiscriminately. MPAs are also a good illustration of why a shift in policy formulation, research, and community engagement efforts is urgently needed, if one is serious about protecting biodiversity and achieving sustainability. Many solutions to the fisheries problems require careful consideration of the pros and cons, as well as meaningful participation of relevant stakeholders in the entire process, beginning at the 'step zero,' in order not to fail (Chuenpagdee and Jentoft 2007; Chuenpagdee et al. 2013).

With the adoption of the Voluntary Guidelines for Securing Sustainable Small-Scale Fisheries in the context of Food Security and Poverty Eradication (SSF Guidelines FAO 2015) in 2014, the timing could not be better to reflect on what the goals of fisheries governance are, expand our thinking about how to organize the fisheries systems, and broaden our perspectives and approaches in dealing with the fisheries problems. While several international instruments have been developed to promote good practices in fisheries and sustainable development since the launch of the Code of Conduct for Responsible Fisheries in 1995 (FAO 1995), the SSF Guidelines is the only instrument aiming specifically at small-scale fisheries. The comprehensiveness of the SSF Guidelines, covering ecological, socio-cultural, technological, and economic aspects of small-scale fisheries, and the grounding on universal principles like human rights, social justice and gender equality and equity, represent a vision for change in the culture of governance, research, grant making, and business development. In the same spirit that they were developed, the SSF Guidelines also invite participation of small-scale fishers, fishing industries, civil society organizations, environment groups and research communities to work with the local, national and regional governments to deliver what is promised to be a major turning point in the history of fisheries governance.

In effect, the SSF Guidelines call for concerted efforts from all stakeholders in three orders of governance. Theorized by Kooiman (2003) and initially applied to fisheries in Kooiman et al. (2005), reflection about values, norms and principles are the meta-order elements that provide foundation for the formulation of policy and the arrangement of institutions in the second-order (Kooiman and Jentoft 2009; Jentoft and Chuenpagdee 2015). Spending time deciphering elements, actions and interactions at these two orders is not a luxury but a necessity for the implementation of the SSF Guidelines, which requires much richer intellectual resources for the governing actors to draw upon. For governing actors and relevant stakeholders to be able to go above and beyond what is expected in conventional practice and norms in the operationalization of an international instrument, they need to develop new visions and capacity that is grounded in the spirit of transdisciplinarity, which is what this book and the Too Big To Ignore – Global Partnership for Small-Scale Fisheries Research (TBTI; toobigtoignore.net) is about (see Chaps. 1 and 2).

The main aim of this chapter is to help facilitate the transition from theory to practice in transdisciplinary governance, research and development in small-scale fisheries. Drawing from the governability assessment framework (Bavinck et al. 2013), the chapter begins with a discussion about how the major problems in small-scale fisheries can be understood, and how some of them make fisheries governance a complex and challenging task. Next, it provides examples of questions that can be used to examine the fisheries systems and to explore what features of the systems that may give rise to governance challenges. The chapter also analyzes the alignment between principles promoted in the SSF Guidelines and other key international instruments with transdisciplinary thinking. Finally, it concludes by introducing the 'open transdisciplinary learning platform' that TBTI has developed.

23.2 Understanding the Problem

Small-scale fisheries are made up of multiple natural and social components, and are also formed by complex relationships and dynamic interactions. The linked and nested nature of small-scale fisheries makes their analysis challenging. As suggested by Rittel and Webber (1973), not only are solutions hard to find, but the problem itself is also hard to diagnose. Governors are confronted with a limited knowledge about how small-scale fisheries work, and how they could or should intervene in ways that steer small-scale fisheries towards a more sustainable path. The latter requires considerations about how change occurs, both from outside and also within the system itself, prior to and after the intervention.

Any intervention, small or large, is likely to have ripple effects that may be hard to anticipate, predict and reverse. Governing actors are also pressured to 'get it right' because of limited resources, and sometimes because of the urgency of the problems. Thus, it is important to know what the 'tension points' are, i.e. those points that have the potential to change the system once they are pushed (see Jentoft, Chap. 16). The problem is that we often cannot know what or who these points are, or where to find them, and we cannot be certain that they are the right ones. We may, for instance, think that a new road will release the development potential of a community, but we cannot be sure that the one road will be sufficient to transform the system. Not knowing where the exact tension points are, interactive governance theory suggests that our search should commence by looking into all layers of the system-to-be-governed, the governing system and the governing interactions (Kooiman et al. 2005), in order to identify potential tension points. This would require empirical research, which due to the multiple facets of small-scale fisheries would need to be transdisciplinary. Individual disciplinary perspectives alone would not be sufficient to capture the entire small-scale fisheries system.

Small-scale fisheries governance requires collective reflection and interactive learning. In search of a governance process, experience-based knowledge of stakeholders is essential, as governance strategies and desired outcomes must be framed from their perspectives. The SSF Guidelines suggest that serious consideration must be given to the normative and ethical aspects of governance because the goals do not always justify the means. 'Good governance' cannot therefore only be defined in functional terms, as it matters how goals are achieved.

As defined by the SSF Guidelines, good governance principles, be they stakeholder participation, justice, gender equity, transparency, rule of law, are qualitative. They are primarily about the process of governing. The normative qualifiers are also why fisheries governance cannot be expert-driven from the top down, but must be inclusive of small-scale fishing people. After all, it is the latter group who lives with the problems the SSF Guidelines aim to address, and who experience the consequences of the governance interventions. Thus, their ideas, values and knowledge must count. On the issue of moral values, scientists cannot claim special authority. The governance principles stated in SSF Guidelines emerged through a process of stakeholder consultation and participation around the world, which in itself gives the Guidelines moral authority (Jentoft, Chap.16). The SSF Guidelines are voluntary and thereby not legally binding. Member states may be pragmatic about how they choose to implement the SSF Guidelines, but by endorsing them, they have committed themselves to the basic meta-principles, like those associated with human rights. If governments and other stakeholders were to refrain from implementing them, they must provide moral justification as to why. Pragmatic or opportunistic arguments would not suffice.

Rittel and Webber (1973) list a number of general criteria of wicked problems; they are hard to define, are part of bigger problems, not solved once and for all, and hard to reverse once implemented. The wickedness of the problems must also be identified in concrete contexts. What exactly makes a particular problem in a specific situation wicked? Wicked problems do not occur in isolation, but in the context of a particular system, which influences the way the problem is structured and how it works. Wicked problems have a non-determined cause and effect, and they may be slow in forming, as with climate change and other global change processes (see

Nayak and Berkes, Chap. 11). Wicked problems may also materialize as incidents, begging for urgent action, individual or collective. However, before such action can take place, the problem must be defined and conceptualized. Does the problem already have a name? Does it belong to a category of problems that already have a prescribed solution? If not, governing actors are facing unknown territory. If the problem has no diagnosis, the action taken to address it would be a shot in the dark.

There may be several wicked problems present, more or less interlinked, and in need of detection and definition. The poverty problem may, for instance, be an entitlement problem, a resource scarcity problem, a health problem, a knowledge problem or a corruption problem. These problems may all be present at the same time, forming a conundrum, but it is not always clear exactly how they are connected. For each problem, a different set of corresponding actors and stakeholders exists. Sometimes problems form layers, also with various degree of severity, which require different sets of stakeholders to cooperate. By addressing one problem, other problems may be easier to solve – or be more difficult as the case may be. As when A does X, B may find it easier/difficult to do Y–or *vice versa*. For instance, forming an organization at the community level, as a measure of empowerment, would make it easier for members to adapt to climate change or exercise environmental stewardship. Introducing conservation measures, like MPAs, may exacerbate the poverty and food insecurity if it undermines livelihoods. Augustave (Chap. 9 in this volume) exemplifies this with the case study of the no-take zone in the Velondriake Locally Managed Marine Area, in south-western Madagascar. While some fishers were supportive of the conservation measure, other fishers, especially those from other villages, felt the need to continue fishing for food security reasons, and thus violated the regulations and the local rules. Interventions from conservation organizations, together with government authorities, helped to bring awareness about the importance of the no- take zone and its contribution to long-term sustainability. Small-scale fisheries governance must therefore be designed to address multiple problems that may be in conflict, as part of a well-planned and coordinated effort.

23.3 Unpacking the Fisheries Systems

Knowing the problem and knowing the system are two different things, but they are related. The governability assessment framework that we have proposed elsewhere (see also Appendix 1) suggests that one should start by examining the wickedness of the problem. By understanding the problem, one also gets an idea about the fisheries natural and social systems, for instance, who are involved, how, the extent of their dependency, and scale. The system is providing context for the problem, and their characteristics may very well be what makes the problem wicked. Notably, there is always something unique about the system that governors need to understand and pay attention to. Therefore, governance solutions have no one-size-fits-all but must be formulated and implemented to fit the concrete challenge as it appears in time and space.

As posited by interactive governance theory, exploring the wickedness of the problems that small-scale fisheries harbour would be the first step. The second step is the analysis of the natural, social and institutional settings within which the problems occur. This involves an examination of the properties and the inherent characteristics and linkages of the natural and social systems, and the governing system. Such analysis is also helpful in thinking about what the possibilities (or tension points) are for intervention, and how to open doors for social, technological and governing innovation that can be part of the new solutions in the fisheries governance toolbox. Just like keys that are designed to open certain doors, tools must be tuned to context, as part of the first- and second-order governance.

To understand the wickedness of the problems, interactive governance theory suggests the unpacking of the social and natural systems-to-be governed, the governing system, and the governing interactions (Kooiman 2003; Bavinck et al. 2013). One cannot know in advance where exactly the problem sits. All three systems are assumed to have four properties, which may provide keys to the problem. The three systems are all, in their own right, characterized by diversity, complexity, and dynamics. The system is also characterized by scale, which is the fourth property. Further, interactive governance posits that their diversity is assessed by the components, complexity by the relationships between components, whereas dynamics by how they interact, forming a process that may change the system over time. Finally, scale is determined by the system boundaries, although these boundaries may not be fixed, but permeable by external drivers and internal leakages, which can make them more or less open (see Appendix 2). For instance, as shown in Bower et al. (Chap. 18), the problems in inland fisheries are not bounded within the lake or the rivers, but have multiple origins from upland areas.

In a governability assessment, these characteristics of each system would need to be explored, and hypotheses framed. For instance, one would assume that for a given wicked problem, the larger the number of system components, be they fish species or stakeholders, and the more different they are, structurally or otherwise, the greater the governability problem. Conversely, the fewer and more homogenous they are, the lesser the governability problem is expected to be. Similarly, the greater the number of relationships between components, and hence the more intricate the system, the greater the governability problem, since problems and problem solving cannot be isolated to one part of the system but must take it all into consideration. For these reasons, involving stakeholders in the assessment of governability through a transdisciplinary process, as done in Trat Province Thailand (see Satumanatpan et al., Chap. 12), can help reveal some nuances that governments or scientists alone cannot observe. It is important to note that engaging stakeholders in discussing the system attributes and functioning also brings about new knowledge about the fisheries that can be useful for improving the governance performance.

The nature of these relationships, like power differentials, is also something to consider as it affects the 'governmentality' within the system, i.e. the degree to which actors are willing to abide with rules, and therefore ultimately the governability of the system (Jentoft and Johnsen 2015). Relations determine the configuration of the system, its architecture so to speak, and thus provide a structure for the interactions and process that occur within it. Yet, the structure does not determine which process would take place. Like a football team is made up of positions, relationships, and rules, football is also about the play, which allows skills, spontaneity, and improvisations to run its course. Thus, any instruction from the coach has a limit in terms of determining the dynamics of the play.

Likewise, for governance, interactions are also bound to have governability impacts, whether goals are realized or not, regardless of instructions. The more dynamic the interactions, the faster the change, and the more likely that governance interventions would misfire. Finally, boundaries define the scale of the system, as the lines on the football arena, albeit for fisheries social and natural systems, these lines are not easily drawn. Consider, for example, an MPA. From a governability perspective, complications may occur because MPA boundaries are difficult to control. A large-scale social and ecological system is likely to be more difficult to manoeuvre than one that is easy to locate within narrow boundaries, as articulated by Scholtens et al. (Chap. 17). Hypotheses can also be related to the relationships and interactions between the system-to-be governed and the governing system. For instance, the lesser the fit between the two systems, the greater the governability challenge, such as when the mandate and responsibility of the governing system only covers parts of the ecosystem or the value chain. One may again think of an MPA whose institutional boundary does not match either the ecosystem or the social system boundaries.

When assessing the relationships and interactions between the governing system and the system-to-be-governed, the examiner is advised to look at the structure and functioning of the modes. For the governability of small-scale fisheries, it matters whether governance is exercised hierarchically, from the top down, or the stakeholders are allowed to sort out the problem for themselves, as within a self-governing mode of a local community. This is partly an issue of power, how power is spread throughout the system, and how well power aligns with the capacity to govern. If power-relations are such that interventions are resisted, and do not lead to change they were meant to elicit, the governability problem would be apparent. Conversely, if local stakeholders are poorly organized, lack the necessary capacity to implement and enforce decisions, delegating governing responsibility risks crashing, as there will be no one at the local level who is powerful or knowledgeable enough to take on the task.

Since there are limitations of governability at both the central and local levels, interactive governance theory emphasizes hybrid forms of governance, as with co-governance, where stakeholders and government interact, cooperate and communicate, both about problem definitions and finding appropriate solutions. Still, co-governance is hardly a panacea, but has its own wicked problems that need to be assessed, both from a functional and a normative perspective. It may not deliver as promised, and even if it does, it may not necessarily be in accordance with the good governance principles recommended in the SSF Guidelines. Chapter 19 by Arias Schreiber et al. provides examples of how co-governance modes may marginalize small-scale fisheries.

23.4 Shifting to High Gears

Interactive governance calls attention to the various forms, frequencies, and qualities of interactions, within and between the system-to-be-governed and the governing system, arguing that these are where the wicked problems are located and solutions found. Shifting from the first- and the second-order governance to the meta-order can happen when governing elements in this higher order, such as values, images and principles, are made more explicit (Kooiman and Jentoft 2009; Song et al. 2013).

As illustrated in Johnson et al. (Chap. 3), value is not a straightforward concept to examine and integrate into the governance system. In addition to the different meaning and types, values can be studied from multiple perspectives. Some value propositions are more influential in the policy discourse than others, like economic value over social and cultural values. Similarly, within economic values, there is a tendency to apply market-based valuation methods to obtain dollar figures, despite the recognized weakness about them. Many studies have revealed that the world-views, perceptions and priorities of small-scale fishing people, large-scale fishing industries, other resource users and decision-makers may not align with each other. The examples presented in Johnson et al. (Chap. 3), e.g. Australia, Brazil, Canada, and in other studies (e.g. Chuenpagdee et al. 2001 for Thailand, Song and Chuenpagdee 2015 for South Korea; and Luomba et al. 2016 for Tanzania), suggest that this is a global phenomenon. Unless these value similarities and differences are properly communicated and explicitly discussed, stakeholder conflicts remain the likely outcomes, which inevitably will hamper the governability of small-scale fisheries, and therefore their ability to deliver on the concerns and expectations that stakeholders have for them, including on those principles promoted by the SSF Guidelines.

The governability assessment (Appendix 1) must address how values are distributed among small-scale fisheries stakeholders, including government and civil society organizations, which are part of the governance system. How, for instance, are use-values relative to conservation values distributed among actors? One should not assume that they are well-aligned with each other. In fact, they may well be in conflict. It should be noted that all conflicts in small-scale fisheries are not about conflicting interests, as would be the case when user-groups compete for scarce resources or limited space. Governability problems are aggravated when conflicts occur at the meta-order, for instance in considering what constitutes justice. This is why fisheries rights, which Pinkerton discusses in Chap. 13, are not just about entitlement (who has a right to what) but also about rectitude (what is right or wrong), which is the basis for human rights principles (Donnelly 2003). Both rights give rise to legal conflicts, which in small-scale fisheries often occur within and between multiple normative orders. The situation can be worsened when the governing system must deal with such 'legal pluralism', as when statutory and customary law exist side by side and apply to the same system-to-be-governed (Jentoft and Bavinck 2014, Jentoft et al. Chap. 14).

The meta-order values, and the norms and principles they bring to governance, may not always align well with how the institutions are structured (at the second governing order) and the daily routine procedures at the first-order. This would inevitably affect governability. If there is disconnect between the orders, the governing actors or the clients may well be induced to call for or activate change. Meta-order norms and principles are not necessarily stable either. They may be adjusted to attune to the lower orders of governance. Incoherence between orders is not always apparent, however, particularly among those who have the power to change the system. But even if it were, those in power may still chose the status quo simply because it is in their interest to maintain the current misalignment. How such alignment and re-alignment would work in different modes of governance is an important research question. It is hypothesized that in a co-governance mode, the system can be more dynamic because of the multiple interactions between the different stakeholders, which enables the alignment and re-alignment to happen more easily because of the two-way process.

Stakeholders, be they government, civil society organizations or industry representatives, may be supportive of the guiding governance principles in the SSF Guidelines, but may still insist on keeping things as they are at the second- and the first-order. This brings the governing system into conflict within itself, which may turn governing interactions into a contentious political issue within the small-scale fisheries sector, as illustrated by Jentoft et al. (Chap. 14) with regard to indigenous peoples and their tenure rights. Indigenous small-scale fisheries may be among the poorest of the poor, and marginalized within a fisheries sector. Correcting this injustice would require interventions that may contradict other stakeholders' interests and perceptions of fairness. One may well support general principles until they are converted into concrete action at the second- and the first-order. It remains to be seen whether the human rights approach to fisheries governance, spearheaded by the SSF Guidelines, will trigger change at the lower orders, and finally on the ground, i.e. within the system-to-be-governed.

23.5 Moving Up the Value Chain

In the global development discussion, small-scale fisheries are explicitly mentioned in the 2015 United Nations Sustainable Development Goals (SDGs) as part of SDG14, target 14b, related to the provision of access to resources and markets (United Nations 2015). Yet, as discussed in McConney et al. (Chap. 10), other SDGs, like SDG1 on 'no poverty' and SDG2 on 'zero hunger', speak to the same key issues in the SSF Guidelines. Linking SSF Guidelines more closely to SDGs 1 and 2, in addition to SDG 14, could expand the approaches and mechanisms to address poverty alleviation and food security among small-scale operators in the fisheries, agriculture, and other food production systems. Other SDGs are also much in line with the SSF Guidelines, such as SDG 16, which talks about the need to "build effective, accountable and inclusive institutions at all levels" (see also

Chuenpagdee, Chap. 2). Such connection will also help governments and development agencies to see the central role that fisheries can play in the political and administrative discourses about food security, which presently emphasizes land-based productions despite the fact that fish are the most traded commodities in the world (Loring et al., Chap. 4; FAO 2014).

Situating the discussion about fisheries in the food system network would not only reveal the importance of fish from the nutritional and cultural aspects, as exemplified in Loring et al. with the case of fermented fish sauce in the Philippines, but also from the economic and gender perspective. The latter is illustrated in Frangoudes and Gerrard (Chap. 7) with an account of the range of activities that women are involved in throughout the fish chain. The examples of the women processing groups in Thailand, for instance, bode well for the use of local knowledge in the utilization of freshwater and marine fishes, turning them into staple products in Thai cooking and new healthy snacks for children and other consumers.

The implementation of the SSF Guidelines, which have a strong emphasis on gender equality and equity, would benefit from engaging with practitioners, researchers and organizations working in the food system domain. This is also what Lowitt and colleagues (Chap. 5) reveal through their research in the Lake Superior region in North America. By applying the food systems framework to examine the different values associated with fisheries in Lake Superior, they reveal the opportunities that can be created by enhancing the thinking around the fisheries value chain, for instance through marketing of local fish using alternative food networks in Northwestern Ontario, Canada. This contribution also demonstrates that food security concern is an issue that affects small-scale fisheries in the north as well as in the south.

Small-scale fisheries are a major provider of food security, and thus an important service to society as a whole. They are also about other things than food. It is hardly imaginable that small-scale fisheries could provide food security unless their own livelihoods and communities are sustainable, which among other things require that their tenure rights are secure, as demonstrated by Wojciechowski et al. (Chap. 15) in the case of Pernambuco estuarine in Brazil. Fishers must be able "to defend their beach." Small-scale fisheries are not just 'seafood industry' but a livelihood and community industry that must work together. Again, that calls for holistic perspectives, a comprehensive assessment of the governance system in its totality, as the template (Appendix 1) aims to facilitate. Strengthening communities would be a necessary condition for them to provide the services that the wider society expects from them.

Helping small-scale fisheries to perform their function is a governance challenge, and a governability problem due to the impoverished and marginalized situation that small-scale fisheries are in. Governing intervention is needed throughout the value chain in the harvest sector and in the post-harvest. As suppliers of food, small-scale fishing people may be victims of exploitation because of low bargaining power, especially when there are few buyers or monopsony. The problem can also be related to poor transportation and lack of other infrastructure, like refrigeration, landing places, water supplies, sanitation, etc. The governability assessment must take the whole value chain into consideration, aiming at detecting where bottlenecks may exist. This is among the areas where government subsidies can play a positive role in the development and wellbeing of small-scale fisheries. This does not necessarily require additional public spending, but a re-direction of subsidies from the industrial fisheries, which at the moment are the primary beneficiaries of government support globally, as shown by Schuhbauer et al. (Chap. 6). The assessment of the vertical axis of the value chain must be followed by the assessment of the horizontal axis, how livelihoods are secured and how communities are functioning as social and economic entities.

The governing actors need to address the distributional conflict that may occur in the system. The fact that small-scale fisheries, while meeting subsistence needs and serving local markets, also supply fish into national and international markets raises distributional issues. More of one means less of the other, and currently, with governments in the global south allowing foreign fleets onto their fishing grounds. While providing earnings for public expenditure, it may easily reduce the supply for local consumption, as there is often no trickle down mechanism that ensures that the local livelihoods and communities benefit. Adding values to local catches through market innovation can help to increase the role of small-scale fisheries in the supply chain, and secure their status within the communities. Pascual-Fernández et al. (Chap. 8) illustrate this with an example from Tenerife where a new marketing strategy, involving branding of locally caught seafood, has created awareness among consumers, in addition to giving fishers higher prices for their catches. Their initiative serves also as a model for transdisiciplinary practice where collaborations between scientists, local government and the industry led to the production of the innovative marketing strategies.

23.5.1 The 'Open Transdisciplinary Learning Platform'

Just like the diverse, complex and dynamic nature of small-scale fisheries, governing the sector and improving its governability requires a broad perspective and a comprehensive and holistic framework. The values and importance of small-scale fisheries are situated throughout the fish chain. They contribute to local stewardship at pre-harvest, provide food security to households and the community, as well as local employment as part of the harvest, and offer income opportunities to fishing families, women, and other business enterprises at post-harvest. Governing interventions need therefore to take place along the value chain, as suggested in the SSF Guidelines, e.g. to support conservation and protection of the ecosystems, to secure access to resources and promote new markets for small-scale fisheries, and to provide adequate infrastructure and facilities for the health and safety of fishers and fish workers. As illustrated by the chapters in this book, issues and concerns affecting small-scale fisheries come from multiple sources, directly from heavy fishing pressure or destructive practices, changes in the environment, or resulting from management, regulations and development policies that limit their ability to have viable and secured livelihoods. These issues and concerns are not situated within the fisheries system alone, but come from other sectors. Thus, addressing fisheries problems needs to involve a broad range of actors and other stakeholders, including those responsible for social and community development. On a whole, the sustainability of small-scale fisheries relies for the most part on building governance and research capacity in all governing orders and modes, throughout the value chain, and with a broad range of stakeholders. The same applies for the implementation of the SSF Guidelines, as well as the SDGs, which must begin with developing major capacity in all aspects of governance.

Mechanisms to encourage involvement of a broad range of stakeholders, especially those not previously engaged like the community of practice in the food production system, need to be established, in order to facilitate collaboration, and foster commitments among all actors to share knowledge and co-create solutions. There are, of course, many barriers to getting together, both practically because of the geographical spread and ideologically because of the different backgrounds and experiences. Recognizing the challenges, TBTI is developing an 'Open Transdisciplinary Learning Platform' (or TD platform, in short) to facilitate the process as well as to contribute to knowledge building and integration, which is one of the expected outcomes.

Many scholars have articulated transdisciplinarity from a theoretical and conceptual perspective and discussed its applications in various fields. Like many concepts today, there may not be general consensus about the definition, but this does not prevent it from gaining popularity and from being useful. In fact, the ambiguity in the term may well be suited for a concept like transdisciplinarity, which requires some organic framing of what it is, what it does, and how to implement it. In some ways, this is similar to the lack of universal definition for small-scale fisheries. The term governance also created confusion among delegates, during the technical consultation on the SSF Guidelines, with some arguing that there was no equivalent meaning of the word in their language. There are certain characteristics about small-scale fisheries that are unique to the sector, as described in the SSF Guidelines. But the specificity of what small-scale fisheries are in each location is something that member states need to determine. Another way to think about a 'hard-to-define' concept is by comparing and contrasting. Small-scale fisheries can be easily described when compared with large-scale fisheries, and governance can be better explained in terms of how it is different from management. Likewise, transdisciplinarity can be explained along with other concepts like multi-disciplinarity and interdisciplinarity, as done in Said et al. (Chap. 22). According to these authors, what differentiates transdisciplinarity from others are the knowledge production process and the expected outcomes. Transdisciplinarity is more demanding than others as it aims to 'transcend' disciplinary boundaries and anticipates innovation and transformation in the way we do sciences and in governance. As suggested by Msomphora and Jentoft (Chap. 21), it is a delicate balance between the disciplinary-based research driven by academic curiosity (Mode 1) and the problem-based research driven by industry's interests (Mode 2). Transdisciplinary research is a mixture of the two modes, not in equal measure but in the best possible mix given the context.

The word 'open' in the TD platform has two meanings. In the transdisciplinary discourse, open transdiscplinary refers to an engagement with non-academics in the process of co-identification and co-creation of the problems and the solutions, respectively. This is what TBTI aspires to do with its partnership structure and activities (see Chuenpagdee, Chap. 2). More can always be done to engage better, with a broader range of stakeholders, and in all steps. Since practicing transdisciplinarity comes with its own challenges (Said et al., Chap. 22), any advancement counts, whether it is large or small. The other use of the word 'open' is in the context of the delivery method of the transdisciplinary training and capacity building materials. Following the massive open online course (MOOC) model, the TD platform is developed as an open access, self-guided study but with interactive features and multi-media contents that enable learning and interaction with other course participants. It will resemble the connectivity philosophy (Prpić et al. 2015), with its aim to connect participants to each other in order to work collaboratively in examining the problems and discussing pathways towards solutions. The course is free for registered participants.

Another key feature of this approach is the crowdsourcing of the materials, which encourages participants to remix and re-purpose the existing contents and contribute new materials to help enhance the learning experience and quality, and for future learning. This includes case study contributions and other real world experiences that participants want to share, as well as suggestions of topics that need to be covered in the course. New data and information can also be gathered through this process, which can then be integrated and synthesized, as suggested by Agapito et al. (Chap. 20), and presented back to the TD platform. Similar to other open platforms, the TD platform will use tools like blog posts and threaded discussion, to stimulate discussion and interaction. While the majority of the instructions are predesigned, 'resident' instructors and guest lecturers will participate in the course, especially as new topics are being added, and also to facilitate the dynamic learning process. Linking the TD platform to the Information System on Small-Scale Fisheries (ISSF; issfcloud.toobigtoignore.net) that TBTI has developed (Chuenpagdee et al. 2017) is another way to share the data and exchange knowledge.

Challenges with the online education range from practical issues such as reliability of Internet access and language barrier to learner issues, like self-motivation. Attrition and drop out rates are generally higher than traditional classroom course offerings, which could be due to many reasons including lack of motivation, lack of time to complete the course, or lack of experience in online learning. These challenges are likely to affect the TD platform, barring it from reaching certain target groups, like fishers and communities in remote areas. Thus, the on-site, in-person learning and on-the-ground capacity building will also be organized using the TD platform materials to supplement the online learning. The duration and the content of the in-situ training will vary depending on the interests of the participants or organizations supporting the training.

Generally speaking, the content of the TD platform is guided by the big questions addressed in TBTI, the thematic areas covered in the research clusters, and the topics presented in the SSF Guidelines. It makes use of the existing case studies within TBTI to portray the wide range of issues, like those described in the chapters of the book, but also incorporates new case study materials that emphasize other aspects not sufficiently covered in the project. In addition to learning how to look at the problems, understand their root causes, and think broadly about what may be possible solutions, the course also offers tools, methods and approaches that help build analytical skill sets and that foster the practices of transdisciplinary principle and philosophy. In addition to interactive governance, which is the theoretical foundation for the learning, the course includes other integrative frameworks like wellbeing (Johnson et al. 2017), I-ADApT (Bundy et al. 2015), and viability assessment (Schuhbauer et al., Chap. 6), to provide exercises and practical applications to facilitate learning.

The official launch of the TD platform and the in-situ training is scheduled to begin in January 2018. The course will be monitored and assessed in order to gauge its reception and effectiveness. It is expected that the training will help form a community of transdisciplinary practitioners, with fishers, fishing industry, community groups, civil society organizations, environmental groups, government officials, policy makers and researchers conversing about small-scale fisheries in ways that transcend disciplinary boundaries and that leads to knowledge integration and capacity building for sustainable small-scale fisheries, and for the sustainability of the world fisheries and oceans.

23.6 Concluding Thoughts

Governance for small-scale fisheries sustainability must begin with the recognition that: (1) small-scale fisheries contribute significantly to local food security and poverty alleviation; (2) small-scale fishing people are important actors in the society, helping to improve the wellbeing and viability of coastal communities; (3) they can play imminent roles in conservation and stewardship; (4) they hold traditional and cultural values, as well as offer additional values, for instance, to enhance education and tourism; and (5) small-scale fishing people need to have secure rights and access to resources and markets. For governing actors and relevant stakeholders to recognize the values, roles and importance of small-scale fisheries, a major investment in transdisciplinary capacity building is required. The urgency of some of the issues facing small-scale fisheries, such as with climate change and globalized markets, means that the window of opportunities maybe closing down. Governments and donor organizations need to allocate funds to support capacity development in transdisciplinary research and governance among policy makers, practitioners, fishers and their organizations, community groups, environmental organizations, and research community, both in-situ and remotely through the TD platform. Providing space and opportunities for people to come together to unpack the problems and broaden the thinking about possible solutions is imperative and should be the priority of all government departments and agencies.

The transdisciplinary perspective enables involved actors and stakeholders to see the problems through multiple lenses. It helps resist the temptation of prescribing solutions too early or applying only familiar tools and approaches. A transdisciplinary process fosters broad thinking around what the problems are, how to define them, and what possible solutions may be. It requires involved actors to listen respectfully to other ideas and constructively contribute their own, in ways that lead eventually to better approaches and outcomes than what they would be able to achieve individually. This process facilitates learning and sharing of knowledge, and through further discussion and careful considerations about options and possibilities, the most feasible solutions, or pathways towards them, may emerge. Given that transdisciplinary capacity development can take place in-situ or remotely through the TD platform, it opens up the opportunity for anyone concerned about the future of small-scale fisheries and willing to support fisheries sustainability, broadly speaking, to make their contribution. With the wicked problems facing global fisheries, all efforts count.

Appendices

Appendix 1: The Governability Assessment Template for a Study of Small–Scale Fisheries (SSF)

How Governable Are SSF?

Section 1: Wicked Problems and Stakeholders

- List problems/issues facing SSF in a location
- For each problem, list stakeholders ‘involved’ (e.g. those creating the problem, those affected by it, and those attempting to address it)
- For each problem, indicate the ‘level of severity’ of the problem (based on its perceived/observed effects)
- For each problem, indicate whether it is perceived differently or similarly by each stakeholders
- For each problem, indicate whether something has been done to address it, by who and to what outcome
- For each problem, indicate whether it is part of a bigger problem
- Once all problems have been evaluated based on the above, provide a rating of 1–5 to each problem in terms of ‘relatively degree of the wickedness.’ (1 = not very wicked, 3 = moderately wicked, 5 = very wicked)

Section 2: System Characterization

- Describe the ‘natural system’ (that is being governed; N-SG) in terms of diversity, complexity, dynamics and scale (DCDS)
- Describe the ‘social system’ (that is being governed; S-SG), in terms of DCDS

- Describe the ‘governing system’ (GS) in terms of DCDS. Note that GS can be governments, non-governments, and private actors. Also,
  - Identify the ‘mode’ of governance, accordingly (e.g. hierarchical/top-down, co-governance/co-management, self-governance).
  - List key rules/norms and regulations (formal and informal) used in GS
- Describe the ‘governing interactions’ (GI) in terms of DCDS. Note that these are interactions between SG and GS, not internal interactions within components of the system.
- For each system, indicate the level of DCDS (high, medium, low)

  Note:

1. Diversity is determined by the number of ‘components’ in the system (e.g. the more components, the higher the diversity)
2. Complexity is related to how the systems’ components ‘relate’ to each other (e.g. a system with only one type of relationship is less complex than a system with several types of relationships)
3. Dynamics is determined by how the components ‘interact’ with each other (e.g. two systems may have the exact same number of components; thus same diversity. The components in the two systems may also have similar relationships; thus same complexity. Yet, System A can still be more ‘dynamics’ than System B if there are a lot of interactions between its components.)
4. Scale is about ‘boundaries’. (e.g. when the system has a clear boundary or is contained in a confined area, then it has little or no scale issue. When a boundary is not well defined or uncertain, then the system may have scale issues).

Section 3: Governing System Quality

The ability of the GS to govern depends on its characteristics (Section 2), and on three other qualities in relation to SG. Rate each quality on a scale of 1–5 after the description (1 = bad, 3 = moderate, 5 = good)

1. Goodness of fit

   Here, we are looking at three elements of governance, i.e. images, instruments and actions, whether there are good fits between GS and SG. For instance, GS may take decisions based a certain images that may or may not match what SG sees. The different image/perception of the problems identified in Part 2 is an indication of mismatch. The institutional arrangements within GS may not match the characteristics of SG. What are the gaps? Do they align well?

2. Responsive of modes

   Describe whether the mode of governance identified above (in the description of GS) is ‘suitable’ to deal with SG and the problem. For instance, is it able to respond appropriately to the problem? Can it resolve conflicts? Is it sensitive to what’s going on in SG? Does it foster learning?

3. Performance of GS in the three orders

   GS needs to operate at three orders. The first-order is about 'day-to-day' operation and daily problem solving. The second-order is about institutional design and arrangement. The third (or meta-) order is about informing/directing the second- and the first-order using a set of 'good governance' principles and agreed-upon societal values. Does GS perform well in all orders of governance? Give examples. For the meta-order, list principles and values that provide direction to the governing structure, institutions, rules and regulations (second-order), and also decisions and actions at the first-order. Also consider the following criteria in rating the performance of GS: consistency, effectiveness, transparency, justice, etc.

Section 4: Governing Interactions

- List types of interactions that take place between SG and GS. Interactions could be in a form of meetings, workshops, and other means of communications and information sharing.
- For each interaction, indicate frequency, who initiates it, and 'rate' its quality on a scale of 1–5 (1 = poor, 3 = moderate, 5 = good). Consider the following criteria in the rating: inclusiveness, informative, representativeness, fairness, interactive, participatory, etc. Note any power relation and power dynamics between SG and GS, as well as among SG.

Appendix 2: The Governability Assessment Matrix for Unpacking SSF Systems

System properties	N-SG	S-SG	GS	GI	System variables
Diversity	What species are there, in what number, and what habitats exist, in what quantity (size)?	Who are the relevant SSF stakeholders, and how are they using the resources and the ecosystem?	What are the relevant organizations and institutions, and what are their mandates, responsibilities and roles?	Which channels and mechanisms of interaction are used, and who are partaking?	Components
Complexity	How do species and habitats relate to each other?	What are the relationships between different SSF stakeholders?	What are the relationships between SSF organizations and institutions?	Are these channels and mechanisms independent of each other or related, and how?	Relations

(continued)

System properties	N-SG	S-SG	GS	GI	System variables
Dynamics	How do species interact with each other and with the habitat, and have these interactions changed over time?	How do stakeholders interact with each other, and have these interactions changed over time?	How do governing organizations and institutions interact, and have these interactions changed over time?	Are there interactions between the chosen channels and mechanisms, and have these interactions changed over time?	Interaction
Scale	What is the natural ecosystem boundary, and how open or closed is it?	What is the social boundary, and how open or closed is it?	What are the organizational and institutional boundaries, and how open or closed are they?	What are the boundaries of the chosen channels and mechanisms, and how open or closed are they?	Boundaries

References

Bavinck M, Chuenpagdee R, Jentoft S et al (eds) (2013) Governability in fisheries: theory and applications. Springer, Dordrecht

Bundy A, Chuenpagdee R, Cooley S et al (2015) A decision support tool for response to global change in marine systems: the IMBER-ADApT framework. Fish Fish 17:1183–1193. https://doi.org/10.1111/faf.12110

Chuenpagdee R, Jentoft S (2007) Step zero for fisheries co-management: what precedes implementation. Mar Policy 31(6):657–668

Chuenpagdee R, Knetsch JL, Brown TC (2001) Environmental damage schedules: community judgments of importance and assessment of losses. Land Econ 77(1):1–11

Chuenpagdee R, Pascual-Fernandez JJ, Szelianszky E et al (2013) Marine protected areas: re-thinking their inception. Mar Policy 39:234–240

Chuenpagdee R, Rocklin D, Bishop D et al (2017) The global information system on small-scale fisheries (ISSF): a crowdsourced knowledge platform. Mar Pol. https://doi.org/10.1016/J.MARPOL.2017.06.018

Degnbol P, Gislason H, Hanna S et al (2005) Painting the floor with a hammer: technical fixes in fisheries management. Mar Policy 30:534–543

Donnelly J (2003) Universal human rights – in theory & practice. Cornell University Press, Ithaca

FAO (1995) Code of conduct for responsible fisheries. FAO, Rome

FAO (2014) The state of world fisheries and aquaculture 2014. FAO, Rome

FAO (2015) Voluntary guidelines for securing sustainable small-scale fisheries in the context of food security and poverty eradication. FAO, Rome

Jentoft S, Bavinck M (2014) Interactive governance for sustainable fisheries: dealing with legal pluralism. Curr Opin Environ Sustain 11:71–77

Jentoft S, Chuenpagdee R (2009) Fisheries and coastal governance as a wicked problem. Mar Policy 33(4):553–560

Jentoft S, Chuenpagdee R (eds) (2015) Interactive governance for small-scale fisheries: global reflections. Springer, Cham
Jentoft S, Johnsen JP (2015) The dynamics of small-scale fisheries in Norway: from adaptamentality to governability. In: Jentoft S, Chuenpagdee R (eds) Interactive governance for small-scale fisheries: global reflections, MARE publication series. Springer, Cham, pp 705–723
Johnson D, Acott TG, Stacey N et al (2017) Social wellbeing and the values of small-scale fisheries. Springer, Cham
Kooiman J (2003) Governing as governance. Sage, London
Kooiman J, Jentoft S (2009) Meta-governance: values, norms and principles, and the making of hard choices. Public Adm 87(4):818–836
Kooiman J, Bavinck M, Jentoft S et al (2005) Fish for life – interactive governance for fisheries. Amsterdam University Press, Amsterdam, pp 25–37
Lang DJ, Wiek A, Bergmann M et al (2012) Transdisciplinary research in sustainability science: practice, principles, and challenges. Sustain Sci 7(Suppl 1):25–43. https://doi.org/10.1007/s11625-011-0149-x
Luomba J, Chuenpagdee R, Song AM (2016) A bottom-up understanding of illegal, unreported, and unregulated fishing in Lake Victoria. Sustainability 8:1062. https://doi.org/10.3390/su8101062
Prpić J, Melton J, Taeihagh A et al (2015) MOOCs and crowdsourcing: massive courses and massive resources. First Monday 20(12). https://doi.org/10.5210/fm.v20i12.6143
Rittel HWJ, Webber MM (1973) Dilemmas in a general theory of planning. Policy Sci 4:155–169
Song AM, Chuenpagdee R (2015) Eliciting values and principles of fishery stakeholders in South Korea: a methodological exploration. Soc Nat Resour 28(10):1075–1091
Song AM, Chuenpagdee R, Jentoft S (2013) Values, images, and principles: what they represent and how they may improve fisheries governance. Mar Policy 40:167–175
United Nations (2015) Rethinking poverty: report on the world social situation 2010 | social perspective on development

About the Authors

Melinda Agapito is a research fellow with the Too Big To Ignore: Global Partnership for Small-Scale Fisheries Research (TBTI). She obtained her PhD in Geography from Memorial University of Newfoundland, St. John's, Canada. Her PhD focused on developing a conservation decision-making method to help identify conservation areas. Currently, her work in small-scale fisheries research relates to food security, global changes, and ecological impacts of fishing gears.

Alfonso Aguilar-Perera works at the Universidad Autónoma de Yucatán, in Mérida, Yucatán, México, where he teaches undergraduate and graduate courses on coastal conservation and management, conservation of reef fishes, and marine protected areas design. He received a PhD from University of Puerto Rico and developed a postdoc at the Caribbean Coral Reef Institute. His scientific expertise includes fish conservation biology incorporating fisher's local ecological knowledge in the Mexican Caribbean and Gulf of Mexico, and lionfish ecology in the Yucatan Peninsula with collaboration of fishers from the Yucatan. He has published more than 30 peer-reviewed, scientific papers on conservation biology.

A. Minerva Arce-Ibarra holds a Researcher/Professor position at the College of Southern Border (ECOSUR) located in Chetumal, Mexico, since 1996. She is a former Fishery Biologist and completed a PhD in Interdisciplinary Studies at Dalhousie University. Her current research areas are Indigenous and Rural fisheries, community-based conservation, ecological economics, and transdisciplinary approaches. She also collaborates in studies that address Social-Ecological Systems, indigenous and rural adaptive strategies to climatic variability, and food security. From 2001 to 2011 she was a member of the Steering Committee of the UNDP Program 'COMPACT- Sian Ka'an Biosphere reserve'. She is also a member (co-applicant) of the Community Conservation Research Network.

R. Chuenpagdee, S. Jentoft (eds.), *Transdisciplinarity for Small-Scale Fisheries Governance*, MARE Publication Series 21,
https://doi.org/10.1007/978-3-319-94938-3

Lovasoa Cédrique Augustave is a national from Madagascar, who is currently pursuing his Master's degree in International Fisheries Management at the University of Tromsø (UiT). Back in 2014, he was an intern at the International Research Institute of Stavanger (IRIS) and undertook his research on ecotoxicology of a treated/untreated oil based drilling waste using a biomarker approach. Nowadays, he works with the European Union project FarFish. His work represents a contribution to the H2020 project 'Improving knowledge and management of EU fisheries outside Europe, while contributing to sustainability and long-term profitability'. His interests focus on fisheries management and sustainability.

Maarten Bavinck is a professor of coastal resource governance at the Department of Human Geography, Planning and International Development Studies of the University of Amsterdam, as well as at the Norwegian College of Fisheries, UiT Arctic University of Norway. He has a PhD in human geography and is specialized in marine and coastal societal dynamics, with special relevance to India and Sri Lanka. Maarten is co-founder and co-director of the social science Centre for Maritime Research (MARE), and has led projects on fisheries conflict, food security, governance, legal pluralism, and coastal transformation in various parts of the world.

Fikret Berkes is Distinguished Professor Emeritus at the Natural Resources Institute, University of Manitoba, Canada. Dr. Berkes' past and current work deals with social-ecological resilience, commons, co-management, and local and traditional ecological knowledge. His 10 books include *Sacred Ecology* (4th edition, Routledge, 2018) and *Coasts for People* (Routledge, 2015). Dr. Berkes is the recipient of the Sustainability Science Award of the Ecological Society of America (2014), the Elinor Ostrom Award for Senior Scholar of the International Association for the Study of Commons (2015), and the International Conservation Union IUCN-CEESP Inaugural Award for Meritorious Research (2016).

Bonnie Bishop is a recent graduate of the Master of Science program at Memorial University of Newfoundland, St. John's, Canada. The focus of her thesis was the human dimensions of the 'Wolf Capital of the World' initiative in Thompson, Manitoba, Canada. She also holds a B.Sc. in Geography and a B.Ed. in adult education. Bishop currently works as an education coordinator for a private organization in Vancouver, British Columbia. Her research interests are tied to public perception of environmental issues and environment-related education. Bonnie worked with TBTI in the transdisciplinary cluster as a research assistant. She participated in the transdisciplinary workshop in Merida, Mexico in 2016 and has collaborated on other works relating to TBTI transdisciplinary work.

Shannon D. Bower is a co-coordinator of the Too Big to Ignore Inland Fisheries Cluster, a Science Ambassador for Keepemwet, and the Director of Recreational Angling for the Mahseer Trust. As a researcher, she primarily studies the growth of recreational fisheries in developing and emerging economies, an undervalued and understudied sector with enormous potential for both benefit and impact. In this research, she uses a social-ecological systems approach to understand the biological, social, and economic dynamics of recreational fisheries, including relationships to other small-scale fishing sectors.

Ratana Chuenpagdee is a Professor at Department of Geography, Memorial University of Newfoundland, St. John's, Canada. She held a position of Canada Research Chair in Natural Resource Sustainability and Community Development at Memorial University from 2006–2016. Her research emphasizes interdisciplinary approaches to coastal fisheries and ocean governance, focusing particularly on small-scale fisheries, marine protected areas, community-based management, and food security. She has worked in several countries including Cambodia, Canada, Malawi, Mexico, Spain, and Thailand. Dr. Chuenpagdee is a project director of Too Big To Ignore: Global Partnership for Small-Scale Fisheries Research (TBTI) and her most recent book — The Small-Scale Fisheries Guidelines: Global Implementation — co-edited with Svein Jentoft, María José Barragán-Paladines, and Nicole Franz was published by Springer in 2017.

Andrés Cisneros-Montemayor is a Program Manager and Research Associate with the Nereus Program at the University of British Columbia, and specializes in applied resource economics. Linking field and theoretic work, he has studied the economics of ecotourism, competing fishing sectors, alternative management strategies and ecosystem approaches to policy, and Indigenous fisheries, in developing and developed regions including Belize, Canada, Central America, East Asia, Mexico, Patagonia, the USA, and West Africa. Andrés holds a Ph.D. and M.Sc. in Resource Management and Environmental Studies (University of British Columbia), and a B.Sc. in Marine Biology (Universidad Autónoma de Baja California Sur).

Steven J. Cooke is a Canada Research Chair and professor in the Fish Ecology and Conservation Physiology Laboratory at Carleton University. His research spans freshwater and marine systems around the globe and encompasses the natural and social sciences. Cooke has particular expertise in the realm of recreational fisheries science and watershed science. He is Editor-in-Chief of the journal *Conservation Physiology* and has received a number of awards including the NSERC E.W.R. Steacie Fellowship and the Medal from the Fisheries Society of the British Isles and is currently serving as President of the International Fisheries Section of the American Fisheries Society.

Alyne E. Delaney is affiliated with Innovative Fisheries Management (IFM), Aalborg University, Aalborg, Denmark and Center for Northeast Asian Studies (CNEAS), Tohoku University, Japan. She has dual appointments as Associate Professor at Aalborg University's IFM and Tohoku's Center for Northeast Asian Studies (CNEAS). Her research foci include resource rights, resilience, and social sustainability, and she has worked on fisheries and coastal community-related projects around the world. She is a member of the MARE Board and is an *ex officio* council member for the International Association for the Study of the Commons. She is currently coordinating the EU H2020 Project 'Preserving and sustainably governing cultural heritage and landscapes in European coastal and maritime regions' (PERICLES).

Rodolphe Devillers is Professor of Geography at Memorial University of Newfoundland, Canada, specialized in marine applications of GISciences and geospatial technologies. His research focuses on marine conservation and marine protected areas, fisheries, marine habitat mapping, and the mapping of species at risk. He is part of the international Too Big To Ignore: Global Partnership for Small-Scale Fisheries Research where he leads the group that developed the Information System on Small-scale Fisheries (ISSF).

David V. Fazzino II is an Assistant Professor of Anthropology at Bloomsburg University of Pennsylvania. He is trained in Law (JD), Anthropology (PhD), and Agroecology (MS). His research interests include food systems, energy policy, intellectual property rights, heritage preservation, health and healing, and structural violence.

Katia Frangoudes is a Senior Researcher in UMR AMURE laboratory of the University of Brest, France. Her main research topics are women's roles in European fisheries, marine resource governance, and fisheries management. She is also the facilitator of AKTEA, a European Network of Women Organizations in Fisheries lobbying for gender equality and women rights in the European Union.

Glenna Gannon is a Master of Environment and Sustainability candidate at the University of Saskatchewan. Her research focuses on how rural fishing dependent communities in Western Alaska respond to fisheries disruptions, and are changing as result of increasing environmental variations.

Catalina García Chavez is a PhD Candidate at the Amsterdam Institute for Social Science Research (AISSR). She holds a Bachelor in Anthropology and an M.Phil. degree in History. As part of Erigaie Research Center in Colombia, Catalina has worked with fluvial and maritime communities and stakeholders in various research projects. Her research interests include maritime culture, contested marine borders, transboundary practices, maritime territory, and ocean governance.

Jennifer Gee is a fishery officer with the Food and Agriculture Organization of the UN. She is the officer responsible for fleet and employment statistics and has project-based work that focuses largely on socioeconomic data for fisheries. Her work interests rest at the nexus of data and the human stories that can be told with this data. Jennifer started her work in fisheries as an onboard trawl fleet observer before moving to work and conduct a Master's degree in aquaculture-focused performance indices still with a focus on bringing together disparate data sources to tell a larger story.

Siri Gerrard is currently professor emerita at The Center for Women and Gender Research. Her main research topics since the 1970s have been and still are women's situation and gender roles in the North Norwegian fisheries, fishing communities and politics. She has arranged and participated in meetings, conferences and political actions focusing on women in coastal areas in Norway as well as internationally.

Miguel González (PhD, York University) is an Assistant Professor in the International Development Studies program at York University. His research examines indigenous self-governance and territorial autonomous regimes in Latin America; he also investigates the governance of small-scale indigenous fisheries in the Global South, with particular attention to the Nicaraguan Caribbean Coast. His most recent publication examines tenure rights and informed consent in Indigenous fisheries of Nicaragua, a piece that is published in the edited volume 'The Small-Scale Fisheries Guidelines, Global Implementation' edited by Svein Jentoft, Ratana Chuenpadgee, Maria J. Barragan-Paladines, and Nicole Franz (Springer 2017). Miguel is affiliated with TBTI and the Centre for Research on Latin America and the Caribbean (CERLAC) at York University.

Tek Bahadur Gurung is working as Principal Scientist (Fisheries), Director of Livestock and Fisheries Research in Nepal Agricultural Research Council. He also serves as the Chief Editor of the Journal of Nepal Agricultural Research Council and Adjunct Professor. He has developed successful farming practices of rainbow trout; breeding methods of pond-reared Himalayan Mahseer, and restoring methods of degraded lakes using carp. His interest also lies on native fish conservation in relation to cross dams obstructing natural flows. Dr. Gurung completed his M.Sc. in India, his Doctorate from Kyoto University, and his Post-doc from University of Missouri, US.

Gordon M. Hickey is Associate Professor and William Dawson Scholar in the Department of Natural Resource Sciences at McGill University. His research examines sustainable natural resource management, policy, and governance across scales.

Helga Hosupeit holds a M.Sc. in mathematics and economics, from the University of Cologne. She worked as a fisheries official with FAO for more than 30 years in the field of international trade of fishery products and in fisheries management. She is now working as an independent consultant and as a Senior Advisor with INFOPESCA, and is participating in the market opportunities cluster of TBTI.

Moenieba Isaacs is a Professor with the Institute for Poverty, Land, and Agrarian Studies (PLAAS) at the University of Western Cape. Her research focus is on understanding the social and political processes of fisheries reform in South Africa and southern Africa, mainly through the lenses of small-scale fisheries policy processes and implementation. She has worked extensively with communities in South Africa to find policy solutions to the problems, which has highlighted the need to deal with social differentiation in fishing 'communities'—unless an understanding of such complexities are part of research processes, they can perpetuate local conflicts. In 2017, she received the Institutional Individual Engagement Excellence Award for her research, policy, and community engagement work.

Svein Jentoft is a sociologist and a Professor at the Norwegian College of Fishery Science at UiT – The Arctic University of Norway. Throughout his career, he has worked extensively on fisheries and coastal issues, including resource management, industrial organization, and community development in his native Norway, as well as in many other countries. Jentoft has published more than 25 books and numerous journal articles on fisheries and coastal governance. He is a founding member and cluster leader within the Too Big To Ignore: Global Partnership for Small-Scale Fisheries Research. His most recent book—The Small-Scale Fisheries Guidelines: Global Implementation—co-edited with Ratana Chuenpagdee, María José Barragán-Paladines, and Nicole Franz was published by Springer in 2017.

Derek Johnson is an Associate Professor of Socio-cultural Anthropology at the University of Manitoba and a Research Associate at the Centre for Maritime Research at the University of Amsterdam. Derek's research integrates political ecology and social wellbeing approaches in the analysis of small-scale fisheries governance and cultural economies of food. Since the mid-1990s, his primary geographical area of interest has been South Asia, and particularly the Indian state of Gujarat. Derek has led work on values and the social and cultural dimensions of small-scale fisheries within the Canadian Social Sciences and Humanities Research Council project Too Big to Ignore: Global Partnership for Small-Scale Fisheries Research.

Andrew F. Johnson is a postdoctoral researcher in the Marine Biology Research Division and the Centre for Marine Biodiversity and Conservation at the Scripps Institution of Oceanography. Andrew received his PhD in marine fisheries ecology from the School of Ocean Sciences at Bangor University, UK. His work focuses on understanding the impacts of fishing on fish populations and marine ecosystems. He invests considerable efforts in integrating fisheries ecology, economics, and social science to better understand how fish populations, seafood markets, and fishers behave. Andrew has worked with many different fisheries in Europe, Central and South America, and Southeast Asia.

Kungwan Juntarashote is an Emeritus Professor at the Faculty of Fisheries, Kasetsart University, Bangkok, Thailand. He has 40 years of teaching and research experience in the field of fishery management and small-scale fisheries development, both in Thailand and in Southeast Asia. He currently serves as an advisor to Department of Fisheries, Ministry of Agriculture and Cooperatives, and also to Fisheries Association of Thailand. Dr. Juntarashote has recently been appointed as a Member of the National Fishery Policy Committee of Thailand.

Jeppe Kolding is Professor in fisheries at University of Bergen, specializing in small-scale fisheries in developing countries and their importance in terms of food security. Previously he was employed by the Institute of Marine Research, Centre for Development Cooperation in Fisheries, where he worked with the R/V Dr. Fridtjof Nansen programme under FAO. He has lived and worked for several years in various African countries and has experience from shorter-term engagements in more than 25 developing countries. His primary research interests are fish stock assessment, ecology, harvest strategies, and management of small-scale tropical fisheries. He has been a member of the IUCN/CEM Fisheries Expert Group (FEG) since 2008 where he initiated the work on 'Balanced Harvest'.

Annie Lalancette has a background in biology (M.Sc.), environmental assessment (graduate diploma), and social sciences (PhD). Her work in the last 10 years has focused on coastal communities' resilience and wellbeing, particularly in the context of fishing. During her PhD with Dr. Monica Mulrennan (Concordia University, Canada), she worked with indigenous Torres Strait Islanders in northern Australia to investigate their perspectives about fisheries management and the potential impacts of proposed fisheries measures in the tropical rock lobster fishery. She is currently a MEOPAR postdoctoral fellow (Saint Mary's University, Canada) working with Dr. Tony Charles on community-based responses to marine hazards in Canada.

Mimi E. Lam, a Marie Curie Fellow at the University of Bergen, Centre for the Study of the Sciences and the Humanities, specializes in seafood values and ethics, policy, and governance. In 2017, Mimi received the Conservation Beacon Award from the Society for Conservation Biology "for pioneering an ethical approach to the conservation of marine resources, both natural and cultural, through interdisciplinary research and community engagement at the science-policy interface." Her Marie Curie project, eSEAS: Enhancing Seafood Ethics and Sustainability (2017–2019), focuses on developing innovative fit-for-purpose ethical decision-support tools to aid decision-makers in resolving resource conflicts and policy trade-offs.

Marta Leite began her career working in fishing communities on Brazil's coast, with her interests in the nexus between humans and natural resources. Her early explorations led her to marine conservation and socioenvironmental NGOs in Brazil, before turning to academia. Her Master's focused on fishers' knowledge and ecosystem-based management, while her PhD at the University of Manitoba, investigates resilience and wellbeing among small-scale fishing people. Currently, Marta's interests lie in expanding our understanding of various interpretations of wellbeing among diverse groups; in particular, how wellbeing is shaped by diverse values found among natural resource dependent cultures.

Marc Léopold is an Institutional Economist in fisheries at the French Research Institute for Sustainable Development (IRD). He is interested in innovative participatory approaches, such as transdisciplinary research and action research, to address governance and management issues of small-scale fisheries in the South. He has leaded several projects in small Pacific island countries and the Indian Ocean during the last 10 years to investigate the production and use of fishery knowledge to design and implement context-specific management rules and multi-level governance systems.

Charles Z. Levkoe is the Canada Research Chair in Sustainable Food Systems and an Assistant Professor at Lakehead University. His research uses a food systems perspective to better understand the importance of, and connections between, social justice, ecological regeneration, regional economies, and active democratic engagement.

Sebastian Linke, School of Global Studies, Gothenburg University, Gothenburg, Sweden, Senior lecturer in Human Ecology and Associate Professor in Environmental Social Sciences at the University of Gothenburg. Previous to the present position he was an assistant professor in theory of science at the University of Gothenburg. He studies the interfaces between science, politics, and society with a focus on fisheries and marine governance. He is interested in the changing relations between scientists' and other stakeholders' knowledge in new forms of environmental governance.

Philip A. Loring is an Associate Professor and Arrell Chair in Food, Policy, and Society at the Department of Geography and Arrell Food Institute, University of Guelph. He is a broadly trained anthropologist with interests in food systems and security, social and environmental justice, and indigenous issues. He received his MA and PhD from the University of Alaska, and has done research with coastal communities in Alaska and British Columbia. You can follow him on Twitter: @ ConserveChange.

Kristen Lowitt is an Assistant Professor in the Department of Geography at Brandon University. Her research program explores the relationships between natural resource governance and food security in rural and coastal contexts.

Sérgio M. Gomes de Mattos works as a Public Manager at the Ministry of Planning, Brazil, and is graduated in Fishing Engineering from the Federal Rural University of Pernambuco, Brazil, with a PhD in Marine Sciences from the Marine Sciences Institute of Barcelona, Spain. His work focuses on the management of small-scale fisheries and fisheries resources assessment, with an emphasis in fisheries bioeconomics and issues relating to the implementation of public policies. This approach highlights the concepts of economic and biological sustainability and calls for careful management that includes participation of fishers and other stakeholders. He is also a coordinator of the Too Big To Ignore (TBTI) 'Global Synthesis' cluster and a contributor to others TBTI clusters.

Ana Isabel Márquez Pérez is an Anthropologist, with a Master's degree in Sustainable Tourism Management and a Ph.D. in Social Science in Development,

Agriculture and Society. She is Assistant Professor in Social Sciences at the Universidad Nacional de Colombia, Caribbean Campus, San Andres Island. Her main interests are related to the diverse links between culture and environment. Particularly, she focuses on social perspectives on aquatic environments, including small-scale fisheries, customary sea tenure, maritime cultures, territories, and territorialities. She has participated in the TBTI 'Transdisciplinarity on Fisheries' research cluster since 2015. She is now starting a project on Social Maritime Cartography in the Colombian Caribbean and teaching a subject in Anthropology of Fishing.

Patrick McConney is a Senior Lecturer in Marine Resource Management Planning at the Centre for Resource Management and Environmental Studies, The University of the West Indies, Cave Hill Campus in Barbados. He has an interdisciplinary PhD in resource management from the University of British Columbia and was a fisheries manager in the Caribbean. His current work covers many aspects of research and outreach in small-scale fisheries and marine protected areas, but especially governance, livelihoods, socioeconomics, and developing adaptive capacity within the contexts of resilience and social-ecological systems.

Rodrigo Pereira Medeiros is a Professor at the Center for Marine Studies, Federal University of Paraná, in Brazil. He is an oceanographer and holds a PhD in Political Sociology. His work and projects explore different aspects for evaluating small-scale fisheries dynamics and in support of marine protected areas governance in Brazil. He has recently been working on a 4-year FAO/GEF project to engage small-scale fishers in reducing bycatch in trawl fisheries in Latin American and the Caribbean. He is the current Regional SocMon Coordinator for Brazil.

Beatriz Mesquita P. Ferreira is a Researcher at the Joaquim Nabuco Foundation (Fundaj) in Brazil, with a PhD in Fisheries Resources from the Federal Rural University of Pernambuco, Brazil. She holds a degree in Business Administration (Católica University of Pernambuco – Unicap) and a Master's in Development and Environment (Federal University of Ceará – UFC). She conducts research focused on environmental issues, governance, marine protected areas, and artisanal fisheries, always targeting sustainability, participation, and social control of public policies. Since 2011, she is a voluntary member of International Collective in Support of Fishworkers (ICSF), where she has been contributing to the construction and implementation of the Voluntary Guidelines for Securing Sustainable Small-Scale Fisheries.

Mbachi Ruth Msomphora has a PhD in fisheries science from UiT, the Arctic University of Norway. She is a Master of Science in Fisheries Management and a Master of Science in Public Health with specialization in Epidemiology from UiT. She also has a Bachelor of Science in Agriculture with a bias in Aquaculture from University of Malawi. Msomphora is a reviewer for international journals. She is currently working as Senior Academic Librarian in Dentistry at UiT. With such a diverse knowledge-background, Msomphora has acquired and mastered multidisciplinary and interdisciplinary research skills. She has expertise in both qualitative and quantitative research as exemplified in the publication of her articles in renowned scientific research journals, in international conferences, and her book chapters.

Prateep K. Nayak is an Associate Professor in the School of Environment, Enterprise, and Development at the University of Waterloo, ON, Canada. His academic background is in political science, environmental studies, and international development, and holds a PhD in Natural Resources and Environmental Management from University of Manitoba. Prateep's research focuses on the understanding of complex human-environment connections (or disconnections) in coastal-marine systems with attention to social-ecological change and its drivers. His main research interests include coastal commons, environmental change, and governance, resilience, social-ecological regime shifts, environmental justice, and political ecology. Prateep is a past SSHRC Banting Fellow, Trudeau Scholar, a Harvard Giorgio Ruffolo Fellow in Sustainability Science, and a recipient of Canada's Governor General Academic Gold Medal.

Connie Nelson is Professor Emeritus at Lakehead University and Director of the Food Security Research Network. Her research focuses on building resilient local food systems through a complexity science approach.

Paul Onyango is a Senior Lecturer at the University of Dar es Salaam, Tanzania, where he lectures on fisheries social science. His research interests are in the areas of natural resource governance, poverty, climate change, and variability effects on small-scale fishing communities. He has been studying small-scale fisheries for over 20 years in Africa. He holds a doctorate degree from the Arctic University of Norway, formerly the University of Tromsø.

Sölmundur K. Pálsson is a PhD candidate at the Department of Anthropology of the University of Manitoba. He conducts his PhD research on contestation of values in environmental politics and deliberations surrounding fisheries on Lake Winnipeg in Manitoba, Canada.

José J. Pascual-Fernández is a *Profesor Titular* of Social Anthropology at the University of La Laguna (Tenerife) since 1993. His research has been related to fisheries and coastal areas, focusing most recently on governability, MPAs, coastal tourism, fisher organizations, and small-scale fisheries markets. In the last 10 years, he has been involved as principal investigator or researcher in a dozen of competitive projects and a similar number of research contracts. Currently he is the Academic Coordinator of the PhD Program on Law, Society, and Tourism at University of La Laguna.

Maria Pena is a Project Officer at the Centre for Resource Management and Environmental Studies, the University of the West Indies, Barbados. She is interested in the socioeconomic aspects of marine resource management, and has been involved in the implementation and management of externally funded marine resource management projects in the wider Caribbean, including: socioeconomic monitoring at marine protected areas (MPAs) and coastal sites; fisheries management planning; MPA management effectiveness evaluation; marine resource governance; climate change adaptation and disaster risk management in fisheries and aquaculture; and, gender in small-scale fisheries. She is the current Regional SocMon Coordinator for the Caribbean.

Graham J. Pierce is a Researcher in the Ecology and Marine Resources Department at Instituto de Investigaciones Marinas (Consejo Superior de Investigaciones Cientificas) in Vigo. He previously worked at the University of Aberdeen for 36 years, latterly as a Professor at the University's Oceanlab and is also Visiting Professor in Marine Science at the University of Aveiro. His research encompasses a range of topics in marine biology, ecology, fisheries, and conservation, including studies on cephalopods, marine mammals, and nematode parasites of marine fish. He has published over 250 peer-reviewed journal papers. He is co-chair of the International Council for the Exploration of the Sea's Working Group on Cephalopod Fisheries and Life History and was a member of the ICES Science Committee during 2012–2016.

Evelyn Pinkerton, educated at Wellesley College, Harvard University, Brandeis University and the Sorbonne (but most of all by talking to fishermen), is a maritime anthropologist who has been theorizing and supporting fisheries co-management efforts for the last five decades. In the last decade, she has focused on the impact of neoliberal policies on these efforts, especially documenting the forms of co-management which are explicit alternatives to or resistance to neoliberal governance models, such as ITQs. She is a professor in an interdisciplinary graduate department, the School of Resource and Environmental Management, Simon Fraser University.

Cristina Pita holds a PhD in Social and Environmental Sustainability, from the University of Aberdeen (UK), and is currently an Assistant Investigator at the Centre for Environmental and Marine Studies (CESAM), University of Aveiro (Portugal). Her primary areas of research focus on small-scale fisheries management and socio-economics, small-scale markets and market initiatives, and sustainable use of marine resources. She is the co-leader of the cluster market opportunities in the global network Too Big To Ignore (TBTI) and is involved in several international projects focused on market initiatives to add value to small-scale fisheries products.

João Garcia Rodrigues works as a Researcher at the University of Santiago de Compostela in Galicia. He is currently doing his PhD studies in Marine Sciences at Campus Do Mar, an international PhD programme. With a background in Biology, he has proceeded with his studies at the University of Kiel where he obtained his

Erasmus Mundus Master's degree in Ecohydrology. Presently his research focuses on social-ecological systems, having published scientific articles on small-scale fisheries governance, seafood value chains, and marine ecosystem services.

Alicia Said is a Post-doctoral Fellow with the Too Big To Ignore: Global Partnership for Small-Scale Fisheries Research. Her research interests include resource governance, socioeconomic and socioecological resilience, and sustainable livelihoods in the context of small-scale and artisanal fisheries. She has recently finished her PhD in Biodiversity Management at the School of Anthropology and Conservation at the University of Kent (UK), within which she investigated the implications of marine policy initiatives on the sustainability of the small-scale fishing sector in Malta. With TBTI, she is now conducting regional and global research on the governance of different fishing communities by fostering the development of a governability index and progressing on a systematic approach for transdisciplinary fisheries research.

Suvaluck Satumanatpan is a Lecturer in the Faculty of Environment and Resource Studies, Mahidol University. She received the Fulbright US-ASEAN Visiting Scholars in 2015. Her Thai textbook, Coastal management: Integration to sustainability, received the Mahidol Excellent Textbook award in 2012, and has been used for teaching on coastal management subject in several Thai universities. She has contributed to several national policy papers for marine and coastal resources in Thailand. She is interested in coastal and fishery governance research (interactive governance), integrated coastal management, marine and environmental protected areas effectiveness, coastal and fishery resilience, and marine spatial planning.

Joeri Scholtens is a Post-doctoral Researcher at the Amsterdam Institute for Social Science Research, University of Amsterdam and the Center for Maritime Research (MARE). He obtained his PhD in 2016 from the University of Amsterdam on the reproduction of marginality of fishers in a transboundary fishing conflict in South Asia. His key areas of interest include political ecology of marine resources, oceans and food security, fisheries conflict and governance, and environmental injustice and inequality.

Milena Arias-Schreiber is a Researcher and Project Leader at the Swedish Institute for the Marine Environment, Gothenburg University, Gothenburg, Sweden. With a background on marine ecology and doctoral studies in the social sciences from the University of Bremen (Germany), her research focuses on institutional challenges for sustainable fisheries management. On this topic, she has worked with commercial anchovy fisheries for fishmeal production in the Peruvian Upwelling Ecosystem and with small-scale fisheries in the Baltic Sea. She is currently doing research and supporting the implementation of the 'social dimension' of ecosystem-based fisheries management in Sweden.

Anna Schuhbauer recently completed a Post-doctoral position at the Fisheries Economics Research Unit at the University of British Columbia. Her research covers ecological, economic, and socioeconomics dimensions of small-scale fisheries. She has published in a variety of scientific journals and technical reports informing fisheries policy and management. Focussing on applied research she currently specializes in the economic viability of fisheries. Anna holds a PhD in Resource Management and Environmental Studies (UBC) and a M.Sc. in Tropical Aquatic Ecology (University of Bremen) and has worked in Peru, the Falkland Islands, the Galapagos Islands, Tanzania, Mexico, and at a global scale.

Andrew M. Song is a Research Fellow at the Australian Research Council Centre of Excellence for Coral Reef Studies, James Cook University, Australia, and holds broad interest in the governance and geography of small-scale fisheries. His current research focuses on trans-boundary, multi-scalar, and inter-organizational issues as they relate to community-based fisheries management, IUU fishing, and human rights based approaches. Also affiliated with WorldFish, he has worked in various inland and marine fishery settings including the Philippines, the African and Laurentian Great Lakes, Atlantic Canada, South Korea, and the Pacific islands.

Merle Sowman is an Associate Professor and Head of the Department of Environmental and Geographical Sciences at the University of Cape Town (UCT). After obtaining a Ph.D. in integrated coastal management from UCT in 1994, she worked for a Development Agency in South Africa before taking on the role of Director of the Environmental Evaluation Unit, a research, training and consultancy unit at UCT until 2013. Her key areas of expertise include integrated coastal management, governance of small-scale fisheries, human dimensions of marine protected areas and more recently assessing vulnerability of coastal communities to climate change to inform adaptation planning.

Natasha Stacey is an applied anthropologist with interests in maritime anthropology in Northern Australia and Indonesia and sea nomadic populations in Southeast Asia. She is currently Associate Professor at the Research Institute for the Environment and Livelihoods, Charles Darwin University, leading a multidisciplinary group of scientists and postgraduate scholars working on natural resource management and livelihood research projects in Australia and Asia. Her recent research projects have included mall-scale fisheries in Indonesia: benefits to households, the roles of women, and opportunities for improving livelihoods; Supporting development of Indigenous coastal fishing enterprises in Arnhem land, Northern Australia; and Improving coastal livelihoods, food security and fisheries management in the cross-border regions of the Arafura-Timor Seas.

Johny Stephen is an Assistant Professor at the Tata Institute of Social Sciences (TISS), Hyderabad. He has an interdisciplinary background. He is a Marine Biologist by training and has completed his Master's degree in International Fisheries Management from the University of Tromso, Norway. He completed his

PhD from the University of Amsterdam in Political Geography. His interest lies in understanding resource governance from the perspective of spatial relations. In this context, he is keen to understand the geographies of resource access and use.

Wichin Suebpala is a Researcher of the Marine Biodiversity Research Group, Ramkhamhaeng University. He received B.Sc. and M.Sc. in Environmental Science from Ramkhamhaeng University and Kasetsart University, Thailand, respectively. He has been involved as a researcher in several projects regarding environmental science, integrated coastal management, and fisheries management. He is interested in applying integrated approaches in managing coastal fisheries and livelihoods to achieve sustainability. As a doctoral student of the Interdisciplinary Program of Environmental Science, Chulalongkorn University, he is currently conducting his Ph.D. research project on assessing ecological impacts of fisheries in Thailand.

Rashid Sumaila is Professor and Director of the Fisheries Economic Research Unit at the University of British Columbia. He specializes in bio-economics, marine ecosystem valuation, and the analysis of global issues, such as fisheries subsidies, illegal fishing, climate change, and oil spills. Sumaila is widely published including *Science*, *Nature* and the *Journal of Environmental Economics and Management*. Sumaila has won many awards such as the 2017 Volvo Environment Prize, 2017 Benchley Award for Science, 2016 Killam Research Prize, American Fisheries Society's 2013 Excellence in Public Outreach Award, the Leopold Leadership Fellowship (Stanford), and the Pew Fellowship for Marine Conservation.

Jackie Sunde is an independent Researcher and Activist working on small-scale fisheries in South Africa. She is a member of ABALOBI, a social enterprise working to enable small-scale fishing communities to use ICTs4Fisheries.

Brice Trouillet (PhD, geography) is Associate Professor (University of Nantes, Fr) since 2007. His research focuses on the relations between human groups and maritime space. Through the lens of STS and critical approaches, he uses the case of fisheries in marine spatial planning to study how power relations and knowledge issues intertwine in (geo)technological devices that ultimately shape 'socio-technical agencements', forming an intermediate with the marine 'environment'. The past decade, he supervised 60 Master and three PhD theses and received four research grants as PI, and has been involved in 11 other projects.

Daniele A. Vila-Nova is a Biologist, with a Master's/PhD in Ecology and Conservation from the Federal University of Paraná, Brazil. Her research focus started with an ecosystem-based approach to design Marine Protected Areas. Later on, she started aggregating to her projects, a more holistic vision towards coastal-marine management, which included local users/beneficiaries, such as small-scale fisheries. For the past years, she has been involved in multiple consulting projects related to small-scale fisheries, at both the local and national scale in Brazil.

Matias John Wojciechowski graduated with a Bachelor in Urban and Regional Planning and a Master's in Local Economic Development from the University of Waterloo (2002) and has a Master's degree in Territorial Planning and Management from the Federal University of ABC (2014) with focus on state interventions in coastal small-scale fisheries' development. Since 2006, he has been working with the World Fisheries Trust with inland and coastal fishing communities focusing on value-chain development, food safety, occupational health, economic development, and solidarity economy. From 2012 to 2017, he participated in the 'Pesces Para la Vida' Project in Bolivia, implementing participatory value-chain assessments and economic feasibility studies. John, born in Poland, is a Canadian citizen residing in Brazil for the past 15 years.

Thamasak Yeemin has worked at the Marine Biodiversity Research Group, Ramkhamhaeng University, in Bangkok since 1992, after completing his D.Sc. in Biology from Kyushu University. He has experience in all aspects of coral reef and related ecosystem management, conservation, research, and administration based on over 20 years of fieldwork in Thailand and other parts of East Asia. He has conducted research on ecology, biology, environmental science, and socioeconomics of coastal ecosystem management and marine protected areas, and provided expertise to several management agencies to design and implement management plans for coastal resources and environment. Currently, he is a president of the Marine Science Association of Thailand.

Appendix: List of Reviewers

Melinda AgapitoShehu Latunji AkintolaMilena Arias Schreiber
Denis Bailly
María José Barragán-Paladines
Leopoldo Cavaleri Gerhardinger
William Cheung
Reade Davis
Maricela de la Torre Castro
María José Espinosa-Romero
Lucia Fanning
Simon Funge Smith
Miguel González
Else Grete
Madeleine Hall-Arbor
Mohammad Mahmudul Islam
Svein Jentoft
Estelle Jones
Ahmed Khan
Danika Kleiber
Mirella de Oliveira Leis
Yinji Li
Lars Lindström
Yajie Liu
Kristen Lowitt
Sérgio Macedo G. de Mattos
Patrick McConney
Knut Mikalsen
Mbachi Ruth Msomphora
Prateep K. Nayak
Paul Onyango

R. Chuenpagdee, S. Jentoft (eds.), *Transdisciplinarity for Small-Scale Fisheries Governance*, MARE Publication Series 21,
https://doi.org/10.1007/978-3-319-94938-3

Sölmundur K. Pálsson
Maria Pena
Rodrigo Pereira Medeiros
Graham Pierce
Evelyn Pinkerton
Alicia Said
Hakan T Sandersen
Andrew M. Song
Jan van Tatenhove
Brice Trouillet

Index

R. Chuenpagdee, S. Jentoft (eds.), *Transdisciplinarity for Small-Scale Fisheries Governance*, MARE Publication Series 21,
https://doi.org/10.1007/978-3-319-94938-3

G

H

MIX
Papier aus verantwortungsvollen Quellen
Paper from responsible sources
FSC® C105338

If you have any concerns about our products,
you can contact us on
ProductSafety@springernature.com

In case Publisher is established outside the EU,
the EU authorized representative is:
Springer Nature Customer Service Center GmbH
Europaplatz 3, 69115 Heidelberg, Germany

Printed by Libri Plureos GmbH
in Hamburg, Germany